Mathematical Optimization and Evolutionary Algorithms with Applications

Mathematical Optimization and Evolutionary Algorithms with Applications

Editors

Antonin Ponsich
Mariona Vila Bonilla
Bruno Domenech

MDPI • Basel • Beijing • Wuhan • Barcelona • Belgrade • Manchester • Tokyo • Cluj • Tianjin

Editors

Antonin Ponsich
Universitat Politècnica de
Catalunya (UPC)
Barcelona
Spain

Mariona Vila Bonilla
EAE Business School
Barcelona
Spain

Bruno Domenech
Universitat Politècnica de
Catalunya (UPC)
Barcelona
Spain

Editorial Office
MDPI
St. Alban-Anlage 66
4052 Basel, Switzerland

This is a reprint of articles from the Special Issue published online in the open access journal *Mathematics* (ISSN 2227-7390) (available at: https://www.mdpi.com/si/mathematics/ Mathematical_Optimization_Evolutionary_Algorithms).

For citation purposes, cite each article independently as indicated on the article page online and as indicated below:

LastName, A.A.; LastName, B.B.; LastName, C.C. Article Title. *Journal Name* **Year**, *Volume Number*, Page Range.

ISBN 978-3-0365-7978-8 (Hbk)
ISBN 978-3-0365-7979-5 (PDF)

Contents

About the Editors . . . . **vii**

Antonin Ponsich, Bruno Domenech and Mariona Vilà
Preface to the Special Issue "Mathematical Optimization and Evolutionary Algorithms with Applications"
Reprinted from: *Mathematics* **2023**, *11*, 2229, doi:10.3390/math11102229 . . . . **1**

Francisco Yuraszeck, Gonzalo Mejía, Jordi Pereira and Mariona Vilà
A Novel Constraint Programming Decomposition Approach for the Total Flow Time Fixed Group Shop Scheduling Problem
Reprinted from: *Mathematics* **2022**, *10*, 329, doi:10.3390/math10030329 . . . . **7**

Min Wang, Guoshan Liu and Xinyu Lin
Dynamic Optimization of the Multi-Skilled Resource-Constrained Project Scheduling Problem with Uncertainty in Resource Availability
Reprinted from: *Mathematics* **2022**, *10*, 3070, doi:10.3390/math10173070 . . . . **33**

Daniela Ambrosino and Carmine Cerrone
The Cost-Balanced Path Problem: A Mathematical Formulation and Complexity Analysis
Reprinted from: *Mathematics* **2022**, *10*, 804, doi:10.3390/math10050804 . . . . **53**

Saúl Zapotecas-Martínez, Abel García-Nájera and Adriana Menchaca-Méndez
Improved Lebesgue Indicator-Based Evolutionary Algorithm: Reducing Hypervolume Computations
Reprinted from: *Mathematics* **2022**, *10*, 19, doi:10.3390/math10010019 . . . . **67**

Akram Belazi, Héctor Migallón, Daniel Gónzalez-Sánchez, Jorge Gónzalez-García, Antonio Jimeno-Morenilla and José-Luis Sánchez-Romero
Enhanced Parallel Sine Cosine Algorithm for Constrained and Unconstrained Optimization
Reprinted from: *Mathematics* **2022**, *10*, 1166, doi:10.3390/ math10071166 . . . . **93**

Jun Wu, Yuanyuan Li, Li Shi, Liping Yang, Xiaxia Niu and Wen Zhang
ReRec: A Divide-and-Conquer Approach to Recommendation Based on Repeat Purchase Behaviors of Users in Community E-Commerce
Reprinted from: *Mathematics* **2022**, *10*, 208, doi:10.3390/math10020208 . . . . **141**

Jin Qin, Xiqiong Li, Kang Yang and Guangming Xu
Joint Optimization of Ticket Pricing Strategy and Train Stop Plan for High-Speed Railway: A Case Study
Reprinted from: *Mathematics* **2022**, *10*, 1679, doi:10.3390/math10101679 . . . . **161**

Jimmy H. Gutiérrez-Bahamondes, Daniel Mora-Melia, Bastián Valdivia-Muñoz, Fabián Silva-Aravena and Pedro L. Iglesias-Rey
Infeasibility Maps: Application to the Optimization of the Design of Pumping Stations in Water Distribution Networks
Reprinted from: *Mathematics* **2023**, *11*, 1582, doi:10.3390/math11071582 . . . . **179**

Bruno Domenech, Laia Ferrer-Martí, Facundo García, Georgina Hidalgo, Rafael Pastor and Antonin Ponsich
Optimizing PV Microgrid Isolated Electrification Projects—A Case Study in Ecuador
Reprinted from: *Mathematics* **2022**, *10*, 1226, doi:10.3390/ math10081226 . . . . **195**

Rosa Galleguillos-Pozo, Bruno Domenech, Laia Ferrer-Martí and Rafael Pastor
Balancing Cost and Demand in Electricity Access Projects: Case Studies in Ecuador, Mexico and
Peru
Reprinted from: *Mathematics* **2022**, *10*, 1995, doi:10.3390/math10121995 . . . . **219**

Mohamed Abdelhamid, Salah Kamel, Emad M. Ahmed and Ephraim Bonah Agyekum
An Adaptive Protection Scheme Based on a Modified Heap-Based Optimizer for Distance and
Directional Overcurrent Relays Coordination in Distribution Systems
Reprinted from: *Mathematics* **2022**, *10*, 419, doi:10.3390/math10030419 . . . . **239**

Luis Miguel Reyes-Barquet, José Octavio Rico-Contreras, Catherine Azzaro-Pantel,
Constantino Gerardo Moras-Sánchez, Magno Angel González-Huerta,
Daniel Villanueva-Vásquez and Alberto Alfonso Aguilar-Lasserre
Multi-Objective Optimal Design of a Hydrogen Supply Chain Powered with Agro-Industrial
Wastes from the Sugarcane Industry: A Mexican Case Study
Reprinted from: *Mathematics* **2022**, *10*, 437, doi:10.3390/math10030437 . . . . **261**

Luis Fernando Grisales-Noreña, Brandon Cortés-Caicedo, Gerardo Alcalá and
Oscar Danilo Montoya
Applying the Crow Search Algorithm for the Optimal Integration of PV Generation Units in DC
Networks
Reprinted from: *Mathematics* **2023**, *11*, 387, doi:10.3390/math11020387 . . . . **303**

Jeewon Park, Oladayo S. Ajani and Rammohan Mallipeddi
Optimization-Based Energy Disaggregation: A Constrained Multi-Objective Approach
Reprinted from: *Mathematics* **2023**, *11*, 563, doi:10.3390/math11030563 . . . . **321**

Denis D. Chesalin and Roman Y. Pishchalnikov
Searching for a Unique Exciton Model of Photosynthetic Pigment–Protein Complexes:
Photosystem II Reaction Center Study by Differential Evolution
Reprinted from: *Mathematics* **2022**, *10*, 959, doi:10.3390/math10060959 . . . . **335**

Xavier Martínez, Jordi Pons-Prats, Francesc Turon, Martí Coma, Lucía Gratiela Barbu and
Gabriel Bugeda
Multi-Objective Multi-Scale Optimization of Composite Structures, Application to an Aircraft
Overhead Locker Made with Bio-Composites
Reprinted from: *Mathematics* **2023**, *11*, 165, doi:10.3390/math11010165 . . . . **353**

About the Editors

Antonin Ponsich

Antonin Ponsich, Ph.D., is a lecturer at the Department of Management of the Technical University of Catalonia (UPC). He is a Process Engineer and a Doctor in Process and Environmental Engineering from the National Polytechnic Institute of Toulouse (INPT, France). He has held various positions as a researcher and teacher at the INPT, Autonomous Technological Institute of Mexico, Center for Research and Advanced Studies of the National Polytechnic Institute (Mexico) and Systems Department of the Autonomous Metropolitan University (UAM, Mexico), Azcapotzalco unit. His research focuses on the adaptation and application of mono- and multi-objective optimization techniques for operations research problems, mainly based on metaheuristics and evolutionary algorithms. He has participated in several research projects and is the author or co-author of books and articles published in international reference journals (ASOC, IEEE TEVC, COR, among others).

Mariona Vila Bonilla

Mariona Vila Bonilla, Ph.D., is a lecturer at EAE Business School and an associate professor at Universitat Politècnica de Catalunya, where she teaches courses regarding business management, supply chain management and project management. She is a Chemical Engineer and Doctor in Business Management from the UPC (Universitat Politècnica de Catalunya), with more than 11 years of experience in higher education teaching, both within public and private institutions. Her research focus is on the application of heuristic and exact algorithms in solving production, logistics and transportation problems.

Bruno Domenech

Bruno Domenech, Ph.D., is an associate professor at the Department of Management of the Technical University of Catalonia (UPC), under the Serra Húnter program of the Generalitat de Catalunya, at the Barcelona School of Industrial Engineering (ETSEIB) and the Barcelona East School of Engineering (EEBE). He is an Industrial Engineer, Scheduling Engineer and Doctor in Industrial Engineering from the UPC (Universitat Politècnica de Catalunya). He has held various positions as a researcher and teacher at the UPC, Open University of Catalonia (UOC) and Pompeu Fabra University (UPF), as well as at the University College Dublin (UCD), co-funded by the Marie Curie FP7-PEOPLE-2013-COFUND program. Within the framework of the Institute of Industrial and Control Engineering of the UPC, his research focuses on the application of quantitative methods of operational research to solve logistics and industrial organization problems, with a practical, applied, social and sustainable approach. Specifically, he has conducted studies in the areas of energy planning, production organization and supply chain design. He has participated in several competitive projects and is the author or co-author of many book chapters and scientific articles published in reference journals (EJOR, RSER, EGY, JEPO, RENE, ESD, IJPDLM, among others).

 *mathematics*

Editorial

Preface to the Special Issue "Mathematical Optimization and Evolutionary Algorithms with Applications"

Antonin Ponsich [1,*], Bruno Domenech [1] and Mariona Vilà [2]

[1] Management Department, Universitat Politècnica de Catalunya—BarcelonaTech, 08028 Barcelona, Spain; bruno.domenech@upc.edu

[2] Academic Department, EAE Business School, 08015 Barcelona, Spain; mariona.vila.bonilla@upc.edu

* Correspondence: antonin.sebastien.ponsich@upc.edu

Citation: Ponsich, A.; Domenech, B.; Vilà, M. Preface to the Special Issue "Mathematical Optimization and Evolutionary Algorithms with Applications". *Mathematics* **2023**, *11*, 2229. https://doi.org/10.3390/math11102229

Received: 27 April 2023
Accepted: 4 May 2023
Published: 10 May 2023

It is recognized that many real-world problems can be interpreted and formulated as optimization problems. This feature has fostered the development of research studies aiming to design and implement efficient optimization methods, able to address the increasing complexity of the applications that are intended to be solved. These research studies have mostly followed two main axes.

The first one focuses on the theoretical development of advanced solution strategies through the perspective of tackling problems of increasing complexity. For instance, multimodal objective functions, highly constrained search spaces, single vs. multi-objective problems, optimization of stochastic systems, among others. In this matter, thanks to both cutting-edge mathematical tools and the increasing power of computational hardware, exact solution methods (in general based on mathematical programming) now enable solving large-size intricate problems. However, many problems have also required the implementation of approximated, heuristic or metaheuristic techniques, which are not affected by the mathematical properties of the tackled problem but, on the other hand, are unable to guarantee result optimality. Within this class of approximated optimization methods, evolutionary algorithms occupy a relevant part of the devoted literature.

On the other hand, a great effort has also been made towards developing problem-devoted techniques that aim to efficiently find high-quality solutions to specific applications drawn from a wide spectrum of areas (engineering, social sciences, biotechnologies, finances, etc.). The corresponding studies do not usually start designing a new solution strategy from scratch, but rather reuse techniques developed in general frameworks and adapt their working mode to the specific feature of the problem that is being tackled. As a consequence, it is necessary to take advantage of the problem structure, conditioning factors or particular characteristics of the considered application for an efficient solution technique to be built.

The Special Issue proposed here illustrates both types of studies. Indeed, as shown in Table 1, 5 out of the 16 published articles tackle the issue of the theoretical development of optimization techniques or the formulation of academic operations research problems. Among these theoretical papers, two of them propose novel mathematical formulations for academic problems, while the other three focus on the development of evolutionary algorithms as a solution technique. The remaining 11 papers propose original and ad hoc solution strategies for different applications. Table 1 provides an overview of the topics addressed in these papers. It is worth highlighting that among these 11 studies, a majority of them use evolutionary algorithms, while four are based on mathematical programming.

The papers will be explained in detail, beginning with the theoretical studies. Yuraszeck et al. [1] propose a novel heuristic procedure to solve the fixed group shop scheduling problem, in which the tasks corresponding to each job have been assigned to stages, and the tasks of each stage share a set of machines. The authors introduce an algorithm that uses both a decomposition-based approach, as well as a constraint programming solver, allowing for the inclusion of extra constraints found in real-life instances. To test the performance of the proposed approach, computational tests are carried out to compare the algorithm with some available solvers; the former obtained the best solution in most instances. The heuristic procedure is also used in a Colombian automotive company case study, in which not only is the scheduling of jobs optimized, but also information about bottlenecks is easily obtained.

Table 1. Classification of the papers included in the Special Issue.

Type	Application	Mathematical Modelling	Evolutionary Algorithms
Theory	Scheduling	[1]	[2]
	Mathematics	[3]	[4,5]
Application	Distribution and Commerce	[6]	[7,8]
	Energy	[9,10]	[11–14]
	Physics and Materials	[15]	[16]

In Zapotecas et al. [4], the authors focus on one of the main paradigms employed for handling multi-objective optimization problems (MOPs) with evolutionary algorithms, which use hypervolume as a performance indicator governing the selection operator. A well-known drawback of this strategy is the complexity of the hypervolume computation when the number of objectives increases. This paper uses the property regularity of continuous MOPs, as well as the locality property of the hypervolume in order to reduce the number of computations of this indicator within a novel and efficient multi-objective evolutionary algorithm (MOEA). Three academic applications, with a number of objectives ranging from 4 to 7, are solved with the new algorithm, and the numerical experiments highlight the benefits of the proposed methodology for identifying efficiently better approximations of the Pareto front (when compared with classical MOEAs based on the hypervolume indicator).

In Ambrosino and Cerrone [3], a variant of the shortest path problem is proposed, considering both negative and positive costs at the edges of a graph. The aim consists of obtaining the Hamiltonian cycle such that the sum of the costs associated with each edge on the chosen path is close to 0. The resulting problem, called the cost-balanced path problem, is proved to be NP-hard since it can be reduced to the Hamiltonian path problem, which is NP-hard. Different versions of this problem are also introduced, so that practical conditions can be included through the appropriate constraints, and their complexity is also studied. Finally, computational experiments empirically confirm the problem complexity and suggest the need for heuristic or metaheuristic solution techniques to address large-size instances.

Belazi et al. [5] introduce an improved version of the sine–cosine algorithm (SCA), which is a population-based metaheuristic recently developed in the area of continuous optimization. The modifications proposed consist of the introduction of a new equation within the algorithm's variation operator, leading to an enhanced intensification effect, which promotes convergence towards the best solutions found. In addition, several parallelization strategies are implemented and tested in order to identify the best performing one. Finally, the new technique proves to significantly outperform the original SCA when both versions are compared over a benchmark, including 30 classical unconstrained text functions and several constrained engineering problems. Additionally, the enhanced SCA obtains very good results when its performance levels are compared with those of a set of state-of-the-art algorithms, such as differential evolution or grey wolf optimizer.

In the last theoretical work of this Special Issue, Wang et al. [2] address the multi-skilled resource-constrained project scheduling problem, which combines a typical scheduling of activities with the skill assignment of resources, taking into account uncertainty in resource availability. The authors formulate the corresponding mathematical model and, given its complexity, propose a genetic algorithm combined with priority rules. A computational experiment is performed comparing dynamic, random and static scheduling, showing the effectiveness of the first option. These results can help project managers in the selection of resources at the beginning of a project and the reinforcement of resources during the execution, especially under uncertain contexts such as COVID-19.

Regarding the 11 studies devoted to the solution of specific applications, Wu et al. [6] present a novel approach for algorithms devoted to community commerce recommendation for repeated purchases. The authors attempt to fill in the perceived gap in these types of purchase recommendation algorithms by accounting not just for past customer behaviors, but also for the repeat purchase behavior of different types of customers. The method uses a divide-and-conquer strategy, separating users into four categories: active users with stable interest, active users with unstable interest, inactive users with stable interest and inactive users with unstable interest. The proposed algorithm is tested on a real dataset and outperforms well-known recommendation algorithms by at least 13.6% in all categories, showing an even greater performance among active users.

Abdelhamid et al. [11] propose an adaptive protection scheme, used to overcome the coordination problems presented by protection relays. The adaptive protection scheme presented is based on both original and modified heap-based optimization. The algorithm proposed is tested using the IEEE 8-bus and the IEEE 14-bus test systems, obtaining better results than the existing algorithms. Specifically, the adaptive protection scheme is able to more reliably investigate the benefits of both directional overcurrent relays and distance relays. Additionally, the modified heap-based optimization makes the algorithm more effective at solving relay coordination.

Reyes-Barquet et al. [12] present a multi-objective genetic algorithm, which, combined with a TOPSIS analysis for multi-criteria decision making, is applied in the design stage of hydrogen supply chain networks. A specific case study is selected, where the hydrogen is obtained using energy generated by the biomass waste produced by Mexican sugar factories. The algorithm uses both the maximization of profit and the minimization of greenhouse gas emissions as optimization objectives. The results of the study highlight the benefits that could be obtained from this unorthodox energy source, as the case study was validated by several economic metrics, such as an internal rate of return of 21.5%, while remaining environmentally respectful.

In Chesalin and Pishchalnikov [15], the optical properties of pigment–protein complexes (PPCs) are investigated due to their major relevance in the study of photosynthetic mechanisms of living species. These properties and, in particular, the spectral response of PPC can be assessed either experimentally or through a simulation. However, the simulation process uses a set of input parameters that should be appropriately tuned in order to produce valid results. In this study, the differences between the experimental and simulated spectral responses of different PPCs are minimized through an evolutionary algorithm, differential evolution (DE), which has proved to perform very well for real-parameter optimization problems. Ten different DE strategies are implemented and their performance levels are compared, showing that the DE/rand-to-best/1/exp version consistently obtains the best results, although the authors recommend the use of self-adaptive implementations to improve the convergence rate.

Domenech et al. [9] deals with the design of autonomous electrification systems in Ecuador's amazon region (RAE), which is an isolated area with communities scattered across the rainforest. This situation involves great practical and economic difficulties for the development of electrification systems promoting the access to power for rural and indigenous local populations. This work introduces a mathematical model for the design of stand-alone rural electrification systems based on photovoltaic technologies, including both microgrid or individual supply configurations. The corresponding mixed integer linear programming (MILP) problem considers economic, technical and social aspects, and it is used to design electrification systems (equipment location and sizing and microgrid configurations) in three real communities, providing relevant insights regarding RAE electrification.

Another original application is presented in Qin et al. [7], which tackles the management of a high-speed railway in China. In particular, the problems of ticket pricing, train stop planning and seat allocation are all addressed in this study. A mathematical model is formulated, with the aim of maximizing the total revenue of the railway company while minimizing passengers' time loss. Due to the complexity of the resulting MILP problem, a simulated annealing algorithm is adopted as a solution technique, with two nested neighborhood structures; the first one deals with the stop plan, and the second focuses on ticket pricing and seat allocation. A solution using the proposed methodology is provided for the case study that is presented in this study, allowing for significant improvements of the chosen performance criteria when compared with those observed in the real system operation mode.

Galleguillos-Pozo et al. [10] develop a fuzzy MILP model to design wind–PV–battery electricity access projects for remote communities of developing countries. It is hard to estimate the electricity needs of the population in those areas, so fuzziness is introduced to balance the project cost vs. the demand supplied within a range of predefined values. Two approaches are considered: maximizing the general satisfaction of the whole community and maximizing the satisfaction of the least satisfied consumption point. The model is used to design electricity access projects in Ecuador, Mexico and Peru. The results achieve a generally better balance between the project cost and the electricity supplied than those that would have been obtained without using a fuzzy MILP model.

In Martínez et al. [16], a multi-objective and multi-scale optimization procedure is designed to improve the structure performance of eco-composites. As objectives, the shelf stiffness and the material cost and weight are optimized by modifying the configuration of the structure at macro and micro levels. The results highlight the importance of considering both the micro and macro structure when designing composite materials. An illustrative example is shown for the design of the cabin stowage bin located above the seats in airplanes. This procedure can be helpful for optimizing the design of eco-composites in many engineering structures, reducing the environmental impact.

Grisales-Noreña et al. [13] propose a mixed integer non-linear programming model to minimize the yearly operation costs of PV generators integrated into DC grids. The problem is solved through a primary—secondary methodology. First, the primary problem is addressed to locate and size the PV modules using a discrete-continuous version of the crow search algorithm. Second, the secondary problem searches the objective function value through the successive approximation power flow method. Test instances are used to validate the proposed methodology, which better performs in terms of applicability and effectiveness in comparison to other literature approaches; lower operation costs of the solution and computation times to solve the problem are achieved.

Park et al. [14] focus on the energy disaggregation problem, which consists of estimating the energy consumption of each device given the aggregated measure from the smart meter. In this perspective, the authors develop a multi-objective model that optimizes sparsity and disaggregation, subject to constraints related to equipment operational characteristics. The model is solved by means of an evolutionary algorithm. The results are compared to those obtained using different formulations from the literature, achieving better performance either on the appliance level or on the disaggregation accuracy.

Finally, Gutiérrez-Bahamontes et al. [8] identify the complexity of designing pump stations in real-size water distribution networks. To address this gap, they propose reducing the problem size through a preprocess where the range of flows that every pump station can manage is calculated, which leads to the construction of infeasibility maps. Then, the problem is optimized by means of a pseudo-genetic algorithm. They later perform a computational experiment showing that the preprocess effectively reduces the solution space, significantly improving the computation time and achieving better solutions in terms of the objective function value obtained.

Finally, it is worth providing a general overview of the Special Issue in terms of the geographical origin of the institutions of the papers' authors in this Special Issue. Figure 1 illustrates the fact that the most represented institutions are from Spain, followed by China and Latin American countries (20, including Mexico, Chile and Colombia). Additionally, it can be mentioned that the mean number of authors per paper is 4.3.

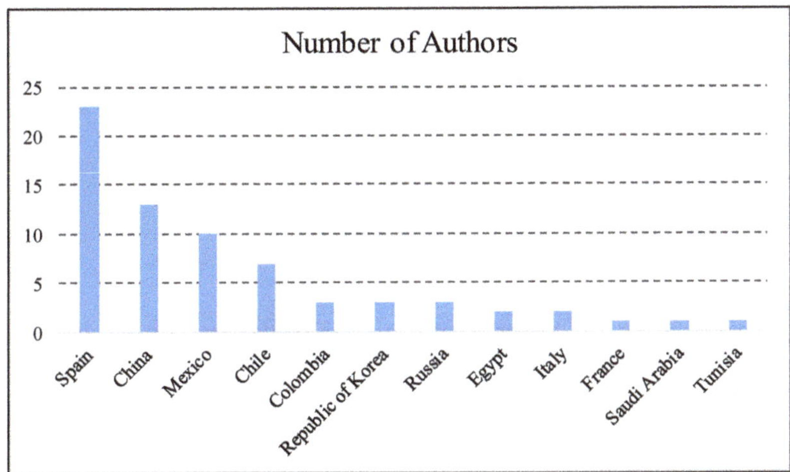

Figure 1. Geographical distribution of the institutions of accepted papers' authors.

In conclusion, the resulting mixture of methods, algorithms and applications for the treatment of complex optimization problems presented in this Special Issue, either through mathematical tools or metaheuristic algorithms, is expected to contribute to the development of research in this area. We also believe that the new knowledge acquired here, as well as the applied results are attractive and useful for young scientists, doctoral students and researchers from various scientific specialties.

Conflicts of Interest: The authors declare no conflict of interest.

References

1. Yuraszeck, F.; Mejía, G.; Pereira, J.; Vilà, M. A Novel Constraint Programming Decomposition Approach for the Total Flow Time Fixed Group Shop Scheduling Problem. *Mathematics* **2022**, *10*, 329. [CrossRef]
2. Wang, M.; Liu, G.; Lin, X. Dynamic Optimization of the Multi-Skilled Resource-Constrained Project Scheduling Problem with Uncertainty in Resource Availability. *Mathematics* **2022**, *10*, 3070. [CrossRef]
3. Ambrosino, D.; Cerrone, C. The Cost-Balanced Path Problem: A Mathematical Formulation and Complexity Analysis. *Mathematics* **2022**, *10*, 804. [CrossRef]
4. Zapotecas-Martínez, S.; García-Nájera, A.; Menchaca-Méndez, A. Improved Lebesgue Indicator-Based Evolutionary Algorithm: Reducing Hypervolume Computations. *Mathematics* **2022**, *10*, 19. [CrossRef]
5. Belazi, A.; Migallón, H.; Gónzalez-Sánchez, D.; Gónzalez-García, J.; Jimeno-Morenilla, A.; Sánchez-Romero, J.L. Enhanced Parallel Sine Cosine Algorithm for Constrained and Unconstrained Optimization. *Mathematics* **2022**, *10*, 1166. [CrossRef]
6. Wu, J.; Li, Y.; Shi, L.; Yang, L.; Niu, X.; Zhang, W. ReRec: A Divide-and-Conquer Approach to Recommendation Based on Repeat Purchase Behaviors of Users in Community E-Commerce. *Mathematics* **2022**, *10*, 208. [CrossRef]
7. Qin, J.; Li, X.; Yang, K.; Xu, G. Joint Optimization of Ticket Pricing Strategy and Train Stop Plan for High-Speed Railway: A Case Study. *Mathematics* **2022**, *10*, 1679. [CrossRef]
8. Gutiérrez-Bahamondes, J.H.; Mora-Melia, D.; Valdivia-Muñoz, B.; Silva-Aravena, F.; Iglesias-Rey, P.L. Infeasibility Maps: Application to the Optimization of the Design of Pumping Stations in Water Distribution Networks. *Mathematics* **2023**, *11*, 1582. [CrossRef]
9. Domenech, B.; Ferrer-Martí, L.; García, F.; Hidalgo, G.; Pastor, R.; Ponsich, A. Optimizing PV Microgrid Isolated Electrification Projects—A Case Study in Ecuador. *Mathematics* **2022**, *10*, 1226. [CrossRef]
10. Galleguillos-Pozo, R.; Domenech, B.; Ferrer-Martí, L.; Pastor, R. Balancing Cost and Demand in Electricity Access Projects: Case Studies in Ecuador, Mexico and Peru. *Mathematics* **2022**, *10*, 1995. [CrossRef]

11. Abdelhamid, M.; Kamel, S.; Ahmed, E.M.; Bonah Agyekum, E. An Adaptive Protection Scheme Based on a Modified Heap-Based Optimizer for Distance and Directional Overcurrent Relays Coordination in Distribution Systems. *Mathematics* **2022**, *10*, 419. [CrossRef]
12. Reyes-Barquet, L.M.; Rico-Contreras, J.O.; Azzaro-Pantel, C.; Moras-Sánchez, C.G.; González-Huerta, M.A.; Villanueva-Vásquez, D.; Aguilar-Lasserre, A.A. Multi-Objective Optimal Design of a Hydrogen Supply Chain Powered with Agro-Industrial Wastes from the Sugarcane Industry: A Mexican Case Study. *Mathematics* **2022**, *10*, 437. [CrossRef]
13. Grisales-Noreña, L.F.; Cortés-Caicedo, B.; Alcalá, G.; Danilo Montoya, O. Applying the Crow Search Algorithm for the Optimal Integration of PV Generation Units in DC Networks. *Mathematics* **2023**, *11*, 387. [CrossRef]
14. Park, J.; Ajani, O.S.; Mallipeddi, R. Optimization-Based Energy Disaggregation: A Constrained Multi-Objective Approach. *Mathematics* **2023**, *11*, 563. [CrossRef]
15. Chesalin, D.D.; Pishchalnikov, R.Y. Searching for a Unique Exciton Model of Photosynthetic Pigment–Protein Complexes: Photosystem II Reaction Center Study by Differential Evolution. *Mathematics* **2022**, *10*, 959. [CrossRef]
16. Martínez, X.; Pons-Prats, J.; Turon, F.; Coma, M.; Gratiela Barbu, L.; Bugeda, G. Multi-Objective Multi-Scale Optimization of Composite Structures, Application to an Aircraft Overhead Locker Made with Bio-Composites. *Mathematics* **2023**, *11*, 165. [CrossRef]

 *mathematics*

Article

A Novel Constraint Programming Decomposition Approach for the Total Flow Time Fixed Group Shop Scheduling Problem

Francisco Yuraszeck [1,2], Gonzalo Mejía [3], Jordi Pereira [4,5] and Mariona Vilà [6,*]

[1] Facultad de Ingeniería, Universidad Andres Bello, Quillota 980, Viña del Mar 2531015, Chile; francisco.yuraszeck@unab.cl

[2] Escuela de Ingeniería Industrial, Pontificia Universidad Católica de Valparaíso, Avenida Brasil 2241, Valparaíso 2362807, Chile

[3] Facultad de Ingeniería, Universidad de La Sabana, Campus Universitario Puente del Común, Km 7 Autopista Norte de Bogotá, Chía 250001, Colombia; gonzalo.mejia@unisabana.edu.co

[4] Facultad de Ingeniería y Ciencias, Universidad Adolfo Ibáñez, Av. Padre Hurtado 750, Viña del Mar 2520001, Chile; jorge.pereira@uai.cl or jordi.pereira@bsm.upf.edu

[5] UPF Barcelona School of Management, Universitat Pompeu Fabra, C. Balmes 132-134, 08008 Barcelona, Spain

[6] Academic Department, EAE Business School, 08015 Barcelona, Spain

* Correspondence: mariona.vila.bonilla@upc.edu

Abstract: This work addresses a particular case of the group shop scheduling problem (GSSP) which will be denoted as the fixed group shop scheduling problem (FGSSP). In a FGSSP, job operations are divided into stages and each stage has a set of machines associated to it which are not shared with the other stages. All jobs go through all the stages in a specific order, where the operations of the job at each stage need to be finished before the job advances to the following stage, but operations within a stage can be performed in any order. This setting is common in companies such as leaf spring manufacturers and other automotive companies. To solve the problem, we propose a novel heuristic procedure that combines a decomposition approach with a constraint programming (CP) solver and a restart mechanism both to avoid local optima and to diversify the search. The performance of our approach was tested on instances derived from other scheduling problems that the FGSSP subsumes, considering both the cases with and without anticipatory sequence-dependent setup times. The results of the proposed algorithm are compared with off-the-shelf CP and mixed integer linear programming (MILP) methods as well as with the lower bounds derived from the study of the problem. The experiments show that the proposed heuristic algorithm outperforms the other methods, specially on large-size instances with improvements of over 10% on average.

Keywords: scheduling; fixed group shop; group shop; constraint programming

Citation: Yuraszeck, F.; Mejía, G.; Pereira, J.; Vilà, M. A Novel Constraint Programming Decomposition Approach for the Total Flow Time Fixed Group Shop Scheduling Problem. *Mathematics* **2022**, *10*, 329. https://doi.org/10.3390/math10030329

Academic Editor: Ripon Kumar Chakrabortty

Received: 20 December 2021
Accepted: 17 January 2022
Published: 21 January 2022

1. Introduction

In the academic world, traditional scheduling problems such as the flow shop scheduling problem, FSSP, the job shop scheduling problem, JSSP, or the open shop scheduling problem (OSSP) have been widely studied (see [1] for a general reference on scheduling problems). However, these scheduling problems may not cover all the requirements for specific manufacturing settings [2]. In this context, the group shop scheduling problem (GSSP) emerges as a generalized shop scheduling problem that includes, among others, the OSSP and the JSSP as special cases [3]. Due to its characteristics, the GSSP is a more flexible model with which address the requirements of multiple challenging real-life scheduling problems often found in the manufacturing industry.

In this paper, we consider a particular case of a GSSP that we denote as fixed group shop scheduling problem (FGSSP) [4]. In a fixed group shop environment, the operations of each job have been divided into stages, and the operations corresponding to each stage share the same set of machines. All jobs must proceed through each stage and perform the associated operations in the stage before proceeding to the next stage. Therefore, the FGSSP generalizes both OSSP and FSSP, and contains common features from many industrial environments in which manufacturing is organized in multiple sequential OSSP stages. By contrast, the classical GSSP formulation contains the OSSP and the JSSP as special cases, as each job may have a different route through the stages.

An example of a FGSSP can be found in mechanical workshops where routine car maintenance operations are performed. The number of operations for each job (car) and their processing times depend on different factors as the odometer count or the time between maintenances. For each car the set of maintenance operations can be divided into stages and the tasks to be performed on each stage must be completed before proceeding to the next stage (i.e., change the air filter, change the motor oil, etc.) with an ordering of stages predefined by the layout of the workshop, but there is no specific order in which operations within a single stage are to be performed (i.e., they have no relationship between them). Additionally, some setup operations may be required between jobs performed in a single machine, and thus sequence-dependent setup times are to be expected.

Another example of the FGSSP can be found in the manufacturing process of leaf springs. Each leaf requires several punching and forming operations that can be performed in any order. Once all these operations have been performed, the leaf is transferred to the heat treatment, sandblasting, painting, and assembly workstations. In the computational experiments section, see Section 6, we present a case study taken from a Colombian automotive company that falls within this specific example of application.

As in any other scheduling problems, multiple objective functions may be considered. In this work, we consider minimizing the total flow time.

To the best of our knowledge, the FGSSP has not been studied before even if the model has practical applications. Due to its computational complexity, it subsumes several well-known hard-to-solve problems. This work presents a novel ad hoc heuristic approach to solving the FGSSP with and without anticipatory sequence-dependent setup times under the total flow time minimization objective. According to the three-field notation proposed in [5], these problems can be denoted as the $FGSSP_s \mid S_{jik} \mid \sum_j C_j$ and the $FGSSP_s \mid\mid \sum_j C_j$, respectively.

The proposed heuristic relies on decomposing the problem into smaller subproblems and solving each subproblem through constraint programming (CP) [6] for a fixed amount of time. The heuristic can be seen as a hybrid metaheuristic [7] or a matheuristic [8] as it combines two optimization solution methods (i.e., a heuristic and an exact method approach). To test the performance of the decomposition approach, we performed extensive computational experiments with small, medium and large instances. We report the results of a computational study in which we compare our approach with off-the-shelf state-of-the-art CP and mixed integer linear programming (MILP) approaches. The results show the validity of our decomposition approach over a traditional method providing significant improvements over commercial solvers, specially for large instances.

The remainder of the paper is organized as follows. Section 2 reviews the literature on problems with similar characterstics to the FGSSP. Section 3 introduces the FGSSP, provides a problem definition, and gives an illustrative example of it. Section 4 puts forward a MILP and a CP formulation for the problem, and provides some lower bounds on the optimal objective value. Section 5 describes the decomposition-based procedure used to solving the problem, including the generation of initial solutions, the local search phase, and a shaking procedure designed to escape from local optima. Section 6 provides the results of the computational experiments conducted to test the method, as well as an industrial case study. Finally, Section 7 concludes and provides some possible research lines.

2. Literature Review

Scheduling corresponds to the allocation of scarce resources (i.e., machines) to perform tasks (i.e., jobs) over time [1]. Due to its generality and broad use, scheduling has become an important area within the operations research (OR) and operations management (OM) communities that focus their contributions on the development of decision-making methods to optimize one or multiple goals.

Among the scheduling problems, we focus our attention on a family known as "shop" problems. Among the different classifications of "shop" problems we are interested in a classification based on (i) their routes, that is, the path that the jobs must follow on the machines and, (ii) the sequence of operations that must be processed in each machine. The most common models in the literature consider that jobs follow a unique route (the FSSP), each job has its own route (the JSSP), or arbitrary routes (the OSSP)—but other cases exist. For these cases more elaborate models are needed to cope with different scheduling conditions.

An early example of these models can be found in [9]. In [9] the authors propose a hybrid model denoted as the mixed shop scheduling problem (MSSP) problem in which some jobs have their own predefined routes (i.e., as in a JSSP) and some jobs do not (i.e., as in an OSSP). Another example of alternative route schemes is the group shop scheduling problem (GSSP), also known as the stage shop problem [10]. The GSSP generalizes the MSSP and considers a set of distinct machines that perform operations on the jobs. Each stage must perform a subset of operations associated to the jobs and can perform these operations in any order within the stage, but the stages to be must be processed in a predetermined order. Note that the MSSP is a special case of GSSP, in which each stage has one operation or there is only one stage that contains all operations [11].

We now proceed to review the literature on the GSSP, as well as some works that make use of CP approaches within the scheduling literature.

The literature of the GSSP is abundant and mainly focuses on the makespan minimization objective [3,12–19]; other optimization objectives for the GSSP have been less studied. An example of other objectives can be found in [20], where the authors propose an application of a chance-constrained version of the GSSP with a total weighted completion time objective.

Sequence-dependent setup times, as well as transportation times have also been studied within the GSSP literature, [13]. In [21] the authors considered the use of a robot to transport material through the multiple processing stages in a GSSP environment.

Additionally, other publications have addressed stochastic and/or fuzzy extensions for the GSSP [12,13,20,22,23].

Regarding solution procedures, most of the literature focuses on metaheuristic approaches. Among them, genetic algorithms [13,18], Tabu Search [11,16,18,24], artificial bee colonies [17], iterated local search [24], Simulated Annealing [24], evolutionary algorithms [24], multi-start multi-level evolutionary local search [15] and ant colony optimization [3,24] are the most common. According to [24] the Tabu search showed the best results among the compared methods (an ant colony optimization, an evolutionary algorithm, an iterated local search, and a simulated annealing approach).

Exact methods, such as constraint programming (CP), have also been used to solve scheduling problems but, to the best of our knowledge, they have not been used to address the GSSP or similar problems. We review the works on exact methods for shop scheduling problems that are relevant to the method proposed in this work.

In [25] the authors proposed a CP approach to solve the JSSP, and in [26] the authors propose a CP approach to solve the OSSP. The approach proposed in [26] uses a new upper bound heuristic combined with constraint propagation and a branching technique to solve the problem. In [27] the authors proposed MILP and CP models to solve the online printing shop scheduling problem (OPSSP). The OPSSP can be seen as a JSSP where there are multiple units of some of the machines (hence, leading to a degree of flexibility within the sequence of operations for each job). The numerical experiments in [26] show that the CP method outperforms the MILP approach by a large extent. In [28], the authors used a CP model as their benchmark to compare the performance of a variable neighborhood search (VNS) for the OSSP with travel/setup times. Their VNS makes use of a probabilistic learning mechanism to self-tune a parameter that balance the generation of active or non-delay solutions. More recently, in [29] the authors proposed four CP formulations to tackle four complex flexible shop scheduling problems (i.e., the no-wait hybrid flow shop scheduling problem, the hybrid flow shop scheduling problem with sequence-dependent setup times, the flexible job shop scheduling problem with worker flexibility and the semiconductor final testing problem). Their experimental results report that the CP models outperform previously proposed solution methods. Authors in [30] address the distributed flexible job shop scheduling problem (an environment with multiple factories in which each factory is a flexible JSSP) comparing the performance of a MILP and a CP approached, showing that the CP method outperforms the CP.

A third type of solution procedure combines exact and heuristic approaches. Such methods are known as hybrid methods or matheuristics. In [31] the authors describe an method that combines constraint programming with a decomposition method and use it to solve the JSSP. Authors in [32] described a hybrid decomposition method to solve the continuous-time scheduling problem of multipurpose batch plants where the assignment of units to tasks is made using a MILP master problem, and CP subproblems are used to check the feasibility of specific assignments as well as to generate cuts for the master problem. Additionally, in [33], a hybrid method based on CP and local search is proposed in order to solve the routing and the scheduling of feeder vessels in multi-terminal ports. The results indicate that the the the variability in solution quality provided by local search heuristics can be decreased by combining of the local search and the CP method. In another study, authors in [34] provided a survey of intelligent scheduling systems. The work categorizes previous contributions according to five solution techniques: fuzzy logic, expert systems, machine learning, stochastic local search optimization algorithms, and CP. Lastly, authors in [35] hybridize a VNS with a CP search strategy for the OSSP with operation repetitions under a makespan criterion, showing good performance on the tested instances.

3. The Fixed Group Shop Scheduling Problem

3.1. Problem Definition

The fixed group shop scheduling problem (FGSSP) is a variant of the group shop scheduling problem (GSSP) in which not only jobs, but also machines are grouped into stages.

The FGSS considers a set of n jobs $J = \{J_1, J_2, \ldots, J_n\}$, each of them consisting of a set of non-preemtive operations $o_{ij} = \{O_{j1}, O_{J_2}, \ldots, O_{jm}\}$ that must be performed on a set of m machines $M = \{M_1, M_2, \ldots, M_m\}$. Each job $j \in J$ must be processed by each machine and must proceed through each stage $S = \{S_1, S_2, \ldots, S_s\}$, wherein a subset of its operations must be performed before advancing to the next stage. The operations of all jobs $j \in J$ that must be processed at stage S require the same set of machines.

As in the GSSP, in the FGSSP all jobs must perform an operation on each machine, and the operations associated to a given job in a given stage can be performed in any order. Unlike the GSSP, in the FGSSP each machine is associated to a given stage and stages are ordered in a fixed route that all jobs perform. Consequently, when the number of machines in each stage is 1, the GSSP becomes a job shop, while the FGSSP becomes a flow shop. The OSSP is both a special case of the GSSP and the FGSSP in which all operations belong to a single stage.

3.2. An Illustrative Example

Table 1 provides a small-size example with 3 jobs and 7 machines for a total of 21 operations. The table details the processing times of each operation associated to each job in the 7 machines.

A solution to the FGSSP can be visualized through a classical disjunctive graph representation or a Gantt chart. Figure 1a provides an arbitrary solution to the example problem with $\sum_j C_j = 60$, $C_1 = 23$, $C_2 = 12$, $C_3 = 24$, where C_j is the completion time of job j. The red dotted arcs in Figure 1a show the sequence at the machines and the black dotted arcs show the groups-permutations (i.e., the route of operations for J_1 at S_1 is O_{12}, O_{13} and

O_{11} then the route within the stage is $M_2 - M_3 - M_1$). A Gantt chart representation of the solution is provided in Figure 1b.

Table 1. Processing times of the small-size instance. For every job, J_1 ,J_2 and J_3, and machine, $M_1, \dots, M_7$, the processing time is provided. Additionally, machines are grouped according to their stage, S_1, S_2 and S_3.

Stages		S_1		S_2		S_3	
	M_1	M_2	M_3	M_4	M_5	M_6	M_7
J_1	1	3	4	1	1	9	1
J_2	2	1	3	2	2	1	1
J_3	3	2	1	2	4	2	3

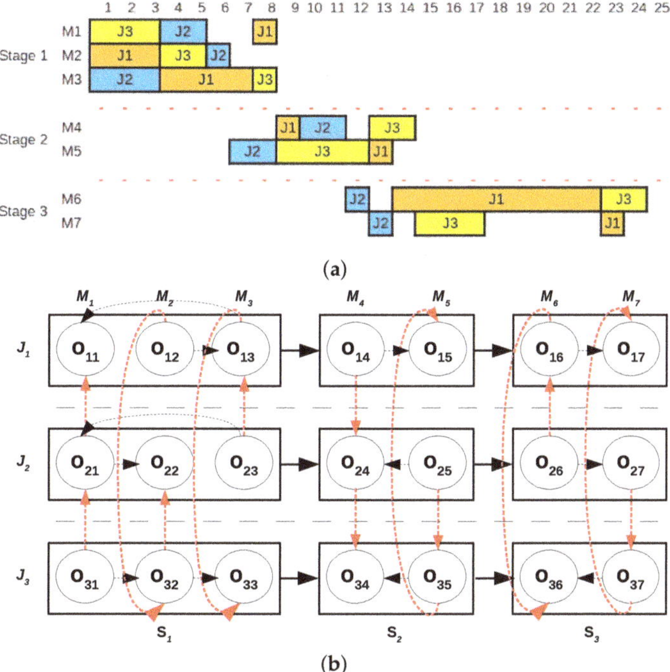

(a)

(b)

Figure 1. Graphical representations of a solution to the example instance provided in Table 1. (**a**) Gantt chart representation of an arbitrary feasible solution for the FGSSP instance presented in Table 1 with $\sum_j C_j = 60$. (**b**) Disjunctive graph representation of an arbitrary feasible solution for the FGSSP instance presented in Table 1 with $\sum_j C_j = 60$.

The representations in Figure 1a,b show the major characteristics of a FGSSP solution. The disjunctive graph representation visualizes the FGSSP as a sequence of serially arranged OSSP subproblems. Once a job finishes all operations in a stage and then job can start its operations in the subsequent stage. The Gantt chart representation also shows the FSSP behavior among stages. While in the GSSP, a job may have different machines in any given stage, in the FGSSP each job has the same machines in each stage. As a result, the machines of later stages remain idle until operations in preceding stages are completed. These differences motivate the need to separately consider resolution procedures for the FGSSP.

4. FGSSP Formulations and Lower Bounds

This section presents an MILP and a CP formulation for the FGSS problem with total flow time minimization objective and anticipatory sequence-dependent setup times ($FGSS_s \mid s_{jik} \mid \sum_j C_j$). The section also introduces three lower bounds on the value of the optimal objective function. The extension of both formulations for the case without setup times is straightforward, and the changes are described after providing the models with sequence-dependent setup times.

4.1. MILP Formulation

The formulation is an adaptation of the formulation provided in [28] for the $OSSP_m \mid S_{jik} \mid \sum_j C_j$. We now proceed to define the parameters, sets, indices and decision variables of the formulation.

Parameters and Indices:

- $nbJobs$: Number of jobs.
- $nbMchs$: Number of machines.
- $nbStgs$: Number of stages.
- j, k: Indices for jobs, $\{1, \ldots, nbJobs\}$.
- i, l: Indices for machines, $\{1, \ldots, nbMchs\}$.
- s: Index for the stages, $\{1, \ldots, nbStgs\}$.
- o_{ji}: Operation associated to job j at machine i.
- p_{ji}: Processing time of operation o_{ji}.
- S_{jik}: Setup time of job j if it is performed immediately after job k on machine i ($j \neq k$).
- B_{is}: 1 if machine i belongs to stage s; and 0 otherwise.
- A_{li}: 1 if machine l belongs to the stage immediately before the stage to which machine i belongs; and 0 otherwise ($i \neq l$).
- M: A sufficiently large number.

Decision variables:

- C_j: Continuous variable that takes the value of the completion time of job j.
- C_{ji}: Continuous variable that takes the value of the completion time of job j at machine i.
- f_{js}: Continuous variable that takes the value of the completion time of job j at stage s.
- x_{jil}: Binary variable that takes value equal to 1 if operation o_{ji} is performed after operation o_{jl}; or 0 in any other case.
- y_{jik}: Binary variable that takes value equal to 1 if operation o_{ji} is performed after operation o_{ki}; or 0 in any other case.

An MILP formulation follows.

$$\min \sum_{j=1}^{nbJobs} C_j \tag{1}$$

$$\text{s.t.} \quad C_{ji} \geq p_{ji} \qquad \forall i,j \tag{2}$$

$$C_{jl} \geq C_{ji} + p_{jl} - Mx_{jil} \qquad \forall i,j,l \mid i \neq l \tag{3}$$

$$C_{ji} \geq C_{jl} + p_{ji} - M\left(1 - x_{jil}\right) \qquad \forall i,j,l \mid i \neq l \tag{4}$$

$$C_{ji} \geq C_{ki} + p_{ji} + S_{jik} - M\left(1 - y_{jik}\right) \qquad \forall i,j,k \mid j \neq k \tag{5}$$

$$C_{ki} \geq C_{ji} + p_{ji} + S_{kij} - My_{jik} \qquad \forall i,j,k \mid j \neq k \tag{6}$$

$$C_{ji} - p_{ji} \geq C_{jl}A_{li} \qquad \forall i,j,l \mid i > l \wedge A_{li} = 1 \tag{7}$$

$$f_{js} \geq B_{is}C_{ji} \qquad \forall i,j,s \tag{8}$$

$$C_j \geq f_{js} \qquad \forall j,\, s = nbStgs \tag{9}$$

$$C_{ji} \in \mathbb{Z}^{\geq 0} \qquad \forall i,j \tag{10}$$

$$f_{js} \in \mathbb{Z}^{\geq 0} \qquad \forall j,s \tag{11}$$

$$x_{jil} \in \{0,1\} \qquad \forall i,j,l \mid i \neq l \tag{12}$$

$$y_{jik} \in \{0,1\} \qquad \forall i,j,k \mid j \neq k \tag{13}$$

The objective (1) minimizes the total flow time, i.e., the sum of completion times of the jobs. Constraint set (2) imposes that the completion time of each operation must be larger than the processing time of the job. Disjunctive constraints sets (3) and (4) ensure that each job is not processed in two machines simultaneously. Constraints sets (5) and (6) consider anticipatory sequence-dependent setup times and ensure that each machine does not perform multiple jobs simultaneously. Constraint set (7) defines that the starting time of job j at machine i must be equal to or greater than the completion time of job j at machine l if and only if machine l belongs to the stage immediately before the stage to which machine i belongs. Constraint set (8) calculates the completion time of a job j in a stage s as the maximum completion time of the job j on the machines belonging to the stage. Constraint set (9) computes the flow time of a job as the completion time in the last stage. Finally, constraint sets (10)–(13) define the domain of the decision variables.

Note that while we do not provide a model for the case without sequence-dependent setup times, removing, or setting to 0 the values of, S_{jik} in constraint sets (5) and (6) constitutes a valid model for the case without setup times.

4.2. CP Formulation

As in the MILP case, we develop a CP formulation for the FGSSP problem with total flow time minimization objective and anticipatory sequence-dependent setup times ($FGSS_s \mid s_{jik} \mid \sum_j C_j$). The formulation makes use of several constructs that are available in many CP modeling languages. Specifically, we use interval and sequence variables as well as specific scheduling constraints that are available in the IBM CP Optimizer solver as it is the one used in our our experimental tests.

An interval variable is a construct defined by two variables (the start value and the end value of the interval) as well as a known parameter, the size, that indicates the difference between the end and the start value. A sequence variable is a construct that encodes an ordering of variables. Here, the sequence variables provide an ordering of interval variables corresponding to jobs and machines.

We now proceed to describe the elements of the proposed model.
Parameters and Indices:

- $nbJobs$: Number of jobs.
- $nbMchs$: Number of machines.

- j, k: Indices for the jobs, $\{1, \ldots, nbJobs\}$.
- i, l: Indices for the machines, $\{1, \ldots, nbMchs\}$.
- o_{ji}: Operation associated to job j at machine i.
- p_{ji}: Processing time of operation o_{ji}.
- A_{li}: 1 if machine l belongs to the stage immediately before the stage to which machine i belongs; and 0 otherwise ($i \neq l$).
- T_i: A transition matrix that reports the minimum delay required by any pair of jobs j, k, to perform in machine i. The transition matrix values equal S_{jik}.

Decision variables:

- $itvs_{ji}$: Interval variables that define the start and the end of the operation of job j at machine i. The interval variable ensures that the difference between the start and the end value equals the processing time p_{ji}.
- $jobs_j$: Sequence of interval variables $itvs_{ji}$ associated to the operations of job j.
- $mchs_i$: Sequence of interval variables $itvs_{ji}$ associated to operations performed in machine i.

The objective function consists of minimizing the total flow time of jobs, which is computed using the end value of the interval variables:

$$\min \sum_{j=1}^{nbJobs} \max_{i=1}^{nbMchs} endOf(itvs_{ji}) \tag{14}$$

where $endOf()$ is an integer expression that reports the end of an interval variable. Consequently, $\max_{i=1}^{nbMchs} endOf(itvs_{ji})$ reports the flowtime of job j and (14) provides the total flowtime.

The model contains three constraint sets, (15)–(17).

$$noOverlap(jobs_j) \qquad\qquad \forall j \tag{15}$$

$$noOverlap(mchs_i, T_i) \qquad\qquad \forall i \tag{16}$$

$$endBeforeStart(itvs_{jl}, itvs_{ji}) \qquad\qquad \forall i, j, l \mid i > l \wedge A_{l,i} = 1 \tag{17}$$

Constraint set (15) ensures that each job j is processed on no more than one machine i at any given time (i.e., since $jobs_j$ is the subset of operations associated to a job, j $noOverlap()$ ensures that the intersection of these intervals is empty).

Constraint set (16) ensures that each machine i does not process more than one job j at a time. Moreover, the transition matrix T_i enforces the setup times between two consecutive operations (the difference between the finalization of an operation and the start of the succeeding operation must be no smaller than their corresponding values in the transition matrix).

Finally, constraint set (17) enforces the stage condition by ensuring that the end of all operations of any given job in a given stage must precede the start of any operation of said job in the next stage.

As in the case of the MILP model, the proposed model can be adapted to the case without sequence-dependent setup times by ignoring setup time values. Here, the change applies to constraint set (16) and the transition matrix of each machine i.

4.3. Lower Bounds

We provide three lower bounds that serve as a basis for comparison of our solution methods. Moreover, as the lower bounds relax some of the conditions of the FGSSP, the gap between the solutions to the FGSSP and the lower bounds may help identify some sources of complexity of the problem, see the results provided in Section 6.

4.3.1. Lower Bound LB_1

This lower bound considers that the completion time of each job must be no smaller than the sum of its processing times at the machines. Consequently, we can obtain a lower bound by summing the operation time of each job on each machine, see (18). We should expect this bound to be tight when routing decisions are not important, and the stages do not play an important role in the instances, that is, problems where it is possible to obtain solutions without idle times.

$$lb_1 = \sum_{j \in J} \sum_{i \in I} p_{ji} \tag{18}$$

4.3.2. Lower Bounds LB_2 and LB_3

LB_2 and LB_3 both build upon the relationship of each stage of the FGSSP with the OSSP. As each stage of the FGSSP is an OSSP instance, we can derive a general lower bound by optimally solving (or finding a lower bound) on a OSSP instance with special characteristics (i.e., release dates and delivery times derived from the operation times in the remaining stages).

Consider any stage $s \in \{1, \dots, nbStgs\}$ and divide the set of stages into three groups, a first group with the stages $\{1, \dots, s - 1\}$ that contains all stages that precede stage s, a second group containing stage s, and a third group with stages $\{s + 1, \dots, s_{nbStgs}\}$ that correspond to the stages following stage s. Clearly, the optimal solution to the OSSP associated to stage s is a lower bound to the objective value of the FGSSP, as it disregards all other stages

Consequently, and to include the remaining stages into the calculation of the lower bound, we estimate the operation times required to complete the operations associated to these stages and associate them to the release dates and delivery times for the OSSP problem in stage s (i.e., we estimate the minimum time unit in which the job can start their operations in stage s and the minimum time required to finish the job once they depart stage s).

The resulting problem corresponds to problem $OSSP \mid r_j, d_j \mid \sum_j C_j$ or to problem $OSSP \mid r_j, d_j, S_{jik} \mid \sum_j C_j$ for case without or with sequence-dependent setup times respectively, and the optimal solution, as well as any lower bound of its value is a lower bound for the original FGSSP instance. In order to calculate the bound, we search for a solution to the resulting OSSP model for a limited amount of time using a CP exact solver, see Section 6, and report the optimal solution, if found, or best-known lower bound reported by the solver when the time limit is reached.

The described method provides $nbStgs$ different lower bounds, but we focus our attention on two of these bounds, i.e., the bounds provided by the first and the last stage, as they related problems are easier for the CP solver, and it is more likely that the solver finds the optimal solution, or a better lower bound, for them.

The lower bound for the first stage, LB_2, corresponds to the optimal resolution of problem $OSSP \mid\mid \sum_j C_j$, or $OSSP \mid S_{jik} \mid \sum_j C_j$ for the case with sequence-dependent setup times, plus the sum of operation times in the remaining stages, see Equations (19) and (20), as it is easy to show that the delivery times are constant values that add to the total flow time of the operations independently of the job they are associated to.

$$lb_2 = lb_{OSSP(s=1)\mid\mid\sum_j C_j} + \sum_{j \in J} \sum_{i \in I : s \geq 2} p_{ji} \tag{19}$$

$$lb_2 = lb_{OSSP(s=1)\mid S_{jik}\mid\sum_j C_j} + \sum_{j \in J} \sum_{i \in I : s \geq 2} p_{ji} \tag{20}$$

The lower bound for the last stage, LB_3 only contains release dates, which may play a role on the optimal schedule of the operations as release dates change the instance where the jobs are available. The resulting bounds correspond to Equation (21), for the case without sequence-dependent setup times, and (22), for the case with sequence-dependent setup times.

$$lb_3 = lb_{OSSP(s=nbStages)|r_j|\Sigma_j C_j} \tag{21}$$

$$lb_3 = lb_{OSSP(s=nbStgs)|r_j, S_{jik}|\Sigma_j C_j} \tag{22}$$

5. Proposed Solution Method

The proposed decomposition-based approach (which we will denote as DEC) exploits the inherent structure of the FGSSP. The structure of a fixed group sShop is similar to the structure of a flow shop but each stage corresponds to an OSSP rather than a single machine. This structure naturally leads to a decomposition in which each Open Shop is individually optimized considering that the sequence of operations on preceding and succeeding stages for each job and on each machine to be known and fixed.

While the approach does not globally optimize the problem, there are intrinsic advantages of the decomposition, specifically, (1) the subproblems do not structurally differ from the original problem and (2) the optimization of each stage allows for minor changes within other stages (i.e., the sequence is fixed but the start time and end time of each operation may vary to accommodate for the changes introduced within the stage under inspection). Moreover, as the sequence of most stages is fixed, the resulting problem is smaller and, supposedly, easier to solve through exact methods. As a result, the proposed method mixes exact and heuristic ideas into a single procedure, a type of method usually referred to as a matheuristic [8] within the literature.

Algorithm 1 provides an outline of the approach. The DEC algorithm creates an initial, incumbent, solution using a constructive heuristic that solves the scheduling problem of each stage sequentially, starting from the first stage, proceeding to the second stage and repeating the process until all *nbStgs* have been solved. After the initial solution is found, the local search phase is initialized. The local search attempts to improve the solution by solving the subproblems associated to each stage in non-sequential order. If an improving solution is found, the incumbent is updated and the local search is repeated. Otherwise, the incumbent is modified in order to escape from local optimality and the local search phase is called again.

Algorithm 1 gives an overview of the procedure. We now provide details of each step of the DEC method, including an example of the behavior of the algorithm solving the example introduced in Section 3.2.

Algorithm 1: Outline of the DEC procedure.

Read instance;
$incumbent \leftarrow \varnothing$;
for $s \in \{1, \ldots, nbStgs\}$ **do**
    $incumbent \leftarrow incumbent \cup solve(OSSP_s \mid r_j \mid \sum_j C_j)$, or
    $solve(OSSP_s \mid r_j, S_{ij} \mid \sum_j C_j)$;
end
$best \leftarrow incumbent$;
$change \leftarrow true$;
while *time limit not exceeded* **do**
    $change \leftarrow false$;
    $pending \leftarrow \{1, \ldots, nbStgs\}$;
    while $pending \neq \varnothing \wedge timelimitnotexceeded$ **do**
        $s \leftarrow random(pending)$;
        $pending \leftarrow pending \setminus \{s\}$;
        $candidate \leftarrow solve(FGSSP \mid\mid \sum_j C_j)$ (or $FGSSP \mid S_{jik} \mid \sum_j C_j$) with
        additional constraints on stages $\{1, \ldots, s-1\} \cup \{s+1, \ldots, nbStgs\}$;
        if $obj(candidate) \leq obj(incumbent)$ **then**
            $incumbent \leftarrow candidate$;
            $change \leftarrow true$;
            if $obj(candidate) \leq obj(best)$ **then**
                $best \leftarrow candidate$;
            end
        end
    end
    if *time limit not exceeded* **then**
        $incumbent \leftarrow shake(incumbent)$;
    end
end
return best;

5.1. Initial Solution

In order to obtain an initial solution to the problem, see lines 3–6 from Algorithm 1, the DEC method starts from an empty solution, and obtains a schedule for the operations of each stage by solving an open shop scheduling problem with release dates and with/without setup times with total flow time objective for each stage (i.e., problem $OSSP_m \mid r_j \mid \sum_j C_j$ or $OSSP_m \mid r_j, S_{ij} \mid \sum_j C_j$, according to [5]).

The procedure starts by obtaining a schedule for the first stage. For this stage, release dates are set to 0. For the remaining stages, stages 2 to $nbStgs$, we solve an OSSP with release dates for each job that are equal to their completion times in their previous stage, These release dates ensure that the operations for any job in a given stage cannot start before the operations of the job finish in previous stages.

Each subproblem is then solved using the model described in Section 4.2 considering only one stage, the stage under consideration, and adding a constraint set, see Equation (23), to impose release dates to the operations associated to each job.

$$startOf(itvs_{j,l}) \geq r_j \qquad\qquad \forall j, l \qquad\qquad (23)$$

Constraint set (23) imposes the release date condition by ensuring that the start of any operation cannot be smaller than the release date of the job. In constraint set (23), r_j stands for the release date of job j in the previous stages.

To illustrate the proposed method, let us consider the example introduced in Section 3.2. The construction procedure would start from Stage 1, solving an $OSSP_m \mid\mid \sum_j C_j$ problem with machines M1, M2 and M3. The completion time of the jobs in the optimal schedule correspond to 8, 6 and 6 time units for job 1, job 2 and job 3, respectively. These completion times constitute the release dates for the problem associated with stage 2. In this case, the optimal solution has completion times equal to 11, 10 and 13 for job 1, job 2 and job 3 respectively. Finally, we solve the problem associated to stage 3. The objective function value of the solution provided by the method is 54, Figure 2a shows the Gantt chart of the solution and Figure 2b its disjunctive graph representation.

As the OSSP is a computationally difficult problem by itself, the CP solver is truncated by imposing a time limit. The time limit given to the solver to solve each stage as well as the overall time devoted to the initialization step is controlled through an algorithmic parameter α %, that limits the total time devoted by the algorithm to the step. The time assigned to this step is then evenly divided into each stage to define the time limit set to the CP solver. Note that the time required to reach and to verify the optimal solution of the problem for any given stage may be smaller than the time limit. In this case, the remaining time is reserved for the local search step of the algorithm.

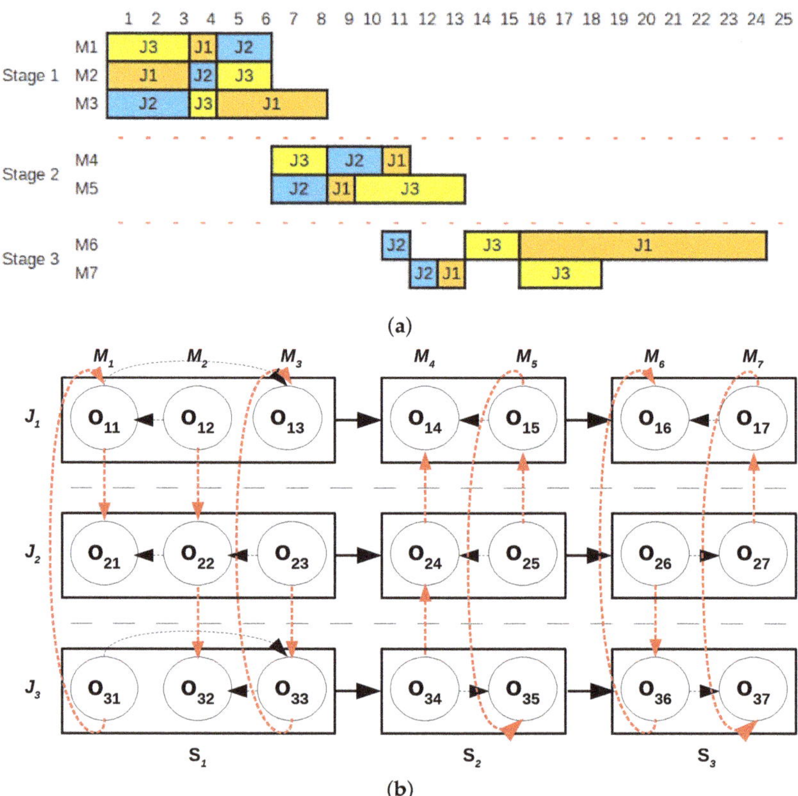

Figure 2. Graphical representation of the constructive heuristic solution of the DEC method for the example instance provided in Table 1. (**a**) Gantt chart representation of the solution provided by the constructive heuristic for the FGSSP instance presented in Table 1. (**b**) Disjunctive graph representation of the solution provided by the constructive heuristic for the FGSSP instance given in Table 1.

5.2. Neighborhood Exploration

The above constructive procedure provides a feasible solution in which greedy decisions in early stages may have a negative impact on later ones. Consequently, and after an initial solution is available, the neighborhood procedure tries to improve the incumbent by reoptimizing stages, taking into account the scheduling decisions from every other stage, see lines 12–24 from Algorithm 1.

The reoptimization stage is performed as follows: first, we add all stages to a list of pending problems. Then, we randomly select a stage from the list, say, stage s, remove it from the list, and fix the sequence of operations for each job and for each machine in the remaining stages, i.e., each stage $q \in S \setminus \{s\}$. The resulting model (i.e., the original model with some fixed variables) is then solved using the CP formulation provided in Section 4.2 truncating the search with a time limit which is a parameter of the method. When the time limit is reached or the solver returns that optimality has been verified the best-found solution is compared to the incumbent and the best ever solution

The neighborhood exploration step ends when the list is empty and no improving solution has been found during the last exploration step, in which case we conclude that a local optimum has been found and proceed to restart the local search by slightly altering the solution using the shaking procedure described in Section 5.3. Otherwise, the exploration step is repeated, i.e., the list is initialized with all stages and an optimization problem is solved for each stage, as described above.

To illustrate the proposed method, let us consider the example introduced in Section 3.2, starting from the solution found in Section 5.1 and depicted in Figure 2.

The neighborhood exploration phase starts by initializing the list of pending problems with the three stages. Then, we randomly select a stage from the list. For the sake of this example, let us suppose that stage 2 is selected. Then, stage 2 is removed from the list and a problem with the routes in stages 1 and 3 fixed is given for the CP solver for resolution. Figure 3a gives a disjunctive graph representation of the problem: the routes in stages 1 and 3 are fixed and the problem is allowed to reoptimize the scheduling decisions for stage 2. The optimal solution to the stage 2 problem improves the incumbent as the objective function value is decreased by two time units from 54 to 52 time units, see Figure 3b. The solution also improves the best found solution, hence it replaces both the incumbent, and the best found.

After updating the incumbent, the method would select another random stage among those still in the list, either stage 1 or 3, and build and solve their respective problems. The solutions to either problem do not provide a better solution and thus a complete iteration of the local search ends. As the method has found an improving solution within the last iteration, another iteration of the local search phase is performed. This second iteration does not lead to improvements, hence we conclude that the incumbent is a local optimum and stops the neighborhood exploration step.

Note that each problem solved in this phase is not theoretically easier than the original problem (i.e., they are NP-hard problems). Consequently, and in order to control the total time used within the resolution of the problems, as well as with the complete local search phase, we control both the total time used by the local search phase, and the time allocated to the CP solver to solve each subproblem. Section 6 gives details on the time allotted to each of these parameters in our computational experiments.

Finally, we attempt to improve the performance of the CP solver by providing a "warmstart" solution to it. In this case, we use the incumbent solution from the procedure as it is a feasible solution for the problem, including the additional constraints. As a result, the solver will never provide a worse solution than the initial one, and it will focus the search of areas that may provide improvements over the initial one.

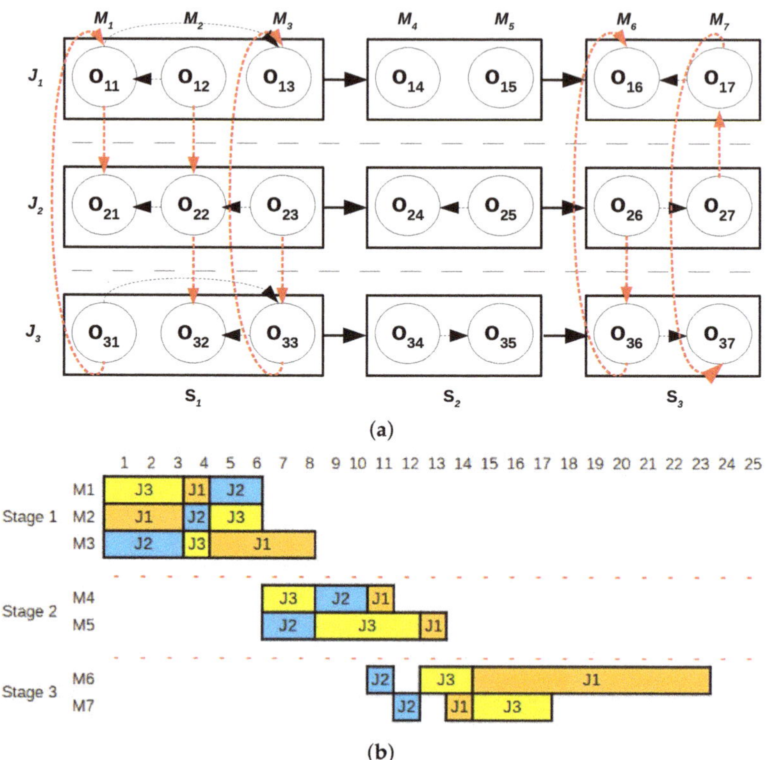

Figure 3. Graphical representation of the neighborhood exploration phase of the DEC method for the example instance provided in Table 1. (**a**) Disjunctive graph representation of the problem associated to the stage 2. The arcs represent the fixed decisions (i.e., the decisions from stage 1 and 3). (**b**) Gantt chart representation of the solution after solving the problem of stage 2. The solution improves the problem by rearranging the order of operations of the stage.

5.3. Shake Procedure

After reaching a local optimum, i.e., the neighborhood exploration step does not improve the incumbent, if the total time limit has not been reached we slightly perturb the solution in order to restart the search from a different position of the solution landscape, i.e., we perform a shake step as in a classical variable neighborhood (VNS) method [36]. Note that the term local optimum in this context is not completely correct, as the truncated nature of our neighborhood exploration scheme may lead us to report that no improving solution has been found when such a solution may exist.

The perturbation scheme considers two randomly-selected consecutive stages and creates an alternative set of routes and assignments for the jobs and the machines in these stages by solving the resulting CP model as in the neighborhood exploration step (i.e., fixing the sequences on the stages that we do not want to modify), but stopping the search when the solver provides a feasible solution and not including the incumbent as a warmstart solution. These decisions help the algorithm to find a solution with enough changes in these stages to move the complete solution away from the current local optimum while the use of the CP model ensures that a solution is found without having to rely on specifically tailored code to ensure feasibility conditions.

To continue with our example of the proposed method, let us continue with the example introduced in Section 3.2 and used in this Section. After reaching a local optimum, see Section 5.2, the incumbent depicted in Figure 2b is modified by selecting two consecutive stages and generating a random feasible solution for these stages. For the sake of this example, consider that stages 1 and 2 are selected. Then, we solve a problem in which the routes of stage 3 is fixed, and stop the search when the CP solver finds a solution. This solution becomes the incumbent and we return to the neighborhood exploration phase. Figure 4 illustrates the new solution.

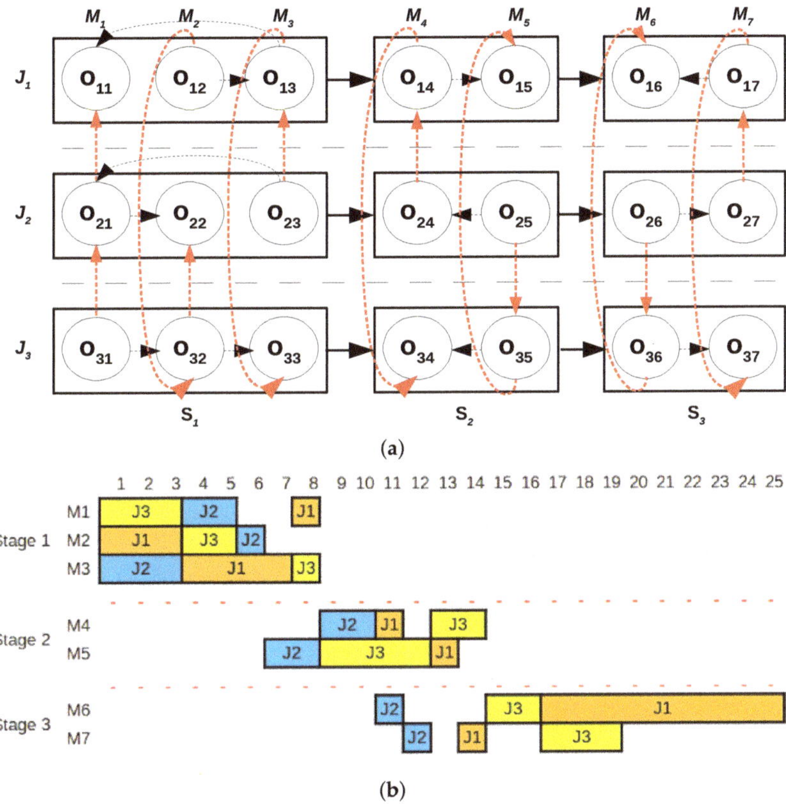

(a)

(b)

Figure 4. Graphical representation of the incumbent solution of the DEC method after the shake is performed for the example instance provided in Table 1. (**a**) Disjunctive graph representation of the incumbent solution after the shake. (**b**) Gantt chart representation of the incumbent solution after the shake.

6. Computational Experiments

All computational experiments were run on an Intel i7-10750H CPU @2.60 GHz with 6 cores and 16 GB of RAM. The code was written and Java and executed in the Java 8 runtime. The IBM ILOG CP and IBM ILOG CPLEX versions 20.1 were used to solve the CP and MILP formulations. The CP model was solved using five different strategies provided by the solver, namely: Auto (a combined search approach automatically controlled by the solver) CP DF (explores the search tree using a depth-first search approach) RS (combines a depth-first search approach with a restart mechanism after a certain number of backtracking decisions), MP (for multi-point search method, an approach that uses some of the characteristics of a population-based metaheuristic) and ID (for iterative diving search method, an approach that resembles a local search-based heuristic).

Each instance was run with the proposed DEC method with a total CPU time limit equal to $\frac{nbJobs\ nbMchs}{4}$ seconds, hence we allocate time proportional to the size of the instances. Consequently, for the exact solvers we provide the total time to the solver, while for the DEC approach, the total time allotted to the solution procedure is divided into an initialization phase, that takes a maximum of $\alpha\%$ of the total time, and the local search phase that takes the remaining time.

During the initialization phase, the resolution of the OSSP associated to each stage is allotted a maximum amount of time equal to $\frac{100}{nbStgs}\%$ of the total time allotted to this phase (i.e., $\frac{\alpha}{nbStgs}\%$). As the allotted time for each subproblem may not be used up (i.e., the CP solver may report that an optimal solution has been found before the time limit has been reached) the initialization phase may take less than the allotted time. Moreover, it is also possible that the CP solver does not find a solution within the time limit. While this never occurred during our experiments, the default implemented strategy allows the algorithm to continue the search until a feasible solution is found.

During the local search phase, each subproblem is allotted a fixed amount of time equal to λ. Parameter λ controls the trade-off between exploitation and exploration within the local search, i.e., a large value of λ has a higher chance to reach an optimal solution for the subproblems at the expense of considering fewer subproblems, while a smaller value of λ leads to considering a larger number of subproblems but the CP solver may fail to reach the optimal solution for the subproblem.

After some preliminary tests we opted for $\alpha = 0.25$ and $\lambda = \frac{100\,nbJobs\,nbMchs}{4nbStgs(1-\alpha)}$. This value of λ should lead to consider each stage no less than three times within the local search phase (our preliminary tests showed that this number usually sufficed to reach the best-found solution and reducing the time to solve each subproblem only lead to degradation in the solution quality).

For the exact methods we impose the following run time limits: For the MILP experiments, we impose a 3600 s time limit, while for the stand-alone CP solver we allocate the same running time as our decomposition approach. Please note that the larger amount of time devoted to the MILP formulation tries to ensure that the performance issues reported in Section 6.2 could not be solved by allocating more computational resources, i.e., running time, to the method.

6.1. Instance Generation

As no previous work for the FGSSP is available in the literature, we generated our own instance set. The generation procedure follows the procedure described in [28] for the $OSSP_m \mid S_{jik} \mid \sum_j C_j$, which extends the procedure described in [37]. Processing times for instances with up to 20 jobs and 20 machines are identical to the processing times used in [37]. For larger instances, the processing times were generated following the indications provided in [37], i.e., they are randomly generated using a discrete uniform distribution $U\,[1,99]$.

As a result, we generated instances with 4 to 80 jobs, 4 to 80 machines and 2 to 8 stages. In each instance, the first $nbMchs - \left\lfloor \frac{nbMchs}{nbStgs} \right\rfloor nbStgs$ stages contain $\left\lceil \frac{nbMchs}{nbStgs} \right\rceil$ machines, while the remaining stages have $\left\lfloor \frac{nbMchs}{nbStgs} \right\rfloor$. We generate 37 groups of instances, each containing 10 instances for a total of 370 instances.

For instances with sequence-dependent setup times we additionally generated setup times as follows: first we generate a random two dimensional Cartesian coordinate (x, y) for each job drawing each coordinate value from a discrete uniform distribution $U\,[0,30]$. Then, the setup time between any pair of jobs, j, k, in a given machine is computed as the rectilinear distance between the coordinates associated to each pair of jobs, $|x_j - x_k| + |y_j - y_k|$. This method ensures that setup times comply with the triangle inequality, hence $S_{jik} \leq S_{jiu} + S_{uik}$ for any triplet of jobs j, k, u and machine i. Finally, initial setup times were set to 0, i.e., we allow the machines to start working on any job without any setup.

As a result, a total of 740 instances were used for the reported experiments, 370 without sequence-dependent setup times and 370 with setup times.

6.2. Results for Small Size Instances

To evaluate the quality of the solutions provided by the lower bounds and the exact methods introduced in Section 4 we perform two sets of experiments using small instances (those with 10 or fewer jobs and machines) both on the instances with and without sequence-dependent setup times.

The first experiment considers the performance of the exact methods and the DEC procedure. The DEC procedure is run ten times with different random seeds and the results report their average performance among different runs as well as the best solution found within the ten runs.

Tables 2 and 3 report the results. For each solution method, we report the average relative gap *rel.gap*, see Equation (24), in which UB stands for the objective function value reported by the method and UB_b corresponds to the best-known objective value among all solution approaches, and, in parentheses, the number of best-known solutions found by the method. For the average performance of the DEC method, UB correspond to the average obtained by the ten independent runs. We also include the results provided by the DEC method using the MILP solver rather than the CP solver for comparison purposes.

$$rel.gap = 100\frac{UB - UB_b}{UB} \tag{24}$$

Table 2. Results for small instances without sequence-dependent setup times (problem $FGSSP_s \parallel \sum_j C_j$). For each instance size (represented by the number of jobs, column *nbJobs*, machines, column *nbMchs*, and stages, column *nbStgs*, we report the average gap to the best known solution and the number of best known solutions (in parentheses) provided by each CP search strategy (columns Auto, DF, RS, MP and ID), the best solution provided by all combined CP approaches (column CP), the results from the MILP approach (column MILP) and the best and the average found among 10 independent runs of the DEC approach (columns, (best) and (av.) respectively) using both the MILP and the CP solvers as their underlying methods to tackle the subproblems required by the approach (columns DEC MILP and DEC CP). The results of the best performing method for each group of instances are highlighted in boldface.

| | | | | | | | | | | DEC MILP | | DEC CP | |
nbJobs	*nbMchs*	*nbStgs*	Auto	DF	RS	MP	ID	CP	MILP	(av.)	(Best)	(av.)	(Best)
4	4	2	**0.0 (10)**	**0.0 (10)**	**0.0 (10)**	**0.0 (10)**	**0.0 (10)**	**0.0 (10)**	**0.0 (10)**	**0.0 (10)**	**0.0 (10)**	**0.0 (10)**	**0.0 (10)**
5	5	2	**0.0 (10)**	1.2 (0)	**0.0 (10)**	0.4 (4)	0.4 (5)	**0.0 (10)**	**0.0 (10)**	**0.0 (10)**	**0.0 (10)**	**0.0 (10)**	**0.0 (10)**
7	7	2	0.3 (4)	10.3 (0)	0.8 (2)	2.4 (0)	3.7 (0)	**0.1 (6)**	2.1 (0)	2.7 (0)	2.3 (0)	0.4 (3)	0.3 (4)
		3	0.1 (8)	13.7 (0)	0.3 (7)	1.6 (0)	2.5 (1)	**0.0 (10)**	0.5 (5)	2.4 (0)	1.9 (1)	2.6 (0)	2.5 (0)
10	10	2	2.2 (1)	14.7 (0)	1.7 (0)	3.7 (0)	3.7 (0)	1.3 (1)	8.9 (0)	11.1 (0)	10.4 (0)	0.5 (3)	**0.1 (9)**
		3	1.0 (2)	15.8 (0)	1.2 (2)	2.9 (1)	3.1 (0)	**0.4 (5)**	6.2 (0)	7.4 (0)	7.4 (0)	0.6 (1)	0.5 (5)
		4	1.0 (2)	15.7 (0)	1.1 (1)	1.6 (2)	2.5 (1)	**0.4 (6)**	3.4 (1)	4.9 (0)	2.9 (2)	1.9 (0)	1.4 (4)

Table 3. Results for small instances with sequence-dependent setup times (problem $FGSSP_s \mid S_{jik} \mid \sum_j C_j$). For each instance size (represented by the number of jobs, column *nbJobs*, machines, column *nbMchs*, and stages, column *nbStgs*, we report the average gap to the best-known solution (in parentheses) provided by each CP search strategy (columns Auto, DF, RS, MP and ID), the best solution provided by all combined CP approaches (column CP), the results from the MILP approach (column MILP) and the best and the average found among 10 independent runs of the DEC approach (columns, (best) and (av.) respectively) using both the MILP and the CP solvers as their underlying methods to tackle the subproblems required by the approach (columns DEC MILP and DEC CP). The results of the best performing method for each group of instances are highlighted in boldface.

| | | | | | | | | | | DEC MILP | | DEC CP | |
nbJobs	*nbMchs*	*nbStgs*	Auto	DF	RS	MP	ID	CP	MILP	(av.)	(Best)	(av.)	(Best)
4	4	2	0.0 (10)	0.0 (10)	0.0 (10)	0.0 (10)	0.0 (10)	0.0 (10)	0.0 (10)	0.0 (10)	0.0 (10)	0.0 (10)	0.0 (10)
5	5	2	0.0 (10)	4.1 (0)	0.0 (10)	0.5 (4)	2.2 (1)	0.0 (10)	0.0 (10)	0.0 (10)	0.0 (10)	0.0 (10)	0.0 (10)
7	7	2	1.4 (1)	15.7 (0)	1.4 (1)	2.3 (1)	4.8 (0)	0.7 (3)	1.4 (2)	**2.6 (0)**	**2.0 (0)**	0.3 (3)	**0.2 (5)**
		3	0.7 (6)	17.3 (0)	0.8 (3)	1.9 (2)	4.2 (0)	**0.1 (8)**	0.6 (5)	**3.1 (0)**	**2.3 (0)**	2.4 (1)	2.3 (1)
10	10	2	4.2 (0)	22.0 (0)	4.3 (0)	5.7 (0)	6.0 (0)	2.8 (0)	9.2 (0)	**13.1 (0)**	**11.3 (0)**	0.5 (3)	**0.0 (10)**
		3	2.2 (3)	23.6 (0)	1.4 (4)	3.9 (1)	5.5 (0)	**0.4 (8)**	5.4 (0)	**7.4 (0)**	**6.1 (0)**	1.5 (0)	1.2 (2)
		4	1.7 (3)	22.3 (0)	1.6 (3)	3.9 (0)	3.6 (1)	**0.3 (7)**	4.0 (1)	**6.1 (0)**	**4.2 (0)**	1.8 (0)	1.3 (2)

The results in Tables 2 and 3 show a similar trend, hence we discuss them together pointing out to the differences when needed:

- If we consider the behavior of the exact methods, i.e., the CP variants as well as the MILP, the results show that each of these methods have difficulties even for moderately small instances with 10 jobs and 10 machines. In fact, we do not report the number of optimal solutions found by any of these methods because they fail to verify optimality even for instances with 7 jobs and 7 machines and up. Note that these methods solve all instances with 4 or 5 jobs and machines to optimality, but the combined effort of all the exact methods only verifies optimality for four additional instances.

- Among the different search strategies available in the CP solver, all methods perform similarly except for DF. If we consider this result together with the difficulty of each exact method to verify optimality, we are led to believe that a depth-first search approach as conducted by the DF strategy fails to backtrack to the initial stages of the problem, leading to suboptimal early decision never being reconsidered.

- When we compare the CP approaches and the MILP approach, the CP outperforms the MILP method in every instance group and metric (either number of best found or relative gap to best known). Moreover, the additional time allocated to the MILP does not result in better solutions and the CP approaches, except for the DF strategy, outperform the MILP. Specifically, for instances with 10 jobs and machines, the MILP fails to find solutions of the quality provided by the CP approaches. Consequently, we recommend the use of a CP strategy for the problem and avoid the use of the MILP approach in larger instances.

- The CP methods do not perform as well on instances with sequence-dependent setup times. Specifically, relative gaps increase and two search strategies, i.e., Auto and RS, tend to provide the best solutions among the five search methods. This result may be attributed to shortcomings of the CP approach that makes use of internal components within its search procedure that are more efficient in problems with fewer features to consider.

- The performance of the DEC approach using a CP solver to tackle the subproblems for small and medium instances is similar to the exact CP methods. The same does not hold true for the DEC method using the MILP solver, as their results are inferior to either the CP or the DEC method using CP.

While the DEC method finds better solutions than the CP methods, specially on instances with fewer stages and the relative gaps are small, it does not outperform the exact methods for these instances. Please note that for small instances, the exact method benefits from considering the problem as a whole, unlike our method that tackles smaller parts of the complete problem. For small sized instances, dividing the problem into part leads to disadvantages in terms of the ability of the method to optimize all stages simultaneously.

The similarity between the results of both methods was statistically checked using a paired *t*-test for statistical significance. The paired *t*-test compares the best solution found by any CP method with the best found among the ten replicates of the DEC method using the CP solver, as well as with each of its individual runs.

The tests between the best solutions show that the results are not statistically different, with a *p*-value of 0.204 for the instances without sequence dependent setup times, and a *p*-value of 0.981 for the case with setup times. Note that Anderson–Darling tests show that the differences among values are not normally distributed, and thus we conduct Wilcoxon signed-rank non-parametric tests that confirm the results from the parametric tests. With regards to the statistical test between individual run of the DEC method when compared to the CP method, similar results are found. For the cases without sequence dependent setup times, six report statistical differences for the parametric test, but after a Bonferroni correction is run to account for multiple comparisons, none of the *p*-values suffice to point to statistically significant differences. For instances with setup times, none of the replicates report statistically significant differences to the best CP solutions.

To conclude, the results show that the exact methods fail to verify optimal solutions even for moderately small instances, being the CP approaches more competitive in terms of solution quality than their MILP counterparts. While the decomposition scheme can reach solutions of similar quality than the combined effort of all CP methods, and it even outperforms the exact methods for instances with a small number of stages, the results suggest that relying on exact methods is the best approach to solve small-sized instances.

To further analyze the performance of the exact methods, we conducted a second experiment considering the lower bounds introduced in Section 4.3 as well as the lower bounds reported by the CP and the MILP methods after reaching their termination condition, either proving optimality of the incumbent or reaching the imposed time limit.

Lower bounds lb_2 and lb_3 require the resolution of an OSSP model which is solved using a CP formulation for a fixed time limit equal to $\frac{nbJobs \cdot nbMchs}{4}$ seconds using the default, i.e., *Auto* strategy, provided by the CP solver. If the time limit is reached without verifying optimality, the lower bound provided by the code is used for the computation of lb_2 and lb_3. Tables 4 and 5 report, respectively, the results for small size instances without and with sequence-dependent setup times.

For each group of instances, we report the results for each lower bound described in Section 4, columns lb_1, lb_2 and lb_3, as well as the best lower bound reported by the CP methods and the MILP model. For each method, we provide two metrics; namely: the optimality gap, calculated as in (25), where ub_b is the best known solution and lb_x corresponds to the lower bound provided by the method and, in parentheses, the number of instances in which the lower bound provides the best bound among all of the methods.

$$opt.gap = 100 \frac{ub_b - lb_x}{ub_b} \tag{25}$$

Table 4. Results for the lower bounds for small-size instances without sequence-dependent setup times (problem $FGSSP_s \parallel \sum_j C_j$). For each combination of instance size (represented by the number of jobs, column *nbJobs*, machines, column *nbMchs*, and stages, column *nbStgs*, and solution method (columns lb_1, lb_2, lb_3, CP and MILP), we report the optimality gap and, in parentheses, the number of instances (out of 10) in which the method reported the best solution. The results of the best performing method for each group of instances are highlighted in boldface.

nbJobs	*nbMchs*	*nbStgs*	lb_1	lb_2	lb_3	CP	MILP
4	4	2	21.5 (0)	3.5 (0)	5.0 (0)	**0.0 (10)**	**0.0 (10)**
5	5	2	21.4 (0)	3.6 (0)	4.9 (0)	**0.0 (10)**	**0.0 (10)**
7	7	2	20.7 (0)	15.7 (2)	**12.7 (4)**	35.4 (0)	**12.7 (4)**
		3	24.2 (0)	16.7 (0)	8.9 (0)	21.1 (2)	**1.6 (10)**
10	10	2	22.1 (0)	**17.9 (10)**	19.5 (0)	46.2 (0)	40.4 (0)
		3	26.1 (0)	**19.3 (10)**	22.4 (0)	35.5 (0)	29.3 (0)
		4	28.9 (0)	**16.4 (6)**	19.7 (3)	32.0 (0)	22.6 (1)

Table 5. Results for the lower bounds for small-size instances with sequence-dependent setup times (problem $FGSSP_s \mid S_{jik} \mid \sum_j C_j$). For each combination of instance size (represented by the number of jobs, column *nbJobs*, machines, column *nbMchs*, and stages, column *nbStgs*, and solution method (columns lb_1, lb_2, lb_3, CP and MILP), we report the optimality gap and, in parentheses, the number of instances (out of 10) in which the method reported the best solution. The results of the best performing method for each group of instances are highlighted in boldface.

nbJobs	*nbMchs*	*nbStgs*	lb_1	lb_2	lb_3	CP	MILP
4	4	2	28.6 (0)	4.7 (0)	7.2 (0)	**0.0 (10)**	**0.0 (10)**
5	5	2	28.6 (0)	5.1 (0)	5.3 (0)	**0.0 (10)**	**0.0 (10)**
7	7	2	28.2 (0)	22.7 (0)	20.5 (2)	41.0 (0)	**13.9 (8)**
		3	31.4 (0)	24.7 (0)	10.9 (0)	34.2 (0)	**3.2 (10)**
10	10	2	30.6 (0)	**26.2 (10)**	27.6 (0)	51.8 (0)	44.4 (0)
		3	34.5 (0)	**28.3 (9)**	30.8 (0)	42.7 (0)	33.9 (1)
		4	36.9 (0)	**25.7 (8)**	30.8 (0)	39.4 (0)	27.6 (2)

The results show that large gaps are common. Specifically, for instances with 7 or 10 jobs and machines, the gap after reaching the time limit is very large, hence the inability of the exact solution methods to verify optimality as it cannot prune the search space through tight bounds and has to rely on enumeration to verify optimality. Moreover, the specially tailored lower bounds outperform the general bounds provided by the off-the-shelf solvers. but they still cannot provide tight bounds and the gaps are still large. Finally, we would also like to discuss the differences between the results provided by lb_2 and lb_3. While theoretically both bounds should report similar results (we try to optimally solve one stage and estimate the contribution of the remaining stages) the experiments show that lb_2 usually outperforms lb_3. We conjecture that this result comes from the performance of the CP solver on the problems solved using this approach. While lb_2 solves a classical OSSP as a subproblem, lb_3 solves an OSSP with release dates. The differences may be attributed to a better ability of the CP solver to solve the said subproblem.

To conclude. These results highlight the computational hardness of the problem and the need to rely on specially tailored heuristics to solve large-size instances.

6.3. Results for Medium and Large Size Instances

In this section, we report the results for medium to large-size instances. Due to the results found for small instances, we focus our analysis on solution methods and do not report lower bounds, as the large gaps found for small instances show the difficulty of finding good lower bounds.

Tables 6 and 7 show average results for these instances grouped according to the number of jobs, the number of machines and the number of stages. The tables compare the results of the best-performing CP strategy, the Auto strategy of the solver, the best solution found among the five search strategies provided by the solver the average result provided by ten independent runs of the DEC method and the best-found solution among these ten independent runs.

Table 6. Results for medium and large instances without sequence-dependent setup times (problem $FGSSP_s \,||\, \sum_j C_j$). For each instance size (represented by the number of jobs, column $nbJobs$, machines, column $nbMchs$, and stages, column $nbStgs$, we report the average gap to the best solution and the number of instances where the best known solution was found (in parentheses) by the best CP search strategy (column Auto), the best solution provided among the CP approaches (column CP), the results from the MILP approach (column MILP) and the best and the average found among 10 independent runs of the DEC approach (columns, DEC (best) and DEC (av.) respectively). The results of the best performing method for each group of instances are highlighted in boldface.

$nbJobs$	$nbMchs$	$nbStgs$	Auto	CP	DEC (av.)	DEC (Best)
15	15	2	2.9 (0)	2.3 (0)	0.3 (0)	**0.0 (10)**
		3	1.3 (2)	1.0 (2)	0.5 (3)	**0.2 (8)**
		4	0.3 (5)	**0.3 (7)**	1.4 (0)	1.1 (3)
		5	0.7 (4)	**0.4 (5)**	1.3 (1)	1.0 (5)
20	20	2	3.8 (0)	3.1 (0)	0.2 (3)	**0.0 (10)**
		3	2.5 (0)	1.9 (0)	0.3 (0)	**0.0 (10)**
		4	1.9 (2)	1.7 (2)	0.7 (0)	**0.1 (8)**
		5	1.3 (2)	1.1 (2)	0.8 (0)	**0.2 (8)**
		6	1.2 (3)	0.8 (4)	1.2 (0)	**0.6 (6)**
		7	1.0 (2)	**0.4 (6)**	1.4 (0)	1.0 (4)
30	30	2	5.7 (0)	4.5 (0)	0.0 (5)	**0.0 (10)**
		3	3.0 (0)	2.9 (0)	0.1 (5)	**0.0 (10)**
		4	3.5 (0)	2.5 (0)	0.2 (3)	**0.0 (10)**
		5	2.2 (0)	1.9 (0)	0.5 (2)	**0.0 (10)**
		6	1.9 (1)	1.6 (1)	0.5 (1)	**0.1 (9)**
		7	2.0 (1)	1.7 (2)	0.6 (0)	**0.0 (8)**
50	50	2	8.6 (0)	6.6 (0)	0.0 (2)	**0.0 (10)**
		3	5.7 (0)	5.4 (0)	0.1 (1)	**0.0 (10)**
		4	4.5 (0)	4.3 (0)	0.3 (0)	**0.0 (10)**
		5	4.0 (0)	3.7 (0)	0.1 (0)	**0.0 (10)**
		6	4.3 (0)	4.0 (0)	0.3 (0)	**0.0 (10)**
		7	3.3 (0)	3.2 (0)	0.0 (0)	**0.0 (10)**
		8	3.9 (0)	3.7 (0)	0.4 (0)	**0.0 (10)**
80	80	2	9.6 (0)	9.6 (0)	0.0 (7)	**0.0 (10)**
		3	6.9 (0)	6.6 (0)	0.1 (1)	**0.0 (10)**
		4	5.2 (0)	5.0 (0)	0.2 (1)	**0.0 (10)**
		5	3.7 (0)	3.6 (0)	0.2 (0)	**0.0 (10)**
		6	3.5 (0)	3.3 (0)	0.2 (0)	**0.0 (10)**
		7	3.3 (0)	3.1 (0)	0.3 (0)	**0.0 (10)**
		8	3.1 (0)	3.1 (0)	0.3 (0)	**0.0 (10)**

Table 7. Results for medium and large instances without sequence-dependent setup times (problem $FGSSP_s \mid S_{jik} \mid \sum_j C_j$). For each instance size (represented by the number of jobs, column *nbJobs*, machines, column *nbMchs*, and stages, column *nbStgs*, we provide the average gap to the best solution and (in parentheses) the number instances where the best CP strategy finds the best known solution (column Auto), the best solution provided among the CP approaches (column CP), the results from the MILP approach (column MILP) and the best and the average found among 10 independent runs of the DEC approach (columns, DEC (best) and DEC (av.) respectively). The results of the best performing method for each group of instances are highlighted in boldface.

nbJobs	*nbMchs*	*nbStgs*	Auto	CP	DEC (av.)	DEC (Best)
15	15	2	4.6 (0)	3.2 (0)	0.7 (0)	**0.0 (10)**
		3	1.6 (1)	1.3 (2)	0.6 (0)	**0.1 (8)**
		4	1.2 (5)	**0.6 (7)**	2.5 (0)	1.6 (3)
		5	2.7 (2)	**0.4 (6)**	1.8 (0)	1.3 (4)
20	20	2	5.0 (0)	4.1 (0)	0.5 (0)	**0.0 (10)**
		3	4.2 (0)	3.1 (2)	1.1 (0)	**0.3 (8)**
		4	2.1 (4)	**1.1 (4)**	2.8 (2)	2.0 (6)
		5	2.7 (1)	**0.8 (6)**	3.0 (2)	2.5 (4)
		6	2.4 (3)	0.8 (5)	1.2 (0)	**0.4 (5)**
		7	2.0 (0)	0.5 (4)	1.3 (0)	**0.7 (6)**
30	30	2	9.6 (0)	7.7 (0)	0.5 (4)	**0.0 (10)**
		3	5.4 (0)	4.2 (0)	0.5 (0)	**0.0 (10)**
		4	6.3 (0)	5.3 (0)	1.8 (0)	**0.0 (10)**
		5	3.4 (1)	3.3 (1)	1.5 (0)	**0.2 (9)**
		6	3.3 (1)	2.8 (2)	1.2 (0)	**0.1 (8)**
		7	3.4 (1)	2.5 (1)	1.0 (0)	**0.0 (9)**
50	50	2	12.6 (0)	8.4 (0)	0.1 (8)	**0.0 (10)**
		3	7.1 (0)	5.4 (0)	1.0 (0)	**0.0 (10)**
		4	9.9 (0)	9.1 (0)	1.9 (0)	**0.0 (10)**
		5	4.8 (0)	4.4 (1)	1.4 (0)	**0.1 (9)**
		6	6.2 (0)	6.0 (0)	1.1 (0)	**0.0 (10)**
		7	4.1 (0)	4.0 (0)	0.0 (0)	**0.0 (10)**
		8	4.7 (1)	4.3 (1)	1.5 (0)	**0.4 (9)**
80	80	2	14.4 (1)	12.2 (2)	2.0 (2)	**1.4 (8)**
		3	24.8 (0)	23.5 (0)	1.9 (0)	**0.0 (10)**
		4	23.2 (0)	21.9 (0)	3.4 (0)	**0.0 (10)**
		5	21.1 (0)	19.9 (0)	4.1 (0)	**0.0 (10)**
		6	16.0 (0)	14.0 (1)	3.0 (0)	**0.3 (9)**
		7	12.4 (0)	11.7 (1)	4.9 (0)	**0.1 (9)**
		8	14.2 (0)	13.1 (0)	1.9 (0)	**0.0 (10)**

The results show similar trends to those found in the small instances (i.e., a deterioration on the performance of the methods when the number of jobs and machines increase). Specifically, for large size instances, the relative gaps increase up to a 9.6% for instances with a small number of stages, and remain above 3% for any number of stages in instances without sequence-dependent setup times. For instances with sequence-dependent setup times, the gaps increase, reporting average relative gaps above 10% on average for any number of stages. These large instances highlight the advantages of the decomposition approach, which is still able to outperform the combined effort of all CP methods in most of the medium-sized and large-sized instances.

While the DEC decomposition approach still relies on a CP solver, the division of the larger problem into smaller subproblems that can be more efficiently tackled in short running times leads to clear improvements over the off-the-shelf method.

The dissimilarity between the results from the CP and the DEC methods were statistically checked using a paired *t*-test. The paired *t*-test compares the best solution found by each of the methods.

The test shows that the results are statistically different, with a *p*-value of 8.79×10^{-31} for the instances without sequence dependent setup times, and a *p*-value of 2.05×10^{-19} for the case with setup times. The Anderson–Darling test for normality showed that the differences between the values are not normally distributed, and thus we conduct Wilcoxon signed-rank non-parametric tests to confirm the results from the parametric test. The results of the Wilcoxon tests confirmed the conclusions reached by the parametric tests with a *p*-value of 7.18×10^{-46} for instances without setup times and a *p*-value of 2.05×10^{-43} for instances with setup times.

6.4. An Industrial Case Study

The case study provided below is taken from a Colombian automotive company. The company is dedicated to the manufacturing and assembling of leaf springs that are part of the suspension systems of cars and trucks. The company has over 200 hundred customers and exports to over ten countries. The customers are the car assemblers and the many car and truck repair shops and dealers of the country. A leaf spring consists of "leaves" that are metal plates that are bolted together. Each batch of springs of the same reference is considered a master production order (MPO). In turn, each batch of plates conforming a leaf spring is defined a single production order (SPO) derived from the MPO. Consider a typical reference with 10 plates. If an MPO for 100 leaf springs is issued, a total of 10 SPOs are generated, each with 100 leaf springs. The manufacturing of leaf springs consists of seven stages: (1) plate cutting, (2) center hole drilling and stamping, (3) tempering and quenching, (4) bending, (5) sand blasting, (6) painting and (7) assembly. The assembly operation was not considered in this research as this stage is not really scheduled. All stages have a single machine except stage two that has five forming operations. Stage two is generally the bottleneck station and, for this reason, the company keeps a buffer equivalent to 4–5 days of demand. Each SPO is transferred between workcenters by lift trucks. In this research, jobs correspond to SPOs. The 73 jobs used in this case study correspond to roughly the production of one week, which is the time lapse at which the schedule is revised and updated.

The total number of machines is 16. At the time of writing, there were one cutting station, one drilling machine, ten stamping presses, one tempering/quenching equipment, one bending and adjustment press, one sand blasting equipment and one painting workcenter.

Although the company has and uses an MRP system, the scheduling task is made manually. This is due to the inherent complexity of the manufacturing process and the constant pressure exerted by the vendors of the sales department. Although the company has implemented the Sales and Operations Planning (S and OP) methodology, frequent changes are common on the agreed schedules. For this reason, the company wanted to implement a scheduling system and wanted to test a prototype computer scheduler.

For the tests, we collected data of processing times and production orders from the MRP system. The processing times ranged between 25 min and 5 h depending on the operation. Setups are also important, but the company does not have exact records of the setups. For this reason, we generated setup times based on the suggestions of the plant personnel. Setups are only important in the stamping and drilling operations.

We ran all CP-based algorithms on the proposed instance. The CP DEC was run ten times. Table 8 summarizes the results.

Table 8. Results for case study. For each algorithm the objective value is reported.

Algorithm	Objective Function Value (Minutes)
CP AUTO	263,366
CP DF	293,522
CP RS	274,561
CP MP	293,442
CP ID	264,043
CP DEC (av.)	261,532
CP DEC (best)	260,349

As expected, the CP-DEC outperformed the other algorithms showing that the method can also performed well on realistic instances. The best performer among the CPs was the CP AUTO. The difference in terms of the objective function between CP AUTO and CP DEC was around 2200 min per week, which translates into an improvement of 30 min per job (2200/73).

After analyzing the schedule resulting from the CP DEC algorithm, we validated that the bottleneck station (as it is called by the plant personnel) was stage 2. The machine utilizations at this stage ranged from 25% to 78% (average 47%) whereas at the other stages, with the exception of tempering/quenching (66%), was around 30%. These figures of utilization are expected to be higher as the machines are always loaded with jobs from the previous week. We did not have such an information, and therefore we assumed that the factory floor was empty for the purpose of this case. In the experience of the authors, not only the better performance of the scheduling algorithms but also the information they provide, justifies its use.

7. Conclusions and Future Work

In this paper, we introduce fixed group shop scheduling problem (FGSSP) without/with sequence-dependent setup time. The FGSSP is a particular case of the group shop scheduling problem (GSSP) in which the machines of a given stage are the same for all jobs. This case can be found in different settings, as mentioned above.

We describe the characteristics of the proposed problem and provide two formulations, one based on mixed integer linear programming (MILP) and one on constraint programming (CP).

To solve the FGSSP, we developed a novel hybrid heuristic procedure based on a decomposition approach (which we denoted as DEC). Our procedure solves sequentially smaller scheduling problems with CP and presents a simple mechanism to escape from local optima. Moreover, the proposed method can accommodates for additional characteristics required in specific settings by introducing additional constraints within the formulations without the need to modify the solution procedure itself.

To test the performance of the approach, we performed computational experiments where we compare our method to the results provided by off-the-shelf CP and MILP solvers. Additionally, we computed several lower bounds for the FGSSP to have a baseline comparison.

The experimental results show that the DEC and all the tested CP are very similar in terms of performance for small and medium-sized instances, especially when the number of stages is small. For medium and large-sized instances, the DEC outperforms the CP methods with independence of the number of stages, finding the best solution in most of the cases.

Future work will be devoted to studying other solution approaches for the problem, to study the application of the proposed method to similar problems with the proposed approach and to study issues related to Industry 4.0 technologies, such as re-scheduling in the presence of real-time information and rework.

Author Contributions: Conceptualization, G.M., J.P. and F.Y.; methodology, G.M., J.P., M.V. and F.Y.; software, J.P. and F.Y.; validation, M.V. and F.Y.; formal analysis, G.M. and J.P.; investigation, F.Y.; resources, G.M., J.P., M.V. and F.Y.; data curation, F.Y.; writing—original draft preparation, G.M., J.P., M.V. and F.Y.; writing—review and editing, J.P.; visualization, M.V. and F.Y.; supervision, G.M. and J.P.; project administration, G.M.; funding acquisition, G.M., J.P. and M.V. All authors have read and agreed to the published version of the manuscript.

Funding: J.P. acknowledges the support of ANID through the grant FONDECYT No. 1191624 "Assembly line balancing for industry 4.0".

Institutional Review Board Statement: Not applicable.

Informed Consent Statement: Not applicable.

Data Availability Statement: Datasets available at https://github.com/yuraszeck/fgssp (accessed on 1 December 2021).

Conflicts of Interest: The authors declare no conflict of interest.

References

1. Pinedo, M.L. *Scheduling: Theory, Algorithms and Systems*, 5th ed.; Springer: New York, NY, USA, 2016.
2. Zobolas, G.I.; Tarantilis, C.D.; Ioannou, G. Exact, Heuristic and Meta-heuristic Algorithms for Solving Shop Scheduling Problems. In *Metaheuristics for Scheduling in Industrial and Manufacturing Applications. Studies in Computational Intelligence, 128*; Xhafa, F., Abraham, A., Eds.; Springer: Berlin/Heidelberg, Germany, 2008.
3. Blum, C.; Sampels, M. An ant colony optimization algorithm for shop scheduling problems. *J. Math. Model. Appl.* **2004**, *3*, 285–308. [CrossRef]
4. Yuraszeck, F.; Mejía, G.; Pereira, J. Modeling and Solving the Total Flow Time Fixed Group Shop Scheduling Problem. In Proceedings of the ICPR Americas, Bahía Blanca, Argentina, 9–11 December 2020; Editorial de la Universidad Nacional del Sur: Bahía Blanca, Argentina, 2020; pp. 2819–2822.
5. Graham, R.L.; Lawler, E.L.; Lenstra, J.K., Kan, A.H.G.R. Optimization and heuristic in deterministic sequencing and scheduling: A survey. *Ann. Discrete Math.* **1979**, *5*, 287–326.
6. Rossi, F.; Beek, V.P.; Walsh, T. *Handbook of Constraint Programming*, 1st ed.; Elsevier Science: Amsterdam, The Netherlands, 2006.
7. Blum, C.; Raidl, G.R. *Hybrid Metaheuristics: Powerful Tools for Optimization*, 1st ed.; Springer: New York, NY, USA, 2016.
8. Maniezzo, V.; Boschetti, M.A.; Stützle, T. *Matheuristics: Algorithms and Implementations (EURO Advanced Tutorials on Operational Research)*, 1st ed.; Springer: New York, NY, USA, 2021.
9. Masuda, T.; Ishii, H.; Nishida, T. The mixed shop scheduling problem. *Discret. Appl. Math.* **1985**, *11*, 175–186. [CrossRef]
10. Nasiri, M.M.; Kianfar, F. A hybrid scatter search for the partial job shop scheduling problem. *Int. J. Adv. Manuf. Syst.* **2011**, *52*, 1031–1038. [CrossRef]
11. Zubaran, T.K.; Ritt, M. An effective heuristic algorithm for the partial shop scheduling problem. *Comput. Oper. Res.* **2018**, *93*, 51–65. [CrossRef]
12. Ahmadizar, F.; Ghazanfari, M.; Mohammad, S.; Fatemi, T. Group shops scheduling with makespan criterion subject to random release dates and processing times. *Comput. Oper. Res.* **2010**, *37*, 152–162. [CrossRef]
13. Ahmadizar, F.; Rabanimotlagh, A. Group shop scheduling with uncertain data and a general cost objective. *Int. J. Adv. Manuf. Technol.* **2014**, *70*, 1313–1322. [CrossRef]
14. Ahmadizar, F.; Shahmaleki, P. Group-shop scheduling with sequence-dependent set-up and transportation times. *Appl. Math. Model.* **2014**, *38*, 5080–5091. [CrossRef]
15. Kemmoé-Tchomté, S.; Féniès, P.; Lamy, D.; Tchernev, N. A Multi-start Multi-level ELS for the Group-Shop Scheduling Problem. *IFAC-PapersOnLine* **2018**, *51*, 1299–1304. [CrossRef]
16. Liu, S.Q.; Ong, H.L.; Ng, K.M. A fast tabu search algorithm for the group shop scheduling problem. *Adv. Eng. Softw.* **2005**, *36*, 533–539. [CrossRef]
17. Nasiri, M.M. A modified ABC algorithm for the stage shop scheduling problem. *Appl. Soft Comput.* **2015**, *28*, 81–89. [CrossRef]
18. Nasiri, M.M.; Kianfar, F. A GA/TS algorithm for the stage shop scheduling problem. *Comput. Ind. Eng.* **2011**, *61*, 161–170. [CrossRef]
19. Nasiri, M.M.; Hamid, M. The stage shop scheduling problem: Lower bound and metaheuristic. *Sci. Iran.* **2020**, *27*, 862–879. [CrossRef]
20. Ahmadizar, F.; Ghazanfari, M.; Fatemi Ghomi, S.M.T. Application of chance-constrained programming for stochastic group shop scheduling problem. *Int. J. Adv. Manuf. Syst.* **2009**, *42*, 321–334. [CrossRef]
21. Nie, X.D.; Chen, Y.P.; Yang, Y.J. The Cyclic Scheduling of Material Transporting Robot in Group Shop. *Appl. Mech. Mater.* **2012**, *263–266*, 634–638. [CrossRef]
22. Ahmadizar, F.; Zarei, A. Minimizing makespan in a group shop with fuzzy release dates and processing times. *Int. J. Adv. Manuf. Syst.* **2013**, *66*, 2063–2074. [CrossRef]

23. Ahmadizar, F.; Rabanimotlagh, A.; Arkat, J. Stochastic group shop scheduling with fuzzy due dates. *J. Intell. Fuzzy Syst.* **2017**, *33*, 2075–2084. [CrossRef]

24. Sampels, M.; Blum, C.; Mastrolilli, M.; Rossi-doria, O. Metaheuristics for Group Shop Scheduling. In Proceedings of the LNCS 2439, PPSN VII, Granada, Spain, 7–11 September 2002; Merelo Guervós, J.J.; Adamidis, P., Beyer, H.-G., Schwefel, H.-P., Fernández-Villacañas, J.-L., Eds.; Springer: Berlin/Heidelberg, Germany, 2002; pp. 631–640.

25. Zhou, J. A Permutation-Based Approach for Solving the Job-Shop Problem. *Constraints* **1997**, *2*, 185–213. [CrossRef]

26. Malapert, A.; Cambazard, H.; Guéret, C.; Jussien, N.; Langevin, A.; Rousseau, L.M. An optimal constraint programming approach to the open-shop problem. *INFORMS J. Comput.* **2012**, *24*, 228–244. [CrossRef]

27. Lunardi, W.T.; Birgin, E.G.; Laborie, P.; Ronconi, D.P.; Voos, H. Mixed Integer Linear Programming and Constraint Programming Models for the Online Printing Shop Scheduling Problem. *Comput. Oper. Res.* **2020**, *123*, 105020. [CrossRef]

28. Mejía, G.; Yuraszeck, F. A self-tuning variable neighborhood search algorithm and an effective decoding scheme for open shop scheduling problems with travel/setup times. *Eur. J. Oper. Res.* **2020**, *285*, 484–496. [CrossRef]

29. Meng, L.; Lu, C.; Zhang, B.; Ren, Y.; Lv, C.; Sang, H.; Li, J.; Zhang, C. Constraint programing for solving four complex flexible shop scheduling problems *IET Collab. Intell. Manuf.* **2021**, *3*, 147–160. [CrossRef]

30. Meng, L.; Zhang, C.; Ren, Y.; Zhang, B.; Lv, C. Mixed-integer linear programming and constraint programming formulations for solving distributed flexible job shop scheduling problem. *Comput. Ind. Eng.* **2020**, *142*, 106347. [CrossRef]

31. Dorndorf, U.; Pesch, E.; Phan-Huy, T. Constraint propagation and problem decomposition: A preprocessing procedure for the job shop problem. *Ann. Oper. Res.* **2002**, *115*, 125–145. [CrossRef]

32. Maravelias, C.T.; Grossmann, I.E. A hybrid MILP/CP decomposition approach for the continuous time scheduling of multipurpose batch plants. *Comput. Chem. Eng.* **2004**, *28*, 1921–1949. [CrossRef]

33. Sacramento, D.; Solnon, C.; Pisinger, D. Constraint Programming and Local Search Heuristic: A Matheuristic Approach for Routing and Scheduling Feeder Vessels in Multi-terminal Ports. *SN Oper. Res. Formu* **2020**, *1*, 32. [CrossRef]

34. Fazel Zarandi, M.H.; Sadat Asl, A.A.; Sotudian, S.; Castillo, O. A state of the art review of intelligent scheduling. *Artif. Intell. Rev.* **2020**, *53*, 501–593. [CrossRef]

35. de Abreu, L.R.; Guimara es Araújo, K.A.; de Athayde Prata, B.; Nagano, B.S.; Moccellin, J.V. A new variable neighbourhood search with a constraint programming search strategy for the open shop scheduling problem with operation repetitions. *Eng. Opt.* **2021**. [CrossRef]

36. Hansen, P.; Mladenović, N. *Variable Neighborhood Search*. In *Handbook of Heuristics*; Martí', R., Pardalos, P.M., Resende, M.G.C., Eds.; Springer: Berlin/Heidelberg, Germany, 2018.

37. Taillard, E. Benchmarks for basic scheduling problems. *Eur. J. Oper. Res.* **1993**, *64*, 278–285. [CrossRef]

 *mathematics*

Article

Dynamic Optimization of the Multi-Skilled Resource-Constrained Project Scheduling Problem with Uncertainty in Resource Availability

Min Wang [1], Guoshan Liu [2] and Xinyu Lin [2,*]

1 College of Business Administration, Fujian Jiangxia University, Fuzhou 350108, China
2 Business School, Renmin University of China, Beijing 100872, China
* Correspondence: 2020000696@ruc.edu.cn

Abstract: Multi-skilled resources have brought more flexibility to resource scheduling and have been a key factor in the research of resource-constrained project scheduling problems. However, existing studies are mainly limited to deterministic problems and neglect some uncertainties such as resource breakdowns, while resource availability may change over time due to unexpected risks such as the COVID-19 pandemic. Therefore, this paper focuses on the multi-skilled project scheduling problem with uncertainty in resource availability. Different from previous assumptions, multi-skilled resources are allowed a switch in their skills, which we call dynamic skill assignment. For this complex problem, a nested dynamic scheduling algorithm called GA-PR is proposed, which includes three new priority rules to improve the solving efficiency. Moreover, the algorithm's effectiveness is verified by an example, and the modified Project Scheduling Problem Library (PSPLIB) is used for numerical experimental analysis. Numerical experiments show that when the uncertainty in resource availability is considered, the more skills the resource has and the more resources are supplied, the better the dynamic scheduling method performs; on the other hand, the higher the probability of resource unavailability and the more skills are required, the worse the dynamic scheduling method performs. The results are helpful for improved decision making.

Keywords: project scheduling; uncertainty in resource availability; multi-skilled resource; dynamic skill assignment

MSC: 90B36

Citation: Wang, M.; Liu, G.; Lin, X. Dynamic Optimization of the Multi-Skilled Resource-Constrained Project Scheduling Problem with Uncertainty in Resource Availability. *Mathematics* **2022**, *10*, 3070. https://doi.org/10.3390/math10173070

Academic Editors: Antonin Ponsich, Mariona Vila Bonilla and Bruno Domenech

Received: 12 July 2022
Accepted: 22 August 2022
Published: 25 August 2022

Publisher's Note: MDPI stays neutral with regard to jurisdictional claims in published maps and institutional affiliations.

1. Introduction

The resource-constrained project scheduling problem (RCPSP) has been an important topic within project management over the past few decades. Extensive focus has been placed on single-skilled resources, while multi-skilled resources are becoming increasingly common with the development of the economy [1]. This extension of the RCPSP is known as the multi-skilled resource-constrained project scheduling problem (MSRCPSP). It was inspired by a problem in the software development industry, where employees had several skills relating to programming, data analysis, debugging and so on [2]. MSRCPSP is suitable in projects with multi-skilled human resources or multi-functional machines. It has been a prevalent topic in recent years and has been gradually applied in production scheduling [3], research and development [4], construction engineering [5] and other projects [6,7]. Although multi-skilled resource increases scheduling flexibility and expands alternatives for project scheduling, it makes the problem more challenging. One needs to decide not only resource scheduling matters but also skill assignments.

In practice, resource unavailability is a frequent occurrence, especially in the wake of COVID-19, such as staff turnover, equipment maintenance, and transportation interruption. In this situation, project managers are forced to take a series of measures to make the project scheduling more efficient and to adapt quickly as possible to uncertainties [8]. Therefore, we focus on the MSRCPSP with uncertainty in resource availability in this paper. Moreover, distinct from previous assumptions, multi-skilled resources are allowed to switch their skills, which we call dynamic skill assignment. This means that when some resources with one skill are unavailable, the impact of resource shortage can be alleviated by skill switching from other idle resources with the same skill. If dynamic skill assignment still fails to make up for the shortage of resources, additional resources will be considered under the constraint of deadlines, such as the recruitment of temporary workers, equipment renting and so on. According to toWEC (Word Employment Confederation), temporary employment accounts for 70 percent of the global HR market, which is worth nearly USD 4 hundred billion. It plays an important role in reducing the cost and relieving the shortage of resources. Methods for optimizing project scheduling and dynamic skill assignment with uncertainty in resource availability so as to achieve the goal of minimum additional resource costs are the focus of this paper.

This paper has the following three contributions. First, we extend the MSRCPSP with uncertainty in resource availability, and the uncertainty is described by the Markov process. Second, dynamic skill assignment is proposed, which allows multi-skilled resources to switch skills. Third, a nested dynamic scheduling algorithm called GA-PR is proposed, which includes three new priority rules to improve the solving efficiency, and the effectiveness of the algorithm is proved by comparing the existing static and random scheduling method.

The remainder of this paper is organized as follows. A literature review is presented in Section 2. Definitions of the MSRCPSP with uncertainty in resource availability are discussed in Section 3. The nested dynamic scheduling algorithm is explained in detail in Section 4. A numerical example is provided in Section 5 to illustrate the new model and the new algorithm. The computational experiments and results analysis are shown in Section 6. Section 7 is the conclusion.

2. Literature Review

Although multi-skilled resource make solving MSRCPSP more flexible, it also renders the scheduling procedure more complex and difficult; thus, modern methods and tools are usually used to improve scheduling processes [9]. Methods for designing more effective scheduling procedures with modern methods have become important topics in the MSRCPSP. Bellenguez and Néron (2007) [2] proposed that each activity needs a specific set of skills, and the resources are staff members who possess fixed skill(s). Moreover, these staff members have unavailable periods. To minimize the makespan, a Branch-and-Bound method is proposed. Generally, the more skills a staff member possesses, the more costs are incurred. To minimize the total costs for multi-skilled personnel, Li and Womer (2009) [10] develop a hybrid Benders decomposition (HBD) algorithm that combines the complementary strengths of mixed-integer linear programming and constraint programming. Correia and Saldanha-da-Gama (2014) [11] consider that the costs associated with resources include fixed and variable costs. The fixed costs are incurred simply by using resources, while variable costs depend on the final makespan of the project. For this problem, a mathematical programming modeling framework is proposed, and a non-linear objective function is included, which can be linearized at the expense of an additional set of continuous variables. For resources considering skills, in addition to the cost, the skill level also directly affects the project scheduling scheme such that the higher the skill, the shorter the task duration. Heimerl and Kolisch (2010) [4] consider the MSRCPSP in a multi-project-environment (i.e., the processing of the projects' external and internal resources with different skills and at different performance levels). Thus, the question is how projects are scheduled and how resources are assigned to a project such that different requirements are met, keeping the costs minimal. To address the complex project-scheduling problem, a mixed-integer linear program with a tight LP-relaxation, which makes solving real-world problems possible, is proposed. A related problem was examined by Fırat and Hurkens (2012) [12]. The authors consider a mixed-integer-based approach for a multi-skill work-load problem and where skill levels are not homogeneous. Each activity has requirements for each skill-level combination. The goal is to maximize the number of tasks processed in each workday. Snauwaert and Vanhoucke (2021) [13] addressed an MSRCPSP with breadth and depth of skills, where the breadth of a resource is perceived as the amount of skills an employee masters and the depth of a skill is the efficiency level at which work can be performed by a resource that masters that skill. After that, in 2022, they studied how hierarchical skills (depth of skills) affect project scheduling from aspects of efficiency, cost, and quality [14].

Although the above studies provide references for the model and algorithmic design of MSRCPSP, they assume that the availability of renewable resources remains constant over time and rarely consider uncertainties in resource availability [1]. This assumption may be too strict. Resource availability might change in response to the availability of labor due to vacation days or varying availability of equipment due to maintenance [15]. A relatively common type of research is the project scheduling problem under uncertainties in project duration, including proactive scheduling [16] and reactive scheduling [17]. Once resources are unavailable (staff turnover, machine failure, etc.), the original scheduling is no longer feasible, especially for key resources. Therefore, the uncertainty in resource availability has gradually become an important and difficult point in project scheduling. Lambrechts et al. (2008) [18] introduced a variant of the RCPSP, for which the uncertainty in resource availability is considered. The objective is to find a robust schedule that minimizes the schedule's instability cost. The schedule's instability cost is the expected weighted sum of the absolute deviations between the planned and the actual starting times of the activity during the execution of proposed proactive and reactive strategies. Furthermore, to determine the impact of unexpected resource breakdowns on activity durations, they developed an approach for inserting explicit idle time into the project schedule. This was also implemented to protect it from disruptions caused by resource unavailability [19].

The literature cited above indicates that regardless of whether project scheduling considers uncertainties in terms of duration or resource availability, the idea is to set a buffer, namely time or resource buffer. The buffer can effectively protect the scheduling benchmark and improve the robustness of the solution. However, considering multi-skilled resources, effectively using this attribute to deal with the disturbance caused by uncertainties in resource availability has become a noteworthy problem. To our knowledge, there are only a few authors that incorporate uncertainty in resource availability in MSRCPSP, thus making this topic an interesting and novel path for research.

Ahmadpour and Ghezavati (2019) [20] provide a fuzzy scheduling model for the RCPSP, which considers fuzzy conditions for the calendar of the project. Multi-skilled human resources are also being considered to cope with the risk of resource shortages and delays in project completion. The results obtained from the fuzzy model for the value of objective function were evaluated under the influence of the resource calendar, consequently showing its benefits. The results provide a research idea for the MSRCPSP with resources uncertainty. However, this study assumes that once resources are assigned to a specific skill, they will be completely unchangeable until the end of the project. However, when a resource is unavailable due to resource uncertainty, other resources with the same skill are often recruited to continue the activity and avoid delay. When the resource is available again, the resource may be required to use other skills and perform other tasks to avoid information asymmetry caused by the resource transition. This multi-skilled resource dynamic skill assignment realizes the rotation of different resources with different skills among different tasks and alleviates the disturbance caused by the uncertainty in resource availability [21]. Compared with the situation that skills cannot be changed, this scheduling is closer to real-world issues, thus making resource scheduling more flexible and effective in dealing with absenteeism [22]. Moreover, as multi-skilled resources are often acquired through cross-training, a worker who does not frequently practice one skill may tend to forget it [23,24]. In the long run, multi-skill can be easily transformed into single-skill. Therefore, in this study, we relax the assumption that resources are allowed to change skills.

Compared to the existing literature, uncertainties in resource availability and dynamic skill assignment of multi-skilled resources are considered on the basis of MSRCPSP in this paper. When a resource is unavailable, it can be replaced by skill switching relative to other multi-skilled resources; alternatively, the external supplement of resources is another method. The uncertainty in resource availability is described by the Markov process. This is mainly because the evolution of availability or unavailability of resources is a discrete stochastic process that evolves with time. In other words, only the present determines the future, and the past is irrelevant. This corresponds to the Markov chain [25]. Then, we designed a nested scheduling algorithm combined with three new priority rules to solve this complex problem to minimize additional resource costs. Finally, the effectiveness of the proposed algorithm is verified by experimental analysis. The results of this research are helpful in the decision making of multi-skilled project scheduling. It is also valuable for expanding the project scheduling research in an uncertain environment, especially since uncertainty appears to be an extremely difficult element to deal with. Although most researchers recognize the importance and ubiquity of uncertainty, it remains the most popular topic of future research sections in many papers [26].

3. Problem Description

In the MSRCPSP, we employ the activity-on-node (AoN) representation and assume a zero time lag for precedence relations. The project consists of $n + 2$ activities. The duration of activity $j (j \in V, V = \{0, 1, 2, \ldots, j, \ldots n+1\})$ is $d_j (d_j \in N_0)$ and $d_0 = d_{n+1} = 0$. We assume that the project needs K types of renewable resources, the set is $R(R = \{1, 2, \ldots, k, \ldots, K\})$, and the availability of each type of resource is $|R_k|$; that is, $R_k = \{r_{k1}, r_{k2}, r_{k3}, \ldots, r_{ki}, \ldots\}$. Moreover, the project needs s_K types of skills, and the set is $S_k (S_k = \{s_1, s_2, \ldots, s_k, \ldots, s_K\})$. The skill(s) mastered by each resource is predefined, and we assume that all resources in the set of R_k master skill s_k, which we call the initial skill of resource r_{ki}. Whether there is mastery of other skills or not is randomly generated, and it is indicated by x_{r_{ki}, s_k}, where $x_{r_{ki}, s_k} = 1$ indicates that resource r_{ki} has mastery of skill s_k. When $\sum_{s_k \in S_k} x_{r_{ki}, s_k} > 1$, resource r_{ki} masters more than one skill. We describe the uncertainty in resource availability by the Markov stochastic process, and we set the state of the resource as $Z_t (Z_t = \{z_t^{r_{11}}, z_t^{r_{12}}, \ldots, z_t^{r_{1i}}, \ldots, z_t^{r_{ki}}, \ldots | t = 1, 2, \ldots, T\})$, where T is the deadline of project and $z_t^{r_{ki}}$ represents whether resource r_{ki} is available or not at time t. If it is available, $z_t^{r_{ki}} = 1$; otherwise, it is 0. We assume that the project starts at time zero; thus, Z_0 denotes the initial available/unavailable status of all resources, and we set $z_0^{r_{ki}} = 1$ for each resource.

The variables and parameters are shown in Table 1.

Table 1. Variables and parameters.

Variables and Parameters	
$x_{j,t}$	Binary variable, if activity j is executed in period $[t, t+1)$ or not.
x_{r_{ki}, j, s_k}	Binary variable, if resource r_{ki} with skill s_k or not.
$x_{r_{ki}, j, s_k, t}$	Binary variable, if resource r_{ki} with skill s_k in period $[t, t+1)$ or not.
$R_{s_k, t}$	The additional amount of the resource with skill s_k at period t.
$V = \{1, 2, \ldots, j, \ldots, n+1\}$	The set of activities.
$SUCC_j$	The immediate successor set of activity j.
$Z_t = \{z_t^{r_{11}}, z_t^{r_{12}}, \ldots, z_t^{r_{1i}}, \ldots, z_t^{r_{ki}}, \ldots | t = 1, 2, \ldots, T\}$	The set of resources state.
$z_t^{r_{ki}}$	Binary parameters, if the resource r_{ki} is available at time t or not.
S_j	The start time of activity j.
d_j	The duration of activity j.

Table 1. *Cont.*

Variables and Parameters	
$R = \{1, 2, \ldots, k, \ldots, K\}$	The set of K types of renewable resources.
$R_k = \{r_{k1}, r_{k2}, r_{k3}, \ldots, r_{ki}, \ldots\}$	The set of R_k.
$|R_k|$	The availability of R_k.
$S_k = \{s_1, s_2, \ldots, s_k, \ldots, s_K\}$	The set of skills.
x_{r_{ki}, s_k}	Binary parameters, if resource r_{ki} has mastery of the skill s_k or not.
r_{j, s_k}	The requirements of skill s_k for executing activity j.
$p_{r_{ki}}$	The unavailability probability of resource r_{ki}
c_{s_k}	The cost of the resource with skill s_k.
T	The deadline of the project.
$P_{r_{ki}}$	The state transition matrix of r_{ki}.
$u_t^{r_{ki}}$	The state probability vector of resource r_{ki} at period t.
e	The degree of infeasibility.
M	A sufficiently large penalty coefficient.

General RCPSP's goal is to study how to schedule activities under the constraint of resources and precedence relations in order to minimize the makespan of the project. While in the MSRCPSP, as resources are multi-skilled, it is necessary to decide not only activity scheduling but also the skill assignment of resources. To ensure that activities remain uninterrupted and to avoid delays in the project, we allow other idle resources with the same skill to replace the unavailable resources when the skill requirements of activities cannot be met. If this still does not work, additional resources with the same skill would be considered (purchasing, renting, or overtime). Adopting additional resources is an easy and popular method to increase flexibility [26]. According to some studies, it is a proper assumption where there is no difference in performance between temporary and permanent resources [27–29]. Given this situation, a problem arises as to which idle resource will be selected and how to assign them, how many additional resources are needed to meet skill requirements. To solve this, we set the goal to minimize additional costs. Thus, the MSRCPSP with uncertainties in resource availability studied in this paper can be described as follows: Under the constraints of precedence relations, resource availability, and project deadline, determine activity scheduling and dynamic skill assignment of multi-skilled resources to minimize additional costs of the project. The research assumptions of this paper is as follows:

(1) There are many types of multi-skilled resources needed by the project. Each resource can possess one or more skills, and the initial skill of the resource has priority in scheduling. Skill levels are homogeneous among resources.
(2) The availability or unavailability of different resources is independent of each other.
(3) The idle resources are allowed to switch skills to replace unavailable resources.
(4) The unavailability probability of resource r_{ki} is p_{ki}, and the state transition matrix of r_{ki} can be denoted as follows:

$$P_{ki} = \begin{pmatrix} 1 & 0 \\ p_{ki} & 1 - p_{ki} \end{pmatrix} \tag{1}$$

p_{ki} cannot be changed in per unit time, and the state $z_t^{r_{ki}}$ is subject to the Bernoulli distribution of p_{ki}. If u_t^{ki} represents the state probability vector of resource r_{ki} at period t, then $u_{t+1}^{ki} = u_t^{ki} P_{ki}$, and the initial state probability vectors of all resources are $(0,1)$.

(5) The initial resource cost of the project is fixed; in other words, it will not decrease or increase because of the unavailability of resources. Thus, the objective function does not consider the initial sunk cost of the resource and only considers additional costs.

Mathematically, the MSRCPSP with uncertainty in resource availability can be conceptually formulated as follows.

$$Min \sum_{k=1}^{K} c_{s_k} R_{s_k,t} + eM \tag{2}$$

$$S_j - S_g \geq d_j \; (j \in V, g \in SUCC_j) \tag{3}$$

$$S_{n+1} \leq T \tag{4}$$

$$\sum_{t \in [S_j, S_j + d_j]} x_{j,t} = d_j \; (j \in V) \tag{5}$$

$$\sum_{t \in (0,S_j) \cup (S_j + d_j, S_{n+1}]} x_{j,t} = 0 \; (j \in V) \tag{6}$$

$$\sum_{t \in T} t(x_{j,t} - x_{j,t-1}) = S_j \; (j \in V) \tag{7}$$

$$\sum_{s_k \in S_k} \sum_{j \in V} x_{r_{ki},j,s_k,t} \leq 1 \; (k \in K, \, t \in T) \tag{8}$$

$$\sum_{j \in V} r_{j,s_k} \leq \sum_{r_{ki} \in R_k} x_{r_{ki},s_k} z_t^{r_{ki}} x_{r_{ki},j,s_k,t} + R_{s_k,t} \; (k \in K, s_k \in S_k, \, t \in T) \tag{9}$$

$$u_{t+1}^{r_{ki}} = u_t^{r_{ki}i} P_{r_{ki}} \; (r_{ki} \in R_k, k \in K, \, t \in T) \tag{10}$$

$$x_{j,t}, x_{r_{ki},j,s_k}, x_{r_{ki},j,s_k,t} \in \{0,1\} \tag{11}$$

$$R_{s_k,t} \geq 0 \tag{12}$$

Objective Function (2) minimizes the cost of additional resources, and a penalty value of eM is considered if scheduling is infeasible. Constraint (3) ensures that the precedence relations among activities need to be satisfied. Constraint (4) ensures that the makespan of the project should not exceed the deadline. Equations (5) and (6) ensure that the activity cannot be interrupted. Equation (7) is a representation of the activity's start time. Constraint (8) ensures that every resource can be assigned only in one activity at any time. Constraint (9) guarantees that the activity's skills need need to be met. Equation (10) is the state probability vector of each resource. Equations (11) and (12) describe the domain of decision variables.

4. The GA-PR Algorithm

As the MSRCPCP is NP-Hard [30], the possibility of solving the problem optimally using exact solution procedures is limited by the size of instances. However, real instances of project scheduling problems are considerably large. Therefore, having efficient heuristics for finding good quality solutions is of great relevance, especially when considering the dynamic skill assignment of multi-skilled resources with uncertainties in resource availability.

The proposed model not only optimizes activity scheduling (the start time of each activity) but also resource scheduling (dynamic skill assignment of multi-skilled resources). To solve the new model, we designed a modified genetic algorithm combined with priority rules, called GA-PR. Based on the characteristics of the model, the algorithm is divided into two layers. The outer layer comprises activity scheduling optimization according to genetic algorithm, and the inner layer comprises resource scheduling optimizations according to priority rules.

4.1. The Outer Algorithm-GA

Step 1. Initialization of activity scheduling

In the outer layer, activity scheduling is the decision maker. Assuming that the project consists of $n+1$ activities, there are $n+1$ genes on each chromosome, representing the start time (ST) of each activity. The earliest start time (ES) of the dummy activity 0 is 1, and the project deadline is set as the latest start time (LS) of the dummy activity $n+1$. According to the forward and backward iteration algorithms in the critical path theory (CPM), the start time interval of activity j is $[ES_j, LS_j]$. The initial ST_j is a random value among $[ES_j, LS_j]$.

Step 2. Calculation of objective function

The objective is to minimize the additional resource cost. First, if the activity scheduling is subjected to precedence constraints, insert it into the inner algorithm. If not, the fitness value is a relatively large penalty value eM, where M is a sufficiently large penalty coefficient, and e reflects the degree of infeasibility—the degree of violating constraints. All individuals in the population are listed in a descending order of fitness value, and the individual with the minimum fitness value is set as the optimal individual.

Step 3. Selection, crossover, and mutation

The binary tournament method is used to select parent individuals from the population. Subsequently, crossover and mutation are carried out to generate the new population. Then the optimal individual is updated. This step is iterated until the maximum number of iterations has been reached, and then the final optimal individual is output. (Parameters, such as crossover probability and mutation probability, are determined after many tests.)

4.2. The Inner Algorithm-PR

Step 1. Identify the unavailability probability $p_{ki}(k = 1, 2, \ldots K, \ i = 1, 2, \ldots)$ of each resource and generate the resource state matrix Z_t based on the Markov process.

Step 2. Internal resources ranking

Assume that the project requires s_K types of skill, and there are K types of resource. For each type of resource, we rank the internal resource's scheduling order according to their skill number in ascending order, and the internal resource's scheduling order is A_k. Here, we define the first resource-scheduling priority rule.

Priority Rule 1: Within each type of resource, the one with the lowest skill number is scheduled preferentially because the one with more skills can replace unavailable resources.

Step 3. External skills ranking

The set of activities that are executed at moment t is O_t, generated based on activity scheduling, which is the outer layer's solution. Calculate the total demand of each type of skill at moment D_{st}, and calculate the total initial supply of each type of skill at moment S_{st} according to the resources' initial skills and their available state. The gap between D_{st} and S_{st} is defined as skill-demand tension L_{st}. The smaller the gap, the smaller the demand tension of s. We rank these skills according to the demand tension in ascending order, and the external skill order at moment t is W_t.

When the scheduling of the skill with the smallest demand tension is completed, the unscheduled resources with this skill can convert its skill to the next skill that needs to be scheduled and so on. Thus, the demand tension of the next skill can be alleviated. Therefore, here, we define the second skill scheduling priority rule.

Priority Rule 2: Among different skills, the one with the smaller demand tension is scheduled preferentially.

Step 4. Feasibility analysis of activity scheduling with resource constraints

There are three cases for resource scheduling with constraints.

Case 1: The skill demand tensions are all negative, indicating that the initial skill provided by resources can satisfy the demand under the given activity scheduling (see Step 5).

Case 2: The skill demand tensions are all positive, indicating that the initial skills provided by resources cannot satisfy the demand under the given activity scheduling. This means that every skill needs to be complemented by additional resources (see Step 6).

Case 3: In other cases, it indicates that the initial skills provided by the resources cannot satisfy the demand under the given activity scheduling, but multi-skilled resources may satisfy skill needs through skill switching (see Step 6).

Step 5. Resource scheduling without multi-skill

First, the skill with the smallest demand tension is scheduled. The scheduling order of resources with initial skill W_{1t} is $A_{W_{1t}}$. For each resource, it is necessary to determine whether it is available. If available, it is removed from $A_{W_{1t}}$ and added to the resource profile of the task in O_t. If it is unavailable, it is removed directly from $A_{W_{1t}}$.

Step 6. Resource scheduling with multi-skill

The cost of skill is used as the basis of the skill scheduling order; the skill with high cost is scheduled first to satisfy its demand as far as possible and to minimize the cost of additional resources as much as possible. Here, we define the third skill scheduling priority rule.

Priority rule 3: When the skill's demand tension is positive at a certain moment, the skill with high costs has priority.

For other cases, based on step 5, after completing the scheduling of skill W_{1t}, as its demand tension is the smallest, if $A_{W_{1t}}$ is a non-empty set, then merge $A_{W_{1t}}$ into $A_{W_{2t}}$ to schedule skill W_{2t} and so on. If it is still unable to satisfy the skills demand, to ensure that activity scheduling is feasible, additional resources should be considered, and the project's cost will increase.

Step 7. Inner iteration

Steps 4–6 are iterated before resource scheduling is completed. Then, output the final fitness, activity scheduling, and resource scheduling.

The flow chart of GA-PR algorithm is shown in Figure 1.

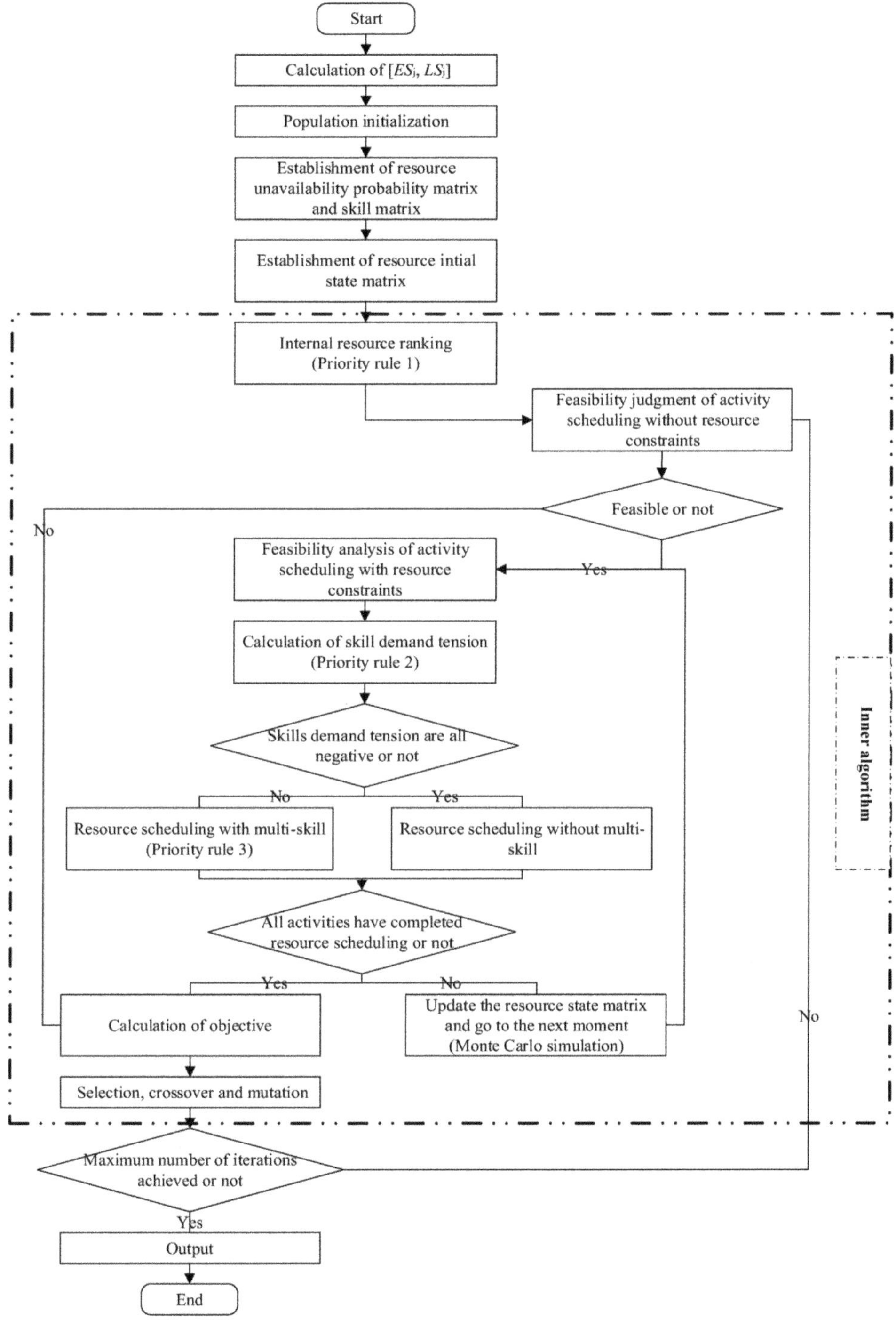

Figure 1. Flowchart of the GA-PR.

5. A Numerical Example

A project with 12 activities is defined to illustrate the new model and the new algorithm. The duration of each activity is known, and precedence relations are shown in Figure 2 as finish-to-start relations. Assume that the project needs three types of renewable resources: $R_1, R_2,$ and R_3, i.e., $K = 3$, and the total number of each type of resources was considered be 10; then, the set of each type of resources can be described as follows: $R_1 = \{r_{11}, r_{12}, r_{13}, r_{14}, r_{15}, r_{16}, r_{17}, r_{18}, r_{19}, r_{110}\}$; $R_2 = \{r_{21}, r_{22}, r_{23}, r_{24}, r_{25}, r_{26}, r_{27}, r_{28}, r_{29}, r_{210}\}$; $R_3 = \{r_{31}, r_{32}, r_{33}, r_{34}, r_{35}, r_{36}, r_{37}, r_{38}, r_{39}, r_{310}\}$. All resources in R_1 master initial skill s_1, all resources in R_2 master initial skill s_2, all resources in R_3 master initial skill s_3, and whether these resources master other skill(s) is generated randomly. The resource state matrix is generated by a Markov process (see Table A2 in Appendix A for details). Each activity demands different skills. Table 2 shows the number of skills required for performing activities and other information.

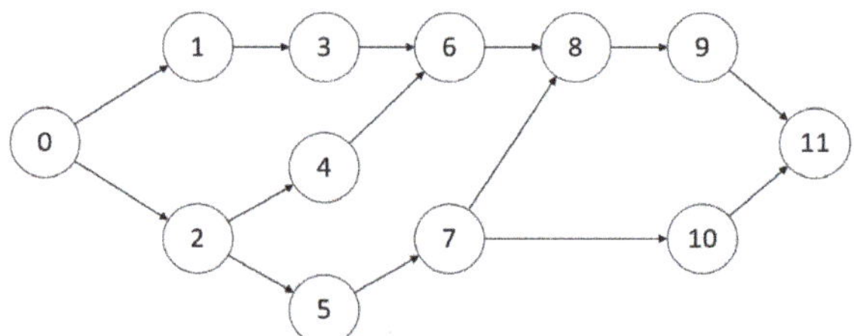

Figure 2. Project Network Example.

Table 2. Project information.

Activists	Successors	d_i	s_1	s_2	s_3
0	1, 2	0	0	0	0
1	3	2	5	6	6
2	4, 5	3	7	5	4
3	6	6	6	4	4
4	6	5	4	6	3
5	7	3	7	7	6
6	8	6	5	6	4
7	8, 10	4	8	5	5
8	9	5	3	4	3
9	11	4	6	5	4
10	11	2	6	3	6
11	-	0	0	0	0
	c_{s_k}		10	8	12

5.1. The Effectiveness of Dynamic Scheduling

According to the critical path method, the shortest makespan of the project is 23. Considering the unavailability of resources, we assume that the deadline of the project is 26 (any number greater than 23 is allowed) (that is, $T = 26$). According to the GA-PR algorithm, the optimized scheduling scheme can be obtained, as shown in Figure 3 and Table 3. Taking scheme 1 as an example, the detailed project scheduling is shown in Figure 4 and the detailed resource scheduling is shown in Table 4.

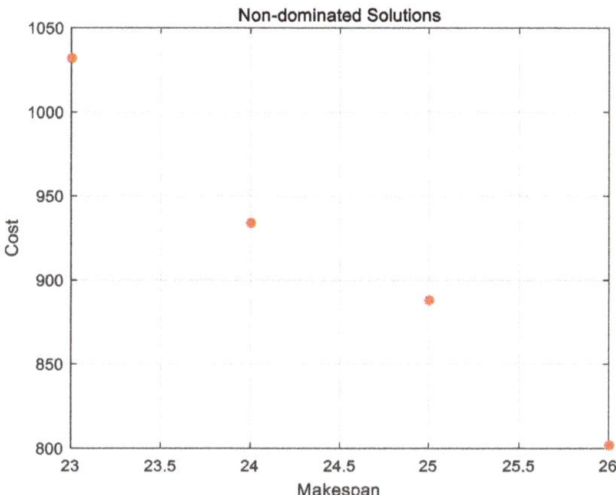

Figure 3. The schedule solution. Note: The red dots indicate the makespan and additional cost of 4 scheduling schemes respectively.

Table 3. Optimization Schedules.

Scheme	Makespan	Objective Value (Costs of Temporary Resources)	Schedule
1	23	1032	[0, 1, 1, 3, 4, 8, 9, 11, 15, 20, 15, 23]
2	24	934	[0, 1, 2, 3, 5, 9, 10, 12, 16, 21, 18, 24]
3	25	888	[0, 3, 1, 5, 4, 9, 11, 12, 17, 22, 21, 25]
4	26	802	[0, 3, 1, 5, 7, 4, 12, 12, 18, 23, 21, 26]

According to Figure 3, it can be found that with the extension of the project deadline, the cost of the additional resources becomes increasingly smaller. In other words, the buffer period plays a role in alleviating resource unavailability. Resource scheduling in Table 3 shows that multiple resources have performed more than one skill. This gives scheduling more flexibility. Therefore, the dynamic scheduling considering multi-skilled resources can effectively alleviate the impact of uncertainty in resource availability on a project's makespan and cost.

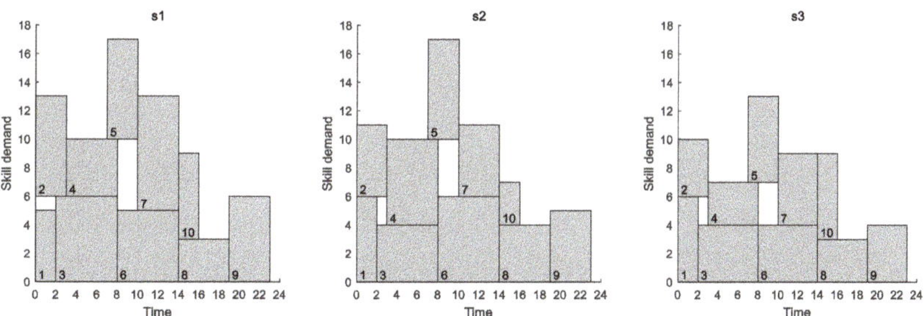

Figure 4. The scheduling of scheme 1.

Table 4. Multi-skilled resource scheduling of scheme 1.

j	t	s_1	s_2	s_3
1	1	r_{11}; r_{12}; r_{14}; r_{15}; r_{110}	r_{21}; r_{22}; r_{23}; r_{24}; r_{27}; r_{29}	r_{32}; r_{33}; r_{34}; r_{36}; r_{37}; r_{39}
	2	r_{14}; r_{17}; r_{110}; 2 temporary	r_{22}; r_{29}; 4 temporary	r_{32}; r_{35}; r_{37}; r_{38}; r_{13}; r_{210}
2	1	r_{13}; r_{16}; r_{17}; r_{18}; r_{19}; 2 temporary	r_{25}; r_{26}; r_{28}; r_{210}; 1 temporary	r_{31}; r_{35}; r_{38}; r_{310}
	2	r_{16}; r_{18}; r_{19}; r_{24}; r_{25}; r_{28}; 1 temporary	5 temporary	r_{33}; r_{36}; r_{310}; r_{11}
	3	r_{12}; r_{14}; r_{17}; r_{110}; 3 temporary	r_{22}; r_{23}; r_{29}; r_{210}; 1 temporary	r_{13}; r_{34}; r_{37}; r_{38}
3	3	r_{11}; r_{15}; r_{18}; r_{19}; r_{25}; r_{28}	r_{27}; 3 temporary	r_{31}; r_{33}; r_{36}; r_{39}
	4	r_{11}; r_{12}; r_{14}; r_{15}; r_{17}; r_{110}	r_{21}; r_{22}; r_{23}; r_{29}	r_{32}; r_{34}; r_{36}; r_{39}
	5	r_{11}; r_{12}; r_{15}; r_{17}; r_{18}; r_{110}	r_{21}; r_{22}; r_{27}; r_{29}	r_{34}; r_{36}; r_{37}; r_{39}
	6	r_{12}; r_{14}; r_{17}; r_{110}; r_{26}; 1 temporary	r_{22}; r_{29}; r_{210}; 1 temporary	r_{23}; r_{32}; r_{37}; r_{38}
	7	r_{12}; r_{13}; r_{14}; r_{15}; r_{18}; r_{19}	r_{21}; r_{22}; r_{27}; r_{29}	r_{32}; r_{36}; r_{37}; r_{39}
	8	r_{13}; r_{16}; r_{110}; r_{28}; 2 temporary	r_{23}; r_{27}; 2 temporary	r_{34}; r_{35}; r_{39}; r_{310}
4	4	r_{13}; r_{18}; r_{19}; r_{38}	r_{25}; r_{26}; r_{28}; r_{210}; r_{310}; 1 temporary	r_{31}; r_{33}; r_{35}
	5	r_{13}; r_{19}; r_{310}; 1 temporary	r_{23}; r_{24}; r_{26}; r_{28}; r_{210}; r_{38}	r_{32}; r_{33}; r_{35}
	6	r_{16}; r_{18}; r_{19}; r_{21}	r_{25}; r_{27}; r_{28}; 3 temporary	r_{31}; r_{33}; r_{36}
	7	4 temporary	r_{23}; r_{25}; r_{26}; r_{28}; r_{38}; r_{310}	r_{31}; r_{33}; r_{35}
	8	r_{14}; r_{17}; r_{19}; r_{24}	r_{29}; r_{210}; 4 temporary	r_{11}; r_{31}; r_{36}
5	8	r_{15}; r_{26}; 5 temporary	r_{22}; 6 temporary	r_{18}; r_{25}; r_{33}; r_{38}; 2 temporary
	9	r_{12}; r_{14}; r_{17}; r_{110}; r_{26}; 2 temporary	r_{22}; r_{29}; r_{210}; 4 temporary	r_{13}; r_{23}; r_{32}; r_{35}; r_{37}; r_{38}
	10	r_{14}; r_{17}; r_{110}; r_{26}; 3 temporary	r_{22}; r_{29}; 5 temporary	r_{13}; r_{23}; r_{37}; r_{38}; 2 temporary
6	9	r_{11}; r_{15}; r_{18}; r_{19}; r_{310}	r_{21}; r_{25}; r_{27}; 3 temporary	r_{31}; r_{33}; r_{36}; r_{39}
	10	r_{15}; r_{18}; r_{19}; r_{24}; r_{28}	r_{25}; r_{27}; 4 temporary	r_{11}; r_{33}; r_{36}; r_{310}
	11	r_{11}; r_{12}; r_{15}; r_{15}; r_{17}	r_{21}; r_{22}; r_{23}; r_{24}; r_{27}; r_{29}	r_{34}; r_{36}; r_{37}; r_{39}
	12	r_{12}; r_{14}; r_{17}; r_{110}; r_{26}	r_{210}; 5 temporary	r_{37}; r_{38}; r_{23}; r_{13}
	13	r_{11}; r_{12}; r_{14}; r_{16}; r_{17}	r_{22}; r_{23}; r_{28}; r_{29}; 2 temporary	r_{32}; r_{34}; r_{36}; r_{37}
	14	r_{12}; r_{14}; r_{15}; r_{17}; r_{110}	r_{21}; r_{22}; r_{23}; r_{24}; r_{27}; r_{29}	r_{34}; r_{36}; r_{37}; r_{39}
7	11	r_{13}; r_{18}; r_{19}; 5 temporary	r_{25}; 4 temporary	r_{31}; r_{32}; r_{35}; r_{38}; r_{310}
	12	r_{11}; r_{15}; r_{16}; r_{18}; r_{19}; r_{21}; r_{24}; r_{25}	r_{27}; 4 temporary	r_{31}; r_{33}; r_{36}; r_{39}; r_{310}
	13	r_{13}; r_{18}; r_{19}; 5 temporary	5 temporary	r_{31}; r_{33}; r_{35}; r_{38}; r_{310}
	14	r_{13}; r_{18}; r_{19}; 5 temporary	r_{25}; r_{210}; r_{310}; 2 temporary	r_{31}; r_{32}; r_{33}; r_{35}; r_{38}
8	15	r_{11}; r_{14}; r_{110}	r_{22}; r_{23}; r_{27}; r_{29}	r_{36}; r_{37}; r_{310}
	16	r_{12}; r_{15}; r_{110}	r_{22}; r_{26}; r_{27}; r_{29}	r_{34}; r_{36}; r_{37}
	17	r_{14}; r_{15}; r_{110}	r_{22}; r_{23}; r_{24}; r_{29}	r_{34}; r_{36}; r_{37}
	18	r_{14}; r_{15}; r_{110}	r_{21}; r_{22}; r_{27}; r_{29}	r_{34}; r_{36}; r_{37}
	19	r_{14}; r_{15}; r_{110}	r_{21}; r_{22}; r_{27}; r_{29}	r_{36}; r_{37}; r_{310}
9	20	r_{13}; r_{15}; r_{17}; r_{18}; r_{19}; r_{25}	r_{21}; r_{22}; r_{23}; r_{24}; r_{27}	r_{32}; r_{34}; r_{36}; r_{37}
	21	r_{11}; r_{12}; r_{14}; r_{15}; r_{16}; r_{17}	r_{21}; r_{22}; r_{23}; r_{24}; r_{29}	r_{31}; r_{32}; r_{33}; r_{37}
	22	r_{11}; r_{12}; r_{14}; r_{15}; r_{16}; r_{110}	r_{21}; r_{22}; r_{23}; r_{24}; r_{29}	r_{34}; r_{36}; r_{37}; r_{39}
	23	r_{11}; r_{12}; r_{15}; r_{16}; r_{17}; r_{110}	r_{21}; r_{22}; r_{23}; r_{27}; r_{29}	r_{34}; r_{36}; r_{37}; r_{39}
10	15	r_{12}; r_{13}; r_{16}; r_{17}; r_{18}; r_{19}	r_{26}; r_{28}; r_{210}	r_{25}; r_{31}; r_{33}; r_{38}; r_{310}; 1 temporary
	16	r_{13}; r_{16}; r_{17}; r_{18}; r_{19}; 1 temporary	r_{25}; r_{210}; r_{310}	r_{31}; r_{32}; r_{33}; r_{35}; r_{38}; r_{39}

5.2. Comparison of Three Scheduling Methods

As shown in Section 4, we know that scheduling multi-skilled resources is mainly determined by the priority rules in the inner algorithm, which we call dynamic scheduling. To further analyze the effectiveness of these priority rules, we set up two groups of experiments to compare with the dynamic scheduling method proposed in this paper—random scheduling and static scheduling. Random scheduling means that the priority rules designed in Section 4.2 and the skill scheduling order and resource scheduling order are random. Static scheduling means that once a resource is assigned a skill, it cannot be changed. The pseudo-codes of dynamic, random and static scheduling are shown in Appendix B, and the results relative to three different scheduling methods are shown in Figure 5a–c, respectively. The figures indicate that the objective of dynamic scheduling is the best: There is minimum additional cost, followed by random scheduling and static scheduling. This is because of the design of scheduling priority rules. This proves the effectiveness of the proposed dynamic scheduling method. The operation times of the three scheduling methods are 19.44 s, 16.39 s, and 0.57 s, respectively.

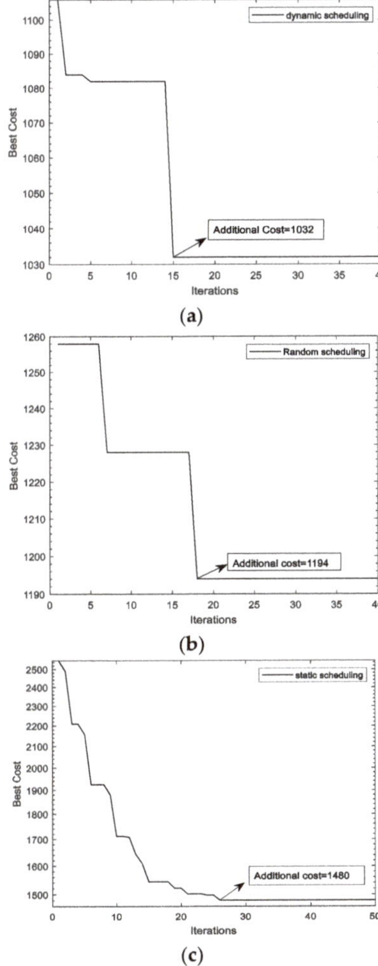

Figure 5. (**a**) Dynamic scheduling. (**b**) Random scheduling. (**c**) Static scheduling.

5.3. The Effect of the Buffer Period

We can know that a project's deadline affects the objective value from the results of Figure 3. Therefore, we define parameter $C_{deadline}$. to represent the buffer coefficient, which is used to reflect the margin of buffer duration interval—$C_{deadline} = T/DueDate$, and the *DueDate* refers to the makspan of the critical path of the project. The larger $C_{deadline}$ is, the larger the buffer period. The objective values of three scheduling methods with different parameters are shown in Table 5.

Table 5. The effect of the buffer period.

Parameter		Dynamic	Random	Static
			Best Cost	
$C_{deadline}$	1.1	952	1154	1076
	1.3	538	802	556
	1.5	364	476	402
	1.7	250	406	270

Table 5 shows that with the increase in buffer period, the objective values decrease for three different scheduling programs. This is because when the activities cannot be started due to the unavailable resources, the larger the buffer period, the more likely that activity can be allowed to delay, and fewer additional resources will be added. Thus, the cost will be reduced.

6. Experiment Analysis

In this section, a computational experiment is designed to assess the performance of the heuristic algorithm proposed in this paper. The algorithm was coded in Matlab R2018b and ran in the environment of Microsoft Windows 10 (CPU 1.68 GHZ, RAM 8 GB).

6.1. Test Data

Considering that the resources are multi-skilled and their availabilities are uncertain, the following changes are made to the dataset from PSPLIB (http://www.om-db.wi.tum.de/psplib/, accessed on 17 June 2020):

(1)   The number of skill types required by the project corresponds to that of resource types required by the project in the original PSPLIB. Each type of resource has a corresponding initial skill. Assuming that the original PSPLIB J30 needs four types of resources $(R_1, R_2, R_3,$ and $R_4)$, the project in this paper needs four types of skills in which all resources in R_1 have initial skill s_1, those in R_2 have initial skill s_1, and so on. Except for the initial skills, whether every resource has other skill(s) is randomly generated.

(2)   The skill requirements of activities correspond to the resource requirements in PSPLIB—assuming that the resource requirements of $R_1, R_2, R_3,$ and R_4 for activity 1 in a case of PSPLIB is 4, 5, 7, and 8, respectively, then the skill requirements of $s_1, s_2, s_3,$ and s_4 for activity 1 is 4, 5, 7, and 8 in this paper.

Network complexity (NC) reflects the precedence relations of activities. Resource Strength (RS) reflects the intensity of resources, where the larger the value is, the more resources are supplied. Resource Factor (RF) reflects the activity's skill requirement, where the larger the value is, the more skills are needed. In addition, we defined Modified Resource Strength (MRS) based on the Resource Strength (RS), which reflects the skill strength mastered by resources. The formula of MRS is shown in Equation (13). The larger the value, the more the skills are mastered by resources. The Rate of Resource Unavailability (RRU) is introduced to reflect the unavailability of resources. The larger the value, the greater the probability that the resource is unavailable.

$$MRS = \frac{\left(\sum_i \sum_{k=1}^{K} X_{kis_k}\right) / \left(\sum_i \sum_{k=1}^{K} r_{ki}\right)}{K} \tag{13}$$

As the problem studied in this paper adds uncertainty in resource availability and dynamic skill assignment to the classical MSRCPSP, this improves scheduling flexibility while increasing the difficulty for solutions. As such, solution times increase exponentially. Therefore, this paper only selected the J30-dataset from the PSPLIB. The parameters of dataset are shown in Table 6 below.

Table 6. Parameters and values.

Parameters	Values				
NC	1.5	1.8	2.1	—	—
RS	0.2	0.5	0.7	1	—
RF	0.25	0.5	0.75	1	—
MRS	0.25	0.625	1	—	—
RRU	0.1	0.2	0.3	0.4	0.5

6.2. Computational Results

In this section, dynamic, random, and static scheduling are compared based on the same dataset. The comparison results under different parameters are shown in Figure 6a–e, where $Opt(\%)$ is the proportion of the optimal solution. The optimal solution refers to the minimum cost of the three scheduling methods (dynamic, stochastic, and static). The proportion of the optimal solution refers to that of the number of optimal solutions in all 480 instances.

As Figure 6a–e indicate, regardless of how parameters change, the performance of dynamic scheduling is always superior than random and static scheduling. Moreover, compared to static scheduling, random scheduling is superior. This is because static scheduling limits the resource's skill switching; thus, more additional resources need to be supplied to reach skill requirements, leading to an increase in costs. This shows that multi-skilled scheduling can effectively alleviate the disturbance caused by resource uncertainty, and the design of priority rules in dynamic scheduling leads to improved optimization results, which further shows the effectiveness of the proposed algorithm.

As shown in Figure 6a, network complexity (NC) has no obvious impact on the results of the three scheduling schemes. This is because the three types of scheduling mainly optimize results from the perspective of resources and skills. Therefore, regardless of how NC changes, the results are relatively stable. With the increase in Resource factor (RF), the proportion of optimal solution of dynamic scheduling gradually decreases, as shown in Figure 6b. With limited resources, when the activity skill requirements increase, project scheduling is more easily affected by resources. With the increase in skill requirements, the flexibility of multi-skilled resources is limited. As such, more additional resources are needed, which leads to an increase in costs. Therefore, compared with random and static scheduling, the optimal proportion of the target value decreases. As observed in Figure 6c, with the increase in RS, the proportion of the optimal solution of dynamic scheduling gradually increases. This is because the greater RS is, the more resources are likely to meet the skill requirements of activities, and the replacement of idle resources is easier. Thus, there is less necessity for additional resources, and costs will be less. In other words, the increase in RS weakens the impact of uncertainty in resource availability. Similarly, MRS reflects the strength of skills. The greater the MRS, the more skills the resources can master. There is more flexibility in resources scheduling, and it is more likely to produce improved solutions compared with static scheduling. Consequently, costs are lower, as shown in Figure 6d. As RRU reflects the state of resources, the higher the value, the greater the probability of resource unavailability. At this time, the role of multi-skilled resources will be weakened, and there are fewer resources that can meet skill requirements. Therefore, with the increase in *RRU*, more additional resources are needed, and the optimal proportion of dynamic scheduling will also decrease, as shown in Figure 6e.

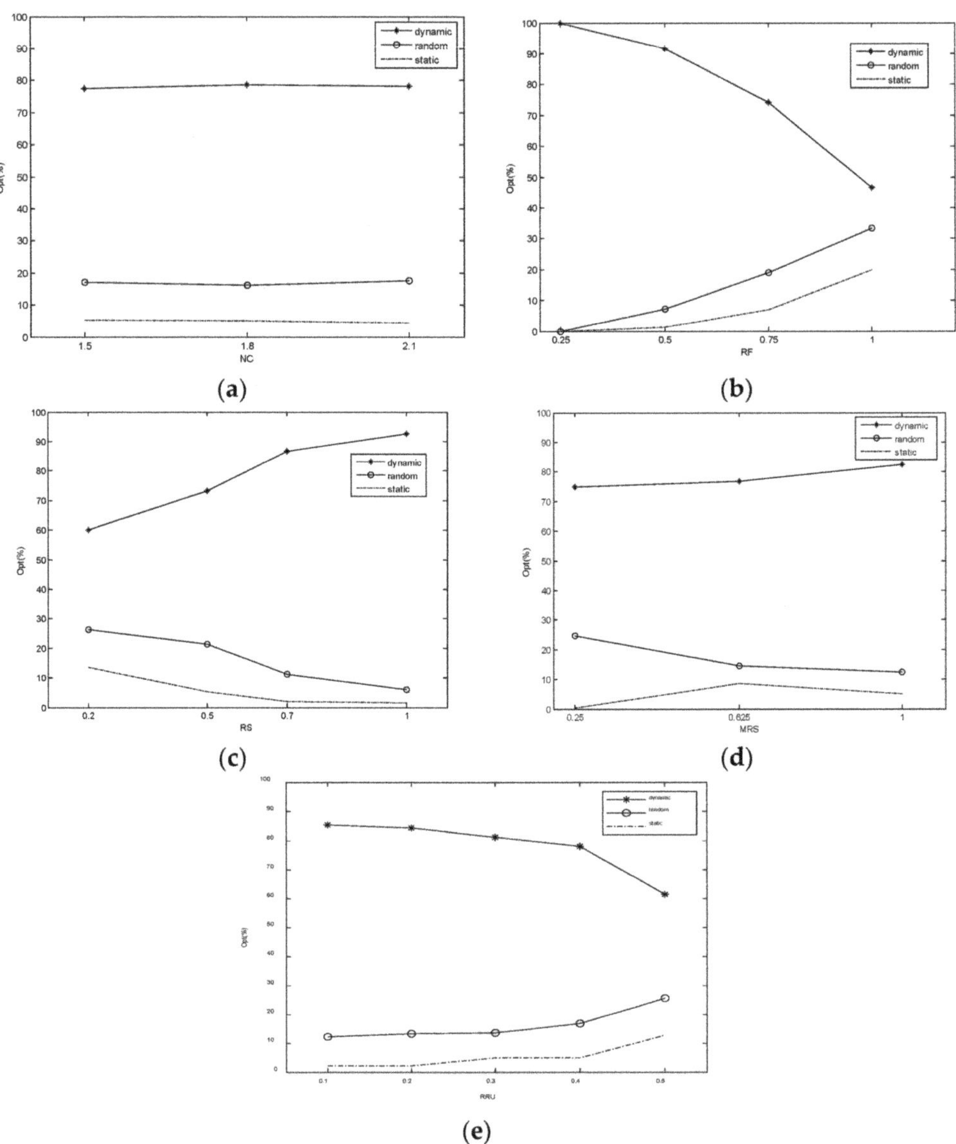

Figure 6. (**a**) Comparison results of *NC*. (**b**) Comparison results of *RF*. (**c**) Comparison results of *RS*. (**d**) Comparison results of *MRS*. (**e**) Comparison results of *RRU*.

Therefore, as a project manager, to reduce the additional cost of the project in an uncertain environment, it is necessary to improve the multi-skilled level of resources, avoid the unavailability of resources, and pay attention to the order in which skills and resources are scheduled, which have certain guiding significance for maximizing project benefits.

7. Conclusions

This paper studies the MSRCPSP with uncertainty in resource availability, and dynamic skill assignment and additional resource replenishment are considered simultaneously when resources are not available. Although extensive research has been conducted for the MSRCPSP, little research focused on uncertainty in resource availability. In an uncertain environment, it is easy to encounter resource shortage or conflict. Idle resources can replace unavailable resources to complete activities through skill switching; thus, dynamic skill assignment has become a method for alleviating resource conflict. If it still does not work, additional resources are considered to ensure that activities are not interrupted. To solve this complex problem, a new model is built and a nested GA-PR dynamic scheduling algorithm is proposed. Finally, an example and numerical experiments are used to verify the performance of the algorithm. Simultaneously, the performance of the algorithm is illustrated by comparing the solution efficiency of static and random scheduling, and the influences of different parameters on the algorithm scheduling are analyzed.

Numerical experiments show that, although the running time of the proposed dynamic scheduling is not optimal, its solution is always superior compared to the other two scheduling methods. When uncertainties in resource availability are considered, the resource has more skills and more resources are supplied, and the dynamic scheduling method has improved performance; on the other hand, the higher the probability of resource unavailability and the more skills are required, the worse the dynamic scheduling method performs. Moreover, by comparing the performance of dynamic scheduling and random scheduling, we can find that the scheduling order has a significant impact on the results and the three new priority rules contribute to the optimization of costs. By comparing the performance of dynamic scheduling and static scheduling, we can find that the skill switching of multi-skill resources also plays an important role.

Our research findings can also provide project managers with some guidance when scheduling projects in an uncertain environment. First, at the start-up stage of the project, mangers should select as many multi-skilled resources as possible when establishing the project team. Second, at the project planning stage, it is important to decide which skill should be scheduled first and which resource should be scheduled first. Moreover, dynamic skill assignment and additional resource replenishment are great methods for alleviating resource shortages. Third, during the project-execution period, managers can take some incentives to encourage single-skilled person to learn from multi-skilled persons to master more skills.

However, it should be noted that this paper has still some limitations. First, this paper assumes that the skill level of each resource is homogeneous. For future research, the heterogeneity of skill level can be considered. Second, when the splitting of activity is allowed, resource conflicts can also be solved by interrupting activities so as to reduce the additional cost of resources. In this case, scheduling will be a more interesting and difficult problem. Finally, in this paper, the costs of resource skill switching are not taken into account, which will render MSRCPSP a trade-off problem between additional resource costs and skill-switching costs.

Author Contributions: All authors contributed to the study conception and design. Material preparation, data collection, and analysis were performed by G.L., M.W. and X.L. The first draft of the manuscript was written by M.W., and all authors commented on previous versions of the manuscript. All authors have read and agreed to the published version of the manuscript.

Funding: This research received no external funding.

Institutional Review Board Statement: Not applicable.

Informed Consent Statement: Not applicable.

Data Availability Statement: The data of PSPLIB can be downloaded from "http://www.omdb.wi.tum.de/psplib/getdata_sm.html" (accessed on 17 June 2015).

Acknowledgments: We are grateful to anonymous reviewers for their thorough reviews and valuable comments.

Conflicts of Interest: The authors declare no conflict of interest.

Appendix A

Table A1. The skill matrix mastered by resources.

Resource	s_1	s_2	s_3	Resource	s_1	s_2	s_3	Resource	s_1	s_2	s_3
r_{11}	1	0	1	r_{21}	1	1	0	r_{31}	1	1	1
r_{12}	1	1	0	r_{22}	0	1	0	r_{32}	1	0	1
r_{13}	1	1	1	r_{23}	0	1	1	r_{33}	1	0	1
r_{14}	1	0	0	r_{24}	1	1	0	r_{34}	0	0	1
r_{15}	1	0	0	r_{25}	1	1	1	r_{35}	1	0	1
r_{16}	1	0	1	r_{26}	1	1	0	r_{36}	0	0	1
r_{17}	1	1	0	r_{27}	0	1	0	r_{37}	0	0	1
r_{18}	1	0	1	r_{28}	1	1	0	r_{38}	1	1	1
r_{19}	1	1	1	r_{29}	0	1	0	r_{39}	0	0	1
r_{110}	1	0	0	r_{210}	0	1	1	r_{310}	1	1	1

Table A2. The resource state matrix.

Resource	t																						
	1	2	3	4	5	6	7	8	9	10	11	12	13	14	15	16	17	18	19	20	21	22	23
r_{11}	1	1	1	1	1	0	0	1	1	1	1	1	1	0	1	0	0	1	1	0	1	1	1
r_{12}	1	0	1	1	1	1	1	0	1	0	1	1	1	1	1	1	1	1	1	0	1	1	1
r_{13}	1	1	1	1	1	0	1	1	1	1	1	1	1	1	1	1	1	1	0	1	1	1	1
r_{14}	1	1	1	1	0	1	1	1	1	1	0	1	1	1	1	0	1	1	1	0	1	1	0
r_{15}	1	0	1	1	1	0	1	1	1	1	1	1	0	1	0	1	1	1	1	1	1	1	1
r_{16}	1	1	0	0	0	1	0	1	0	0	1	1	1	0	1	1	1	0	1	0	1	1	1
r_{17}	1	1	1	1	1	1	0	1	1	1	1	1	1	1	1	1	0	1	1	1	1	1	1
r_{18}	1	1	1	1	1	1	1	1	1	1	1	1	1	1	1	1	1	1	1	1	1	1	1
r_{19}	1	1	1	1	1	1	1	1	1	1	1	1	1	1	1	1	1	1	1	1	1	1	1
r_{110}	1	1	1	1	1	1	0	1	1	1	0	1	0	1	1	1	1	1	0	0	0	1	1
r_{21}	1	0	0	1	1	1	1	0	1	0	1	1	0	1	0	0	0	1	1	1	1	1	1
r_{22}	1	1	1	1	1	1	1	1	1	1	1	0	1	1	1	1	1	1	1	1	1	1	1
r_{23}	1	0	1	1	1	1	1	1	1	1	1	1	1	1	1	0	1	1	1	1	1	1	1
r_{24}	1	1	0	0	1	0	0	1	0	1	1	1	0	1	0	0	1	0	0	1	1	1	1
r_{25}	1	1	1	1	0	1	1	1	1	1	1	1	0	1	1	1	1	1	1	1	1	1	1
r_{26}	1	0	0	1	1	1	1	1	1	1	0	1	0	0	1	1	1	1	1	0	1	1	0
r_{27}	1	0	1	0	1	1	1	1	1	1	1	1	0	1	1	1	0	1	1	1	0	0	1
r_{28}	1	1	1	1	1	1	1	1	0	1	0	0	1	0	1	0	0	0	1	0	0	1	1

Table A2. *Cont.*

Resource	t																						
	1	2	3	4	5	6	7	8	9	10	11	12	13	14	15	16	17	18	19	20	21	22	23
r_{29}	1	1	1	1	1	1	1	1	1	1	1	0	1	1	1	1	1	1	1	0	1	1	1
r_{210}	1	1	1	1	1	1	0	1	1	0	0	1	0	1	1	1	0	1	1	1	1	1	1
r_{31}	1	0	1	1	0	1	1	1	1	0	1	1	1	1	1	1	0	0	1	0	1	0	0
r_{32}	1	1	0	1	1	1	1	0	1	0	1	0	1	1	0	1	0	0	1	1	1	0	1
r_{33}	1	1	1	1	1	1	1	1	1	1	0	1	1	1	1	1	1	1	1	1	1	1	1
r_{34}	1	0	1	1	1	0	0	1	0	0	1	0	1	1	0	1	1	0	0	1	0	1	1
r_{35}	1	1	0	1	1	0	1	1	1	0	1	0	1	1	0	1	1	1	1	0	0	0	0
r_{36}	1	1	1	1	1	1	1	1	1	1	1	1	1	1	1	1	1	1	1	1	0	1	1
r_{37}	1	1	1	0	1	1	1	0	1	1	1	1	1	1	1	1	1	1	1	1	1	1	1
r_{38}	1	1	1	1	1	1	1	1	1	1	1	1	1	1	1	1	1	1	1	0	1	1	0
r_{39}	1	0	1	1	1	0	1	1	1	0	1	1	0	1	1	1	1	0	1	0	0	1	1
r_{310}	1	1	0	1	1	0	1	1	1	1	1	1	1	1	1	1	0	1	0	1	1	1	1

Appendix B

Algorithm for Dynamic Scheduling	Algorithm for Random Scheduling	Algorithm for Static Scheduling
Begin	**Begin**	**Begin**
Input d_j, r_{js_k}, R, S, p_{ki}, $Z_{t=1}$	Input d_j, r_{js_k}, R, S, p_{ki}, $Z_{t=1}$	Input d_j, r_{js_k}, R, S, p_{ki}, $Z_t = 1$
While it < MaxIt	**While** it < MaxIt	**While** it < MaxIt
for $j = 1 : n$	**for** $j = 1 : n$	**for** $j = 1 : n$
$\quad$ Calculate $\left[ES_j, LS_j\right]$	$\quad$ Calculate $\left[ES_j, LS_j\right]$	$\quad$ Calculate $\left[ES_j, LS_j\right]$
end	**end**	**end**
Initialize ST_j	Initialize ST_j	Initialize ST_j, SK_i
Resource ranking (Priority rule 1)	**If** $ST_j + d_j \leq ST_g$, $g \in SUCC_j$, $\forall j$	**If** $ST_j + d_j \leq ST_g$, $g \in SUCC_j$, $\forall j$
If $ST_j + d_j \leq ST_g$, $g \in SUCC_j$, $\forall j$	$\quad$ Then	$\quad$ Then
$\quad$ Then	$\quad\quad$ **for** $t = 1 : T$	$\quad$ Calculate R_s
$\quad\quad$ **for** $t = 1 : T$	$\quad\quad$ Update O_t	**Else**
$\quad\quad\quad$ Calculate L_{st} (Priority rule 2)	$\quad\quad$ Resource scheduling (No order)	$\quad$ Output Result
$\quad\quad\quad$ Update O_t, W_t	$\quad\quad$ Update Z_{t+1}	$\quad$ Update ST_j, SK_i (Select, Crossover,
$\quad\quad\quad$ **Case 1** $L_{st} \leq 0$, $\forall K$	$\quad\quad$ **end**	$\quad$ Mutate)
$\quad\quad\quad$ Resource scheduling	$\quad$ Calculate R_s and Output Result	**End**
$\quad\quad\quad$ **Case 2** $L_{st} > 0$, $\forall K$ (Priority rule 3)	**Else**	
$\quad\quad\quad$ Resource scheduling	$\quad$ Output Result	
$\quad\quad\quad$ **Case 3** others (Priority rule 3)	$\quad$ Update ST_j (Select, Crossover, Mutate)	
$\quad\quad\quad$ Resource scheduling	**End**	
$\quad\quad\quad$ Update Z_{t+1}		
$\quad\quad$ **end**		
$\quad$ Calculate R_s and Output Result		
Else		
$\quad$ Output Result		
Update ST_j (Select, Crossover, Mutate)		
End		

References

1. Afshar-Nadjafi, B. Multi-skilling in scheduling problems: A review on models, methods and applications. *Comput. Ind. Eng.* **2021**, *151*, 107004.
2. Bellenguez, O.; Néron, E. A Branch-and-Bound method for solving Multi-Skill Project Scheduling Problem. *RAIRO-Oper. Res.* **2007**, *41*, 155–170. [CrossRef]

3. Benavides, A.J.; Ritt, M.; Miralles, C. Flow shop scheduling with heterogeneous workers. *Eur. J. Oper. Res.* **2014**, *237*, 713–720. [CrossRef]

4. Heimerl, C.; Kolisch, R. Scheduling and staffing multiple projects with a multi-skilled workforce. *OR Spectr.* **2010**, *32*, 343–368. [CrossRef]

5. Isah, M.A.; Kim, B.S. Integrating cchedule risk analysis with multi-skilled resource scheduling to improve resource-constrained project scheduling problems. *Appl. Sci.* **2021**, *11*, 650. [CrossRef]

6. Chen, R.; Liang, C.Y.; Gu, D.X.; Joseph, Y.-T.L. A multi-objective model for multi-project scheduling and multi-skilled staff assignment for IT product development considering competency evolution. *Int. J. Prod. Res.* **2017**, *55*, 6207–6234. [CrossRef]

7. Polo-Mejía, O.; Artigues, C.; Lopez, P.; Basini, V. Mixed-integer/linear and constraint programming approaches for activity scheduling in a nuclear research facility. *Int. J. Prod. Res.* **2020**, *58*, 7149–7166. [CrossRef]

8. Filip, D. Applying to the mathematical methods to optimize the launching process in manufacturing. *Acta Tech. Napoc. Ser.-Appl. Math. Mech. Eng.* **2018**, *61*, 585–592.

9. Filip, D. Modern methods and tools to improve the production processes from small series and unique production. *Acta Tech. Napoc. Ser.-Appl. Math. Mech. Eng.* **2018**, *61*, 575–584.

10. Li, H.; Womer, K. Scheduling projects with multi-skilled personnel by a hybrid MILP/CP benders decomposition algorithm. *J. Sched.* **2009**, *12*, 281–298. [CrossRef]

11. Correia, I.; Saldanha-da-Gama, F. The impact of fixed and variable costs in a multi-skill project scheduling problem: An empirical study. *Comput. Ind. Eng.* **2014**, *72*, 230–238. [CrossRef]

12. Fırat, M.; Hurkens, C. An improved MIP-based approach for a multi-skill workforce scheduling problem. *J. Sched.* **2012**, *15*, 363–380. [CrossRef]

13. Snauwaert, J.; Vanhoucke, M. A new algorithm for resource-constrained project scheduling with breadth and depth of skills. *Eur. J. Oper. Res.* **2021**, *292*, 43–59. [CrossRef]

14. Snauwaert, J.; Vanhoucke, M. Mathematical formulations for project scheduling problems with categorical and hierarchical skills. *Comput. Ind. Eng.* **2022**, *169*, 108147. [CrossRef]

15. Buddhakulsomsiri, J.; Kim, D.S. Priority rule-based heuristic for multi-mode resource-constrained project scheduling problems with resource vacations and activity splitting. *Eur. J. Oper. Res.* **2007**, *178*, 374–390. [CrossRef]

16. Goldratt, E. *Critical Chain*; North River Press: Great Barrington, MA, USA, 1997.

17. Herroelen, W.; Demeulemeester, E.; De Reyck, B. A classification scheme for project scheduling. In *International Series in Operations Research and Management Science*; Project Scheduling: Recent Models, Algorithms and Applications; Weglarz, J., Ed.; Kluwer Academic: Boston, MA, USA, 1998.

18. Lambrechts, O.; Demeulemeester, E.; Herroelen, W. Proactive and reactive strategies for resource-constrained project scheduling with uncertain resource availabilities. *J. Sched.* **2008**, *11*, 121–136. [CrossRef]

19. Lambrechts, O.; Demeulemeester, E.; Herroelen, W. Time slack-based techniques for robust project scheduling subject to resource uncertainty. *Ann. Oper. Res.* **2011**, *186*, 443–464.

20. Ahmadpour, S.; Ghezavati, V. Modeling and solving multi-skilled resource-constrained project scheduling problem with calendars in fuzzy condition. *J. Ind. Eng. Int.* **2019**, *15*, 179–197. [CrossRef]

21. Azizi, N.; Liang, M. An integrated approach to worker assignment, workforce flexibility acquisition, and task rotation. *J. Oper. Res. Soc.* **2013**, *64*, 260–275. [CrossRef]

22. Wongwai, N.; Malaikrisanachalee, S. Augmented heuristic algorithm for multi-skilled resource scheduling. *Autom. Construct.* **2011**, *20*, 429–445.

23. Kher, H.V.; Malhotra, M.K.; Philipoom, P.R.; Fry, T.D. Modelling simultaneous worker learning and forgetting in dual resource constrained systems. *Eur. J. Oper. Res.* **1999**, *115*, 158–172. [CrossRef]

24. Yue, H.; Slomp, J.; Molleman, E.; Vanderzee, D.J. Worker flexibility in a parallel dual resource constrained job shop. *Int. J. Prod. Res.* **2008**, *46*, 451–467. [CrossRef]

25. Gans, N.; Zhou, Y. Managing Learning and Turnover in Employee Staffing. *Oper. Res.* **2002**, *50*, 991–1006. [CrossRef]

26. De Bruecker, P.; Ven den Bergh, J.; Beliën, J.; Demeulemeester, E. Workforce planning incorporating skills: State of the art. *Eur. J. Oper. Res.* **2015**, *243*, 1–16. [CrossRef]

27. Bard, J.F.; Purnomo, H.W. Preference scheduling for nurses using column generation. *Eur. J. Oper. Res.* **2005**, *164*, 510–534. [CrossRef]

28. Lagodimos, A.G.; Leopoulos, V. Greedy heuristic algorithms for manpower shift planning. *Int. J. Prod. Econ.* **2000**, *68*, 95–106. [CrossRef]

29. Lagodimos, A.G.; Mihiotis, A.N. Overtime vs. regular shift planning decisions in packing shops. *Int. J. Prod. Econ.* **2006**, *101*, 246–258. [CrossRef]

30. Correia, I.; Lourenço, L.L.; Saldanha-da-Gama, F. Project scheduling with flexible resources: Formulation and inequalities. *OR Spectr.* **2012**, *34*, 635–663. [CrossRef]

 *mathematics*

Article

The Cost-Balanced Path Problem: A Mathematical Formulation and Complexity Analysis

Daniela Ambrosino *,† and Carmine Cerrone †

Department of Economics and Business Studies, University of Genoa, 16126 Genoa, Italy; carmine.cerrone@unige.it
* Correspondence: ambrosin@economia.unige.it
† These authors contributed equally to this work.

Abstract: This paper introduces a new variant of the Shortest Path Problem (SPP) called the Cost-Balanced Path Problem ($CBPP$). Various real problems can either be modeled as $BCPP$ or include $BCPP$ as a sub-problem. We prove several properties related to the complexity of the $CBPP$ problem. In particular, we demonstrate that the problem is NP-hard in its general version, but it becomes solvable in polynomial time in a specific family of instances. Moreover, a mathematical formulation of the $CBPP$, as a mixed-integer programming model, is proposed, and some additional constraints for modeling real requirements are given. This paper validates the proposed model and its extensions with experimental tests based on random instances. The analysis of the results of the computational experiments shows that the proposed model and its extension can be used to model many real applications. Obviously, due to the problem complexity, the main limitation of the proposed approach is related to the size of the instances. A heuristic solution approach should be required for larger-sized and more complex instances.

Keywords: shortest path problem; mixed-integer linear programming; cost-balanced paths

Citation: Ambrosino, D.; Cerrone, C. The Cost-Balanced Path Problem: A Mathematical Formulation and Complexity Analysis. *Mathematics* **2022**, *10*, 804. https://doi.org/10.3390/math10050804

Academic Editor: Aleksandr Rakhmangulov

Received: 9 January 2022
Accepted: 1 March 2022
Published: 3 March 2022

Publisher's Note: MDPI stays neutral with regard to jurisdictional claims in published maps and institutional affiliations.

1. Introduction

This paper introduces a new variant of the Shortest Path Problem (SPP) called the Cost-Balanced Path Problem ($CBPP$). The $CBPP$ is defined on a directed graph $G(N, A)$, where N is the set of nodes and A is the set of directed arcs. For each arc $(i, j) \in A$ is also defined a cost $c_{ij} \in \mathbb{R}$. Let the nodes $o, d \in N$, respectively, called origin and destination. A feasible solution of this problem is an acyclic path $p = ((o, n_i), (n_i, n_j), \ldots, (n_h, n_k), (n_k, d))$ in the graph G from node o to node d. Let $c(p) = \sum_{(i,j) \in p} c_{ij}$ the cost of the pah p, i.e., the sum of the cost of all the arcs used in the path p; the objective of $CBPP$ is the minimization of the absolute value of this cost: $MIN|c(p)|$.

Various real problems that present some elements in common with the cost-balanced path problem can either be modeled as $BCPP$ or include $BCPP$ as a sub-problem; some are here briefly discussed. The first mentioned problem is related to the path of an electric vehicle. The route choices of drivers of battery electric vehicles are affected by the many factors related to the battery recharge [1]. The cost-balanced path problem can be solved when defining the path that an electric vehicle has to perform for going from an origin point to a destination one, with the aim of maintaining the same level of electric charge. Suppose that a vehicle starts its trip in the origin node with a charge of 80% and has to arrive at a destination node with the same charge. During the trip, the vehicle can recharge the battery on the downhill roads, while the vehicle reduces its charge on the roads that go uphill. This problem can be formulated on a direct graph G where the weights of arcs represent the charge consumption (negative arc costs) and the recharge (positive costs). In this case, additional constraints are required, such as the level of electric charge to maintain along the whole path that can range from 0 to 100%.

In the bike-sharing systems [2], a particular problem linked to bikes management can be modeled as a $CBPP$. Suppose to have to re-locate bikes among a set of points that can be modeled as arcs in a direct graph G, with weights on the arcs representing the number of bikes to deliver (negative costs) and the number of bikes to pick up (positive costs). A vehicle has to perform a path in such a way to redistribute the bikes. In this example, the additional constraints are related to the number of bikes on the vehicle, that can range from zero to the vehicle capacity; moreover, additional requirements can be added for obtaining a constrained path concerning the duration, the number of arcs visited, the length, that, for example, should be maintained within a given range. Alternatively, suppose to have a depot and a vehicle that has to deliver a certain number of bikes to some points that are the locations for bikes. The vehicle has to deliver bikes to some locations, while eventually re-locate some bikes, that is, to pick up bikes from some locations. In the end, the vehicle has to finish its trip, possibly without bikes on board; the same additional requirements cited here above can be added.

This paper introduces a new problem, thus it can be helpful to summarize the novelties of this work in the following list: (i) definition of a new problem, the $CSPP$; (ii) proof that $CSPP$ is NP-hard; (iii) proof that it is possible to solve the $CSPP$ in polynomial time under specific configurations of the arc costs; (vi) first mathematical formulation for $CSPP$.

The remaining of the paper is organized as follows. Section 2 summarizes the literature related to $BCPP$. Section 3 presents an evaluation of the $BCPP$ complexity, the $BCPP$ mathematical formulation and some model extensions, while Section 4 reports the computational experiments for the validation of the proposed model and its extensions. Section 5 gives some conclusions and perspectives.

2. Literature Review

To the authors' knowledge, the $BCPP$ has never been studied in the literature, although there are many variants of the classic SPP, and there is a paper related to the Traveling Salesman Problem (TSP) that introduces the same objective function of $BCPP$ [3]. The authors of [3] introduce the cost-balanced TSP, in which the main objective is to find a Hamiltonian cycle with total travel cost as close as possible to 0. The authors assumed a cost/length matrix, while negative costs are allowed. To solve the cost-balanced TSP, they proposed a variable neighborhood search algorithm. A similar problem is the balanced TSP, which is related to the uniform (equitable) distribution of resources [4]. In [5], the multiple balanced traveling salesmen problem is proposed to model and optimize the problems with multiple objectives (salesmen). The goal is to find m Hamiltonian cycles in G by minimizing the difference between the highest edge cost and the smallest edge cost in the tours. The SPP [6,7] and many variants have been proposed in the literature for facing problems arising in various fields, together with ever more efficient algorithms (see, for example, in [8–11]). Although the SPP can be solved in polynomial time using various algorithms, many of its variants are known to be NP-hard. Among these variants of the SPP, in the k-Color Shortest Path Problem proposed by Ferone et al. [12,13], the classic SPP is solved on graphs with colored arcs. In the recent Steiner bi-objective Shortest Path Problem introduced in [14], the authors present this new variant of the SPP capable of preprocessing data to solve the well-known vehicle routing problem. SPP in which the cost of the arcs is not known in advance has been studied in the recent literature [15,16]. Stochastic shortest path (SSP) dealing with applications in routing problems and in road networks can be found in [17,18]. Another problem on graphs linked to the balance concept is the balanced trees [19], which are the appropriate structures (balanced tree structures) for managing networks with the aim of balancing two objectives. The constrained path has been studied in [20]; the authors proposed a robust formulation for the Resource-Constrained Shortest Path Problem that is the problem of determining a path p from an origin to a destination with the smallest cost, such that the consumption of a given resource for that path is lower or equal to the maximum amount of available resource.

In the following section the complexity of $CSPP$ is investigated and a mathematical formulation is proposed.

3. Problem Complexity and Mathematical Formulation for $CBPP$

Many variants of SPP are known to be NP-hard; thus, in this section, before presenting a mathematical formulation for modeling and solving $CBPP$, the problem complexity is investigated. In particular, thanks to a reduction algorithm [21,22], it has been proved that the problem is NP-hard in its general form.

3.1. Problem Complexity

Theorem 1. *If not all costs have the same sign, the Cost-Balanced Path Problem is NP-hard.*

Proof of Theorem 1. To prove the theorem, we will describe a reduction algorithm which in polynomial time, reduces the classic Hamiltonian Path Problem (HPP) in an instance of the Cost-Balanced Path Problem. The HPP is a classic problem belonging to the NP-complete complexity class [23]. Let a directed graph $G(N, A)$ where N is a set of nodes, and A is a set of arcs. Let us suppose we want to compute the Hamiltonian path that goes from node h to node k of the graph G. We create the graph $G'(N', A')$ such that $N' = N \cup \{o\}$, $A' = A \cup \{(o, h)\}$. We create the cost c_{ij} such that $c_{ij} = 1 \ \forall (ij) \in A$ and $c_{oh} = 1 - |N|$. Considering that $|N'| = |N| + 1$ and that the longest path in G' can contain $|N|$ arcs. Considering that in every feasible solution the arc (o, h) will always be present and that the value of a solution with k arcs will be equal to $1 - |N| + (k - 1)$. If the value of the solution of $CBPP$ on the graph G' is equal to zero, then in G, there is a Hamiltonian path between the nodes h and k (see Figure 1). $\square$

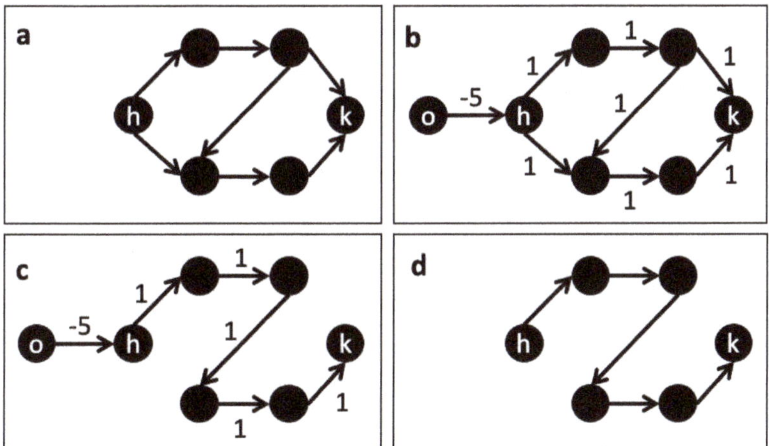

Figure 1. (**a**) An example of a graph G. (**b**) Graph G' derived from G. (**c**) A solution of $CBPP$ with cost zero. (**d**) The Hamiltonian path.

Figure 1 shows an example useful to understand Proof of Theorem 1. In Figure 1a, a direct graph (i.e., G) with six nodes is depicted, while in Figure 1b, a graph G' derived from G is represented: node o has been added together with the weights for the arcs. In Figure 1c, the solution for the $CBPP$ is shown. Finally, the Hamiltonian path from node h to node k, is depicted in Figure 1d).

Corollary 1 (Corollary of Theorem 1). *Given a generic instance of CBPP, if even one cost has the opposite sign to the others, then the problem is NP-hard.*

Proposition 1. *If all costs are non-negative $c_{ij} \geq 0$, the problem is equivalent to the classic Shortest Path Problem, therefore it can be solved in polynomial time.*

Proposition 2. *If all costs are non-positive $c_{ij} \leq 0$, the problem is equivalent to the classic Shortest Path Problem, therefore it can be solved in polynomial time.*

Proof of Proposition 2. By inverting the sign of each cost, we will obtain a scenario in which all costs are positive. Proposition 1 assures us that we can solve the resulting problem in polynomial time. The obtained solution is an optimal solution also for the initial problem with the unique difference that the value of the objecting function is negative. $\square$

A graph with particular characteristics for *CBPP* is the graph in which the cost of the arc is a function of the elevation difference of the two nodes associated with the arc. This graph could represent points positioned at different altitudes (see Figure 2).

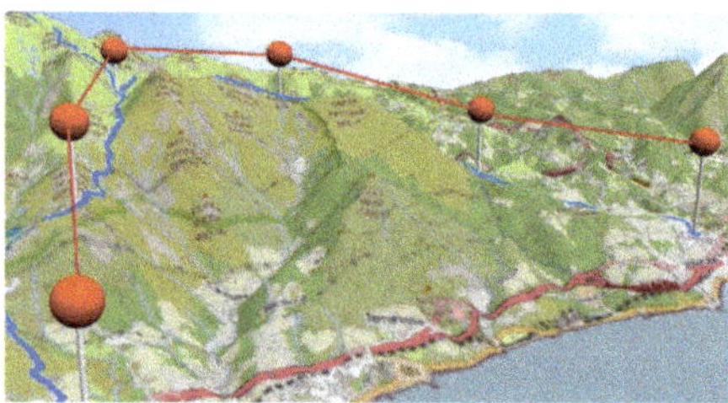

Figure 2. Example of an altimetric graph.

Proposition 3 (Elevation difference). *Given a directed graph $G(N, A)$ in which for each node $i \in N$ is defined a value $v_i \in \mathbb{R}$, and such that $c_{ij} = v_j - v_i \; \forall (ij) \in A$. The Cost-Balanced Path Problem is solvable in polynomial times.*

Proof of Proposition 3. Given a graph G created as described in Proposition 3, let be p a path in G, such that $p = ((n_1, n_2), (n_2, n_3), (n_3, n_4), \ldots, (n_h, n_k))$, $c(p) = c_{12} + c_{23} + c_{34} + \ldots + c_{hk} = (v_2 - v_1) + (v_3 - v_2) + (v_4 - v_3) + \ldots + (v_k - v_h) = -v_1 + (v_2 - v_2) + (v_3 - v_3) + (v_4 - v_4) + \ldots + (v_h - v_h) + v_k \implies c(p) = v_k - v_1$. This implies that the cost of a path depends only on the starting node and the destination node, so each path is also optimal (see Figure 3). $\square$

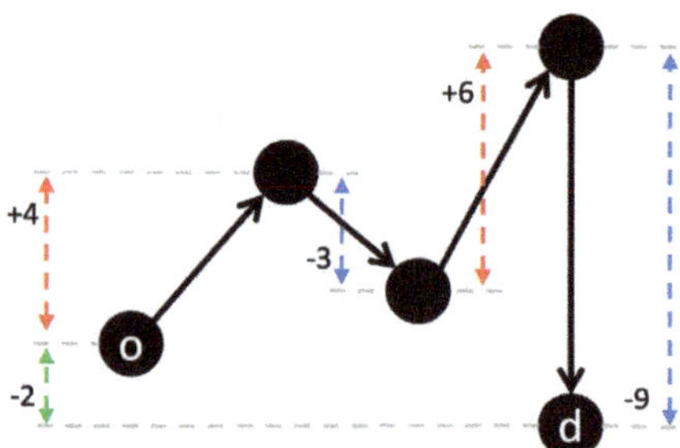

Figure 3. Example of a graph in which the cost of an arc depends on the elevation of the nodes.

3.2. Mathematical Model

In this sub-section, a model for solving *CBPP* is presented; it is a mixed integer linear programming with binary variables (MILP). Let us introduce the following decision variables:

$$x_{ij} \in \{0,1\}, \forall i, j \in N: x_{ij} = 1 \text{ if and only if arc } (i, j) \text{ is included in the problem solution.}$$

$t_i, \forall i \in N$ that represents the flow leaving node i and is used to prevent the creation of loops in the solution.

b represents the costs (in absolute value) of the optimal path.

The resulting model is the following:

$$z = Min\ b \tag{1}$$

subject to:

$$\sum_{(i,j) \in A} c_{ij} x_{ij} \leq b \tag{2}$$

$$-\sum_{(i,j) \in A} c_{ij} x_{ij} \leq b \tag{3}$$

$$\sum_{(o,j) \in A} x_{oj} = 1 \tag{4}$$

$$\sum_{(i,d) \in A} x_{id} = 1 \tag{5}$$

$$\sum_{(j,i) \in A} x_{ji} - \sum_{(i,j) \in A} x_{ij} = 0 \qquad \forall i \in N \setminus \{o,d\} \tag{6}$$

$$\sum_{(i,j) \in A} x_{ij} \leq 1 \qquad \forall j \in N \setminus \{o,d\} \tag{7}$$

$$\sum_{(i,j) \in A} x_{ij} \leq 1 \qquad \forall i \in N \setminus \{o,d\} \tag{8}$$

$$t_o = 0 \tag{9}$$

$$t_j - t_i \geq 1 - |N|(1 - x_{ij}) \qquad \forall (i,j) \in A \tag{10}$$

$$x_{ij} \in \{0,1\} \qquad \forall (i,j) \in A \tag{11}$$

$$t_i \in [0, |N| - 1] \qquad \forall i \in N \tag{12}$$

Equation (1) minimizes variable b that represents the cost of the selected path in absolute value. Variable b is defined thanks to Equations (2) and (3). Equations (4) and (5) impose that one arc leaves the origin node o, and one arc enters the destination node d. (6) impose, for each node of the network that is different from either the origin or the destination node, that the number of arcs entering the node is equal to the number of arcs exiting it. Equations (7) and (8) impose that at most one arc can enter in and exit from each node, except for the origin and the destination ones. Equations (9) and (10) defines variables t_i; t_o is set to zero (i.e., from the origin node the outflow is equal to zero), while t_j is set greater than the flow leaving node i, if arc (i,j) is selected. Finally, in (11) and (12) the decision variables are defined.

3.3. Model Extensions

In the introduction, some real applications that can be solved by the here above proposed model have been briefly described. Unfortunately, some additional constraints should be required, and thus in this sub-section, some of the additional constraints for model (1)–(12) are described.

CBPP has the objective of cost balancing instead of cost minimization. The cost that it is necessary to balance may represent a measure of a level of a particular element that has to be maintained near a pre-defined value (for example, the electric charge, the load of a vehicle, etc.). Each decision, expressed in the graph by the selection of an arc, may increase or decrease the level of the considered element. The scope is to take a sequence of decisions in such a way to have at the end of the process the same starting level that, in particular cases, can be zero.

In real applications, there is often the necessity to maintain this level within given upper and lower limits after each decision, that is, along the selected path. This means that, i.e., in the example of the electric car, the charge must always be within a lower and an upper bound. The same can be required for the cargo loaded on a vehicle.

Constraints for limiting the variance of the level within the given interval (a^{min}, a^{max}) are based on flow variables defined as follows:

$f_{ij}, \forall i, j \in N$: f_{ij} represents the level reached at node i, that will leave node i for reaching node j if and only if arc (i, j) is included in the selected path, i.e., $x_{ji} = 1$.

Thanks to Equations (13) and (14), the flow (f_{ij}) on each selected arc must be less or equal to its maximum value and greater or equal to the minimum required, while thanks to Equation (15) the outflow from the origin node (f_{oj}) is fixed equal to the starting level (a_o).

Equation (16) gives the flow conservation constraints. For each node i, different from either the origin node or the destination one, the flow that leaves node i is equal to the flow that leaves the initial node of the arc entering in i, plus the cost of arc entering in node i.

$$f_{ij} \leq a^{max} x_{ij}, \forall (i, j) \in A \tag{13}$$

$$f_{ij} \geq a^{min} x_{ij}, \forall (i, j) \in A \tag{14}$$

$$\sum_{(d,j) \in A} f_{dj} = a_o \tag{15}$$

$$\sum_{(j,i) \in A} f_{ji} + \sum_{(j,i) \in A} c_{ji} x_{ji} = \sum_{(i,j) \in A} f_{ij}, \forall i \in N \setminus \{o, d\} \tag{16}$$

Thanks to the above constraints, we are able to balance the cost and maintain it in the given required interval along the whole path.

Sometimes, together with the aim of cost balancing some other objectives must be included in the problem. In fact, when dealing with paths, the most common problem is the shortest path problem. For example, in the problem of the electric car, it should be required to have a path not too expensive in terms of either kilometers traveled or times. In this case, it is possible to insert an additional constraint that permits to find a path from origin to destination no longer than a given % of the shortest path.

Let d_{ij} be the distance associated to the arc (i, j), and c^{SP} the distance associated to the shortest path from the origin node to the destination one in the graph under investigation, α the percentage of deterioration accepted, the resulting constraint is the following:

$$\sum_{(i,j) \in A} d_{ij} x_{ij} \leq (1 + \alpha) c^{SP} \tag{17}$$

In other cases, the limitations may concern the length of the path in terms of number of arcs belonging to the path; the model can be extended by simply adding the following constraints that are related, respectively, to the maximum number of arcs that can build the path (b^{max}) and the minimum number of arcs to select (b^{min}).

$$\sum_{(i,j) \in A} x_{ij} \leq b^{max} \tag{18}$$

$$\sum_{(i,j) \in A} x_{ij} \geq b^{min} \tag{19}$$

4. Results

In this section, the computational results obtained by applying the proposed mathematical model (1)–(12) are described. Some computational experiments related to the extended model are presented too. The computational campaigns are based on some generated instances described in the following subsection.

The *MILP* model has been implemented in Java, using CPLEX version 12.8 as a solver. The computational tests were performed on a MacBook Pro, with a 2.9 GHz Intel i9 processor and 32 GB of RAM. Figure 4 shows the flow chart of the proposed approach. Step 1 load the input graph $G(N, A)$ from the text file. Step 2 creates the mathematical model. Its complexity depends on the number of constraints created, which is in the order of $|A|$. Using the *MILP* solver of the *CPLEX* software, in Step 3, the problem is solved. The computation time for Step 3 is exponential [24] as stated in Theorem 1. In Step 4, all the values associated with the model decision variables are extracted to create a textual representation of the solution.

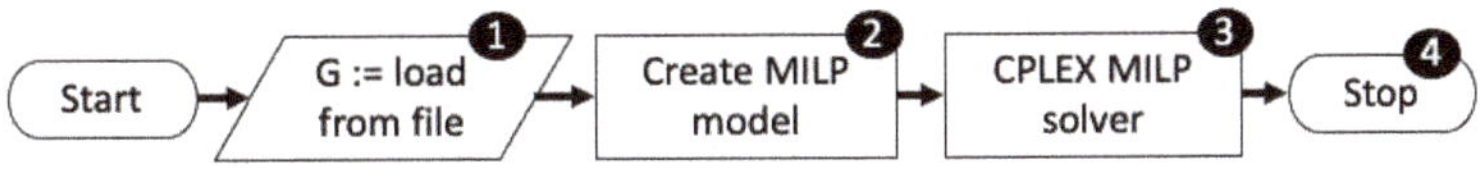

Figure 4. Algorithm flowchart.

4.1. Instances

The proposed mathematical model has been validated with two sets of generated instances for the above described problem. The first set of instances, named *Grid*, is characterized by complete square grids where each node is connected to its four neighbors. In the name of these instances, the first value represents the number of nodes; the second value represents the size of the grid. The second set of instances named *Rand* is characterized by connected graphs in which each node is randomly connected to other nodes until the desired density is reached. In the name of these instances, the first value represents the number of nodes; the second value represents the percentage of arcs incident on each vertex. The costs associated with the arcs of each instance were generated following 5 different schemes.

[−10, 10] Random homogeneous distribution of costs in the range $[−10, 10]$.
[−100, 100] Random homogeneous distribution of costs in the range $[−100, 100]$.
[−1k, 1k] Random homogeneous distribution of costs in the range $[−1000, 1000]$.
EL After associating a random height to each node, the cost of the arc represents the displacement in height.
P-EL Perturbation of the 1% random of the *EL* costs.

4.2. Computational Results for the Proposed Model

Tables 1–3 show the results of the computational tests performed on the grid instances. Each row reports the average of five instances. The last row of each table AVG is the average of all solved scenarios. We used 1800 s as the time limit for the CPLEX solver (Step 3). This implies that in the event of a higher running time, the optimality of the obtained solution is not guaranteed. Table 1 shows the computational times. It is interesting to note that the running time is mainly related to the costs associated with the arcs rather than to the size of the graph. Table 2 shows the number of zero-value solutions identified. For instances with random costs, in the scenarios with homogeneous weight distribution ($[−10, 10], [−100, 100], [−1k, 1k]$), it is always possible to obtain a solution with a cost equal to zero. Using the EL policy to create the costs, all paths will have a cost equal to the difference in height between the source node and the destination node. It is interesting to see that analyzing the P-EL policy is sufficient a 1% perturbation of the EL policy cost in order to identify solutions with a cost equal to zero, in particular as the graph size increases. Table 3 shows the obtained objective function values. The analysis of Table 3 shows that as the size of the graph increases, the solution for the P-EL policy will tend to approach zero.

Table 1. Running times for the grid instances.

Instance	$[-10, 10]$	$[-100, 100]$	$[-1k, 1k]$	EL	P-EL
Grid_100_10	0.27	0.79	1.71	0.02	38.23
Grid_225_15	0.43	0.95	3.60	0.07	1502.42
Grid_400_20	0.69	1.66	4.30	0.11	961.81
AVG	**0.46**	**1.13**	**3.20**	**0.07**	**834.15**

Table 2. Number of solutions with an objective function value equal to zero.

Instance	$[-10, 10]$	$[-100, 100]$	$[-1k, 1k]$	EL	P-EL
Grid_100_10	5	5	5	0	0
Grid_225_15	5	5	5	0	0
Grid_400_20	5	5	5	0	3
AVG	**5**	**5**	**5**	**0**	**1**

Table 3. Objective function value for the grid instances.

Instance	$[-10, 10]$	$[-100, 100]$	$[-1k, 1k]$	EL	P-EL
Grid_100_10	0	0	0	4467	3806
Grid_225_15	0	0	0	4809	3405
Grid_400_20	0	0	0	2920	878
AVG	**0**	**0**	**0**	**4065**	**2696**

Tables 4–6 show the results of the computational tests performed on the random generated instances. As before, each row reports the average of five solved instances, while the two rows AVG refer to the average of all solved scenarios, respectively, for the instances with 100 and 200 nodes. Table 4 shows the computational times. In this test, the running time remains highly dependent on the used cost scheme, but the P-EL scheme is much easier to solve than the $[-1k, 1k]$ scheme. Table 5 shows the number of zero-value solutions identified. For instances with random costs, in the scenarios with homogeneous weight distribution ($[-10, 10]$,$[-100, 100]$,$[-1k, 1k]$), it is always possible to obtain a solution with a cost equal to zero. This test confirms the results obtained previously for the grid instances. In this scenario, it becomes even more evident that the cost scheme P-EL tends as the graph grows to produce instances with cost zero solution (see also Table 6). Table 6 shows the obtained objective function values. Considering that the execution of the model stops reaching a solution equal to zero (Lower Bound), we justify the computational times shown in Table 4.

Table 4. Running times for the random instances.

Instance	$[-10, 10]$	$[-100, 100]$	$[-1k, 1k]$	EL	P-EL
Rand_100_02	0.06	0.09	1.28	0.01	0.05
Rand_100_03	0.13	0.29	1.17	0.02	0.13
Rand_100_04	0.18	0.29	2.40	0.02	0.82
Rand_100_05	0.12	0.48	2.37	0.03	0.50
Rand_100_10	0.13	0.42	2.25	0.05	0.23
Rand_100_20	0.18	0.68	4.14	0.10	0.46
AVG	**0.13**	**0.38**	**2.27**	**0.04**	**0.36**
Rand_200_02	0.30	1.18	3.91	0.05	0.50
Rand_200_03	0.22	1.11	4.53	0.06	0.71
Rand_200_04	0.16	0.35	6.63	0.09	1.30
Rand_200_05	0.25	0.59	6.28	0.11	0.66
Rand_200_10	0.23	3.88	4.95	0.18	1.01
Rand_200_20	0.15	12.66	23.08	0.38	6.66
AVG	**0.22**	**3.30**	**8.23**	**0.14**	**1.81**

Table 5. Number of solutions with an objective function value equal to zero.

Instance	$[-10, 10]$	$[-100, 100]$	$[-1k, 1k]$	EL	P-EL
Rand_100_02	5	5	5	0	1
Rand_100_03	5	5	5	0	0
Rand_100_04	5	5	5	0	2
Rand_100_05	5	5	5	0	1
Rand_100_10	5	5	5	0	2
Rand_100_20	5	5	5	0	2
AVG	**5**	**5**	**5**	**0**	**1**
Rand_200_02	5	5	5	0	1
Rand_200_03	5	5	5	0	1
Rand_200_04	5	5	5	0	1
Rand_200_05	5	5	5	0	2
Rand_200_10	5	5	5	0	2
Rand_200_20	5	5	5	0	4
AVG	**5**	**5**	**5**	**0**	**2**

Table 6. Objective function value for the random instances.

Instance	$[-10, 10]$	$[-100, 100]$	$[-1k, 1k]$	EL	P-EL
Rand_100_02	0	0	0	3634	1435
Rand_100_03	0	0	0	3040	6005
Rand_100_04	0	0	0	3358	434
Rand_100_05	0	0	0	2343	2982
Rand_100_10	0	0	0	3330	1888
Rand_100_20	0	0	0	2107	1762
AVG	**0**	**0**	**0**	**2969**	**2418**
Rand_200_02	0	0	0	2959	2835
Rand_200_03	0	0	0	4109	2559
Rand_200_04	0	0	0	4240	1545
Rand_200_05	0	0	0	2550	1762
Rand_200_10	0	0	0	2512	1676
Rand_200_20	0	0	0	2721	74
AVG	**0**	**0**	**0**	**3182**	**1742**

4.3. Results for the Extended Model

In this section, some results related to the extended model are presented. In particular, these tests are based on the 50 instances named Grid_100_10 and Grid_225_15. In all experiments, a^{max} is equals to $-a^{min}$. Looking at Table 7, it is possible to note that by decreasing the value of $alpha$, the computational time decreases, according to the decrease in the dimension of the admissible region. On the other hand, by introducing in the model the constraints associated with the parameter a^{max}, the computational time increases, according to the increase in the number of decision variables and constraints associated with the problem. As in Table 8, we can see that obviously, as the number of constraints increases, it becomes increasingly challenging to identify zero-sum solutions. Table 9 shows that as the number of constraints increases, it becomes even more difficult to identify feasible solutions: the values reported in round brackets indicate the number of unfeasible solutions. Analyzing Tables 8 and 9, it is possible to state that as the size of the graph increases, the quality of the solutions worsens less by adding further constraints. This is probably due to the increase in alternative paths between source and destination nodes.

Table 7. Running times in seconds.

$|N|$	α	a^{max}	$[-10, 10]$	$[-100, 100]$	$[-1k, 1k]$	EL	P-EL
	∞	∞	0.27	0.79	1.71	0.02	38.23
	0.2	∞	0.14	0.54	1.43	0.03	0.04
	0.1	∞	0.13	0.27	0.29	0.03	0.04
100	0.1	4C	0.27	0.77	1.12	0.15	0.16
	0.1	3C	0.39	0.60	0.88	0.14	0.16
	0.2	3C	0.73	0.86	2.38	0.22	0.20
	∞	∞	0.43	0.95	3.60	0.07	1502.42
	0.2	∞	0.99	10.15	113.64	0.13	0.50
	0.1	∞	1.47	5.09	11.57	0.16	0.24
225	0.1	4C	4.94	30.00	30.21	0.75	0.49
	0.1	3C	5.81	9.16	12.77	0.90	0.79
	0.2	3C	13.15	19.65	144.18	0.85	0.60

Table 8. Number of solutions with an objective function value equal to zero.

$|N|$	α	a^{max}	$[-10, 10]$	$[-100, 100]$	$[-1k, 1k]$	EL	P-EL
	∞	∞	5	5	5	0	0
	0.2	∞	3	2	1	0	0
	0.1	∞	2	0	1	0	0
100	0.1	4C	2	0	0	0	0
	0.1	3C	2	0	0	0	0
	0.2	3C	2	2	0	0	0
	∞	∞	5	5	5	0	0
	0.2	∞	5	4	2	0	0
	0.1	∞	4	2	0	0	0
225	0.1	4C	2	1	0	0	0
	0.1	3C	1	0	0	0	0
	0.2	3C	4	2	1	0	0

Table 9. Objective function value—values in round brackets indicate the number of unfeasible solutions.

$|N|$	α	a^{max}	$[-10, 10]$	$[-100, 100]$	$[-1k, 1k]$	EL	P-EL
	∞	∞	0	0	0	4467	3806
	0.2	∞	6	15	13	4467	4379
	0.1	∞	12	95	233	4467	4402
100	0.1	4C	3 (2)	15 (2)	260	4467	4402
	0.1	3C	3 (2)	15 (2)	179 (1)	4467 (1)	5003 (1)
	0.2	3C	3 (1)	4 (2)	100	4467	4379
	∞	∞	0	0	0	4809	3405
	0.2	∞	0	1	10	4809	4870
	0.1	∞	1	4	175	4809	4900
225	0.1	4C	3	5 (1)	5 (2)	4809	4900
	0.1	3C	1 (2)	17 (2)	18 (3)	4809	4900
	0.2	3C	1 (1)	26	35	4809	4870

5. Conclusions

This paper deals with the Cost-Balanced Path Problem ($CBPP$), a variant of the classic Shortest Path Problem introduced in this paper for the first time. The characteristic of this problem is that it can be used as a sub-problem to model many real scenarios. Using the mixed-integer linear programming model introduced in Section 3.2, we computed the optimal solution for many test instances. It is interesting to note that analyzing the results shown in Section 4, in the case of uniform distribution of the costs of the arcs, there is always an optimal solution with an objective function value equal to zero. To prevent or make the presence of solutions with an objective function value equal to zero rarer, smart methods for defining the cost of the edges $(EL, P - EL)$ have been used. Note that when the model reaches an objective function value equal to zero, it stops instantaneously having reached its lower bound; this implies that the computational time for instances that do not have zero as an optimal solution is significantly higher. Considering these observations, the future developments for this work are manifold. First of all, it would be interesting to develop instance generators capable of preventing or minimizing the presence of zero solutions in order to create computationally complex instances. Using more complex instances, realizing heuristic or meta-heuristic approaches for this problem would become necessary. A constructive approach based on the Dijkstra algorithm [25] improved through the Carousel Greedy, an enhanced Greedy algorithm proposed in [26,27], might identify a feasible solution to the problem. According to the authors' experience, the tabu search, a technique introduced by Glover [28] and widely used in the literature, also by the authors of this work, for example, in [29], might be used to improve the Greedy solution.

Author Contributions: All authors contributed equally to this work. All authors have read and agreed to the published version of the manuscript.

Funding: This research received no external funding.

Institutional Review Board Statement: Not applicable.

Informed Consent Statement: Not applicable.

Data Availability Statement: Data available on request.

Conflicts of Interest: The authors declare no conflict of interest.

Abbreviations

The following abbreviations are used in this manuscript:

SPP	Shortest Path Problem
CBPP	Cost-Balanced Path Problem
HPP	Hamiltonian Path Problem
MILP	Mixed Integer Linear Programming

References

1. He, F.; Yin, Y.; Lawphongpanich, S. Network equilibrium models with battery electric vehicles. *Transp. Res. Part B Methodol.* **2014**, *67*, 306–319. [CrossRef]
2. Dell'Amico, M.; Hadjicostantinou, E.; Iori, M.; Novellani, S. The bike sharing rebalancing problem: Mathematical formulations and benchmark instances. *Omega* **2014**, *45*, 7–19. [CrossRef]
3. Akbay, M.; Kalayci, C. A Variable Neighborhood Search Algorithm for Cost-Balanced Travelling Salesman Problem. In *Metaheuristics Summer School*; Springer: Cham, Switzerland, 2018; pp. 23–36.
4. Larusic, J.; Punnen, A.P. The balanced traveling salesman problem. *Comput. Oper. Res.* **2011**, *38*, 868–875. [CrossRef]
5. Dong, X.; Xu, M.; Lin, Q.; Han, S.; Li, Q.; Guo, Q. IT algorithm with local search for large scale multiple balanced traveling salesmen problem. *Knowl.-Based Syst.* **2021**, *229*, 107330. [CrossRef]
6. Gallo, G.; Pallotino, S. Shortest path algorithms. *Ann. Oper. Res.* **1988**, *13*, 3–79. [CrossRef]
7. Cherkassy, B.V.; Goldberg, A.V.; Radzik, T. Shortest path algorithms: Theory and experimental evaluation. *Math. Program.* **1996**, *73*, 129–174. [CrossRef]
8. Fu, L.; Sun, D.; Rilett, L.R. Heuristic shortest path algorithms for transportation applications: State of the art. *Comput. Oper. Res.* **2006**, *33*, 3324–3343. [CrossRef]
9. Raith, A.; Ehrgott, M. A comparison of solution strategies for bi-objective shortest path problems. *Comput. Oper. Res.* **2009**, *36*, 1299–1331. [CrossRef]
10. Panda, M.; Mishra, A. A survey of shortest-path algorithms. *Int. J. Appl. Eng. Res.* **2018**, *13*, 6817–6820.
11. Yuan, H.; Hu, J.; Song, Y.; Li, Y.; Du, J. A new exact algorithm for the shortest path problem: An optimized shortest distance matrix. *Comput. Ind. Eng.* **2021**, *158*, 107407. [CrossRef]
12. Ferone, D.; Festa, P.; Pastore, T. The k-color shortest path problem. In *Advances in Optimization and Decision Science for Society, Services and Enterprises*; Springer: Cham, Switzerland, 2019; pp. 367–376.
13. Ferone, D.; Festa, P.; Fugaro, S.; Pastore, T. A dynamic programming algorithm for solving the k-Color Shortest Path Problem. *Optim. Lett.* **2021**, *15*, 1973–1992. [CrossRef]

14. Ticha, H.B.; Absi, N.; Feillet, D.; Quilliot, A. The Steiner bi-objective shortest path problem. *EURO J. Comput. Optim.* **2021**, *9*. [CrossRef]
15. Ketkov, S.S.; Prokopyev, O.A.; Burashnikov, E.P. An approach to the distributionally robust shortest path problem. *Comput. Oper. Res.* **2021**, *130*, 1105212. [CrossRef]
16. Zhang, D.; Wallace, S.W.; Guo, Z.; Dong, Y.; Kaut, M. On scenario construction for stochastic shortest path problems in real road networks. *Transp. Res. Part E Logist. Transp. Rev.* **2021**, *152*, 102410. [CrossRef]
17. Ehmke, J.F.; Campbell, A.M.; Thomas, B.W. Data-driven approaches for emissions-minimized paths in urban areas. *Comput. Oper. Res.* **2016**, *67*, 34–47. [CrossRef]
18. Prakash, A.A. Pruning algorithm for the least expected travel time path on stochastic and time-dependent networks. *Transp. Res. Part B Methodol.* **2018**, *108*, 127–147. [CrossRef]
19. Moharam, R.; Morsy, E. Genetic algorithms to balanced tree structures in graphs. *Swarm Evol. Comput.* **2017**, *32*, 132–139. [CrossRef]
20. Di Puglia Pugliese, L.; Guerriero, F.; Poss, M. The Resource Constrained Shortest Path Problem with uncertain data: A robust formulation and optimal solution approach. *Comput. Oper. Res.* **2019**, *107*, 140–155. [CrossRef]
21. Carrabs, F.; Cerrone, C.; Cerulli, R.; Silvestri, S. On the complexity of rainbow spanning forest problem. *Optim. Lett.* **2018**, *12*, 443–454. [CrossRef]
22. Carrabs, F.; Cerrone, C.; Cerulli, R.; Silvestri, S. The rainbow spanning forest problem. *Soft Comput.* **2018**, *22*, 2765–2776. [CrossRef]
23. Garey, M.; Johnson, D. *Computers and Intractability*, 3rd ed.; Freeman: San Francisco, CA, USA, 1979.
24. Schrijver, A. *Theory of Linear and Integer Programming*; John Wiley & Sons: Hoboken, NJ, USA, 1998.
25. Dijkstra, E.W. A note on two problems in connexion with graphs. *Numer. Math.* **1959**, *1*, 269–271. [CrossRef]
26. Cerrone, C.; Cerulli, R.; Golden, B. Carousel greedy: A generalized greedy algorithm with applications in optimization. *Comput. Oper. Res.* **2017**, *85*, 97–112. [CrossRef]
27. Carrabs, F.; Cerrone, C.; D'Ambrosio, C.; Raiconi, A. Column generation embedding carousel greedy for the maximum network lifetime problem with interference constraints. In *International Conference on Optimization and Decision Science*; Springer: Cham, Switzerland, 2017; pp. 151–159.

28. Glover, F. Tabu search—Part I. *ORSA J. Comput.* **1989**, *1*, 190–206. [CrossRef]
29. Carrabs, F.; Cerrone, C.; Cerulli, R. A tabu search approach for the circle packing problem. In Proceedings of the 17th International Conference on Network-Based Information Systems, Salerno, Italy, 10–12 September 2014; pp. 165–171.

Article

Improved Lebesgue Indicator-Based Evolutionary Algorithm: Reducing Hypervolume Computations

Saúl Zapotecas-Martínez [1,*], Abel García-Nájera [1] and Adriana Menchaca-Méndez [2]

[1] Departamento de Matemáticas Aplicadas y Sistemas, Universidad Autónoma Metropolitana Unidad Cuajimalpa, Av. Vasco de Quiroga 4871, Col. Santa Fe Cuajimalpa, Ciudad de México 05348, México; agarcian@cua.uam.mx

[2] Licenciatura en Tecnologías para la Información en Ciencias, ENES, Campus Morelia, Universidad Nacional Autónoma de México, Morelia 58190, México; amenchaca@enesmorelia.unam.mx

[*] Correspondence: szapotecas@cua.uam.mx

Abstract: One of the major limitations of evolutionary algorithms based on the Lebesgue measure for multi-objective optimization is the computational cost required to approximate the Pareto front of a problem. Nonetheless, the Pareto compliance property of the Lebesgue measure makes it one of the most investigated indicators in the design of indicator-based evolutionary algorithms (IBEAs). The main deficiency of IBEAs that use the Lebesgue measure is their computational cost which increases with the number of objectives of the problem. On this matter, the investigation presented in this paper introduces an evolutionary algorithm based on the Lebesgue measure to deal with box-constrained continuous multi-objective optimization problems. The proposed algorithm implicitly uses the regularity property of continuous multi-objective optimization problems that has suggested effectiveness when solving continuous problems with rough Pareto sets. On the other hand, the survival selection mechanism considers the local property of the Lebesgue measure, thus reducing the computational time in our algorithmic approach. The emerging indicator-based evolutionary algorithm is examined and compared versus three state-of-the-art multi-objective evolutionary algorithms based on the Lebesgue measure. In addition, we validate its performance on a set of artificial test problems with various characteristics, including multimodality, separability, and various Pareto front forms, incorporating concavity, convexity, and discontinuity. For a more exhaustive study, the proposed algorithm is evaluated in three real-world applications having four, five, and seven objective functions whose properties are unknown. We show the high competitiveness of our proposed approach, which, in many cases, improved the state-of-the-art indicator-based evolutionary algorithms on the multi-objective problems adopted in our investigation.

Keywords: multi-objective optimization; Lebesgue measure; indicator-based evolutionary algorithms

Citation: Zapotecas-Martínez, S.; García-Nájera, A.; Menchaca-Méndez, A. Improved Lebesgue Indicator-Based Evolutionary Algorithm: Reducing Hypervolume Computations. *Mathematics* **2022**, *10*, 19. https://doi.org/10.3390/math10010019

Academic Editor: David Greiner

Received: 30 October 2021
Accepted: 15 December 2021
Published: 21 December 2021

Publisher's Note: MDPI stays neutral with regard to jurisdictional claims in published maps and institutional affiliations.

1. Introduction

In several engineering and sciences applications, some problems require the simultaneous optimization of a number of objective functions. In the specialized literature, such problems are referred to as multi-objective optimization problems (MOPs). The optimization of a multi-objective problem involves determining the best compensation alternatives considered in a set of conflicting objective functions. Therefore, instead of an optimal solution, as in single-objective optimization, a set of solutions manifesting the best trade-offs among objectives is reached. The population on which evolutionary algorithms are based makes these algorithms a practical tool to solve these types of problems. For this reason, evolutionary multi-objective algorithms (EMOAs) have become a flexible and popular instrument to deal with MOPs. In the specialized literature, a variety of investigations concerning the development of evolutionary approaches for multi-objective optimization can be found. See the extensive review of such approaches presented in [1,2]. According to their conceptual foundations, EMOAs are categorized into three main groups: Pareto-based, decomposition-based, and indicator-based approaches. These approaches incorporate different search strategies that define by themself the performance of a particular EMOA. Distinctly, EMOAs based on indicators—the topic investigated in this work—explicitly optimize a quality indicator (e.g., $R2$ [3], Lebesgue measure [4], ϵ indicator [5], IGD [6], among others) to approximate the Pareto front of a MOP. In this manner, since its origin in the early 2000s, the indicator-based evolutionary algorithm (IBEA) [7] traced a new research line investigated to date.

IBEAs adopting indicators that use reference sets (e.g., $R2$, ϵ indicator, IGD, etc.) are a design challenge since the optimal solutions are unknown. Consequently, reference sets cannot be adequately pre-established. Despite this, some researchers have studied diverse techniques to predict the reference set for these IBEAs [8,9]. In the evolutionary multi-objective optimization (EMOO) literature, the Lebesgue measure, also referred to as hypervolume indicator or S-metric, was introduced by Zitzler and Thiele [4] to evaluate the performance of EMOAs. This quality indicator possesses an attractive property—it is Pareto compliant [5]—that has called the attention of several researchers working on IBEAs. In particular, IBEAs adopting the Lebesgue measure benefit from not requiring reference sets because they exclusively employ reference vectors that are much simpler to state. Therefore, these IBEAs have been a practical approach to solving real-life applications where the characteristics of the problems are not known. Although IBEAs based on the Lebesgue measure are highly docile solving MOPs, their application is restricted by the computational cost of the Lebesgue measure, which grows with the number of objective functions. As pointed out in [10], this indicator cannot be calculated in polynomial time concerning the number of objectives except that $P = NP$. In addition, the complex characteristics of multi-objective problems (for example, multimodality, bias, non-separability, etc.) faced by an IBEA, further increase the computational cost in the search process for such algorithms. In other words, IBEAs use many more iterations (computational efforts) to approximate the real Pareto front of a problem. As a consequence, extensive investigations concerning the design of IBEAs using the Lebesgue measure as a quality indicator have been studied in the last few years [11–15]. To date, the development of EMOAs based on the hypervolume indicator is recognized as an actual area of investigation within the EMOO community, and this is precisely the topic of the investigation presented in this work.

This paper introduces an improved Lebesgue indicator-based evolutionary algorithm for multi-objective optimization. The introduced approach can be seen as an improvement of the Lebesgue indicator-based evolutionary algorithm (LIBEA) [15]. Analogous to LIBEA, the proposed algorithm addresses the notion of IBEA [7] in the sense of optimizing a quality indicator. Nevertheless, it is directed at maximizing the Lebesgue measure of non-dominated solutions obtained through the search. In contrast to several Lebesgue indicator-based EMOAs, the introduced algorithm implicitly applies the regularity property of continuous MOPs advised to approximate continuous MOPs with complicated characteristics [16–18]. Additionally, in order to reduce the computational time, the local property of the Lebesgue measure is considered in the survival mechanism of the proposed algorithm [19]. We hypothesize that an algorithm considering the Lebesgue measure, the regularity property of continuous MOPs, and the local property of the Lebesgue measure can solve problems with difficult features more efficiently than traditional EMOAs based on the Lebesgue measure.

The proposed IBEA is tested by solving a set of artificial test problems known to be challenging in the EMOO literature. As discussed by some researchers [20], algorithms able to solve test problems having different difficulties can be candidates to deal properly with real-life problems. Consequently, we present an analysis of the proposed algorithm solving three real-life applications where the fitness landscapes and Pareto fronts are unknown. A comparison is carried out to analyze the performance of the suggested IBEA versus three state-of-the-art IBEAs based on the Lebesgue measure. We show that the algorithmic proposal outperforms the state-of-the-art IBEAs in most test problems, including the three real-life problems considered in our study. The obtained results are statistically validated over a number of experiments performed as part of our experimental research.

The rest of the manuscript is organized as follows. Section 2 introduces the fundamental concepts to understand the content of this work. Section 3 introduces an overview of the related work to this investigation. Section 4 introduces the proposed algorithm and details its components. Section 5 presents an experimental study of performance on a set of test problems with complicated features. Section 6 introduces three real-life applications from practice in which the suggested algorithm is tested and analyzed with other IBEAs. Lastly, Section 7 presents our outcomes and describes some paths for future investigation.

2. General Background

This section provides the foundations of multi-objective optimization, introduces the indicator-based multi-objective evolutionary algorithms, and presents some concepts related to performance quality indicators.

2.1. Multi-Objective Optimization

Using standard notation and terminology, a multi-objective optimization problem (MOP) can be defined as follows.

Definition 1 (Multi-objective optimization problem). *Without loss of generality, assuming minimization in all the objective functions, a multi-objective optimization problem can be defined as:*

$$\begin{aligned} \underset{\mathbf{x} \in \mathcal{X}}{\text{minimize:}} \quad & \mathbf{F}(\mathbf{x}) = (f_1(\mathbf{x}), \ldots, f_m(\mathbf{x}))^T \\ \text{subject to} \quad & g_i(\mathbf{x}) \leq 0 \ , \ \forall\, i \in \{1, \ldots, p\} \ , \\ & h_j(\mathbf{x}) = 0 \ , \ \forall\, j \in \{1, \ldots, q\}. \end{aligned} \tag{1}$$

where $\mathbf{x} \in \mathcal{X} \subset \mathbb{R}^n$ is a solution to the problem, $\mathcal{X}$ is the solution space, and $f_i : \mathcal{X} \to \mathbb{R}$, for all $i \in \{1, \ldots, m\}$, are m objective functions. The constraint functions $g_i, h_j : \mathcal{X} \to \mathbb{R}$ restrict $\mathbf{x}$ to a feasible region $\mathcal{X}' \subseteq \mathcal{X}$.

In multi-objective optimization, a set of trade-off solutions are normally aimed for, because the minimization of one objective function could lead to the deterioration of the others. To describe the concept of optimality in which we are interested in, the following definitions are presented.

Definition 2 (Pareto dominance). *Let $\mathbf{x}, \mathbf{y} \in \mathcal{X}'$. We say that $\mathbf{x}$ weakly dominates $\mathbf{y}$ ($\mathbf{x} \preceq \mathbf{y}$) if $f_i(\mathbf{x}) \leq f_i(\mathbf{y})$ for all $i \in \{1, \ldots, m\}$. If, in addition $\mathbf{y} \npreceq \mathbf{x}$, we say that $\mathbf{x}$ strictly dominates $\mathbf{y}$ ($\mathbf{x} \prec \mathbf{y}$). If $f_i(\mathbf{x}) < f_i(\mathbf{x})$ for all $i \in \{1, \ldots, m\}$, we say that $\mathbf{x}$ strongly dominates $\mathbf{y}$ ($\mathbf{x} \prec\prec \mathbf{y}$).*

Definition 3 (Pareto optimality). *Let $\mathbf{x}^\star \in \mathcal{X}'$. We say that $\mathbf{x}^\star$ is a Pareto optimal solution if there is no other solution $\mathbf{y} \in \mathcal{X}$ such that $\mathbf{y} \preceq \mathbf{x}^\star$.*

Definition 4 (Pareto optimal set). *The Pareto optimal set $\mathcal{PS}$ of a multi-objective problem is defined by $\mathcal{PS} = \{\mathbf{x} \in \mathcal{X}' \mid \mathbf{x} \text{ is Pareto optimal solution}\}$.*

Definition 5 (Pareto optimal front). *The Pareto optimal front $\mathcal{PF}$ of a multi-objective problem is stated by the image of the Pareto optimal set, that is, $\mathcal{PF} = \{\mathbf{F}(\mathbf{x}) \mid \mathbf{x} \in \mathcal{PS}\}$.*

An interesting property observed in continuous multi-objective problems that has been considered when designing multi-objective algorithms is presented below.

Property 1 (Regularity property of continuous MOPs). *From the Karush–Kuhn–Tucker conditions, it can be induced that under certain assumptions, the $\mathcal{PS}$ ($\mathcal{PF}$) of a continuous MOP with*

*m objectives defines an (m − 1)-dimensional piecewise continuous manifold in the decision space
(objective space)* [21,22].

The regularity property of continuous MOPs defined above was firstly employed
by Hillermeier [22] in the well-known continuation methods for multi-objective optimization. As identified by some authors [18], multi-objective solvers should take into account
this property explicitly or implicitly.

A critical condition of a multi-objective optimization problem is the conflict among its
objectives. If there is no conflict among the objectives, then the problem could be solved
by the optimization of each objective function independently. Although several authors
have given distinct definitions for the relation between pairs of objectives, the following
definition will be used in this paper.

Definition 6 (Conflict relation). *Let $S \subset \mathcal{X}'$. According to Carlsson and Fullér [23], two
objective functions f_i and f_j can be related in the following three ways (assuming minimization):*

1. *f_i is in conflict with f_j on S if $f_i(\mathbf{x}) \leq f_i(\mathbf{y}) \implies f_j(\mathbf{x}) \geq f_j(\mathbf{y}) \; \forall \, \mathbf{x}, \mathbf{y} \in S$;*
2. *f_i supports f_j on S if $f_i(\mathbf{x}) \geq f_i(\mathbf{y}) \implies f_j(\mathbf{x}) \geq f_j(\mathbf{y}) \; \forall \, \mathbf{x}, \mathbf{y} \in S$;*
3. *f_i and f_j are independent on S, otherwise.*

2.2. Indicator-Based Evolutionary Algorithms for Multi-Objective Optimization

Quality indicators have been introduced to compare the outcomes of multi-objective
algorithms in a quantitative manner. They map a Pareto front approximation to a scalar
number that quantifies the performance of a multi-objective approach.

Definition 7 (Quality indicator). *An n-ary quality indicator $\mathcal{I}$ is a function $\mathcal{I} : \mathcal{F}^n \to \mathbb{R}$, which
assigns each vector $(\mathcal{A}_1, \ldots, \mathcal{A}_n)$ of n approximation sets (which can be singletons) a real value
$I(\mathcal{A}_1, \ldots, \mathcal{A}_n)$.*

Currently, we can find a large number of quality indicators for multi-objective optimization. A comprehensive compilation of them can be found in [5,24–26]. Quality
indicators can assess convergence and diversity of solutions along the Pareto front of a
given MOP. However, some indicators require certain knowledge of the problem which, in
many cases, is not available. For example, quality indicators based on reference sets (e.g.,
R2, ε indicator, IGD, etc.) require a discretization of the entire Pareto front.

Although quality indicators were initially employed for comparison purposes of
multi-objective solvers, their use has been extended to guide the optimization process in
EMOAs. In this way, with its emergence in the early 2000s, the indicator-based evolutionary
algorithm (IBEA) [7] posed the possibility to optimize a quality indicator to approximate
the Pareto front of a MOP.

Let us consider the $(\mu + \lambda)$-selection scheme of an EMOA and the combined population Q_t of μ parents and λ offspring. In order to choose the best μ solutions for the next
population (i.e., the updated set of parents), the fitness assignment to each individual is
necessary. Traditionally, EMOAs employ the Pareto ranking and a diversity indicator to
update the parents set. The selection mechanism in IBEAs consists of finding the solution
that contributes the least to the indicator under consideration. Making allowance for the
fitness value $\varphi(\mathbf{q}^i)$ of an individual $\mathbf{q}^i \in Q_t$, Algorithm 1 shows the survival selection
mechanism in IBEAs.

Algorithm 1: IBEAs survival selection mechanism.

Input:
    Q_t: Combined population of μ parents and λ offspring.
Output:
    P_{t+1}: Updated population of μ parents.
1 **while** $|Q_t| > \mu$ **do**
2   Compute $\varphi(\mathbf{q}^1), \ldots, \varphi(\mathbf{q}^{|Q_t|})$ where $\mathbf{q}^i \in Q_t$ for all $i \in \{1, \ldots, |Q_t|\}$;
3      $\mathbf{q}_{worst} \leftarrow \arg\min_{\mathbf{q}^i \in Q_t} \ \max\{\varphi(\mathbf{q}^i)\}$;
4      $Q_t \leftarrow Q_t \setminus \{\mathbf{q}_{worst}\}$;
5 **return** Q_t;

In the sense of evolutionary algorithms, the higher the fitness value $\varphi(\mathbf{q}^i)$, the better the individual $\mathbf{q}^i$. Note, however, that according to their conceptual foundations, quality indicators can be either maximized or minimized. Therefore, the adequate fitness values assignment depends directly on the concerned quality indicator. Note that besides the computational time required to estimate the worst solution in IBEAs could be considerably costly (line 2 in Algorithm 1). To date, in the EMOO literature, we can find several evolutionary algorithms based on various quality indicators. A comprehensive review of these types of algorithms can be found in [27].

2.3. Performance Quality Indicators

As pointed out, performance indicators are employed in IBEAs and are also employed to compare performance between algorithms. In the follows, we present some relevant performance indicators in this study.

2.3.1. Hypervolume Performance Indicator

The hypervolume performance indicator, as well known as Lebesgue measure or S metric, has been employed to guide the search in evolutionary algorithms for multi-objective optimization. The follow definitions are relevant in this work [28,29].

Definition 8 (Hypervolume indicator). *Let $S \subset \mathbb{R}^m$ and $\mathbf{r} \in \mathbb{R}^m$ be a point set and a reference point, respectively. The hypervolume indicator of S is the measure of the region weakly dominated by S and bounded by vector $\mathbf{r}$. Formally:*

$$\mathcal{H}(S, \mathbf{r}) = \mathcal{L}(\{\mathbf{q} \in \mathbb{R}^m \mid \exists\, \mathbf{p} \in S : \mathbf{p} \preceq \mathbf{q} \text{ and } \mathbf{q} \preceq \mathbf{r}\}) \tag{2}$$

where $\mathcal{L}(\cdot)$ refers to the Lebesgue measure.

Definition 9 (Hypervolume Contribution). *The exclusive hypervolume contribution of a solution $\mathbf{q} \in \mathbb{R}^m$ to a set $S \in \mathbb{R}^m$ respect to the reference vector $\mathbf{r}$, is defined as:*

$$\mathcal{H}_c(S, \mathbf{q}, \mathbf{r}) = \mathcal{H}(S \cup \{\mathbf{q}\}, \mathbf{r}) - \mathcal{H}(S \setminus \{\mathbf{q}\}, \mathbf{r}) \tag{3}$$

The hypervolume contribution of a point is sometimes referred to as Lebesgue contribution, incremental, or exclusive hypervolume contribution. In this regard, some contributions to the state of the art on this topic can be found in [30–32].

In the specialized literature, we can find several issues addressed by investigators in relation to the hypervolume indicator, see the comprehensive review on this topic presented in [33]. However, one of the most important challenges in this research area is the exact computation of the hypervolume indicator on a point set S. In this regard, some researchers have designed algorithms that are efficient in a few dimensions, see the works reported in [31,34,35]. The computational complexity of the hypervolume computation is exponential to the number of points in S [19]. An interesting property observed in the two-dimensional objective space that has been exploited for a fast hypervolume computation is presented below.

Property 2 (Locality property of the hypervolume indicator). *Given three consecutive points on the Pareto front, moving the middle point will only affect the hypervolume contribution that is solely dedicated to this point, but the joint hypervolume contribution of the other points remains fixed [19].*

Nonetheless, the challenges presented in high-dimensional objective spaces have motivated a vast research in the design of algorithms for the efficient hypervolume computation. As a flavor of approaches devoted to the exact hypervolume calculation generalized in the number of dimension, Table 1 presents some algorithms known by the EMOO community and their complexities for an m-dimensional set of n points.

Table 1. Algorithms for the exact hypervolume computation on an m-dimensional set of n points.

Algorithm	Dimension	Computational Complexity
HSO [36]	$m \geq 2$	$O(n^{m-1})$
LebMeasure [37]	$m \geq 2$	$O(n^m)$
FPL [38]	$m \geq 2$	$O(n^{m-2} \log n)$
HOY [39]	$m \geq 2$	$O(n^{d/2} \log n)$
WFG [40]	$m \geq 2$	$\Omega(n^{m/2} \log n)$
QHV [41]	$m \geq 2$	$O(2^{m(n-1)})$

2.3.2. Normalized Hypervolume Indicator

The $\mathcal{H}$ indicator (stated in Definition 8) can quantify convergence and distribution of solutions on the $\mathcal{PF}$ of a given problem. The normalized hypervolume can be defined as follows.

Definition 10 (Normalized hypervolume indicator). *Let $S \subset \mathbb{R}^m$, $\mathbf{u} \in \mathbb{R}^m$ and $\mathbf{r} \in \mathbb{R}^m$ be a point set, an ideal point, and a reference point, respectively, such that $\mathbf{u} \preceq \mathbf{s} \preceq \mathbf{r}$ (for all $\mathbf{s} \in S$). The normalized hypervolume indicator of S is the measure of the region weakly dominated by S and bounded by vector $\mathbf{u}$ and $\mathbf{r}$. Mathematical it can be stated as:*

$$\mathcal{H}_n(S, \mathbf{u}, \mathbf{r}) = \frac{\mathcal{H}(S, \mathbf{r})}{\Pi_{i=1}^{m} |r_i - u_i|} \tag{4}$$

where $\mathcal{H}(S, \mathbf{r})$ denotes the hypervolume indicator of S with reference vector $\mathbf{r}$.

The $\mathcal{H}_n$ indicator value is in the range $[0, 1]$. In this way, a large $\mathcal{H}_n$ value indicates that the set of solutions S has a suitable approximation and spread on the real $\mathcal{PF}$.

2.3.3. IGD^+ Indicator

The inverted generational distance plus (IGD^+) [42] is an extension of the IGD indicator [6]. This quality indicator is weakly Pareto compliant and it can quantify how far a given approximation set is from the real Pareto front. Formally, the IGD^+ indicator is stated as follows.

Definition 11 (Inverted Generational Distances plus). *Let $F \in \mathbb{R}^m$ and $S \in \mathbb{R}^m$ be a discretization of the real Pareto front of a given MOP and a set of objective vectors given by an algorithm, respectively. The IGD^+ quality indicator is stated as:*

$$IGD^+(F, S) = \left(\frac{1}{|F|} \sum_{\mathbf{r} \in F} \min_{\mathbf{s} \in S} d^+(\mathbf{r}, \mathbf{s}) \right)^{1/p} \tag{5}$$

where $p = 2$ and d^+ is defined by,

$$d^+(\mathbf{r}, \mathbf{s}) = \sqrt{\sum_{i=1}^{m} (\max\{s_i - r_i, 0\})^2} \tag{6}$$

where m is the number of objective functions of a given MOP.

A value of zero of the IGD^+ indicator notices that all the objective vectors obtained by an algorithm are on the true $\mathcal{PF}$.

3. Previous Related Work

The hypervolume indicator ($\mathcal{H}$), as well known as Lebesgue or S metric, is a quality indicator widely used to assess the performance of evolutionary multi-objective algorithms [26]. Its peculiar property—it is strictly Pareto compliant [5]—has motivated several investigators working on the design of IBEAs. It has been proved that given a finite search space and a reference point, maximizing the hypervolume indicator is equivalent to finding the Pareto optimal set of a given problem [37]. For this reason, several IBEAs have incorporated this indicator in their survival selection mechanism (see the comprehensive survey of approaches presented in [43]). Lebesgue indicator-based EMOAs need to compute the hypervolume contribution ($\mathcal{H}_c$) of non-dominated objective vectors to estimate the worst solution in the current population. As pointed out before, the main disadvantage of the hypervolume indicator is its computation cost which increases exponentially with the number of objectives of the problem. Traditionally, EMOAs based on the $\mathcal{H}$ indicator need to compute the $\mathcal{H}_c$ of each individual in the population per iteration. Examples of these algorithms are SIBEA [44], SMS-EMOA [11], MO-CMA-ES [45], HMOPSO [46], FV-MOEA [14], LIBEA [15], among others. These approaches become impractical when dealing with many objective functions (more than three), employing large populations, or requiring a significant number of generations. Consequently, some authors have focused their investigation on reducing the computational complexity of methods to compute either the $\mathcal{H}$ or $\mathcal{H}_c$ [33,43]. Other alternatives studied by some researchers are the approximation methods to estimate the $\mathcal{H}$ or $\mathcal{H}_c$ [33,43]. In this regard, some authors have incorporated into their IBEAs, approximation methods to calculate $\mathcal{H}_c$. A pioneering study adopting this idea is the HypE algorithm introduced in [47]. Another example of these types of approaches is the R2HCA-EMOA [48], which works similar to SMS-EMOA, but it uses the R2-based hypervolume contribution approximation method [49]. Experimental results presented by the authors show that it outperforms the HypE algorithm in terms of $\mathcal{H}$. Although the approximation methods have decreased the computational cost of IBEAs based on the Lebesgue measure, the performance quality in these algorithms is compromised. This is, in effect, the price to compensate for efficiency in these types of IBEAs.

In this paper, we are interested in designing IBEAs based on the exact hypervolume computation. In this regard, Menchaca-Mendéz and Coello [50] presented an improved version of SMS-EMOA called iSMS-EMOA. iSMS-EMOA generates an offspring per iteration. After that, the nearest individual to the offspring (measured by the Euclidean distance in the objective space) and another randomly selected individual compete with the offspring to survive (comparing their $\mathcal{H}_c$). Therefore, iSMS-EMOA only needs to compute three hypervolume contributions per iteration, unlike SMS-EMOA that calculates n contributions, where n is the population size. The core idea of iSMS-EMOA is to move a solution within its neighborhood to improve its $\mathcal{H}_c$. This idea is based on the locality property stated in [19] (see Property 2). iSMS-EMOA significantly improves the efficiency of SMS-EMOA, and it achieves comparable performance to SMS-EMOA. In [51], the authors studied the behavior of iSMS-EMOA using the approximation method to calculate $\mathcal{H}_c$ proposed by Bringmann and Friedrich [52]. The experimental results show that this version of iSMS-EMOA outperforms HypE. In [53], the authors studied the behavior of iSMS-EMOA if it does not use the randomly selected individual in the competition always. Rostami and Neri [54] proposed the algorithm CMA-PAES-HAGA, which incorporates a fast hypervolume-driven selection mechanism for many-objective optimization called HAGA to CMA-ES. HAGA divides the objective space into grids. Then, it separates the population into subpopulations (each grid contains one subpopulation). When a new individual is created, it only competes with the individuals in its grid. Experimental results show that CMA-PAES-HAGA is able to solve problems with more than three objective functions. Recently, Zapotecas-Martínez et al. [15] introduced a novel Lebesgue-based IBEA (LIBEA) adopting the regularity property of continuous MOPs (see Property 1). The introduced LIBEA employs different neighborhoods for the mating selection mechanism. In this way, if a solution is close to the $\mathcal{PS}$ of a problem, it is possible to create new solutions close to the $\mathcal{PS}$ recombining with neighboring solutions. The authors show the effectiveness of LIBEA when solving continuous MOPs with roughed Pareto optimal sets.

In contrast to the related work, we introduce an improved multi-objective solver considering the Lebesgue measure, the regularity property of continuous MOPs, and the local property of the Lebesgue measure. We investigate a new framework to solve problems with difficult features and unknown fitness landscapes more efficiently than traditional IBEAs based on the Lebesgue measure. In the next section, we describe the components of the proposed algorithm thoroughly.

4. Improved Evolutionary Multi-Objective Algorithm Based on the Lebesgue Indicator

The proposed algorithm presented in this paper is an improvement of the Lebesgue indicator-based evolutionary algorithm (LIBEA) [15] for multi-objective optimization. In analogy to LIBEA, the suggested algorithm addresses the notion of IBEA [7] regarding the optimization of a quality indicator. Nevertheless, it is directed at maximizing the Lebesgue indicator of non-dominated solutions obtained through the search. The differences are clearly observed between our algorithmic proposal and IBEAs adopting the Lebesgue measure. This section introduces details of the new algorithm and its components to be compared against state-of-the-art IBEAs.

4.1. Framework of the Improved Lebesgue Indicator-Based Evolutionary Algorithm

Analogous to its predecessor (LIBEA [15]), the proposed algorithm implicitly adopts the regular property of continuous MOPs to approximate solutions towards the Pareto front of a given problem. The framework of the improved LIBEA (namely here LIBEA-II) is presented in Algorithm 2. Initially, a set $P_t = \{\mathbf{x}^1, \ldots, \mathbf{x}^N\}$ $(t = 0)$ of N candidate solutions is generated randomly (Algorithm 2, line 2). A matrix $\mathbf{D}$ allocating the distances between pairs of objective vectors is calculated and used in the parent selection mechanism of LIBEA-II (Algorithm 2, line 3). At each iteration, for each candidate solution $\mathbf{x}^i \in P$, a parent solution $\mathbf{y}$ is selected according to the *mating selection* mechanism (Algorithm 2, line 5). Thus, the *recombination* procedure is performed by employing the current solution $\mathbf{x}^i$ and the parent solution $\mathbf{y}$ (Algorithm 2, line 6). Section 4.3 illustrates different recombination models that could be adopted into LIBEA-II. Finally, in line 8 of Algorithm 2, a the new population P_{t+1} is updated employing the current population P_t and the offspring solution $\mathbf{y}'$ according to the *survival selection* mechanism described in Section 4.4. In the following, the rest of the components of LIBEA-II are described.

Algorithm 2: LIBEA-II Framework.

Input:
    A stopping criterion;
    N: Population size;
Output:
    P_t: PF approximation of a given MOP.

1   $t \leftarrow 0$;
2   $P_t \leftarrow InitializePopulation()$;
3   $D_{ij} \leftarrow ComputeEuclideanDistances(\mathbf{F}(\mathbf{x}^i), \mathbf{F}(\mathbf{x}^j)), \forall i, j \in \{1, \ldots, N\}$;
4   **while** *stopping criterion is not satisfied* **do**
5   **for** $i \in \{1, \ldots, N\}$ **do**
6          $\mathbf{y} \leftarrow MatingSelection(P_t, D_{ij})$;
7          $\mathbf{y}' \leftarrow Recombination(\mathbf{x}^i, \mathbf{y})$;
8          $P_{t+1} \leftarrow SurvivalSelection(P_t \cup \{\mathbf{y}'\})$;
9          $t \leftarrow t + 1$;

4.2. Mating Selection Mechanism

The regularity property of continuous MOPs, establish that, under certain conditions, the $\mathcal{PS}$ ($\mathcal{PF}$) of a continuous MOP with m objectives defines an $(m - 1)$-dimensional piecewise continuous manifold in the decision space (objective space) [21,22]. Although this property was firstly introduced by Hillermeier [22] to solve multi-objective problems, its use has been adopted by several EMOAs based on different natures (see for example, the approaches reported in [15–17,55,56]). LIBEA-II adopts the regularity property of continuous MOPs in an implicit form by promoting the recombination between neighboring solutions. In this way, if a solution $\mathbf{x}^i$ and its neighbors are close to the $\mathcal{PS}$ ($\mathcal{PF}$), the new offspring solution should also be close to the $\mathcal{PS}$ ($\mathcal{PF}$). In other words, the local manifold approximated by solution $\mathbf{x}^i$ and its neighbors should generate a new solution also close to the $\mathcal{PS}$ ($\mathcal{PF}$). In the following, the mating selection mechanism of LIBEA-II is described.

Let $C_i \subset P_t$ be the solutions set of the closest solutions to $\mathbf{x}^i$ (in the space of the objective functions). LIBEA-II uses a probability δ to select the solutions set (β) to be taken into account in the recombination procedure. In the proposed approach, the solutions set β is stated by either the neighboring solutions to $\mathbf{x}^i$ or the solutions in P_t according to a probability δ. More precisely:

$$\beta = \begin{cases} C_i, & rand(0,1) < \delta \ , \\ P_t, & otherwise. \end{cases} \tag{7}$$

In this way, the parameter δ denotes the probability of picking a neighboring solution to be recombined with solution $\mathbf{x}^i$. Otherwise, with probability $1 - \delta$, any other solution taken from the whole population P_t can be chosen for recombination.

Once the solutions set β is stated, a parent solution $\mathbf{y} \neq \mathbf{x}^i$ is chosen randomly from β. It is worth noticing that LIBEA-II keeps a distance matrix $\mathbf{D}$ updated during the search process (we refer to Section 4.4 for more details). Therefore, the solutions set C_i can be computed by employing the partial sorting algorithm [57] with a computational complexity of $O(N + T \log N)$, such that T denotes the number of desirable solutions in C_i and N represents the number of solutions in P_t.

4.3. Recombination Mechanism

LIBEA-II can be seen as a framework that allows incorporating any recombination mechanism available in the evolutionary computation research area. Nonetheless, it is worth mentioning that for certain recombination operators coming from some metaheuristics (e.g., PSO [58], DE [59], etc.), consider more than one solution. In such cases, the mating selection mechanism should produce more than one solution for the concerned recombination operator. That is, instead of picking one solution $\mathbf{y}$ from β, various solutions $\mathbf{y}^1, \mathbf{y}^2, \ldots \in \beta$, such that $\mathbf{y}^1 \neq \mathbf{y}^2 \neq \ldots \neq \mathbf{x}^i$ have to be selected. In order to exemplify the mating selection and recombination procedures in the LIBEA-II framework, two popular operators coming from the evolutionary computing field are illustrated below.

Operators from Genetic Algorithms.

Genetic algorithms employ crossover and mutation operators to create offspring solutions throughout the search process. In LIBEA-II, an offspring solution $\mathbf{y}'$ can be created employing such operators according to following equation

$$\mathbf{y}' = MUT(CX(\mathbf{x}^i, \mathbf{y})) \tag{8}$$

where $\mathbf{y}$ is a solution randomly chosen from β, CX, and MUT are the crossover and mutation operators, respectively. Therefore, LIBEA-II could adopt operators for combinatorial, integer, or mixed optimization.

Operators from Differential Evolution.

A recombination method employed for solving MOPs with difficult $\mathcal{PS}$ [17] exhibiting good results, is the differential evolution (DE) operator [58]. In LIBEA-II, an offspring solution $\mathbf{y}'$ can be created employing DE operator according to the following equation:

$$\mathbf{y}' = CX(\mathbf{x}^i, \mathbf{u}) \tag{9}$$

such that CX is the DE crossover, $\mathbf{u} = \mathbf{y}^1 + F(\mathbf{y}^2 - \mathbf{y}^3)$ is the perturbed vector, where $\mathbf{y}^1, \mathbf{y}^2$ and $\mathbf{y}^3$ are solutions chosen randomly from β with, $\mathbf{y}^1 \neq \mathbf{y}^2 \neq \mathbf{y}^3 \neq \mathbf{x}^i$, and F denotes the differential factor, respectively. After performing the DE crossover, a mutation mechanism can be applied to improve the search capabilities, as has been employed by a few researchers [17,18].

4.4. Survival Selection Mechanism

The survival selection mechanism in LIBEA-II (line 8 in Algorithm 2) chooses the best solutions from $Q = P_t \cup \{\mathbf{y}\}$ considering either the Pareto dominance relation or the Lebesgue measure. Since the number of solutions in Q is $N + 1$, it is necessary to remove one solution from Q to make way for the next iteration.

Let $d(\mathbf{q}, Q)$ and $P^\star$ be the number of solutions from Q that dominate solution $\mathbf{q} \in Q$, and the set of non-dominated solutions in Q, respectively. That is,

$$
\begin{aligned}
d(\mathbf{q}, Q) &= |\{\mathbf{p} \in Q \mid \mathbf{F}(\mathbf{p}) \prec \mathbf{F}(\mathbf{q}), \mathbf{F}(\mathbf{p}) \neq \mathbf{F}(\mathbf{q})\}| & (10) \\
P^\star &= \{\mathbf{q} \in Q \mid \nexists\, \mathbf{p} \in Q : \mathbf{F}(\mathbf{p}) \prec \mathbf{F}(\mathbf{q})\} & (11)
\end{aligned}
$$

Traditionally, IBEAs based on the Lebesgue indicator employ Pareto ranking [60] followed by computing the exclusive Lebesgue contribution (see Definition 2) of each solution allocated in the last rank (see for example the algorithmic proposals introduced in [11,15,61,62]). Therefore, if the last rank contains a large number of solutions, the computational complexity to estimate the solution to be removed becomes extremely high. In evolutionary many-objective optimization, it can be observed that a population constituted by non-dominated solutions can be preserved for several iterations of an algorithm. Therefore, a high computational time is required to decide which solution should be removed from the population. LIBEA-II considers the following two scenarios in the survival selection mechanism.

- If $P^\star \neq Q$, there are solutions in Q dominated by some solution in $P^\star$. In such a case, we shall remove the solution with the largest $d(\mathbf{q}, Q)$ value avoiding the Lebesgue measure computation and reducing the computational cost of LIBEA-II;

- If $P^\star = Q$, all solutions in Q are non-dominated, and all of them are equally acceptable in terms of the Pareto dominance relation. In such a case, we shall remove the solution $x^j \in S \subseteq Q$ that maximizes the contribution to the Lebesgue measure. In other words, the solution to be removed is the one such that:

$$x^j = \arg\min_{s \in S} \mathcal{H}_c(S, \mathbf{s}, \mathbf{r}) \tag{12}$$

Therefore, a total of $|S|$ Lebesgue measures are required to identify the worst solution (i.e., the solution that contributes the least to the Lebesgue measure) in the population.

Note that in the case of $S = Q$, the worst solution is found after $N + 1$ Lebesgue measures. This is, in fact, computationally expensive and impractical in many-objective optimization problems. LIBEA-II saves Lebesgue measures by reducing the number of candidate solutions in the set S.

A problem observed in IBEAs based on the Lebesgue indicator is the overestimation of the reference vector, which could divert the search. Although the correct estimation of the reference vector for certain types of $\mathcal{PF}$ has been discussed [63], there is no strategy to correctly define this vector for all $\mathcal{PF}$ forms. In such a case, a reference vector close to the nadir vector (the vector opposite the ideal vector) should properly measure the coverage and distribution of solutions along the $\mathcal{PF}$, including the extreme portions of it. Therefore, the solutions that provide information on the nadir vector should be considered in the survival selection mechanism. In other words, the solutions whose objectives vectors are the farthest to the ideal vector should be considered.

The following criteria to define the set S (line 5 in Algorithm 3) are based on the locality property of the hypervolume indicator studied in [19] (the first three criteria) and the problem to estimate the reference vector for computing the Lebesgue indicator discussed in [63] (the last criterion). The set of candidate solutions S is composed by:

1. The offspring solution $\mathbf{y}$;
2. The solution $\mathbf{q} \in Q$ such that the objective vector $\mathbf{F}(\mathbf{q})$ is the closest to the objective vector $\mathbf{F}(\mathbf{y})$;
3. A percentage (ρ_c) of solutions $\mathbf{q} \in Q$ which objective vectors are the closest;
4. A percentage (ρ_n) of solutions $\mathbf{q} \in Q$ which objective vectors are the farthest to the ideal vector $\mathbf{z}$, where the ideal vector $\mathbf{z} = (z_1, \ldots, z_m)^T$ is estimated by $z_j = \min_{\mathbf{x} \in Q} f_j(\mathbf{x})$ for all $j \in \{1, \ldots, m\}$.

In the case that the offspring solution $\mathbf{y}$ is accepted, it shall replace solution $x^j \in Q$ ($x^j \leftarrow \mathbf{y}$) and the distance matrix $\mathbf{D}$ has to be updated calculating the Euclidean distances between the objective vector $\mathbf{F}(x^j)$ and each objective vector $\mathbf{F}(x^i)$, that is:

$$D_{ij} \leftarrow ||\mathbf{F}(x^i) - \mathbf{F}(x^j)||_2, \ \forall x^i \in P_{t+1}$$

In order to deal with different scales (in the objective space), LIBEA-II considers objective vectors normalized in the hypercube bounded by the ideal ($\mathbf{z}$) and the nadir ($\mathbf{n}$) vectors. In such cases, the ideal and nadir vectors are defined by the smallest and the largest values of each objective function found in the set of solutions $Q \cup \{\mathbf{y}\}$. Therefore, the Lebesgue measure is computed employing the normalized objective vectors and considering the reference vector $\mathbf{r} = (1.1, \ldots, 1.1)^T$. In Algorithm 3, we show the complete survival selection mechanism of LIBEA-II.

> **Algorithm 3:** *EnvironmentalSelection(Q).*
>
> ---
> **Input:**
> Q: the population to be truncated;
> **Output:**
> $Q^\star$: the population updated;
> 1 $P^\star \leftarrow NondominatedSolutions(Q);$
> 2 **if** $P^\star \neq Q$ **then**
> 3 $\quad$ $x^j \leftarrow \arg\min_{\mathbf{q} \in Q} d(\mathbf{q}, Q);$
> 4 **else**
> 5 $\quad$ $S \leftarrow CandidateSolutions(Q);$ // Defining the set S
> 6 $\quad$ $; x^j \leftarrow \arg\min_{\mathbf{s} \in S} \mathcal{H}_c(S, \mathbf{s}, \mathbf{r});$
> 7 $Q^\star \leftarrow Q \setminus \{x^j\};$
> 8 **return** $Q^\star;$

---

5. Experimental Study

This section presents the experimental setup and the analysis of results. Firstly, the IBEAs and the benchmark problems adopted for comparison are introduced. Then, the experimental settings are given. Finally, the results on the benchmark problems are analyzed.

5.1. IBEAs Considered for Comparison of Performance

The performance of the proposed LIBEA-II is compared with respect to state-of-the-art IBEAs based on the hypervolume quality indicator. In the first instance, we adopt the S-metric selection EMOA (SMS-EMOA) [11] for performance comparison. SMS-EMO employs a survival mechanism based on Pareto ranking joined by the exclusive Lebesgue contribution of each solution located in the last rank. This evolutionary algorithm has shown its effectiveness and has become popular among state-of-the-art IBEAs. Another algorithm contemplated in this investigation is the improved version of the SMS-EMOA (iSMS-EMOA) [50], which adopts ideas coming from the local property of the hypervolume indicator. Finally, the Lebesgue indicator-based algorithm (LIBEA) [15] for multi-objective optimization is selected. As noticed before, LIBEA adopts the regularity property of continuous MOPs in the mating selection mechanism. Since the proposed LIBEA-II also employs this regularity property, its predecessor, LIBEA, is an obvious competing algorithm.

5.2. Adopted Test Problems

The study presented in this investigation considers the continuous box-constrained MOPs with difficult Pareto sets introduced in [64]. These problems are part of the CEC'2009 competition related to the performance of evolutionary multi-objective algorithms. The adopted test problems have been formulated to assess the performance of EMOAs solving continuous MOPs that exhibit the property of complicated $\mathcal{PS}$ topologies. Since this property has been seen in real-life problems [17], this test suite is a challenge to evaluate the performance of our algorithmic proposal. The adopted test problems (as well known as UFs) offer diverse characteristics regarding separability, multi-modality, and include different $\mathcal{PF}$ shapes, incorporating discontinuities, concavity, convexity, etc. More precisely, we consider the two-objective problems UF1–UF7 and the three-objective problems UF8–UF10.

5.3. Experimental Settings

As pointed out, the results achieved by our proposed algorithm (i.e., LIBEA-II) are analyzed versus those obtained by SMS-EMOA, iSMS-EMOA, and LIBEA on the test problems with roughed $\mathcal{PF}$ (UF1–UF10). As discussed by some authors, MOPs exhibiting complicated $\mathcal{PF}$ shapes shall test specific components of EMOAs, such as the parent selection mechanism and the recombination operators [15,17]. In this work, we use the DE operator whose effectivity has been proved in MOPs with this singular characteristic (see the studies reported in [17]). Therefore, all the IBEAs adopted for performance comparison employ DE operator as their main recombination procedure, such as described in Section 4.3. In order to improve the search capabilities after performing the DE operator, the Polynomial-based mutation [65] is implemented. However, note that for other test problems such as ZDT [66] and DTLZ [67], the performance of IBEAs could be improved by using recombination operators from genetic algorithms, for example, Simulated Binary Crossover (SBX) and Polynomial-based mutation (PBM) [65]. The parameters used by the IBEAs are presented in Table 2, where N denotes the population size. G_{max} represents the maximum number of generations, in our study $G_{max} = 2000$. Therefore, the search process was limited to performing 200,000 fitness function evaluations. In the case of the DE operator, F and CR denote the differential amplitude factor and the crossover rate, which were set as guested in [17] to solve MOPs with complicated $\mathcal{PS}$. P_m and η_m are the mutation probability and the mutation index used by PBM, respectively. For LIBEA and LIBEA-II, T and δ are the neighborhood size and the probability of picking solutions from a determined neighborhood (see Section 4.2), respectively. Note that the smaller the δ value, the effect of the regular property of continuous MOPs is diluted. For LIBEA-II, ρ_c and ρ_n are parameters that define the percentage of solutions to be considered in the survival selection mechanism (see Section 4.4). It is worth emphasizing that the smaller these parameters values (ρ_c and ρ_n), the more efficient LIBEA-II is.

Table 2. Parameter settings for SMS-EMOA, iSMS-EMOA, LIBEA, and LIBEA-II.

Parameter	SMS-EMOA	iSMS-EMOA	LIBEA	LIBEA-II
N	100	100	100	100
G_{max}	2000	2000	2000	2000
F	0.5	0.5	0.5	0.5
CR	1	1	1	1
P_m	$1/n$	$1/n$	$1/n$	$1/n$
η_m	20	20	20	20
T	—	—	20	20
δ	—	—	0.9	0.9
ρ_c	—	—	—	0.1
ρ_n	—	—	—	0.1

For each IBEA, 30 executions were independently performed on each MOP. The IBEAs were assessed employing the $\mathcal{H}_n$ and IGD^+ quality indicators presented in Section 2.3. For each test problem, a statistical analysis was performed over the final approximation produced by the IBEAs in all the experiments using the concerned quality indicator. Since the properties of the UF test functions are known, the $\mathcal{H}_n$ quality indicator was calculated by employing the reference vector $\mathbf{r} = (1.1,\dots,1.1)^T$ and the ideal vector $\mathbf{u} = (0,\dots,0)^T$. Therefore, a reliable measure of approximation and distribution of solutions obtained by the algorithms along the Pareto front is reported. The IGD^+ indicator was calculated by employing the reference sets provided by the authors of the UF test functions.

5.4. Analysis of Results on the UF Test Problems

The non-dominated solutions found by LIBEA-II, and those from SMS-EMOA, iSMS-EMOA, and LIBEA, to each UF test function, were subjected to the $\mathcal{H}_n$ and IGD^+. Tables 3 and 4 show the average $\mathcal{H}_n$ and IGD^+ values, respectively, over 30 repetitions for each UF problem. These tables have five columns: the first identifies the UF test function, and the remaining four correspond to each of the four algorithms under comparison. The best average $\mathcal{H}_n$ and IGD^+ values for each UF problem are in **boldface**. Moreover, in order to distinguish if there is a statistically significant difference among the average $\mathcal{H}_n$ and IGD^+ values for each test function, the Mann–Whitney–Wilcoxon [68] non-parametric statistical test, employing a p-value of 0.05, and Bonferroni correction [69] were applied on them. In this manner, an algorithm can be considered the best regarding the test function and quality indicator if it statistically surpasses the other three. If this is the case, the value presented in the Table is underlined.

Table 3. Average $\mathcal{H}_n$ values for the non-dominated solutions found by the IBEAs to each UF test problem.

MOP	SMS-EMOA	iSMS-EMOA	LIBEA	LIBEA-II
UF1	**0.6674** $\pm$ 0.010	0.6654 $\pm$ 0.010	0.6654 $\pm$ 0.009	0.6669 $\pm$ 0.008
UF2	0.7048 $\pm$ 0.001	0.7039 $\pm$ 0.002	0.7051 $\pm$ 0.001	**0.7054** $\pm$ 0.002
UF3	0.6057 $\pm$ 0.028	0.6050 $\pm$ 0.029	**0.6117** $\pm$ 0.023	0.6108 $\pm$ 0.025
UF4	0.2779 $\pm$ 0.007	**0.2792** $\pm$ 0.006	0.2781 $\pm$ 0.006	0.2761 $\pm$ 0.006
UF5	0.0087 $\pm$ 0.015	0.0143 $\pm$ 0.034	0.0154 $\pm$ 0.024	**0.0175** $\pm$ 0.021
UF6	0.3137 $\pm$ 0.054	0.3001 $\pm$ 0.047	**0.3329** $\pm$ 0.037	0.3319 $\pm$ 0.043
UF7	0.5595 $\pm$ 0.004	0.5511 $\pm$ 0.039	**0.5608** $\pm$ 0.004	0.5489 $\pm$ 0.039
UF8	0.3539 $\pm$ 0.041	0.3563 $\pm$ 0.038	0.3641 $\pm$ 0.054	<u>**0.3667**</u> $\pm$ 0.052
UF9	**0.7582** $\pm$ 0.012	0.7464 $\pm$ 0.011	0.7550 $\pm$ 0.015	0.7366 $\pm$ 0.025
UF10	0.0000 $\pm$ 0.000	0.0000 $\pm$ 0.000	0.0000 $\pm$ 0.000	0.0000 $\pm$ 0.000

Table 4. Average IGD^+ values for the non-dominated solutions found by the IBEAs to each UF test problem.

MOP	SMS-EMOA	iSMS-EMOA	LIBEA	LIBEA-II
UF1	0.0341 $\pm$ 0.006	0.0354 $\pm$ 0.005	0.0350 $\pm$ 0.005	**0.0340** $\pm$ 0.005
UF2	0.0142 $\pm$ 0.001	0.0148 $\pm$ 0.001	0.0141 $\pm$ 0.001	**0.0140** $\pm$ 0.001
UF3	0.0709 $\pm$ 0.025	0.0696 $\pm$ 0.021	0.0666 $\pm$ 0.018	**0.0665** $\pm$ 0.019
UF4	0.1232 $\pm$ 0.006	**0.1223** $\pm$ 0.005	0.1230 $\pm$ 0.005	0.1252 $\pm$ 0.005
UF5	0.7429 $\pm$ 0.101	0.7470 $\pm$ 0.133	0.7232 $\pm$ 0.109	**0.7106** $\pm$ 0.098
UF6	0.2163 $\pm$ 0.048	0.2221 $\pm$ 0.061	**0.2005** $\pm$ 0.036	0.2028 $\pm$ 0.045
UF7	0.0149 $\pm$ 0.002	0.0229 $\pm$ 0.035	**0.0142** $\pm$ 0.002	0.0241 $\pm$ 0.036
UF8	0.1542 $\pm$ 0.029	0.1528 $\pm$ 0.027	0.1468 $\pm$ 0.038	<u>**0.1454**</u> $\pm$ 0.037
UF9	0.0518 $\pm$ 0.007	0.0563 $\pm$ 0.008	**0.0517** $\pm$ 0.007	0.0642 $\pm$ 0.017
UF10	1.3045 $\pm$ 0.222	1.2999 $\pm$ 0.227	**1.2335** $\pm$ 0.203	1.3563 $\pm$ 0.207

In Table 3, we can see the average results for the $\mathcal{H}_n$ indicator. As we can see, the performance of the four algorithms was very similar: SMS-EMOA obtained the best average results for two test problems, solutions from iSMS-EMOA were the best for one test problem, LIBEA found solutions that were the best to three test problems, and the solutions found by LIBEA-II were the best for three test problems. Actually, these results were expected since all four algorithms are based on the $\mathcal{H}$ indicator. However, something remarkable is that LIBEA-II was able to find statistically better solutions than those from the other three algorithms for test instance UF8.

$\mathcal{H}_n$ quality indicator assesses, to some extent, the closeness and spreading of the non-dominated solutions obtained by an EMOA. Nevertheless, quality indicators based on reference sets could provide more information regarding how distant a set of solutions is from the real $\mathcal{PF}$. To this end, the IGD^+ indicator was selected to further evaluate the performance of the IBEAs. Table 4 presents the obtained results of the proposed LIBEA-II and those reached by the adopted IBEAs in terms of the IGD^+ indicator. It can be observed that the results achieved by LIBEA-II exceeded those obtained by SMS-EMOA, iSMS-EMOA, and LIBEA in five out of the ten test problems in terms of $\mathcal{H}_n$ indicator. LIBEA obtained the best average results in four test problems, while iSMS-EMOA was the best in only one. More importantly, LIBEA-II obtained results that are statistically better than the results from the other three algorithms in problem UF8.

Additionally to the quality indicators, Figure 1 shows the average convergence of the four algorithms under comparison. This figure contains ten plots, one for each UF test function, that show the convergence of the IGD^+ indicator for each algorithm. We can see that the convergence of the IGD^+ indicator is very similar for all four algorithms in each test problem, except for test problem UF7.

After these results, regarding the $\mathcal{H}_n$ and IGD^+ quality indicators, we can say that LIBEA-II performs slightly better on the UF test problems than SMS-EMOA, iSMS-EMOA, and LIBEA, since, despite solutions from LIBEA-II, have comparable hypervolumes, they are closer to the $\mathcal{PF}$ in more benchmark functions. Moreover, for the ρ_c and ρ_n parameters adopted in this study, LIBEA-II reduces up to approximately 80% the hypervolume indicator calculations, as shown in Figure 2. This means that LIBEA-II is more efficient than the other three algorithms since, with fewer computational resources, it can find solutions with as high quality as those found by SMS-EMOA, iSMS-EMOA, and LIBEA.

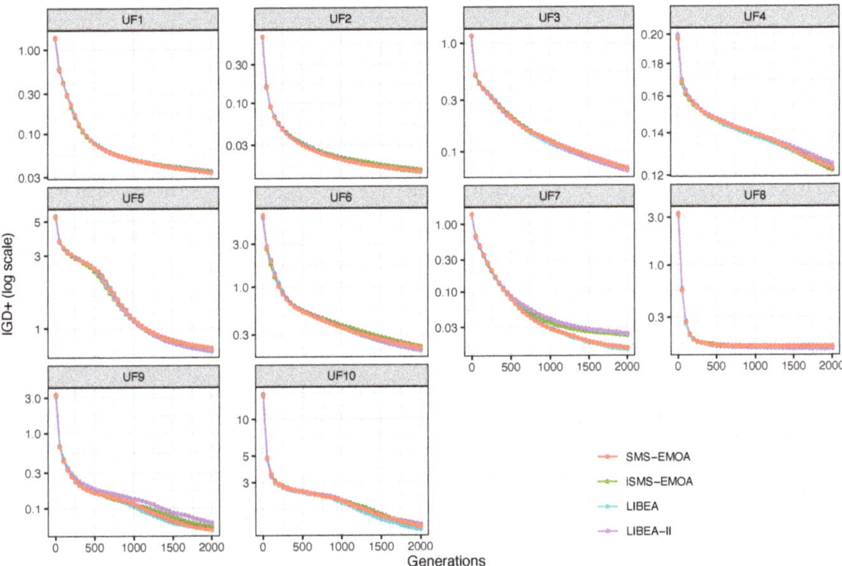

Figure 1. Convergence plots of the $IGD+$ quality indicator on the UF test problems.

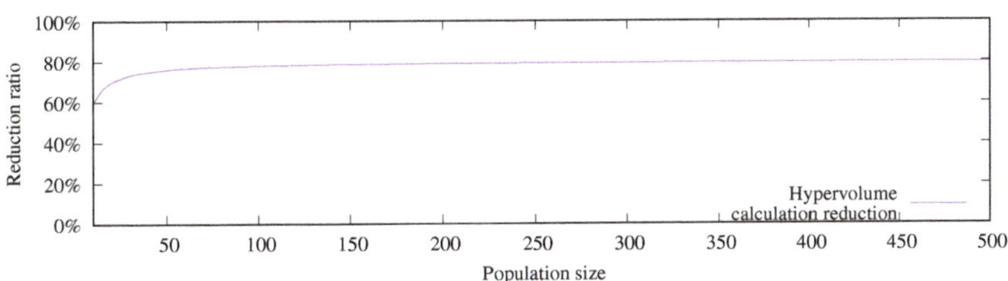

Figure 2. Reduction of $\mathcal{H}_c$ computations in LIBEA-II adopting $\rho_c = 0.1$ and $\rho_n = 0.1$.

In order to illustrate the Pareto approximations obtained by the algorithms, Figure 3 presents the non-dominated solutions found by the four algorithms under consideration to the test problems UF1, UF3, and UF7. It is clear that no algorithm could find a proper approximation set to any of the three test problems. However, solutions from SMS-EMOA and LIBEA-II show the best approximations to the Pareto front.

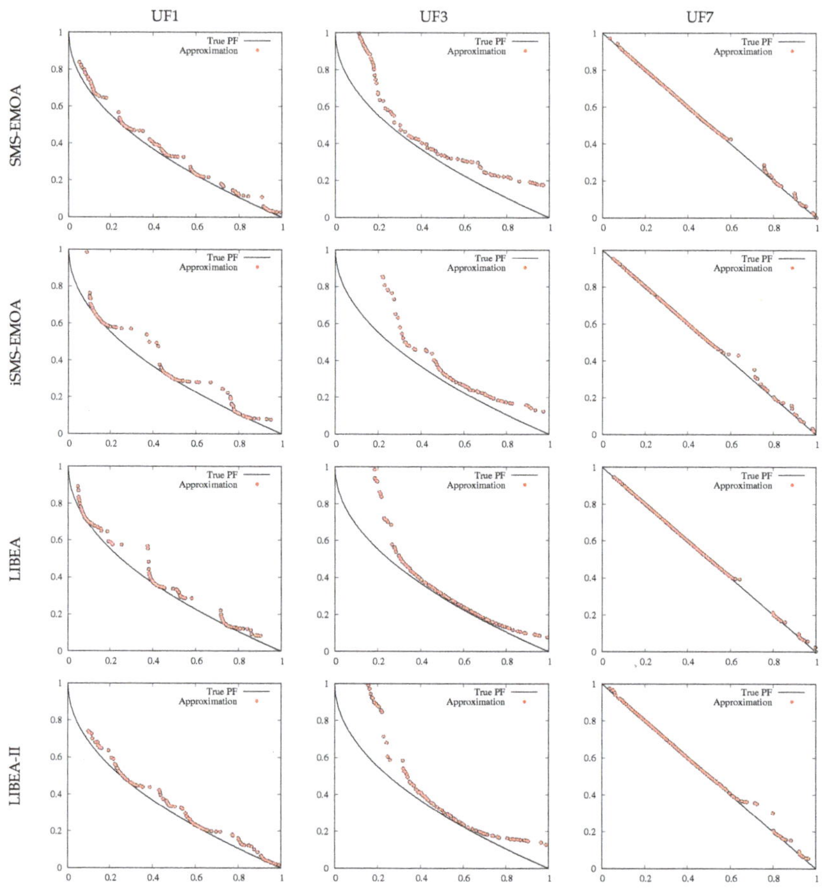

Figure 3. Non-dominated solutions found by the four algorithms under study to the test problems UF1, UF3, and UF7.

6. Three Real-World Applications from Practice

After testing LIBEA-II on the UF benchmark test problems, it is now tested on three real-world applications. This section introduces, firstly, the three real-world applications under consideration. Secondly, the experimental setup is presented. Thirdly, the results are analyzed. Finally, the correlation between pairs of objectives are analyzed.

6.1. Description of the Real-World Applications

The three real-world applications considered in this study are introduced next.

6.1.1. RWA1: Liquid-rocket single element injector design

The design of a liquid-rocket single element injector aims at improving its performance and enlarging its life [70]. Vaidyanathan et al. [71] states that, in order to optimally design such an injector, four objectives should be considered: the maximum injector face temperature (TF_{max}), the wall temperature at a distance of three inches from the injector face (TW_4), the maximum oxidizer post tip temperature (TT_{max}), and the centerline axial location where the combustion is 99% complete (X_{cc}). Specifically, this multi-objective optimization problem can be written as:

$$\begin{aligned}
\text{minimize:} \quad & f_1(\mathbf{x}) = X_{cc} \\
\text{minimize:} \quad & f_2(\mathbf{x}) = TF_{max} \\
\text{minimize:} \quad & f_3(\mathbf{x}) = TW_4 \\
\text{minimize:} \quad & f_4(\mathbf{x}) = TT_{max}
\end{aligned} \tag{13}$$

where $\mathbf{x} = (\alpha, \Delta HA, \Delta OA, OPTT)^T$, such that $0° \leq \alpha \leq 20°$ is the *hydrogen flow angle*, $0\% \leq \Delta HA \leq 25\%$ is the *hydrogen area increment* with respect to the baseline cross-section area (0.0186 in^2), $-40\% \leq \Delta OA \leq 0\%$ is the *oxygen area decrement* with respect to the baseline cross-section area (0.0423 in^2) of the tube carrying oxygen, and $X'' \leq OPTT \leq 2X''$ is the *oxidizer post tip thickness*, where X'' denotes the tip thickness with a baseline value 0.01 in. The mathematical definition of this problem can be seen in [71].

6.1.2. RWA2: Ultra-wideband antenna design

In order to design an ultra-wideband antenna with two stopbands within the WiMAX and WLAN bands, besides achieving the expected impedance features, gain uniformity and high fidelity are also required [72]. Such antenna comprises a planar rectangular patch and a pair of notches at the two lower corners. Two U-shaped thin slots are carved in the monopole patch for the two stopbands. In order to design this antenna, ten parameters have to be considered and five objective functions, which are the voltage standing wave ratio (VSWR) over the passband (f_1), the VSWR among the WiMAX (f_2) and WLAN (f_3) bands, respectively, the fidelity factors in the E-plane and H-plane (f_4), and the relatively uniform peak gains over the passband (f_5) [73]. Hence, the multi-objective optimization problem is stated as:

$$\begin{aligned}
\text{minimize:} \quad & f_1(\mathbf{x}) \\
\text{maximize:} \quad & f_2(\mathbf{x}) \\
\text{maximize:} \quad & f_3(\mathbf{x}) \\
\text{maximize:} \quad & f_4(\mathbf{x}) \\
\text{minimize:} \quad & f_5(\mathbf{x})
\end{aligned} \tag{14}$$

where $\mathbf{x} = (a_1, a_2, b_1, b_2, d_1, d_2, l_1, l_2, w_1, w_2)^T$, such that $5 \leq a_1 \leq 7$, $10 \leq a_2 \leq 12$, $5 \leq b_1 \leq 6$, $6 \leq b_2 \leq 7$, $3 \leq d_1 \leq 4$, $11.5 \leq d_2 \leq 12.5$, $17.5 \leq l_1 \leq 22.5$, $2 \leq l_2 \leq 3$, $17.5 \leq w_1 \leq 22.5$, and $2 \leq w_2 \leq 3$. The mathematical formulation of this problem is presented in [73].

6.1.3. RWA3: Development of oil and water repellent fabric

In the textile industry, one aim is to produce fabrics with added high value. Particularly, the hydrophobicity effect, that is, water, oil, and stain repellence, is one of the most widely used textile surface modification [74]. Hydrophobicity depends on several process parameters, such as the concentration of oil and water repellent (O-CPC) finish, the concentration of the crosslinking agent (K-FEL), and the curing temperature (C-Temp) [75]. The hydrophobicity effect can be measured by means of the following seven responses [76]: the contact angle of a water (WCA) and oil (OCA) droplet touching a surface; the air permeability (AP), which is the comfort property of a woven fabric used to measure the flow of air through it; the crease recovery angle (CRA), which measures the textiles ability to recover from creasing; the *stiffness*, which is a comfort property of cotton fabric; the *tear strength* of the finished fabric, which depends on the chemical finishing treatment applied to the fabric; and the *tensile strength*, which describes the behavior of the fabric under axial stretching load. These seven responses can be considered as objective functions as follows:

$$\begin{aligned}
\textbf{maximize:} \quad & f_1(\mathbf{x}) = WCA \\
\textbf{maximize:} \quad & f_2(\mathbf{x}) = OCA \\
\textbf{maximize:} \quad & f_3(\mathbf{x}) = AP \\
\textbf{maximize:} \quad & f_4(\mathbf{x}) = CRA \\
\textbf{minimize:} \quad & f_5(\mathbf{x}) = Stiffness \\
\textbf{maximize:} \quad & f_6(\mathbf{x}) = Tear\ strength \\
\textbf{maximize:} \quad & f_7(\mathbf{x}) = Tesile\ strength
\end{aligned} \tag{15}$$

where $\mathbf{x} = (\text{O-CPC}, \text{K-FEL}, \text{C-Temp})^T$, such that $10\ \text{g/L} \leq \text{O-CPC} \leq 50\ \text{g/L}$, $10\ \text{g/L} \leq \text{K-FEL} \leq 50\ \text{g/L}$, and $150\ °\text{C} \leq \text{C-Temp} \leq 170\ °\text{C}$. The mathematical description of the problem is presented in [76].

6.2. Experimental Setup

In order to analyze the results achieved by LIBEA-II versus those achieved by SMS-EMOA, iSMS-EMOA, and LIBEA, the following experimental setup was carried out. Since the characteristics of the real-world applications(RWA) described above are not known, the reference $\mathcal{PF}$ had to be constructed to compute the quality indicator IGD^+.

1. The non-dominated solutions obtained by all four IBEAs from the 30 executions were recorded;
2. The maximin fitness function [77] was applied to choose 5000 from these non-dominated solutions and they were considered as the reference set for the IDG^+ quality indicator.

Regarding the $\mathcal{H}_n$ quality indicator, for each RWA problem, the ideal point $\mathbf{u} = (u_1, \ldots, u_m)^T$ was calculated by finding the minimum value for each objective function in the reference $\mathcal{PF}$. On the other hand, the reference vector $\mathbf{r} = (r_1, \ldots, r_m)^T$ was stated by finding the maximum value for each objective function in the reference $\mathcal{PF}$ and scaling its magnitude (with respect to the ideal point) by 10% for each dimension. More precisely, $r_j = 1.1 \times |f_j^{max} - u_j|$, such that f_j^{max} is the maximum value of each objective function in the reference $\mathcal{PF}$, for all $j \in \{1, \ldots, m\}$. Hence, the $\mathcal{H}_n$ indicator will consider, in a more appropriate scope, the boundaries of the $\mathcal{PF}$ approximation found by each IBEA. Due to the computational cost of the original SMS-EMOA and LIBEA when dealing with more than four objectives, the exact calculation of the $\mathcal{H}_c$ was replaced by the HypE indicator [13] employing $1000 \times m$ samples for the $\mathcal{H}_c$ approximation, where m denotes the number of objectives in the problem. It is worth noticing that the computational cost of LIBEA-II and the other IBEAs depends directly on the population size and on the number of objectives. Our experimental study adopts $N = 100$ solutions to solve the three real-world applications. With this number of solutions, LIBEA-II could deal with problems with up to seven objectives in approximately five days. However, LIBEA-II spent less than 24 h performing a single run for problems having four and five objective functions. The experimental study presented in this work was carried out on a desktop PC with a 32-core 2.6 GHz processor and 64GB of RAM.

6.3. Analysis of Results

The results achieved by LIBEA-II were examined versus those obtained by SMS-EMOA, iSMS-EMOA, and LIBEA. Tables 5 and 6 show the results achieved by the algorithms in the three real-world applications described above, for the $\mathcal{H}_n$ and IGD^+ quality indicators, respectively. The structure of these tables is similar to that of Tables 3 and 4. That is, the best average $\mathcal{H}_n$ and IGD^+ values for each real-world application are in **bold-**

face, while the best algorithm regarding the concerned real-world application and quality indicator is underlined.

Regarding the $\mathcal{H}_n$ indicator, we can see from Table 5 that LIBEA-II found solutions that cover a larger hypervolume for the three real-world problems and, remarkably, for problem RWA3, the difference is statistically significant. Concerning the IGD^+ indicator, Table 6 shows that solutions from LIBEA-II obtained, on average, the smallest values for all three real-world problems. In this case, there is a tie between LIBEA-II and iSMS-EMOA for problem RWA1.

Table 5. Average $\mathcal{H}_n$ values for the non-dominated solutions found by the IBEAs to each real-world application.

MOP	SMS-EMOA	iSMS-EMOA	LIBEA	LIBEA-II
RWA1	0.5495 ± 0.001	$0.5550 \pm_{0.001}$	0.5487 ± 0.001	$\mathbf{0.5557} \pm 0.001$
RWA2	0.6883 ± 0.005	$0.6947 \pm_{0.007}$	0.6906 ± 0.005	$\mathbf{0.6969} \pm 0.005$
RWA3	0.1982 ± 0.007	0.1951 ± 0.012	0.1963 ± 0.009	$\underline{\mathbf{0.2052}} \pm 0.007$

Table 6. Average IGD^+ values for the non-dominated solutions found by the IBEAs to each real-world application.

MOP	SMS-EMOA	iSMS-EMOA	LIBEA	LIBEA-II
RWA1	0.0157 ± 0.001	$\mathbf{0.0126} \pm 0.000$	0.0160 ± 0.001	$\mathbf{0.0126} \pm 0.000$
RWA2	4318.4205 ± 952.777	4163.9243 ± 1770.699	3939.6710 ± 539.990	$\mathbf{3829.6227} \pm 1063.089$
RWA3	3.3550 ± 0.871	3.9980 ± 1.691	3.5676 ± 1.213	$\mathbf{3.0701} \pm 0.797$

Figure 4 shows the average convergence for the IGD^+ indicator. This Figure contains three plots, one for each real-world application. It is evident that the convergence of LIBEA-II is faster than those of the other three algorithms for the two real-world applications RWA1 and RWA3. The convergence for problem RWA2 is similar for all three algorithms.

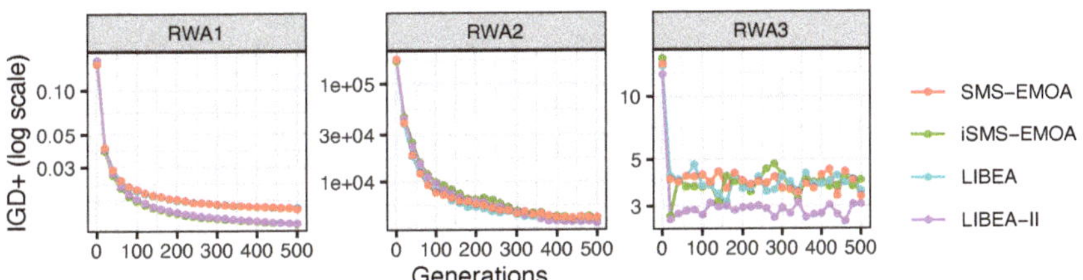

Figure 4. Convergence of the $IGD+$ quality indicator on the three real-world applications.

After these results, it is clear that the performance of LIBEA-II on the three real-world applications is higher than that of SMS-EMOA, iSMS-EMOA, and LIBEA.

6.4. Analysis of the Conflict Relation Between Pairs of Objectives

In addition to studying the performance of the four IBEAs on real-world applications, it is also of interest to know the conflict relation (see Definition 5) between the objective functions for each RWA. To this end, Figure 5 presents the parallel coordinates plots for the three real-world applications considered in this study.

Parallel coordinates are a handy tool for identifying conflict, support, or independence between pairs of objectives. Even though they do not specify information regarding the characteristics of the approximation sets, they are generally utilized for realizing the correlation, whether positive, negative, or neutral, between pairs of objectives [78]. As mentioned earlier, the reference $\mathcal{PF}$ was obtained, for each RWA, by recording all non-dominated solutions found by all four algorithms. For each of these solutions, the values for each objective function were plotted. In Figure 5, the conflict between pairs of objectives is illustrated. This Figure shows three boxes, one for each RWA. For each problem, the objective values are normalized in the vertical axis in the range $[0, 1]$ for a straightforward interpretation, while the objective functions are on the horizontal axis. Lines are plotted from one objective function f_i to the next adjacent f_{i+1} to reflect the correlation between the pair of objectives (f_i, f_{i+1}). If a line depicts a significant slope (whether negative or positive) from one objective to the next, it can be interpreted that those objectives are in conflict, and the longer the slope, the greater the conflict. On the contrary, if a line is horizontal, i.e., it has no slope at all, the objectives support each other.

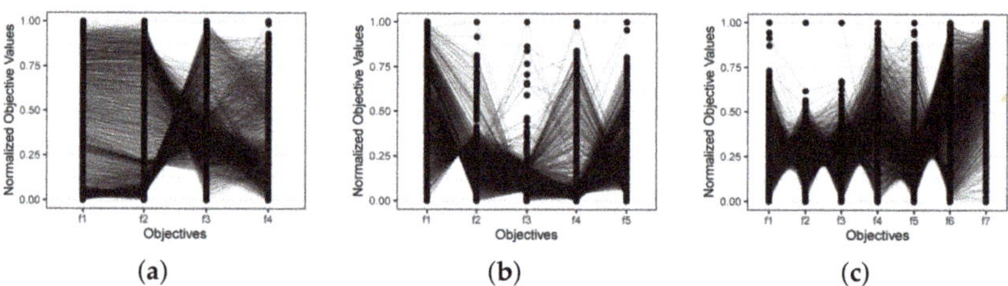

(a)   (b)   (c)

Figure 5. Parallel coordinate plots for the three real-world applications. (a) RWA1; (b) RWA2; (c) RWA3.

In the case of RWA1, we can see that most of the lines between objectives (f_1, f_2) are nearly horizontal or with a slight slope, which indicates that those objectives support each other. For the pair of objectives (f_2, f_3), a considerable number of lines have a more significant slope, whether positive or negative, from which we can infer that those objective functions are in conflict. For the last pair (f_3, f_4), we see that some lines have a slope while others are almost horizontal. Hence, there is no clear conclusion for this pair of objectives.

Following the same analysis as in RWA1, in the case of RWA2, we see that objectives (f_1, f_2) and (f_4, f_5) are clearly in conflict since nearly all the lines present a significant slope and only a few are horizontal or with a slight slope. For the pair of objectives (f_2, f_3) and (f_3, f_4), on the contrary, most of the lines have a slight slope or are horizontal, while the rest of the lines present a significant slope. For these cases, nothing can be said from these plots.

Finally, for problem RWA3, it is evident that there is conflict for the pairs of objectives (f_1, f_2), (f_2, f_3), (f_3, f_4), and (f_5, f_6), since the vast majority of the lines presents a significantly large slope, whether positive or negative. For the other two pairs of objectives, (f_4, f_5) and (f_6, f_7), there is no clear indication whether the objectives are in conflict, support each other, or are independent.

In order to complete the conflict relation analysis between objectives, a numerical analysis of the reference $\mathcal{PF}$s is presented next. Figure 6 contains three matrices, one for each RWA problem. Each matrix shows: in the upper triangular matrix, the Pearson correlation coefficients [79] between pairs of objectives; in the lower triangular matrix, the projection of the objective function values of the non-dominated solutions between pairs of objectives; and in the diagonal, the densities of each objective function.

For problem RWA1, we can see that the pairs of objectives (f_1, f_2) and (f_3, f_4) are positively correlated. Remarkably, the correlation for (f_1, f_2) is approximately 1.0, which means that optimizing one of them, whether f_1 or f_2, will lead to the optimization of the other and vice versa. These results are in accordance with the analysis of the parallel coordinates plots. The remaining four pairs of objectives, i.e., (f_1, f_3), (f_1, f_4), (f_2, f_3), and (f_2, f_4), present a negative correlation. This means that there is a conflict between the objectives in each pair. That is, the optimization of one objective function deteriorates the other and vice versa.

In the case of problem RWA2, we see a positive correlation for pairs of objectives (f_1, f_5), (f_2, f_3), (f_2, f_4), and (f_3, f_4), which means that, to some extent, the objectives in each pair support each other. Particularly for the pair (f_1, f_5), we can observe that the correlation is nearly 1.0. The other six pairs of objective functions, that is (f_1, f_2), (f_1, f_3), (f_1, f_4), (f_2, f_5), (f_3, f_5), and (f_4, f_5), present a negative correlation. From these last pairs of objectives, the conflict that exists in (f_1, f_2) and (f_4, f_5) is in agreement with the observed in the parallel coordinates plots.

Finally, for the RWA3 problem, we can confirm what was noticed from the parallel coordinates plots, that is, the pairs of objective functions (f_1, f_2), (f_2, f_3), (f_3, f_4), and (f_5, f_6) presents a negative correlations, which means the objectives in each pair are in conflict with each other. Other pairs of objectives that present a negative correlation are (f_1, f_6), (f_1, f_7), (f_2, f_4), (f_2, f_5), (f_3, f_6), (f_3, f_7), (f_4, f_6), (f_4, f_7), and (f_5, f_7). The remaining eight pairs of objectives show a positive correlation, however, this does not mean that they can be removed from the problem since they show conflict with other objective functions.

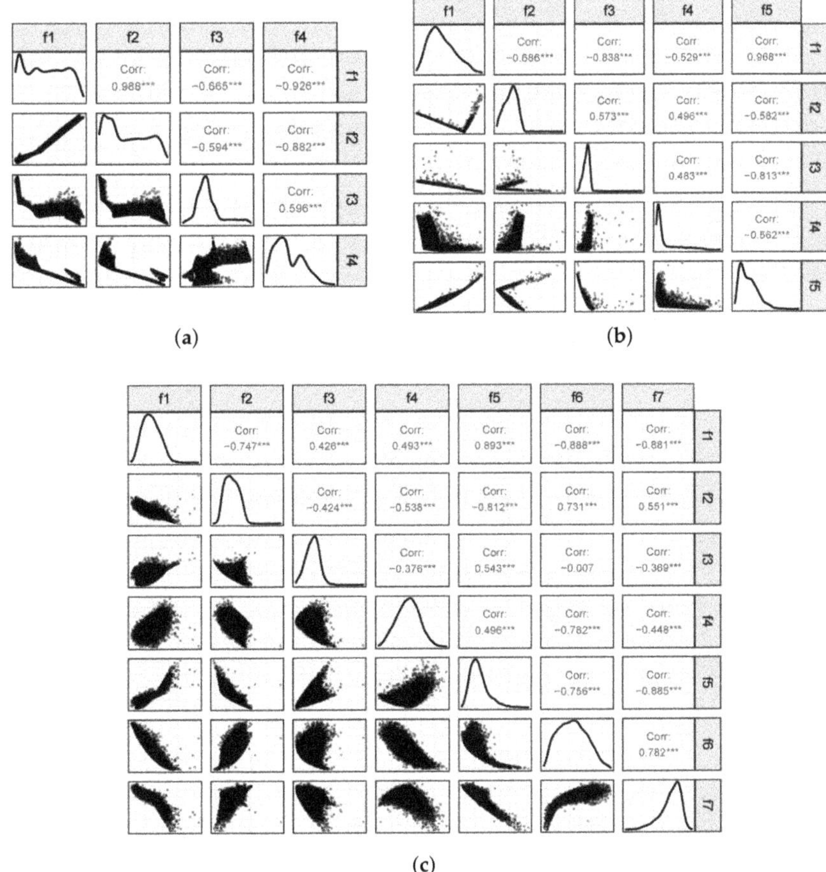

Figure 6. Pearson correlation coefficients between pairs of objectives for the three RWA problems. (**a**) RWA1; (**b**) RWA2; (**c**) RWA3.

7. Conclusions

This paper introduced an improved Lebesgue indicator-based evolutionary algorithm (LIBEA-II) for solving multi-objective optimization problems. The hypothesis put forward in this paper about the efficiency of IBEAs considering the Lebesgue measure, the regularity property of continuous MOPs, and the local property was held. In terms of results, the proposed LIBEA-II and the other three IBEAs, namely SMS-EMOA, iSMS-EMOA, and LIBEA, were tested on the well-known UF benchmark set. These test functions have properties that have been seen in real-life optimization problems in terms of separability, multi-modality, and different $\mathcal{PF}$ shapes, including convexity, concavity, discontinuities, etc. The non-dominated solutions achieved by LIBEA-II and by the other three algorithms, for each test function, were applied the two quality indicators: normalized hypervolume ($\mathcal{H}_n$) and inverted generational distance plus (IGD^+). Results from the $\mathcal{H}_n$ indicator showed that the performance of the four algorithms is similar. Given that all four algorithms are based on the $\mathcal{H}_n$ quality indicator, this result was rather expected. Regarding the IGD^+ indicator, the performance of LIBEA-II was slightly better than the other three algorithms since it obtained the best average results for five out of the ten test functions, and the difference was statistically significant for one of them. In general, LIBEA-II is an efficient algorithm since it can find solutions with the same quality as those found by the other three algorithms but using only 20% of the computing resources.

LIBEA-II was also tested on three real-world applications, precisely: the liquid-rocket single element injector design (RWA1), which has four objective functions and four variables, the ultra-wideband antenna design (RWA2), which considers five objective functions and ten variables, and the development of oil and water repellent fabric (RWA3), which optimizes seven objective functions with three variables. In this case, LIBEA-II was also compared with the same IBEAs used for the UF test instances. Remarkably, LIBEA-II was able to obtain non-dominated solutions with higher quality than those found by the other three IBEAs, since the average value from both quality indicators, i.e., $\mathcal{H}_n$ and IGD^+, was the best. The superiority of LIBEA-II was demonstrated in real-world applications since it obtained higher-quality non-dominated solutions and saved up to approximately 80% of the hypervolume calculations.

As part of our future research, we are interested in extending the applicability of LIBEA-II to deal with constrained MOPs. This line of research has been slightly explored, and it is the course of our outcoming investigations. On the other hand, we would like to test the performance of the proposed LIBEA-II in other real-life applications in order to identify insights that allow us to understand the main weaknesses of IBEAs based on the Lebesgue measure. On the other hand, the hybridization of these types of approaches with mathematical programming is certainly a good path that deserves to be investigated. These are, in fact, part of our future program of investigations.

Author Contributions: Conceptualization, S.Z.-M.; methodology, S.Z.-M.; software, S.Z.-M. and A.M.-M.; validation, S.Z.-M., A.G.-N.; formal analysis, S.Z.-M. and A.G.-N.; investigation, S.Z.-M., A.M.-M., and A.G.-N.; data curation, S.Z.-M.; writing—original draft preparation, S.Z.-M., A.M.-M., and A.G.-N.; writing—review and editing, S.Z.-M. and A.G.-N.; visualization, S.Z.-M. All authors have read and agreed to the published version of the manuscript.

Funding: This research received no external funding.

Institutional Review Board Statement: Not applicable.

Informed Consent Statement: Not applicable.

Conflicts of Interest: The authors declare no conflicts of interest.

References

1. Zhou, A.; Qu, B.Y.; Li, H.; Zhao, S.Z.; Suganthan, P.N.; Zhang, Q. Multiobjective evolutionary algorithms: A survey of the state of the art. *Swarm Evol. Comput.* **2011**, *1*, 32–49. [CrossRef]
2. Nedjah, N.; de Macedo Mourelle, L. Evolutionary multi-objective optimisation: A survey. *Int. J. -Bio-Inspired Comput.* **2015**, *7*, 1–25. [CrossRef]
3. Hansen, M.P.; Jaszkiewicz, A. *Evaluating the Quality of Approximations to the Non-Dominated Set*; Technical Report IMM-REP-1998-7; Technical University of Denmark: Kongens Lyngby, Denmark, 1998.

4. Zitzler, E.; Thiele, L. Multiobjective Optimization Using Evolutionary Algorithms—A Comparative Study. In *Parallel Problem Solving from Nature V*; Eiben, A.E., Ed.; Springer: Amsterdam, The Netherlands, 1998; pp. 292–301.

5. Zitzler, E.; Thiele, L.; Laumanns, M.; Fonseca, C.M.; da Fonseca, V.G. Performance Assessment of Multiobjective Optimizers: An Analysis and Review. *IEEE Trans. Evol. Comput.* **2003**, *7*, 117–132. [CrossRef]

6. Coello, C.A.C.; Reyes Sierra, M. A Study of the Parallelization of a Coevolutionary Multi-Objective Evolutionary Algorithm. In *Proceedings of the Third Mexican International Conference on Artificial Intelligence (MICAI'2004), Mexico City, Mexico, 26–30 April 2004*; Monroy, R., Arroyo-Figueroa, G., Sucar, L.E., Sossa, H., Eds.; Lecture Notes in Artificial Intelligence; Springer: Amsterdam, The Netherlands, 2004; Volume 2972, pp. 688–697.

7. Zitzler, E.; Künzli, S. Indicator-based Selection in Multiobjective Search. In *Parallel Problem Solving from Nature—PPSN VIII*; Lecture Notes in Computer Science; Springer: Amsterdam, The Netherlands, 2004; Volume 3242, pp. 832–842.

8. Rodríguez Villalobos, C.A.; Coello, C.A. A new multi-objective evolutionary algorithm based on a performance assessment indicator. In Proceedings of the GECCO'2012, Philadelphia, PA, USA, 7–11 July 2012; ACM: New York, NY, USA, 2012; pp. 505–512.

9. Zapotecas Martínez, S.; Sosa Hernández, V.A.; Aguirre, H.; Tanaka, K.; Coello Coello, C.A. Using a Family of Curves to Approximate the Pareto Front of a Multi-Objective Optimization Problem. In *Parallel Problem Solving from Nature—PPSN XIII. Proceedings of the 13th International Conference, Ljubljana, Slovenia, 13–17 September 2014*; Bartz-Beielstein, T., Branke, J., Filipič, B., Smith, J., Eds.; Lecture Notes in Computer Science; Springer: Amsterdam, The Netherlands, 2014; Volume 8672, pp. 682–691.

10. Bringmann, K.; Friedrich, T. Approximating the Least Hypervolume Contributor: NP-Hard in General, But Fast in Practice. In *Evolutionary Multi-Criterion Optimization, Proceedings of the 5th International Conference, EMO 2009, Nantes, France, 7–10 April 2009*; Ehrgott, M., Fonseca, C.M., Gandibleux, X., Hao, J.K., Sevaux, M., Eds.; Lecture Notes in Computer Science; Springer: Amsterdam, The Netherlands, 2009; Volume 5467, pp. 6–20.

11. Beume, N.; Naujoks, B.; Emmerich, M. SMS-EMOA: Multiobjective selection based on dominated hypervolume. *Eur. J. Oper. Res.* **2007**, *181*, 1653–1669. [CrossRef]

12. Tsukamoto, N.; Sakane, Y.; Nojima, Y.; Ishibuchi, H. Incorporation of Hypervolume Approximation with Scalarizing Functions into Indicator-based Evolutionary Multiobjective Optimization Algorithms. *Trans. Inst. Syst. Control Inf. Eng.* **2010**, *23*, 165–177.

13. Bader, J.; Deb, K.; Zitzler, E. Faster Hypervolume-Based Search Using Monte Carlo Sampling. In *Multiple Criteria Decision Making for Sustainable Energy and Transportation Systems*; Ehrgott, M., Naujoks, B., Stewart, T.J., Wallenius, J., Eds.; Lecture Notes in Economics and Mathematical Systems; Springer: Amsterdam, The Netherlands, 2010; Volume 634, pp. 313–326.

14. Jiang, S.; Zhang, J.; Ong, Y.S.; Zhang, A.N.; Tan, P.S. A Simple and Fast Hypervolume Indicator-Based Multiobjective Evolutionary Algorithm. *IEEE Trans. Cybern.* **2015**, *45*, 2202–2213. [CrossRef]

15. Zapotecas-Martínez, S.; García-Nájera, A.; López-Jaimes, A. LIBEA: A Lebesgue Indicator-Based Evolutionary Algorithm for multi-objective optimization. *Swarm Evol. Comput.* **2019**, *44*, 404–419. [CrossRef]

16. Zhang, Q.; Li, H. MOEA/D: A Multiobjective Evolutionary Algorithm Based on Decomposition. *IEEE Trans. Evol. Comput.* **2007**, *11*, 712–731. [CrossRef]

17. Li, H.; Zhang, Q. Multiobjective Optimization Problems With Complicated Pareto Sets, MOEA/D and NSGA-II. *IEEE Trans. Evol. Comput.* **2009**, *13*, 284–302. [CrossRef]

18. Zhou, A.; Zhang, Q.; Zhang, G. A multiobjective evolutionary algorithm based on decomposition and probability model. In Proceedings of the 2012 IEEE Congress on Evolutionary Computation (CEC'2012), Brisbane, QLD, Australia, 10–15 June 2012; IEEE Press: Brisbane, Australia, 2012; pp. 3151–3158.

19. Auger, A.; Bader, J.; Brockhoff, D.; Zitzler, E. Theory of the Hypervolume Indicator: Optimal $\{\mu\}$-Distributions and the Choice Of The Reference Point. In Proceedings of the FOGA '09: Tenth ACM SIGEVO Workshop on Foundations of Genetic Algorithms, Orlando, FL, USA, 9–11 January 2009; ACM: Orlando, FL, USA, 2009; pp. 87–102.

20. Chand, S.; Wagner, M. Evolutionary many-objective optimization: A quick-start guide. *Surv. Oper. Res. Manag. Sci.* **2015**, *20*, 35–42. [CrossRef]

21. Miettinen, K. *Nonlinear Multiobjective Optimization*; Kluwer Academic Publishers: Boston, MA, USA, 1999.

22. Hillermeier, C. *Nonlinear Multiobjective Optimization—A Generalized Homotopy Approach*; Birkhäauser: Basel, Switzerland, 2001.

23. Carlsson, C.; Fullér, R. Multiple criteria decision making: The case for interdependence. *Comput. Oper. Res.* **1995**, *22*, 251–260. [CrossRef]

24. Okabe, T.; Jin, Y.; Sendhoff, B. A Critical Survey of Performance Indices for Multi-Objective Optimization. In Proceedings of the 2003 Congress on Evolutionary Computation (CEC'2003), Canberra, Australia, 8–12 December 2003; IEEE Press: Canberra, Australia, 2003; Volume 2, pp. 878–885.

25. Jiang, S.; Ong, Y.S.; Zhang, J.; Feng, L. Consistencies and Contradictions of Performance Metrics in Multiobjective Optimization. *IEEE Trans. Cybern.* **2014**, *44*, 2391–2404. [CrossRef]

26. Li, M.; Yao, X. Quality Evaluation of Solution Sets in Multiobjective Optimisation: A Survey. *ACM Comput. Surv.* **2019**, *52*, 1–38. [CrossRef]

27. Falcón-Cardona, J.G.; Coello, C.A.C. Indicator-Based Multi-Objective Evolutionary Algorithms: A Comprehensive Survey. *ACM Comput. Surv.* **2020**, *53*, 1–35. [CrossRef]

28. Knowles, J.; Corne, D.; Fleischer, M. Bounded archiving using the lebesgue measure. In Proceedings of the 2003 Congress on Evolutionary Computation, 2003 CEC '03, Canberra, ACT, Australia, 8–12 December 2003; Volume 4, pp. 2490–2497. [CrossRef]

29. While, L.; Bradstreet, L. Applying the WFG algorithm to calculate incremental hypervolumes. In Proceedings of the 2012 IEEE Congress on Evolutionary Computation, Brisbane, QLD, Australia, 10–15 June 2012; pp. 1–8. [CrossRef]

30. Emmerich, M.T.; Fonseca, C.M. Computing hypervolume contributions in low dimensions: Asymptotically optimal algorithm and complexity results. In *International Conference on Evolutionary Multi-criterion Optimization*; Springer: Berlin/Heidelberg, Germany, 2011; pp. 121–135.

31. Guerreiro, A.P.; Fonseca, C.M. Computing and updating hypervolume contributions in up to four dimensions. *IEEE Trans. Evol. Comput.* **2017**, *22*, 449–463. [CrossRef]

32. Russo, L.M.; Francisco, A.P. Extending quick hypervolume. *J. Heuristics* **2016**, *22*, 245–271. [CrossRef]

33. Guerreiro, A.P.; Fonseca, C.M.; Paquete, L. The Hypervolume Indicator: Computational Problems and Algorithms. *ACM Comput. Surv.* **2021**, *54*, 119. [CrossRef]

34. Beume, N.; Fonseca, C.M.; Lopez-Ibanez, M.; Paquete, L.; Vahrenhold, J. On the complexity of computing the hypervolume indicator. *IEEE Trans. Evol. Comput.* **2009**, *13*, 1075–1082. [CrossRef]

35. Guerreiro, A.P.; Fonseca, C.M.; Emmerich, M.T. A Fast Dimension-Sweep Algorithm for the Hypervolume Indicator in Four Dimensions. In Proceedings of the Canadian Conference on Computational Geometry, Charlottetown, PEI, Canada, 8–10 August 2012; pp. 77–82.

36. Knowles, J.D. Local-Search and Hybrid Evolutionary Algorithms for Pareto Optimization. Ph.D. Thesis, University of Reading, Reading, UK, 2002.

37. Fleischer, M. The measure of Pareto optima applications to multi-objective metaheuristics. In *International Conference on Evolutionary Multi-Criterion Optimization*; Springer: Berlin/Heidelberg, Germany, 2003; pp. 519–533.

38. Fonseca, C.M.; Paquete, L.; López-Ibánez, M. An improved dimension-sweep algorithm for the hypervolume indicator. In Proceedings of the 2006 IEEE International Conference on Evolutionary Computation, Vancouver, BC, Canada, 16–21 July 2006; pp. 1157–1163.

39. Beume, N. S-metric calculation by considering dominated hypervolume as Klee's measure problem. *Evol. Comput.* **2009**, *17*, 477–492. [CrossRef]

40. While, L.; Bradstreet, L.; Barone, L. A fast way of calculating exact hypervolumes. *IEEE Trans. Evol. Comput.* **2011**, *16*, 86–95. [CrossRef]

41. Russo, L.M.; Francisco, A.P. Quick hypervolume. *IEEE Trans. Evol. Comput.* **2013**, *18*, 481–502. [CrossRef]

42. Ishibuchi, H.; Masuda, H.; Tanigaki, Y.; Nojima, Y. Modified Distance Calculation in Generational Distance and Inverted Generational Distance. In *Evolutionary Multi-Criterion Optimization, Proceedings of the 8th International Conference, EMO 2015, Guimarães, Portugal, 29 March–1 April 2015*; Gaspar-Cunha, A., Antunes, C.H., Coello Coello, C., Eds.; Lecture Notes in Computer Science; Springer: Amsterdam, The Netherlands, 2015; Volume 9019, pp. 110–125.

43. Shang, K.; Ishibuchi, H.; He, L.; Pang, L.M. A Survey on the Hypervolume Indicator in Evolutionary Multiobjective Optimization. *IEEE Trans. Evol. Comput.* **2021**, *25*, 1–20. [CrossRef]

44. Zitzler, E.; Brockhoff, D.; Thiele, L. The Hypervolume Indicator Revisited: On the Design of Pareto-compliant Indicators Via Weighted Integration. In *Evolutionary Multi-Criterion Optimization*; Obayashi, S., Deb, K., Poloni, C., Hiroyasu, T., Murata, T., Eds.; Springer: Berlin/Heidelberg, Germany, 2007; pp. 862–876.

45. Igel, C.; Hansen, N.; Roth, S. Covariance Matrix Adaptation for Multi-objective Optimization. *Evol. Comput.* **2007**, *15*, 1–28. [CrossRef]

46. Mostaghim, S.; Branke, J.; Schmeck, H. Multi-Objective Particle Swarm Optimization on Computer Grids. In Proceedings of the 9th Annual Conference on Genetic and Evolutionary Computation, London, UK, 7–11 July2007; Association for Computing Machinery: New York, NY, USA, 2007; pp. 869–875. [CrossRef]

47. Bader, J.; Zitzler, E. HypE: An Algorithm for Fast Hypervolume-Based Many-Objective Optimization. *Evol. Comput.* **2011**, *19*, 45–76. [CrossRef]

48. Shang, K.; Ishibuchi, H. A New Hypervolume-Based Evolutionary Algorithm for Many-Objective Optimization. *IEEE Trans. Evol. Comput.* **2020**, *24*, 839–852. [CrossRef]

49. Shang, K.; Ishibuchi, H.; Ni, X. R2-Based Hypervolume Contribution Approximation. *IEEE Trans. Evol. Comput.* **2020**, *24*, 185–192. [CrossRef]

50. Menchaca-Mendéz, A.; Coello, C.A.C. A new selection mechanism based on hypervolume and its locality property. In Proceedings of the 2013 IEEE Congress on Evolutionary Computation, Cancún, México, 20–23 June 2013; pp. 924–931. [CrossRef]

51. Menchaca-Mendez, A.; Coello Coello, C.A. An alternative hypervolume-based selection mechanism for multi-objective evolutionary algorithms. *Soft Comput.* **2017**, *21*, 861–884. [CrossRef]

52. Bringmann, K.; Friedrich, T. Approximating the least hypervolume contributor: NP-hard in general, but fast in practice. *Theor. Comput. Sci.* **2012**, *425*, 104–116. [CrossRef]

53. Menchaca-Méndez, A.; Montero, E.; Zapotecas-Martínez, S. An Improved S-Metric Selection Evolutionary Multi-Objective Algorithm With Adaptive Resource Allocation. *IEEE Access* **2018**, *6*, 63382–63401. [CrossRef]

54. Rostami, S.; Neri, F. Covariance matrix adaptation pareto archived evolution strategy with hypervolume-sorted adaptive grid algorithm. *Integr. Comput.-Aided Eng.* **2016**, *23*, 313–329. [CrossRef]

55. Zhang, Q.; Zhou, A.; Jin, Y. RM-MEDA: A Regularity Model-Based Multiobjective Estimation of Distribution Algorithm. *IEEE Trans. Evol. Comput.* **2008**, *12*, 41–63. [CrossRef]

56. Schütze, O.; Coello Coello, C.A.; Mostaghim, S.; Talbi, E.G.; Dellnitz, M. Hybridizing evolutionary strategies with continuation methods for solving multi-objective problems. *Eng. Optim.* **2008**, *40*, 383–402. [CrossRef]
57. Chambers, J.M. Algorithm 410: Partial Sorting. *Commun. ACM* **1971**, *14*, 357–358. [CrossRef]
58. Kennedy, J.; Eberhart, R.C. Particle swarm optimization. In Proceedings of the IEEE International Conference on Neural Networks, Perth, WA, Australia, 27 November–1 December 1995; pp. 1942–1948.
59. Storn, R.M.; Price, K.V. *Differential Evolution—A Simple and Efficient Adaptive Scheme for Global Optimization over Continuous Spaces*; Technical Report TR-95-012; International Computer Science Institute: Berkeley, CA, USA, 1995.
60. Goldberg, D.E. *Genetic Algorithms in Search, Optimization and Machine Learning*; Addison-Wesley Publishing Company: Reading, MA, USA, 1989.
61. Knowles, J.; Corne, D. Properties of an Adaptive Archiving Algorithm for Storing Nondominated Vectors. *IEEE Trans. Evol. Comput.* **2003**, *7*, 100–116. [CrossRef]
62. Huband, S.; Hingston, P.; White, L.; Barone, L. An Evolution Strategy with Probabilistic Mutation for Multi-Objective Optimisation. In Proceedings of the 2003 Congress on Evolutionary Computation (CEC'2003), Canberra, Australia, 8–12 December 2003; IEEE Press: Canberra, Australia, 2003; Volume 3, pp. 2284–2291.
63. Ishibuchi, H.; Imada, R.; Setoguchi, Y.; Nojima, Y. Reference Point Specification in Hypervolume Calculation for Fair Comparison and Efficient Search. In Proceedings of the Genetic and Evolutionary Computation Conference, Berlin, Germany, 15–19 July 2017; Association for Computing Machinery: New York, NY, USA, 2017; pp. 585–592. [CrossRef]
64. Zhang, Q.; Zhou, A.; Zhao, S.; Suganthan, P.N.; Liu, W.; Tiwari, S. Multiobjective Optimization Test Instances for the CEC 2009 Special Session and Competition. In *Special Session on Performance Assessment of Multi-Objective Optimization Algorithms*; Technical Report CES-487; University of Essex: Colchester, UK; Nanyang Technological University: Singapore, 2008; Volume 264.
65. Deb, K.; Pratap, A.; Agarwal, S.; Meyarivan, T. A Fast and Elitist Multiobjective Genetic Algorithm: NSGA–II. *IEEE Trans. Evol. Comput.* **2002**, *6*, 182–197. [CrossRef]
66. Zitzler, E.; Deb, K.; Thiele, L. Comparison of Multiobjective Evolutionary Algorithms: Empirical Results. *Evol. Comput.* **2000**, *8*, 173–195. [CrossRef]
67. Deb, K.; Thiele, L.; Laumanns, M.; Zitzler, E. Scalable Multi-Objective Optimization Test Problems. In Proceedings of the Congress on Evolutionary Computation (CEC'2002), Honolulu, HI, USA, 12–17 May 2002; IEEE Service Center: Piscataway, NJ, USA, 2002; Volume 1, pp. 825–830.
68. Wilcoxon, F. Individual Comparisons by Ranking Methods. *Biom. Bull.* **1945**, *1*, 80–83. [CrossRef]
69. Bonferroni, C.E. Teoria statistica delle classi e calcolo delle probabilita. *Pubbl. R Ist. Super. Sci. Econ. Commer. Firenze* **1936**, *8*, 3–62.
70. Goel, T.; Vaidyanathan, R.; Haftka, R.T.; Shyy, W.; Queipo, N.V.; Tucker, K. Response surface approximation of Pareto optimal front in multi-objective optimization. *Comput. Methods Appl. Mech. Eng.* **2007**, *196*, 879–893. [CrossRef]
71. Vaidyanathan, R.; Tucker, P.K.; Papila, N.; Shyy, W. Computational-fluid-dynamics-based design optimization for single-element rocket injector. *J. Propuls. Power* **2004**, *20*, 705–717. [CrossRef]
72. Chen, Y.S. Performance enhancement of multiband antennas through a two-stage optimization technique. *Int. J. Microw.-Comput.-Aided Eng.* **2017**, *27*, e21064. [CrossRef]
73. Chen, Y.S. Multiobjective optimization of complex antenna structures using response surface models. *Int. J. Microw. Comput.-Aided Eng.* **2016**, *26*, 62–71. [CrossRef]
74. Genzer, J.; Marmur, A. Biological and synthetic self-cleaning surfaces. *MRS Bull.* **2008**, *33*, 742–746. [CrossRef]
75. Sun, T.; Feng, L.; Gao, X.; Jiang, L. Bioinspired surfaces with special wettability. *Accounts Chem. Res.* **2005**, *38*, 644–652. [CrossRef] [PubMed]
76. Ahmad, N.; Kamal, S.; Raza, Z.A.; Hussain, T. Multi-objective optimization in the development of oil and water repellent cellulose fabric based on response surface methodology and the desirability function. *Mater. Res. Express* **2017**, *4*, 035302. [CrossRef]
77. Balling, R. The maximin fitness function for multiobjective evolutionary optimization. In *Optimization in Industry*; Springer: Berlin/Heidelberg, Germany, 2002; pp. 135–147.
78. Tušar, T.; Filipič, B. Visualization of Pareto Front Approximations in Evolutionary Multiobjective Optimization: A Critical Review and the Prosection Method. *IEEE Trans. Evol. Comput.* **2015**, *19*, 225–245. [CrossRef]
79. Pearson, K. Notes on regression and inheritance in the case of two parents. *Proc. R. Soc. Lond.* **1895**, *58*, 240–242.

 *mathematics*

Article

Enhanced Parallel Sine Cosine Algorithm for Constrained and Unconstrained Optimization

Akram Belazi [1,†], Héctor Migallón [2,*,†], Daniel Gónzalez-Sánchez [2,†], Jorge Gónzalez-García [2,†], Antonio Jimeno-Morenilla [3,†] and José-Luis Sánchez-Romero [3,†]

[1] Laboratory RISC-ENIT (LR-16-ES07), Tunis El Manar University, Tunis 1002, Tunisia; akram.belazi@enit.utm.tn

[2] Department of Computer Engineering, Miguel Hernández University, 03202 Elche, Spain; daniel.gonzalez07@goumh.umh.es (D.G.-S.); jorge.gonzalez11@goumh.umh.es (J.G.-G.)

[3] Department of Computer Technology, University of Alicante, 03071 Alicante, Spain; jimeno@dtic.ua.es (A.J.-M.); sanchez@dtic.ua.es (J.-L.S.-R.)

* Correspondence: hmigallon@umh.es; Tel.: +34-966-65-8390

† These authors contributed equally to this work.

Citation: Belazi, A.; Migallón, H.; Gónzalez-Sánchez, D.; Gónzalez-García, J.; Jimeno-Morenilla, A.; Sánchez-Romero, J.-L. Enhanced Parallel Sine Cosine Algorithm for Constrained and Unconstrained Optimization. *Mathematics* **2022**, *10*, 1166. https://doi.org/10.3390/math10071166

Academic Editors: Antonin Ponsich, Mariona Vila Bonilla and Bruno Domenech

Received: 8 March 2022
Accepted: 30 March 2022
Published: 3 April 2022

Publisher's Note: MDPI stays neutral with regard to jurisdictional claims in published maps and institutional affiliations.

Abstract: The sine cosine algorithm's main idea is the sine and cosine-based vacillation outwards or towards the best solution. The first main contribution of this paper proposes an enhanced version of the SCA algorithm called as ESCA algorithm. The supremacy of the proposed algorithm over a set of state-of-the-art algorithms in terms of solution accuracy and convergence speed will be demonstrated by experimental tests. When these algorithms are transferred to the business sector, they must meet time requirements dependent on the industrial process. If these temporal requirements are not met, an efficient solution is to speed them up by designing parallel algorithms. The second major contribution of this work is the design of several parallel algorithms for efficiently exploiting current multicore processor architectures. First, one-level synchronous and asynchronous parallel ESCA algorithms are designed. They have two favors; retain the proposed algorithm's behavior and provide excellent parallel performance by combining coarse-grained parallelism with fine-grained parallelism. Moreover, the parallel scalability of the proposed algorithms is further improved by employing a two-level parallel strategy. Indeed, the experimental results suggest that the one-level parallel ESCA algorithms reduce the computing time, on average, by 87.4% and 90.8%, respectively, using 12 physical processing cores. The two-level parallel algorithms provide extra reductions of the computing time by 91.4%, 93.1%, and 94.5% with 16, 20, and 24 processing cores, including physical and logical cores. Comparison analysis is carried out on 30 unconstrained benchmark functions and three challenging engineering design problems. The experimental outcomes show that the proposed ESCA algorithm behaves outstandingly well in terms of exploration and exploitation behaviors, local optima avoidance, and convergence speed toward the optimum. The overall performance of the proposed algorithm is statistically validated using three non-parametric statistical tests, namely Friedman, Friedman aligned, and Quade tests.

Keywords: constrained optimization; metaheuristic; heuristic algorithm; OpenMP; parallel algorithms; SCA algorithm; unconstrained optimization

MSC: 49M99; 68Q10

1. Introduction

Metaheuristic optimization methods are widely used. Many of these algorithms are based on populations that evolve towards the optimal through an iterative process. In many cases, this iterative process is governed by rules based on natural phenomena, physical processes, or mathematical functions. Depending on both the evolutionary process of the populations (i.e., the algorithm used) and the characteristics of the function to be optimized (single-objective or multi-objective), the use of these methods may not be feasible, either because of the high computing cost or because of the poor quality of the result.

Some of the well-known metaheuristic optimization algorithms are based on natural phenomena. The most common algorithms are the ant colony optimization (ACO) algorithm [1], which imitates the foraging behavior of ant colonies; the evolutionary strategy (ES) algorithm [2], which is based on the processes of mutation and selection seen in evolution; the evolutionary programming [3] uses techniques for evolving programs based on the selection of individuals for reproduction (crossover) and mutation, as well the genetic programming [4]; the particle swarm optimization (PSO) algorithm [5], which is based on the social behavior of fish schooling or bird flocking; the shuffled frog leaping [6] algorithm, which imitates the collaborative behavior of frogs; and the artificial bee colony (ABC) algorithm [7], which was inspired by the foraging behavior of honey bees. Some algorithms are based on physical phenomena, for instance, the simulated annealing (SA) algorithm [8], which is based on the annealing process in metallurgy. Some algorithms based on human or non-human physiological processes have been proposed, such as genetic algorithms (GA) [9], which reflects the process of natural selection; the differential evolution (DE) [10–12] optimizes a problem by iteratively working to promote an agent concerning a given measure of quality; and the artificial immune algorithm (AIA) [13], which is based on the behavior of the human immune system. Some algorithms based on human social processes have also been proposed, such as the harmony search algorithm (HSA) [14] inspired by the process of musical performance. Finally, there are proposed algorithms based on mathematical processing, such as the SCA algorithm [15], which is based on the sine and cosine trigonometric functions.

Almost all of the algorithms mentioned require configuration parameters for an optimal optimization process. An incorrect setting of these parameters can cause either a poor quality solution or that the computational cost drastically increases as more generations are required to be processed. For example, ABC needs the number of bees and limits to be defined, HSA needs the harmony memory consideration rate, the number of improvisations, etc., to be adjusted. However, some of these algorithms do not require parameter tunings, such as teaching-learning based optimization algorithm (TLBO), Jaya, and SCA algorithms. The latter is employed in this paper.

The SCA algorithm has been proven to be efficient in various applications. In [16], SCA is used to train feed foreword neural network to breast cancer classification. Authors in [17] employ SCA algorithm to improve an adaptive fuzzy logic PID (proportional integral derivative) controller for the load frequency control of an autonomous power generation system. In addition, it is used to optimize the parameters of a fractional-order proportional integral differential controller for coordinated control of power consumption in heat pumps [18]. In [19], the unified power quality conditioner is formulated as a single objective problem optimized using SCA. The application spectrum of the SCA algorithm is too large, see for example [20–27]. However, its convergence speed is a bit slow, especially when considered multimodal objectives functions. Indeed, it maintains high global searchability even at the end of iterations. This paper aims to improve the SCA algorithm optimization behavior by intensifying the current solution's refinement with a promising diversification level during the course of the algorithm, speeding it up both in terms of optimization and computational cost.

The major findings of the work are:

- A new optimization algorithm is proposed, dubbed the Enhanced Sine Cosine Algorithm (ESCA), which improves the SCA algorithm and offers better performance than a set of state-of-the-art algorithms. The outstanding optimization performance of the ESCA algorithm is based on the embedding of a best-guided approach along with the local search capability already existing in the SCA algorithm, leading to a decrease in the diversification behavior at the end of the iterations.
- To improve the computational performance of the proposed algorithm, synchronous and asynchronous parallel algorithms have been designed based on parallelization,

initially at an outer, i.e., at a coarse-grained level. Since this level of parallelization is related to subpopulations, the number of subpopulations cannot increase indefinitely. These synchronous and asynchronous one-level parallel ESCA algorithms decrease the computing time by 87.4% and 90.8%, respectively, using 12 processing cores.

- To improve parallel scalability without harming the optimization performance and increasing the number of processes, two-level parallel algorithms have been designed. The parallel strategy includes two levels, namely the outer level and the internal level. The outer level corresponds to coarse-grained parallelization, while the internal level corresponds to fine-grained parallelization. Accordingly, the parallel scalability of the proposed algorithms is extremely improved. The experimental results show significant reductions in the computing time of 91.4%, 93.1%, and 94.5% with 16, 20, and 24 processes mapped on 12 physical cores. These time reductions correspond to speed-ups of $x12.5$, $x15.9$, and $x19.0$ with 16, 20, and 24 processes correctly mapped on 12 physical cores, i.e., using hyperthreading.

The rest of the paper is organized as follows. The preliminaries, including the sine cosine algorithm (SCA) and the related works, are provided in Section 2. The proposed enhanced SCA algorithm (ESCA) along with the proposed parallel algorithms based on multi-population are described in Section 3. Section 4 lists the benchmark functions and the engineering problems employed for testing the performance of the proposed algorithm. The experimental results of these algorithms are discussed in Section 5. Finally, Section 6 concludes the paper.

2. Related Work

The SCA algorithm, on which our ESCA proposal is based, is described in Section 2.1. Other proposals based on the SCA algorithm are listed and briefly described in Section 2.2.

2.1. Sine Cosine Algorithm

The SCA algorithm is an optimization algorithm based on an initial population that evolves in search of a function's optimum, called a cost function. This evolution, i.e., the generation of consecutive new populations (the typical procedure of population-based algorithms), is mainly based on (1) and (2).

$$Pop_m^k = Pop_m^k + (r_1 * \sin(r_2^k)\left|r_3^k * BestPop^k - Pop_m^k\right|) \tag{1}$$

$$Pop_m^k = Pop_m^k + (r_1 * \cos(r_2^k)\left|r_3^k * BestPop^k - Pop_m^k\right|) \tag{2}$$

As can be seen, (1) and (2) differ only in the use of the mathematical functions sine or cosine. In these equations, it has been adopted that each population is composed of m individuals, each individual consists of k variables (this parameter depends on the cost function), and finally, the best current individual is denoted by $BestPop$. Each individual is generated based on both the current individual (Pop_m) and the current best individual ($BestPop$). However, the generation of each variable of each new individual is tuned by using three random values that define the magnitude of the sine or cosine range (r_1), the sine or cosine domain (r_2^k), and the magnitude of the contribution of the target ($BestPop$) in defining the new position of the solution (r_3^k).

In practice, the random numbers r_1 divide the search space into two sub-spaces based on the current individual and the best individual in the current population. Thus, if r_1 is greater than 1, the candidate solutions vacillate outwards the destination, else they fluctuate inwards the destination (see Figure 1).

Both exploration and exploitation phases of the SCA optimization algorithm depend on the capabilities provided by (1) and (2). This selection is decided at random with the same probability.

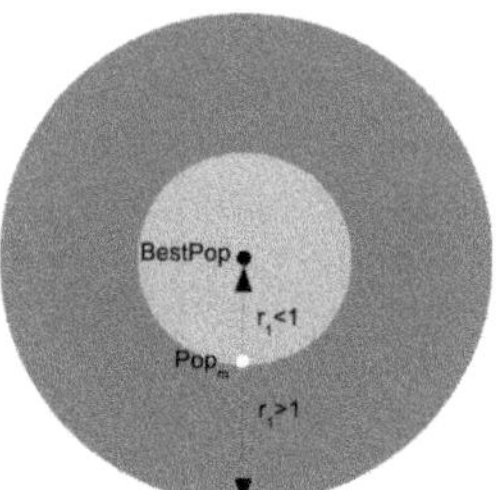

Figure 1. Searching spaces of SCA depending on r_1.

In heuristic optimization algorithms, which are iterative, the exploration phase is usually more decisive in the iterative procedure's final phase. The SCA algorithm prioritizes the exploration phase as more iterations are performed through r_1 (see Equation (3)).

$$r_1 = iniValue_{r_1} - current_{IT}\frac{iniValue_{r_1}}{max_{ITs}} \tag{3}$$

From an initial value ($iniValue\_r_1$), the value of r_1 decreases as the number of iterations performed increases, towards the r_1 minimum value when the last iteration is performed (max_ITs). The initial value of r_1 is set to 2. The number of iterations to be performed (max_ITs) is necessary for all population-based heuristic optimization algorithms. In practice, the value of r_1 modifies the range of values of the terms associated with the sine and cosine, from the original range $[-1, 1]$ to the decreasing variable range, this variables range starts at $[-iniValue\_r_1, iniValue\_r_1]$. These variables' contribution can be seen in Algorithm 1, which shows the steps of the SCA algorithm. The computed new individual is $newPop_m$, the number of individuals in the population is $popSize$ and the number of cost function design variables is $numDesignVars$.

Algorithm 1 The SCA optimization algorithm.

1: Set $iniValue_{r_1} = 2$
2: Set max_{ITs} variable
3: Set population size (m - iterator for individuals)
4: Define function cost (k - iterator for design variables)
5: Generate initial population Pop_0
6: **for** $iterator = 1$ to max_{ITs} **do**
7:   Search for the current $BestPop$
8:   $r_1 = iniValue_{r_1} - iterator\frac{iniValue_{r_1}}{max_{ITs}}$
9:   **for** $m = 0$ to $popSize$ **do**
10:     **for** $k = 1$ to $numDesignVars$ **do**
11:       $r_2 = 2 * \pi * rand_{0..1}$
12:       $r_3 = 2 * rand_{0..1}$
13:       $r_4 = rand_{0..1}$
14:       **if** $r_4 < 0.5$ **then**
15:         $newPop_m^k = Pop_m^k + \left(r_1 * \sin(r_2)\left|r_3 * BestPop^k - Pop_m^k\right|\right)$
16:       **else**
17:         $newPop_m^k = Pop_m^k + \left(r_1 * \cos(r_2)\left|r_3 * BestPop^k - Pop_m^k\right|\right)$
18:       **end if**
19:     **end for**
20:     $Pop_m = newPop_m$
21:   **end for**
22: **end for**
23: Search for the current $BestPop$

2.2. SCA-Based Proposals

Thanks to its simplicity, the SCA algorithm was widely adopted and refined in many research proposals. In [28], the authors proposed a modified SCA algorithm in which the linear transition rule was substituted by a non-linear transition to guarantee a better transition from exploration to exploitation. Second, the best guidance based on the elite candidate solution was entered in the SCA's search equations. Third, to escape from local optimums, a mutation operator is utilized to produce a new position during the course of the algorithm. An improved alternative of SCA named HSCA for train multilayer perceptrons was reported in [29]. The HSCA adjusted the search mechanism of SCA by combining the leading guidance and the simulated quenching algorithm. In [30], a novel SCA based on orthogonal parallel information was presented. It is based on two approaches; multiple-orthogonal parallel information and experience-based opposition direction strategy. The former enabled the algorithm to save the solution diversification and search around promising regions simultaneously. The latter serves to guard the exploration ability of the SCA algorithm. Authors in [31] proposed an improved sine cosine algorithm (ISCA) for feature selection of text categorization. In addition to the position of the leading solution, the ISCA worked with random positions from the search space. That alteration of the solution's position mitigated premature convergence and submitted adequate performance. Ref. [32] suggested an improved sine cosine algorithm in which a couple of new mechanisms are provided. One is the mixing of the exploitation abilities of crossover with the personal lead position of individual solutions. The other is the combination of self-learning and global search tools. Zhiliu et al. proposed a modified SCA algorithm based on vicinity search and greedy levy mutation [33]. It suggests three optimization tactics. Firstly, it mixed the exponential decreasing of conversion parameter and the linear decreasing of inertia weight, which yielded an equilibrium between the algorithm's global and local search abilities. Secondly, to escape from local optimums, a random strategy for search agents around the best one is performed. Thirdly, the greedy Levy mutation strategy is adopted for the best individuals to intensify the algorithm's local searchability. A hybrid modified SCA algorithm was studied in [34]. It was benefited from the ability of random populations through the Latin hypercube sampling method. Next, it was used for hybridization with the cuckoo search algorithm. The algorithm showed sufficient local and global search skills. Mohamed et al. presented an improved SCA algorithm based on opposition-based learning (OBL) [35]. Indeed, OBL is a machine learning approach usually utilized to boost the performance of metaheuristic optimization algorithms. It allowed better accuracy of the obtained solutions by promoting the exploration skills of the algorithm. Since OBL elected the leading element falling between a given solution and its opposite, better solutions are afforded accordingly. An enhanced SCA algorithm for feature selection was described in [36]. It embedded an elitism strategy and a new strategy of best solution updating, yielding better accuracy for pattern classification. In [37], the authors proposed an improved SCA algorithm for solving high-dimensional global optimization problems. The equation for renovating the position of the current solution and the linearly decreasing parameter were modified. In the former, inertia weight was introduced to speed up the convergence rate and avoid local optimums. The latter was replaced by a Gaussian function-based strategy that enabled a non-linear decrease of the parameter. Therefore, a promising exploration-exploitation balance was yielded. Other good attempts for improving the SCA algorithm can be found in [38–42]. In this subsection, some SCA-based algorithms have been reviewed. The motivation for the improvements in each of them is briefly described.

3. Proposed Work

In Section 3.1 our proposed optimization algorithm based on the SCA algorithm, called ESCA, is presented. Then in Section 3.2, the parallel algorithms developed to computationally accelerate the ESCA algorithm are presented.

3.1. Enhanced Sine Cosine Algorithm

The proposed enhanced sine cosine algorithm (ESCA) aims to improve the optimization behavior of the original SCA algorithm. For this purpose, we enhance the exploration and exploitation phases of the SCA optimization algorithm. Indeed, they depend on the capabilities provided by (1) and (2). These capacities are boosted by introducing a new alternative, defined by (4), to generate each new individual.

$$Pop_m^k = BestPop^k + r_5^2(Pop_m^k - r_6 * BestPop^k) \tag{4}$$

When using (4), the new individual is generated based on the current individual and the distance between that individual and the best individual in the current population. Both the magnitude of the best individual and the magnitude of the distance are tuned using two random numbers, r_5 (which is squared) and r_6 respectively, as shown in (4).

The probability of using the sine-based equation, i.e., (1), remains at 50%. While the probability of using the cosine-based equation, i.e., (2), decreases to only 20%. The new equation uses neither sine nor cosine, and it has a 30% chance of being used. The proposed enhanced sine cosine algorithm (ESCA) is described in Algorithm 2.

Algorithm 2 Enhanced SCA (ESCA) optimization algorithm

1: Set $iniValue_{r_1} = 2$
2: Set max_{ITs} variable
3: Set population size (m - iterator for individuals)
4: Define function cost (k - iterator for design variables)
5: Generate initial population Pop_0
6: **for** $iterator = 1$ to max_{ITs} **do**
7:     Search for the current $BestPop$
8:     $r_1 = iniValue_{r_1} - iterator \frac{iniValue_{r_1}}{max_{ITs}}$
9:     **for** $m = 0$ to $popSize$ **do**
10:       **for** $k = 1$ to $numDesignVars$ **do**
11:         $r_2 = 2 * \pi * rand_{0..1}$
12:         $r_3 = 2 * rand_{0..1}$
13:         $r_4 = rand_{0..1}$
14:         **if** $r_4 < 0.5$ **then**
15:           $newPop_m^k = Pop_m^k + \left(r_1 * \sin(r_2) \left| r_3 * BestPop^k - Pop_m^k \right| \right)$
16:         **else if** $r_4 < 0.7$ **then**
17:           $newPop_m^k = Pop_m^k + \left(r_1 * \cos(r_2) \left| r_3 * BestPop^k - Pop_m^k \right| \right)$
18:         **else**
19:           $r_5 = rand_{0..1}$
20:           $r_6 = round(1 + rand_{0..1})$
21:           $newPop_m^k = BestPop^k + r_5^2(Pop_m^k - r_6 * BestPop^k)$
22:         **end if**
23:       **end for**
24:       $Pop_m = newPop_m$
25:     **end for**
26: **end for**
27: Search for the current $BestPop$

In more detail, in the SCA algorithm two equations can be used to obtain a new individual, as can be seen in Algorithm 1 (lines 14–18), the first based on the sine function and the second based on the cosine function. Both equations have the same probability of being used, as can be seen in line 14 of Algorithm 1. In contrast, in our proposal up to three equations can be used, the first two coincide with the functions of the SCA algorithm, and the third is shown in Equation (4). The probability of using the equation based on the sine of the SCA algorithm remains unchanged. The probability of using the cosine-based equation of the SCA algorithm is reduced to 20%, while the new equation proposed in the ESCA algorithm has a 30% probability of being used, as can be seen in Algorithm 1 (lines 18–22).

To compare search agents' behavior of the SCA and ESCA algorithms, the two-dimensional versions of the benchmark functions are solved by 30 search agents. The search maps of the search agents under 300 function evaluation times are shown in Figures 2–4. Similarly, the distributions of all possible solutions over the entire search space are depicted in Figures 5–7. These figures reveal that the ESCA algorithm searches around thoroughly narrow regions from the promising regions of the search space, which means reaching the optimum faster. In contrast, the SCA algorithm searches in dispersed areas of the entire space, so more time is required to attain the promising regions. In addition, the obtained solutions by the ESCA algorithm are almost distributed around the global optimum. This proves that it efficiently exploits the previous solutions to improve the current one and bypass significant jumps in the search space. The SCA algorithm's weakness is that it favors exploration even at the end of iterations. An efficient optimization algorithm should hit an equilibrium of exploitation and exploration. Indeed, it should maintain a high level of diversification at the beginning and a lower one at its end to avoid falling on local optimums. Simultaneously, the algorithm refines the current solution progressively. Briefly, the algorithm should promote exploration in the beginning and exploitation at the end. In this context, the ESCA algorithm is guided by the current best solution (see Equation (4)) to converge toward the optimum and sustain a high level of intensification at the end of the algorithm. Accordingly, a better balance between local search and global search is guaranteed over the course of iterations.

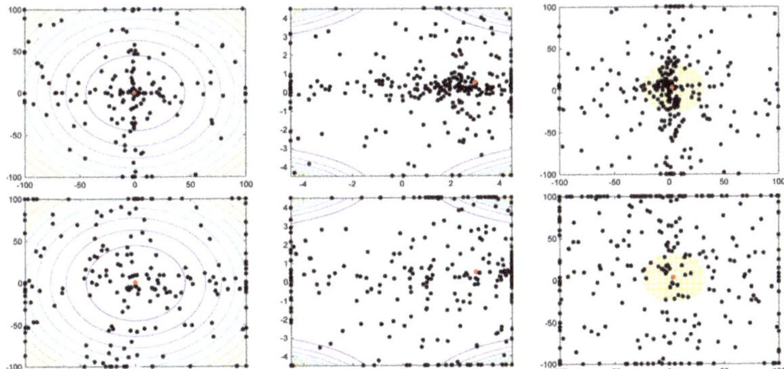

Figure 2. Search maps of search agents when solving functions f_1, f_3, and f_4; by the ESCA algorithm (**first row**); and the SCA algorithm (**second row**).

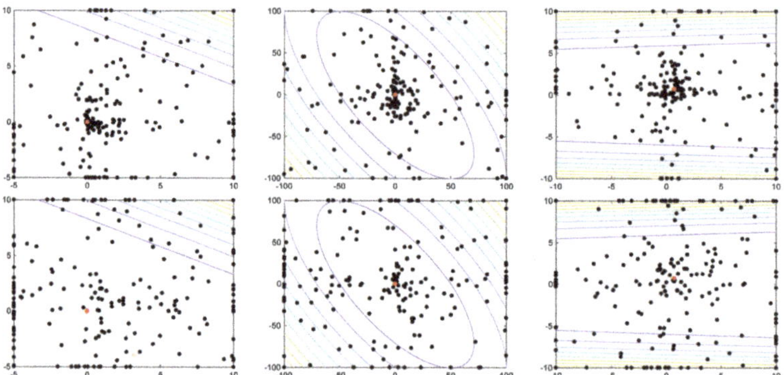

Figure 3. Search maps of search agents when solving functions f_9, f_{10}, and f_{12}; by the ESCA algorithm (**first row**); and the SCA algorithm (**second row**).

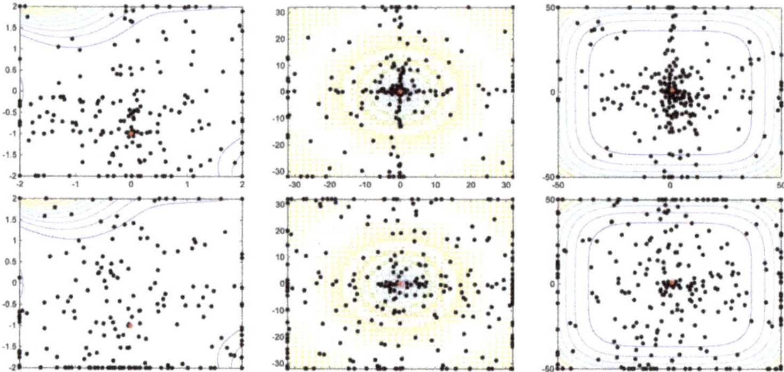

Figure 4. Search maps of search agents when solving functions f_{21}, f_{24}, and f_{25}; by the ESCA algorithm (**first row**); and the SCA algorithm (**second row**).

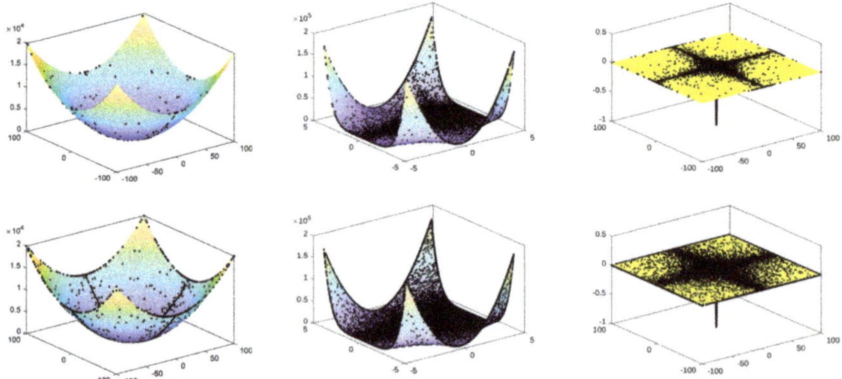

Figure 5. Obtained solutions in the search space of functions f_1, f_3, and f_4; by the ESCA algorithm (**first row**); and the SCA algorithm (**second row**).

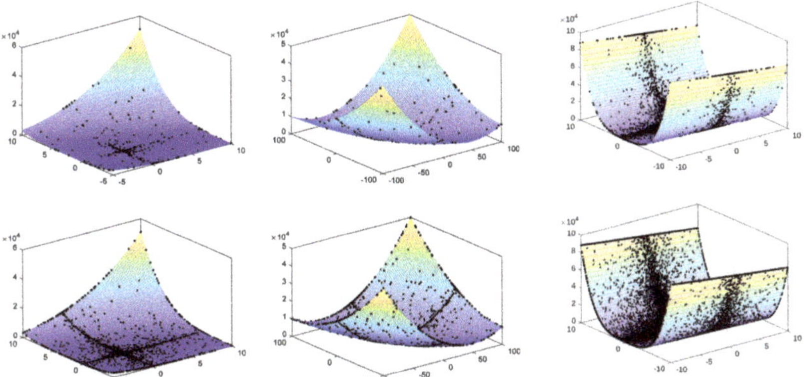

Figure 6. Obtained solutions in the search space of functions f_9, f_{10}, and f_{12}; by the ESCA algorithm (**first row**); and the SCA algorithm (**second row**).

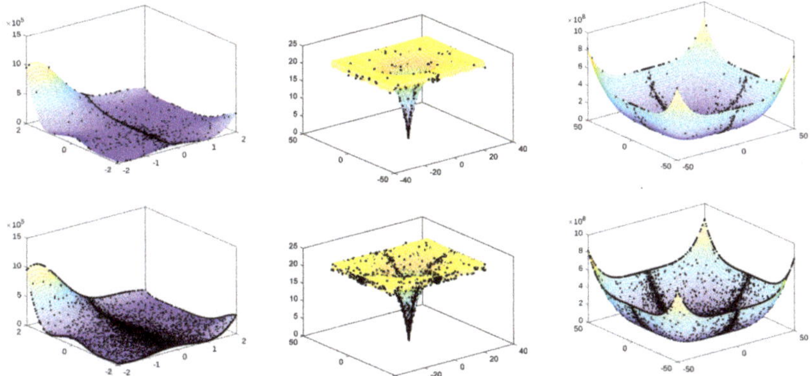

Figure 7. Obtained solutions in the search space of functions f_{21}, f_{24}, and f_{25}; by the ESCA algorithm (**first row**); and the SCA algorithm (**second row**).

3.2. Proposed Parallel Algorithms

Almost all newer computing platforms, regardless of their computing power, are parallel. The main trends to increase the platforms' computing power are (i) increasing the number of processing units (physical cores and/or logical threads) and (ii) including hardware accelerators (GPUs, FPGAs, etc.). We propose parallel algorithms based on multicore platforms to efficiently use the computational resources available on shared memory parallel platforms.

First, two parallel coarse-grained algorithms based on multi-population are developed. Similar strategies applied to different heuristic optimization are presented in [43,44] and some other well-known algorithms. In both, the SCA and the proposed ESCA algorithms, only the population size and the stop criterion need to be established. Since the proposed parallel algorithms are based on multi-populations, the selected population size is that of the initial population, i.e., before it is partitioned. The stop criterion is the number of new generations to be computed. Note that the number of generations and the population's size implicitly determine the number of cost function evaluations to be performed.

The initial population is divided into subpopulations of equal or similar size. The size of the subpopulations depends on the number of used processing units as shown in Algorithm 3 (line 4). If the size of the initial population is not divisible by the number of processing units, the sizes of some subpopulations are increased by one as exhibited in lines 5–9 of Algorithm 3.

Algorithm 3 Multi-population sizes computing

1: Initial population size: $popInitSize$
2: Number of cores (or processes): $NoCs$
3: Process ID: $idPr \in [0, NoCs - 1]$
4: $subpopSize = \frac{popInitSize}{NoCs}$
5: **if** $(subpopSize\%NoCs)! = 0$ **then**
6:   **if** $idPr < (PopulationSize\%NoCs)$ **then**
7:     $subpopSize = subpopSize + 1$
8:   **end if**
9: **end if**

Once the size of the subpopulations is determined according to the size of the initial population and the number of processes, as can be seen in Algorithm 3, each subpopulation is processed by a single process. The required communications between these concurrent processes depend on the operating algorithm. The asynchronous approach reduces these communications with respect to the synchronous algorithm. Note that when hyperthreading is not used, each core runs only one process. In our case, hyperthreading is used when more than 12 processes are required.

As stated, the proposed parallel algorithms are suitable for shared memory platforms. In both algorithms, to efficiently exploit shared-memory platforms, private memory has been used preferably. The first proposed parallel algorithm, shown in Algorithm 4, is asynchronous, i.e., communications between processes are not needed. Algorithm 4 shows the parallel processing implemented in the asynchronous parallel method, i.e., the processing performed once each sequential thread has spawned the parallel region. A new subpopulation individual ($newSP_m$) is computed based on the current subpopulation individual (SP_m) and the best subpopulation individual ($subpopBest$).

It is worth mentioning that the concurrent processing shown in Algorithm 4 lacks synchronization points. This strategy allows having populations of significantly different sizes and leads to balancing the computing load through the number of generations processed by each thread and thus not degrading parallel efficiency.

Algorithm 5 presents the second parallel strategy in which the concurrent processes share data to obtain the best individual from the whole population, i.e., the best of all subpopulations. This process is done both at the beginning (line 7) and after computing each new generation by each parallel process (line 29). To ensure that all concurrent processes use the best individual from the whole population ($wholepopBest$) in each new generation, a synchronization point is needed after the critical section (line 35).

As shown in Algorithms 4 and 5, the population size assigned to each process depends on the size of the whole population ($popInitSize$) and the number of computing processes $NoCs$ (see Algorithm 3). That is, as the number of processes increases, the size of the subpopulations decreases. When tiny populations are used in population-based heuristic optimization algorithms, the optimization behavior can be significantly degraded. To further increase the number of processes and thus further reduce the computing time without drastically reducing the subpopulation sizes, we propose a two-level parallel algorithm. The parallel second level (fine-grained level) is applied to obtain a new generation of each subpopulation (see lines 10 and 26 of Algorithm 4).

Algorithm 4 Asynchronous parallel algorithm.

1: Allocate private memory for subpopulation: $SP_{[0,subpopSize]}$
2: Allocate private memory for best individual: $subpopBest$
3: Set $iniValue_{r_1} = 2$
4: Generation counter: $genIt = 0$
5: Generate initial subpopulation SP_0
6: **while** $genIt < numGenerations$ **do**
7:     Search for the current subpop best $subpopBest$
8:     $genIt = genIt + 1$
9:     $r_1 = iniValue_{r_1} - genIt \dfrac{iniValue_{r_1}}{numGenerations}$
10:     **for** $m = 1$ to $subpopSize$ **do**
11:       **for** $k = 1$ to $numDesignvars$ **do**
12:         $r_2 = 2 * \pi * rand_{0..1}$
13:         $r_3 = 2 * rand_{0..1}$
14:         $r_4 = rand_{0..1}$
15:         **if** $r_4 < 0.5$ **then**
16:           $newSP_m^k = SP_m^k + \left(r_1 * \sin(r_2) \left| r_3 * subpopBest^k - SP_m^k \right| \right)$
17:         **else if** $r_4 < 0.7$ **then**
18:           $newSP_m^k = SP_m^k + \left(r_1 * \cos(r_2) \left| r_3 * subpopBest^k - SP_m^k \right| \right)$
19:         **else**
20:           $r_5 = rand_{0..1}$
21:           $r_6 = round(1 + rand_{0..1})$
22:           $newSP_m^k = subpopBest^k + r_5^2(SP_m^k - r_6 * subpopBest^k)$
23:         **end if**
24:       **end for**
25:       $SP_m = newSP_m$
26:     **end for**
27: **end while**

In the two-level algorithm the subpopulations are not calculated as a function of the total number of processes, since a single process will not process each subpopulation. The total number of processes in the two-level algorithm is equal to the number of subpopulations multiplied by the number of processes that will process each subpopulation. The number of subpopulations will be equal to the number of external processes ($NoCs$), while the number of processes that will process each subpopulation will be denoted by $inCs$. Therefore, the total number of processes equals to $NoCs \times inCs$.

Important modifications in Algorithm 4 are required that could degrade the parallel performance of the two-level parallel algorithm given in Algorithm 6. Since several threads will process each subpopulation, it must be stored in shared memory (line 1 of Algorithm 6), instead of being stored in private memory as in Algorithm 4. Moreover, before processing each subpopulation, the best individual must be available for all the processes involved in processing each subpopulation. This implies a synchronization point (line 9 of Algorithm 6) that determine the best individual. Thereafter each process checks if the current best individual stored in its private memory ($subpopBest$) should be updated.

Algorithm 5 Parallel algorithm with data sharing.

1: Shared memory: $wholepopBest$
2: Allocate private memory for: $SP_{[0,subpopSize]}$ and $subpopBest$
3: Set $iniValue_{r_1} = 2$
4: Generation counter: $genIt = 1$
5: Generate initial subpopulation SP_0
6: Search for the current subpopulation best $subpopBest$
7: $wholepopBest = Bestof(subpopBest_{NoCs})$
8: **while** $genIt < numGenerations$ **do**
9:    $genIt = genIt + 1$
10:    $r_1 = iniValue_{r_1} - genIt \dfrac{iniValue_{r_1}}{numGenerations}$
11:    **for** $m = 1$ to $subpopSize$ **do**
12:      **for** $k = 1$ to $numDesignvars$ **do**
13:        $r_2 = 2 * \pi * rand_{0..1}$
14:        $r_3 = 2 * rand_{0..1}$
15:        $r_4 = rand_{0..1}$
16:        **if** $r_4 < 0.5$ **then**
17:          $newSP_m^k = SP_m^k + \left(r_1 * \sin(r_2) \left| r_3 * subpopBest^k - SP_m^k \right| \right)$
18:        **else if** $r_4 < 0.7$ **then**
19:          $newSP_m^k = SP_m^k + \left(r_1 * \cos(r_2) \left| r_3 * subpopBest^k - SP_m^k \right| \right)$
20:        **else**
21:          $r_5 = rand_{0..1}$
22:          $r_6 = round(1 + rand_{0..1})$
23:          $newSP_m^k = subpopBest^k + r_5^2(SP_m^k - r_6 * subpopBest^k)$
24:        **end if**
25:      **end for**
26:      $SP_m = newSP_m$
27:    **end for**
28:    Search for the current subpopulation best $subpopBest$
29:    **CRITICAL parallel section:**
30:      **if** $F_{eval}(subpopBest) < F_{eval}(wholepopBest)$ **then**
31:        $wholepopBest = subpopBest$
32:      **else**
33:        $subpopBest = wholepopBest$
34:      **end if**
35:    **end CRITICAL**
36: **end while**

Note that, in Algorithm 6 the total number of processes is increased from $NoCs$ to $NoCs \times inCs$, using the same subpopulation size. There are several options to implement the second level of parallelism (lines 13–29 of Algorithm 6), which will be discussed in Section 5.

Algorithm 6 Two-level parallel algorithm.

1: Allocate shared memory for $NoCs$ subpopulations: $SP_{[0,subpopSize]}$
2: Total number of processes: $NoCs \times inCs$ processes.
3: Allocate private memory for best individual: $subpopBest$
4: Set $iniValue_{r_1} = 2$
5: Generation counter: $genIt = 0$
6: Generate initial subpopulation SP_0
7: **while** $genIt < numGenerations$ **do**
8:     Search for the current subpopulation best $subpopBest$
9:     {Synchronization point}
10:     $genIt = genIt + 1$
11:     $r_1 = iniValue_{r_1} - genIt \frac{iniValue_{r_1}}{numGenerations}$
12:     {FOR processed in PARALLEL using $inCs$ processes}
13:     **for** $m = 1$ to $subpopSize$ **do**
14:       **for** $k = 1$ to $numDesignvars$ **do**
15:         $r_2 = 2 * \pi * rand_{0..1}$
16:         $r_3 = 2 * rand_{0..1}$
17:         $r_4 = rand_{0..1}$
18:         **if** $r_4 < 0.5$ **then**
19:           $newSP_m^k = SP_m^k + \left(r_1 * \sin(r_2) \left| r_3 * subpopBest^k - SP_m^k \right| \right)$
20:         **else if** $r_4 < 0.7$ **then**
21:           $newSP_m^k = SP_m^k + \left(r_1 * \cos(r_2) \left| r_3 * subpopBest^k - SP_m^k \right| \right)$
22:         **else**
23:           $r_5 = rand_{0..1}$
24:           $r_6 = round(1 + rand_{0..1})$
25:           $newSP_m^k = subpopBest^k + r_5^2(SP_m^k - r_6 * subpopBest^k)$
26:         **end if**
27:       **end for**
28:       $SP_m = newSP_m$
29:     **end for**
30: **end while**

---

4. Benchmark Test

The benchamark test used in this work is composed of 30 well-known unconstrained functions shown in Section 4.1, and three constrained engineering design problems shown in Section 4.2.

4.1. Benchmark Functions

A total of 30 well-known unconstrained functions used for the performance analysis are listed and described in Tables 1 and 2.

4.2. Engineering Optimization Problems

The proposed algorithms' optimization performance will be further examined through three constrained engineering design problems.

4.2.1. Pressure Vessel Design Problem

The structural design problem of pressure vessels is shown in Figure 8. In this design problem, four variables have to be computed: the thickness of the shell (d_s), the thickness of the heads (d_h), the internal radius (R), and the length (L) of the cylindrical section. These variables should minimize the financial cost by meeting the non-linear stress constraints and yield criteria. Note that d_s and d_h are not continuous variables. Indeed, from 0.0625 inches, the possible values are calculated in steps of 0.0625 inches. The pressure vessel design problem is formulated as in (5).

Pressure vessel design problem:

$$f = 0.6224x_1x_3x_4 + 1.7781x_2x_3^2 +$$
$$3.1661x_1^2x_4 + 19.84x_1^2x_3$$
$$x_1 = d_s, x_2 = d_h, x_3 = R, x_4 = L$$

Constraints:

$$g_1 = -x_1 + 0.0193x_3 \leq 0$$
$$g_2 = -x_2 + 0.00954x_3 \leq 0$$
$$g_3 = -\pi x_3^2 x_4 - (4/3)\pi x_3^3 + 1296000 \leq 0$$
$$g_4 = x_4 - 240 \leq 0$$
$$0.0625 \leq x_1, x_2 \leq 99 * 0.0625$$
$$10 \leq x_3, x_4 \leq 240 \tag{5}$$

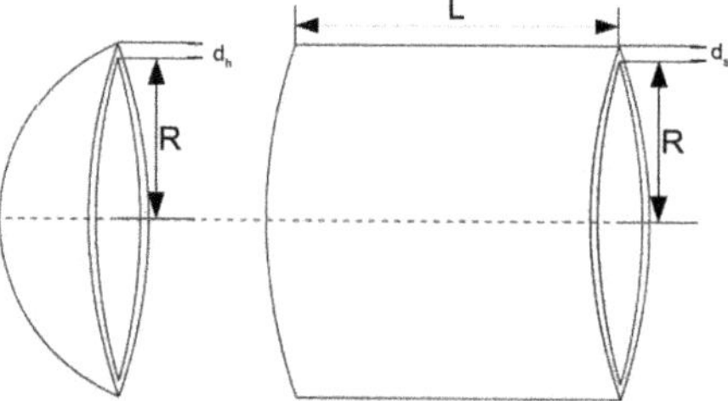

Figure 8. Pressure vessel design problem.

Table 1. Benchmark functions: dimensions and domain.

Id.	Name	Dim. (V)	Domain (Min, Max)
f_1	Sphere	30	$-100, 100$
f_2	SumSquares	30	$-10, 10$
f_3	Beale	2	$-4.5, 4.5$
f_4	Easom	2	$-100, 100$
f_5	Matyas	2	$-10, 10$
f_6	Colville	4	$-10, 10$
f_7	Trid 6	6	$-V^2, V^2$
f_8	Trid 10	10	$-V^2, V^2$
f_9	Zakharov	10	$-5, 10$
f_{10}	Schwefel_1.2	30	$-100, 100$
f_{11}	Rosenbrock	30	$-30, 30$
f_{12}	Dixon-Price	5	$-10, 10$
f_{13}	Foxholes	2	$-2^{16}, 2^{16}$
f_{14}	Branin	2	$x_1 : -5, 10$ $x_2 : 0, 15$
f_{15}	Bohachevsky_1	2	$-100, 100$
f_{16}	Booth	2	$-10, 10$
f_{17}	Michalewicz_2	2	$0, \pi$
f_{18}	Michalewicz_5	5	$0, \pi$
f_{19}	Bohachevsky_2	2	$-100, 100$
f_{20}	Bohachevsky_3	2	$-100, 100$
f_{21}	GoldStein-Price	2	$-2, 2$
f_{22}	Perm	4	$-V, V$
f_{23}	Hartman_3	3	$0, 1$
f_{24}	Ackley	30	$-32, 32$
f_{25}	Penalized_2	30	$-50, 50$
f_{26}	Langermann_2	2	$0, 10$
f_{27}	Langermann_5	5	$0, 10$
f_{28}	Langermann_10	10	$0, 10$
f_{29}	Fletcher-Powell_5	5	$x_i, \alpha_i : -\pi, \pi$ $a_{ij}, b_{ij} : -100, 100$
f_{30}	Fletcher-Powell_10	10	$x_i, \alpha_i : -\pi, \pi$ $a_{ij}, b_{ij} : -100, 100$

Table 2. Benchmark functions: Definitions.

Id.	Function
f_1	$f = \sum\limits_{i=1}^{V} x_i^2$
f_2	$f = \sum\limits_{i=1}^{V} i x_i^2$
f_3	$f = (1.5 - x_1 + x_1 x_2)^2 + (2.25 - x_1 + x_1 x_2^2)^2 + (2.625 - x_1 + x_1 x_2^3)^2$
f_4	$f = -\cos(x_1)\cos(x_2)\exp\left(-(x_1 - \pi)^2 - (x_2 - \pi)^2\right)$
f_5	$f = 0.26(x_1^2 + x_2^2) - 0.48 x_1 x_2$
f_6	$f = 100(x_1^2 - x_2)^2 + (x_1 - 1)^2 + (x_3 - 1)^2 + 90(x_3^2 - x_4)^2 + 10.1\left((x_2 - 1)^2 + (x_4 - 1)^2\right) + 19.8(x_2 - 1)(x_4 - 1)$
f_7 f_8	$f = \sum\limits_{i=1}^{V} (x_i - 1)^2 - \sum\limits_{i=2}^{V} x_i x_{i-1}$
f_9	$f = \sum\limits_{i=1}^{V} x_i^2 + \left(\sum\limits_{i=1}^{V} 0.5 i x_i\right)^2 + \left(\sum\limits_{i=1}^{V} 0.5 i x_i\right)^4$
f_{10}	$f = \sum\limits_{i=1}^{V} \left(\sum\limits_{j=1}^{i} x_j\right)^2$
f_{11}	$f = \sum\limits_{i=1}^{V-1} \left(100(x_{i+1} - x_i^2)^2 + (x_i - 1)^2\right)$
f_{12}	$f = (x_1 - 1)^2 + \sum\limits_{i=2}^{V} i\left(2x_i^2 - x_{i-1}\right)^2$
f_{13}	$f = \left[\frac{1}{500} + \sum\limits_{j=1}^{25} \frac{1}{j + \sum\limits_{i=1}^{2}(x_i - a_{ij})^6}\right]^{-1}$
f_{14}	$f = \left(x_2 - \frac{5.1}{4\pi^2} x_1^2 + \frac{5}{\pi} x_1 - 6\right)^2 + 10\left(1 - \frac{1}{8\pi}\right)\cos x_1 + 10$
f_{15}	$f = x_1^2 + 2x_2^2 - 0.3\cos(3\pi x_1) - 0.4\cos(4\pi x_2) + 0.7$
f_{16}	$f = (x_1 - 2x_2 - 7)^2 + (2x_1 + x_2 - 5)^2$
f_{17} f_{18}	$f = -\sum\limits_{i=1}^{V} \sin x_i \left(\sin\left(\frac{i x_i^2}{\pi}\right)\right)^{20}$
f_{19}	$f = x_1^2 + 2x_2^2 - 0.3\cos(3\pi x_1)\cos(4\pi x_2) + 0.3$
f_{20}	$f = x_1^2 + 2x_2^2 - 0.3\cos(3\pi x_1 + 4\pi x_2) + 0.3$
f_{21}	$f = \left[1 + (x_1 + x_2 + 1)^2(19 - 14x_1 + 3x_1^2 - 14x_2 + 6x_1 x_2 + 3x_2^2)\right] \left[30 + (2x_1 - 3x_2)^2(18 - 32x_1 + 12x_1^2 + 48x_2 - 36x_1 x_2 + 27x_2^2)\right]$
f_{22}	$f = \sum\limits_{j=1}^{V} \left[\sum\limits_{i=1}^{j}(i^j + \beta)\left(\left(\frac{x_i}{i}\right)^j - 1\right)\right]^2$
f_{23}	$f = -\sum\limits_{i=1}^{4} c_i \exp\left[-\sum\limits_{j=1}^{3} a_{ij}(x_j - p_{ij})^2\right]$

Table 2. *Cont.*

Id.	Function
f_{24}	$f = -20\exp\left(-0.2\sqrt{\frac{1}{V}\sum_{i=1}^{V}x_i^2}\right) - \exp\left(\frac{1}{V}\sum_{i=1}^{V}\cos(2\pi x_i)\right) + 20 + e$
f_{25}	$f = 0.1\left\{\sin^2(3\pi x_1) + \sum_{i=1}^{V-1}(x_i-1)^2\left[1+\sin^2(3\pi x_{i+1})\right] + (x_V-1)^2\left[1+\sin^2(2\pi x_V)\right]\right\}$ $+ \sum_{i=1}^{V}u(x_i,5,100,4),$ $u(x_i,a,k,m) = k(x_i-a)^m, x_i > a; 0, -a \le x_i \le a; k(-x_i-a)^m, x_i < -a.$
f_{26} f_{27} f_{28}	$f = -\sum_{i=1}^{5}c_i\left[\exp\left(-\frac{1}{\pi}\sum_{j=1}^{V}(x_j-a_{ij})^2\right)\cos\left(\pi\sum_{j=1}^{V}(x_j-a_{ij})^2\right)\right]$
f_{29} f_{30}	$f = \sum_{i=1}^{V}(A_i-B_i)^2;\ A_i = \sum_{j=1}^{V}(a_{ij}\sin\alpha_j + bij\cos\alpha_j),\ B_i = \sum_{j=1}^{V}(a_{ij}\sin x_j + bij\cos x_j)$

4.2.2. Welded Beam Design Problem

The welded beam design problem is depicted in Figure 9. The cost of manufacturing and assembling the welded beams must be minimized by considering the welding work, material, and labor cost. The variables to be computed are the thickness of the weld (h), the length of the welded joint (l), the width of the beam (t), and the thickness of the beam (b). The optimization problem is formulated as in (6), where $\tau(x)$ is the shear stress in the weld, τ_{max} is the allowable shear stress of the weld, $\sigma(x)$ is the normal stress in the beam, σ_{max} is the allowable normal stress for the beam material, $P_c(x)$ is the bar buckling load, P is the load, $\delta(x)$ is the beam end deflection, and δ_{max} is the allowable beam end deflection. Some auxiliary functions and constant values used to solve the welded beam design problem are given in (7).

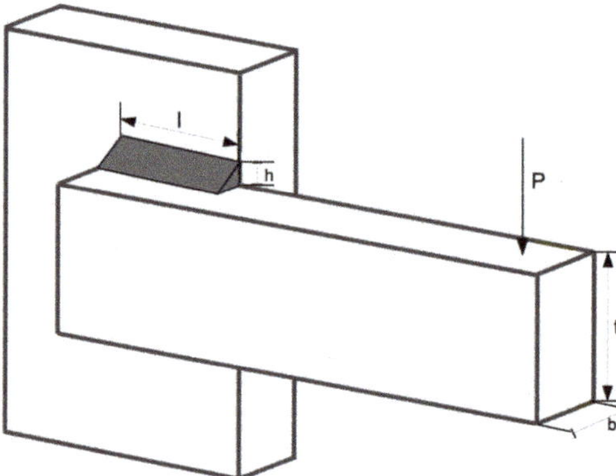

Figure 9. Welded beam design problem.

Welded beam design problem:

$$F = 1.10471x_1^2 x_2 + 0.04811x_3 x_4(14.0 + x_2)$$
$$x_1 = h, x_2 = l, x_3 = t, x_4 = b$$

Constraints:

$$g_1 = \tau(x) - \tau_{max} \leq 0$$
$$g_2 = \sigma(x) - \sigma_{max} \leq 0$$
$$g_3 = x_1 - x_4 \leq 0$$
$$g_4 = 0.10471x_1^2 + 0.04811x_3 x_4(14.0 + x_2) - 5.0 \leq 0$$
$$g_5 = 0.125 - x_1 \leq 0$$
$$g_6 = \delta(x) - \delta_{max} \leq 0$$
$$g_7 = P(x) - P_c(x) \leq 0$$
$$0.1 \leq x_1, x_4 \leq 2.0$$
$$0.1 \leq x_2, x_3 \leq 10.0 \tag{6}$$

Functions and constants of welded beam problem:

$$\tau(x) = \sqrt{(\tau')^2 + 2\tau'\tau''\frac{x_2}{2R} + (\tau'')^2};$$

$$\tau' = \frac{P}{\sqrt{2}x_1 x_2}; \tau'' = \frac{MR}{J}$$

$$M = P\left(L + \frac{x_2}{2}\right); R = \sqrt{\frac{x_2^2}{4} + \left(\frac{x_1 + x_3}{2}\right)^2}$$

$$J = 2\left\{\sqrt{2}x_1 x_2\left[\frac{x_2^2}{12} + \left(\frac{x_1 + x_3}{2}\right)^2\right]\right\}$$

$$\sigma(x) = \frac{6PL}{x_4 x_3^2}$$

$$\delta(x) = \frac{4PL^3}{Ex_3^3 x_4}$$

$$P_c(x) = \frac{4.013E\sqrt{\frac{x_3^2 x_4^6}{36}}}{L^2}\left(1 - \frac{x_3}{2L}\sqrt{\frac{E}{4G}}\right)$$

$$P = 6000lb; L = 14in; \delta_{max} = 0.25in$$
$$E = 30e^{+6}psi; G = 12e^{+6}psi$$
$$\tau_{max} = 13,600psi; \sigma_{max} = 30,000psi \tag{7}$$

4.2.3. Rolling Element Bearing Design Problem

The rolling element bearing design problem is a maximization problem aimed to maximize the dynamic load capacity of a rolling element bearing. This problem, depicted in Figure 10, has five decision variables, namely pitch diameter (D_m), ball diameter (D_b), number of balls (Z), curvature radius coefficient of inner raceway groove ($f_i = r_i/D_b$), curvature radius coefficient of outer raceway groove ($f_o = r_o/D_b$), and the inner and outer ring groove curvature ratio r_i and r_o, respectively. In addition, it has five constraints constants, K_{Dmin}, K_{Dmax}, ϵ, e and ψ. This problem can be formulated as in (8).

Rolling element bearing design problem:

$$f = f_c x_3^{2/3} x_2^{1.8}; \text{if } x_2 \leq 25.4$$

$$f = 3.647 f_c x_3^{2/3} x_2^{1.4}; \text{if } x_2 > 25.4$$

$$x_1 = D_m, x_2 = D_b, x_3 = Z, x_4 = f_i, x_5 = f_o$$

Constraints:

$$g_1 = \frac{\phi_0}{2\sin^{-1}\frac{x_2}{x_1}} - x_3 + 1 \geq 0$$

$$g_2 = 2.0x_2 - x_6(D - d) \geq 0$$

$$g_3 = x_7(D - d) - 2.0x_2 \geq 0$$

$$g_4 = x_{10}B_w - x_2 \geq 0$$

$$g_5 = x_1 - 0.5(D + d) \geq 0$$

$$g_6 = (0.5 + x_9)(D + d) - x_1 \geq 0$$

$$g_7 = 0.5(D - x_1 - x_2) - x_8 x_2 \geq 0$$

$$g_8 = x_4 - 0.515 \geq 0$$

$$g_9 = x_5 - 0.515 \geq 0$$

$$x_6 = K_{Dmin}, x_7 = K_{Dmax}, x_8 = \epsilon, x_9 = e, x_5 = \psi \tag{8}$$

Auxiliary functions and constant values of rolling problem:

$$\gamma = \frac{D_b \cos\alpha}{D_m}$$

$$f_c = 37.91 \times \left\{ 1 + \left[1.04 \left(\frac{1 - \gamma}{1 + \gamma} \right)^{1.72} \left(\frac{f_i(2f_o - 1)}{f_o(2f_i - 1)} \right)^{0.41} \right]^{10/3} \right\}^{-0.3} \times \left\{ \left(\frac{\gamma^{0.3}(1 - \gamma)^{1.39}}{(1 + \gamma)^{1/3}} \right) \left(\frac{2f_i}{2f_i - 1} \right)^{0.41} \right\}$$

$$T = D - d - (2.0x_2)$$

$$\phi_0 = 2\pi - 2\cos^{-1} \left(\frac{\left(\frac{D-d}{2} - \frac{3T}{4} \right)^2 + \left(\frac{D}{2} - \frac{T}{4} - x_2 \right)^2 - \left(\frac{d}{2} + \frac{T}{4} \right)^2}{2\left(\frac{D-d}{3} - \frac{3T}{4} \right)\left(\frac{D}{2} - \frac{T}{4} - x_2 \right)} \right)$$

$$D = 160; d = 90; B_w = 30; \alpha = 0$$

$$90.0 \leq x_1 \leq 150.0$$

$$10.5 \leq x_2 \leq 31.5$$

$$4 \leq x_3 \leq 50$$

$$0.515 \leq x_4, x_5 \leq 0.6$$

$$0.4 \leq x_6 \leq 0.5$$

$$0.6 \leq x_7 \leq 0.7$$

$$0.3 \leq x_8 \leq 0.4$$

$$0.02 \leq x_9 \leq 1.0$$

$$0.6 \leq x_{10} \leq 0.85 \tag{9}$$

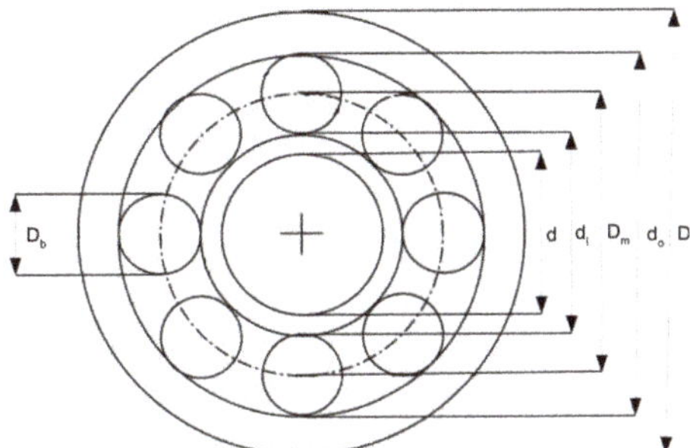

Figure 10. Rolling element bearing design problem.

5. Numerical Experiments

All the numerical experiments have been obtained in Fujitsu Server PRIMERGY TX300 S8 Tower Server. This platform is a multicore platform equipped with a D2949-B1 motherboard with two CPU sockets. In each CPU the processor installed is an Intel(R) Xeon(R) CPU E5-2620 v2 @ 2.10 GHz, with 15 MB Intel Smart Cache. Each processor is composed of 6 physical cores, resulting in a total number of 12 physical cores in the system. The Intel Hyper-Threading Technology is enabled, the number of threads per physical core is 2, therefore the maximum number of processes (or threads) should not exceed 24, in order to obtain the best possible computational performance. The main memory size is 32 GB of DDR3. All the developments, both sequential and parallel, were implemented in the C programming language, using the GCC v.4.4.7 [45]. The OpenMP API v3.1 [46] has been used to develop parallel algorithms. Therefore, all the data in tables and figures included in this section have been obtained running simulations in this platform. In addition, for the computational results to be reliable, the Sun Grid Engine queuing system has been used.

5.1. Comparative Analysis ESCA vs. SCA

First, the computational costs of the SCA algorithm and the proposed ESCA algorithm are examined in Table 3. This table shows the computing time cost when optimizing the benchmark test reported in Section 4 with population sizes of 240, 120, and 60. The number of generations was 50,000, and the number of independent runs was 30. The results in Table 3 point that the proposed ESCA algorithm does not increase the computing cost compared to the SCA algorithm. On the contrary, in more than 80% of the experiments conducted, the computational cost decreases.

Table 3. Computational times (s.) for sequential SCA and ESCA algorithms.

	\multicolumn{6}{c}{**Population Size**}					
	\multicolumn{2}{c}{**60**}	\multicolumn{2}{c}{**120**}	\multicolumn{2}{c}{**240**}				
	SCA	**ESCA**	**SCA**	**ESCA**	**SCA**	**ESCA**
f_1	349.3	308.8	683.5	711.5	1432.5	1325.7
f_2	388.5	312.9	739.0	668.4	1474.9	1405.0
f_3	25.2	23.4	50.4	46.8	100.6	93.6
f_4	27.8	26.4	55.6	52.8	111.2	105.6
f_5	33.9	31.4	68.8	70.2	131.7	138.5
f_6	31.3	28.6	62.4	57.1	124.7	114.1
f_7	48.6	44.1	96.9	87.9	193.7	179.3
f_8	80.5	69.8	160.0	140.8	321.4	279.2
f_9	144.7	132.7	280.6	268.0	558.3	530.3

Table 3. *Cont.*

| | Population Size | | | | | |
| | 60 | | 120 | | 240 | |
	SCA	ESCA	SCA	ESCA	SCA	ESCA
f_{10}	374.5	413.7	773.6	815.8	1562.5	1657.5
f_{11}	223.1	208.3	442.6	416.4	884.5	833.1
f_{12}	38.4	36.4	77.9	72.5	154.9	145.5
f_{13}	461.7	466.7	923.8	933.9	1845.8	1867.0
f_{14}	19.7	18.9	39.5	37.8	79.1	75.6
f_{15}	17.8	17.1	33.9	33.9	70.7	68.0
f_{16}	15.5	14.6	31.2	29.1	62.0	58.1
f_{17}	72.3	55.6	144.7	111.0	291.4	221.9
f_{18}	174.7	125.1	309.0	280.3	620.4	493.1
f_{19}	18.7	16.5	36.1	32.2	72.2	70.2
f_{20}	17.6	16.8	35.2	31.2	69.9	59.7
f_{21}	16.3	15.3	32.6	30.6	65.1	61.0
f_{22}	105.5	101.8	212.0	205.5	419.7	409.7
f_{23}	36.3	36.5	72.0	73.2	146.1	146.3
f_{24}	125.6	123.2	251.6	246.3	501.5	493.8
f_{25}	406.4	321.7	812.3	674.4	1707.8	1321.8
f_{26}	56.4	57.1	113.3	113.5	225.6	227.2
f_{27}	82.0	82.0	164.3	164.1	331.8	328.8
f_{28}	130.6	118.5	262.2	236.1	523.4	473.2
f_{29}	174.0	168.7	346.9	339.4	700.0	675.1
f_{30}	583.4	568.9	1165.5	1134.6	2334.1	2290.9

Once it has been proven that the proposed method decreases the computational cost of the SCA algorithm, the optimization behavior is investigated by comparing both methods in Table 4. This table shows the number of function evaluations for an error of less than $<1 \times 10^{-3}$ (for functions marked with * an error less than $<1 \times 10^2$), with population sizes of 240, 120, and 60. Fewer function evaluations are required when the ESCA method is used instead of the SCA method. The dramatic decrease, particularly for the functions that require more evaluations, is higher than $100\times$, demonstrating the significant improvement of the SCA's optimization behavior.

To perform a parallel efficiency analysis of both parallel proposals, experimental tests are conducted using the same parameters as those used so far, i.e., population sizes of 240, 120, and 60. The number of generations is equal to 50,000, and the number of independent runs is 30. The parallel speed-up values for the data sharing parallel algorithm, depending on the total population size (*popInitSize*) and the number of processes (*NoCs*), are exhibited in Table 5. The obtained speed-up values are close to ideal ones for the largest population size. These values slightly decrease, in most cases, as the population size decreases. However, the values significantly degrade when 12 parallel processes are used for the smaller population size and lower computing cost functions.

The parallel asynchronous algorithm's speed-up values, shown in Table 6, remain close to the ideal values when the number of concurrent processes is increased or when the population size is decreased. Note that this behavior implies outstanding parallel scalability.

Considering the outstanding parallel performance results obtained for the parallel asynchronous algorithm using the 12 available physical cores (see Table 6), it can be concluded that the parallel scalability of the asynchronous algorithm allows increasing the number of processes efficiently. However, the results shown in Table 4 confirm that the size of the subpopulations requires a minimum dimension, which depends on the optimization algorithm and the problem under consideration. Algorithm 6 has been proposed to increase the number of processes without reducing the size of the subpopulations. To implement the inner level of parallelism of Algorithm 6, nested parallelism can be applied using OpenMP features. This strategy has been discarded due to poor experimental results that excessively degrade parallel scalability. When using nested parallelism the generation of each nested parallel region involves computational overhead [47]. The poor experimental results are due to many nested regions ($numGenerations \times NoCs$) and the insufficient computational cost of each nested parallel region. Note that this computational cost depends on the considered algorithm (quasi-non-variable cost) and the objective function.

Table 4. Number of function evaluations for an error $< 1 \times 10^{-3}$ ($* < 1 \times 10^2$).

	Population Size					
	240		120		60	
	SCA	ESCA	SCA	ESCA	SCA	ESCA
f_1	3,639,144	75,384	1,842,864	48,504	971,802	28,074
f_2	3,596,880	73,464	1,808,004	43,500	988,380	24,888
f_3	24,000	2136	24,888	3072	13,878	2082
f_4	306,912	4152	218,220	3432	239,166	2088
f_5	1584	840	756	564	540	312
f_6	–	9,627,227	–	4,450,577	–	2,654,280
f_7 *	3888	960	5724	612	3222	354
f_8 *	5,031,792	317,376	2,684,760	190,053	1,565,184	196,337
f_9	1,528,656	16,848	848,544	9708	490,854	6420
f_{10}	5,048,616	739,296	2,623,800	462,456	1,400,712	311,640
f_{11} *	3,677,160	78,720	1,906,380	45,828	–	32,424
f_{12}	–	6,186,240	–	4,982,240	–	2,624,640
f_{13}	571,008	14,088	547,320	6288	236,148	36,126
f_{14}	70,392	1920	118,296	2256	52,782	1998
f_{15}	5928	2352	2964	1380	2262	762
f_{16}	187,560	3120	236,952	2508	131,712	2400
f_{17}	401,688	3888	419,220	1812	236,400	2910
f_{18} *	480	480	240	240	120	120
f_{19}	6120	2448	3624	1392	1896	882
f_{20}	5160	2112	4560	1296	2340	834
f_{21}	26,856	2040	28,596	1080	15,924	912
f_{22}	–	3,966,264	–	3,528,912	–	1,907,900
f_{23}	7920	57,090	3,739,200	81,345	123,720	27,760
f_{24}	2,290,464	30,408	1,207,956	17,940	668,790	8304
f_{25} *	3,591,960	46,032	1,952,616	33,756	951,708	17,022
f_{26}	20,400	9672	21,744	4848	10,193	2208
f_{27}	–	6,840,528	–	5,366,040	–	2,930,112
f_{28}	480	480	252	252	120	120
f_{29} *	1,127,832	24,168	1,148,604	27,672	840,288	26,940
f_{30} *	–	9,787,467	–	5,338,960	–	2,939,910

Table 5. Parallel speed-up for parallel data sharing algorithm.

	Population Size								
	240			120			60		
	NoCs								
	2	6	12	2	6	12	2	6	12
f_1	2.0	5.7	10.4	2.0	5.7	11.4	1.8	5.0	9.7
f_2	2.0	5.8	10.9	1.8	5.5	10.5	2.0	4.9	9.3
f_3	1.9	4.9	6.9	1.9	4.6	4.5	1.9	3.7	2.5
f_4	2.0	5.5	10.9	1.9	5.5	10.3	1.6	5.3	3.8
f_5	1.9	5.0	9.0	1.8	4.9	7.4	1.6	4.2	3.7
f_6	2.0	5.5	10.9	1.9	5.4	10.4	1.9	5.4	3.6
f_7	1.3	3.3	4.5	1.9	4.7	5.1	1.9	4.1	3.4
f_8	2.0	5.4	9.9	2.0	5.3	9.0	1.9	5.1	6.8
f_9	1.9	5.2	10.2	1.8	5.3	10.1	1.9	5.1	9.2
f_{10}	2.0	5.5	11.0	1.9	5.4	10.6	2.0	5.5	10.4
f_{11}	2.0	5.5	11.0	2.0	5.5	10.9	2.0	5.5	10.9
f_{12}	2.0	5.3	9.1	1.9	5.1	7.2	1.9	4.6	4.0
f_{13}	2.0	5.5	11.1	2.0	5.5	11.0	2.0	5.5	10.8
f_{14}	2.0	5.5	10.7	2.0	5.4	8.8	1.9	5.2	2.4
f_{15}	1.9	5.4	10.2	2.0	5.7	6.3	2.0	5.2	2.2
f_{16}	1.9	5.5	10.4	1.9	5.3	5.2	1.9	5.1	1.8
f_{17}	2.0	5.4	9.5	2.0	5.2	8.2	1.9	4.8	5.8
f_{18}	1.9	5.4	9.3	1.9	6.0	8.7	1.7	4.8	6.0
f_{19}	2.0	5.1	7.1	1.7	4.3	3.8	1.5	3.7	1.9
f_{20}	1.7	4.8	6.4	1.8	4.5	3.8	1.6	4.1	1.9
f_{21}	1.9	5.1	6.4	1.9	4.6	3.5	1.9	3.6	1.7
f_{22}	1.9	5.2	10.2	1.9	5.3	9.9	1.9	5.1	8.8
f_{23}	2.0	5.5	10.6	2.0	5.4	10.2	1.9	5.4	8.9
f_{24}	2.0	5.5	10.4	1.9	5.4	9.6	2.0	5.2	8.5
f_{25}	2.0	5.6	10.4	2.0	5.5	10.0	2.0	5.2	9.1
f_{26}	2.0	5.2	8.4	1.9	5.0	7.2	1.9	4.8	5.6
f_{27}	2.0	5.4	9.8	2.0	5.1	8.7	2.0	5.0	6.9
f_{28}	2.0	5.5	10.9	1.9	5.5	10.9	1.9	5.4	10.8
f_{29}	2.0	5.5	11.0	2.0	5.5	10.6	2.0	5.4	10.4
f_{30}	2.0	5.6	11.1	2.0	5.5	11.0	2.0	5.5	10.9

Table 6. Parallel speed-up for asynchronous parallel algorithm.

	Population Size								
	240			120			60		
	NoCs								
	2	6	12	2	6	12	2	6	12
f_1	2.0	5.7	11.6	2.0	5.8	11.6	1.9	5.7	11.2
f_2	1.9	5.5	11.4	1.9	5.6	11.2	1.9	5.2	10.5
f_3	1.9	5.5	11.0	1.9	5.5	11.0	1.9	5.5	8.2
f_4	1.9	5.5	11.0	2.0	5.5	11.0	2.0	5.5	11.0
f_5	1.9	5.5	11.0	2.0	5.5	11.1	1.8	5.5	11.0
f_6	1.9	5.5	11.0	2.0	5.5	11.1	2.0	5.5	11.0
f_7	1.3	3.6	7.3	2.0	5.5	11.0	1.9	5.5	11.0
f_8	1.9	5.5	11.0	2.0	5.5	11.1	2.0	5.5	11.1
f_9	1.9	5.6	11.1	2.1	5.3	10.6	1.9	5.2	10.7
f_{10}	1.9	5.4	10.9	2.0	5.6	11.1	2.0	5.5	10.9
f_{11}	1.9	5.5	10.9	2.0	5.5	11.1	1.9	5.5	11.1
f_{12}	1.9	5.5	10.7	2.0	5.5	11.0	2.0	5.5	10.9
f_{13}	1.9	5.5	11.0	2.0	5.5	11.0	2.0	5.5	11.1
f_{14}	1.9	5.5	10.9	2.0	5.5	11.0	1.9	5.5	10.9
f_{15}	1.9	5.4	10.9	1.8	5.2	10.3	1.8	5.4	10.9
f_{16}	1.9	5.5	11.0	1.9	5.5	10.9	1.9	5.5	10.8
f_{17}	1.9	5.5	11.1	2.0	5.5	10.7	2.0	5.5	10.9
f_{18}	1.9	5.6	11.3	2.0	5.6	11.1	2.0	5.6	11.3
f_{19}	1.8	5.1	9.9	1.9	5.4	10.5	1.9	5.0	9.9
f_{20}	1.9	5.3	10.4	2.0	5.5	11.0	1.9	5.5	10.9
f_{21}	1.9	5.5	11.0	2.0	5.5	10.2	1.9	5.5	10.9
f_{22}	1.9	5.2	10.4	1.9	5.2	10.5	1.9	5.3	10.3
f_{23}	1.9	5.4	10.7	2.0	5.5	10.9	2.0	5.5	10.8
f_{24}	1.9	5.5	11.0	2.0	5.6	11.2	2.0	5.5	11.0
f_{25}	1.9	5.4	10.9	2.0	5.5	11.1	1.9	5.6	11.0
f_{26}	1.9	5.5	11.0	2.0	5.5	11.0	1.9	5.5	11.0
f_{27}	1.9	5.5	10.9	2.0	5.4	11.0	2.0	5.5	11.0
f_{28}	1.9	5.5	11.0	2.0	5.5	11.1	1.9	5.6	11.1
f_{29}	1.9	5.5	11.0	2.0	5.5	11.0	2.0	5.5	11.0
f_{30}	1.9	5.5	10.9	2.0	5.6	11.0	2.0	5.6	11.1

The two-level parallel algorithm generates a parallel region of $NoCs \times inCs$ processes, organized into $NoCs$ groups of $inCs$ processes each. In each group, only one process works outside the inner parallel region, while all the processes in the group cooperate in the processing associated with the inner level of parallelism (lines 13–29 of Algorithm 6).

As mentioned above, the used parallel platform has two processors with six physical cores each. Hyperthreading can be enabled, allowing to run two processes (or threads) per core efficiently. Thus, it can be run up to 24 concurrent processes without excessively degrading the computer platform's efficiency. Using hyperthreading and fine-grained parallelism, such as the proposed two-level algorithm, the strategy of thread placement on the cores may be relevant. To control the strategy of process placement in the cores, OpenMP affinity features are used. Figure 11a shows that the platform's architecture is equipped with two processors of six physical cores and twelve logical cores each. An example of thread placement of 5 processes when no affinity is used is shown in Figure 11b, in which the operating system decides the process placement. There is no problem in this thread placement if neither hyperthreading nor fine-grained parallelism are used.

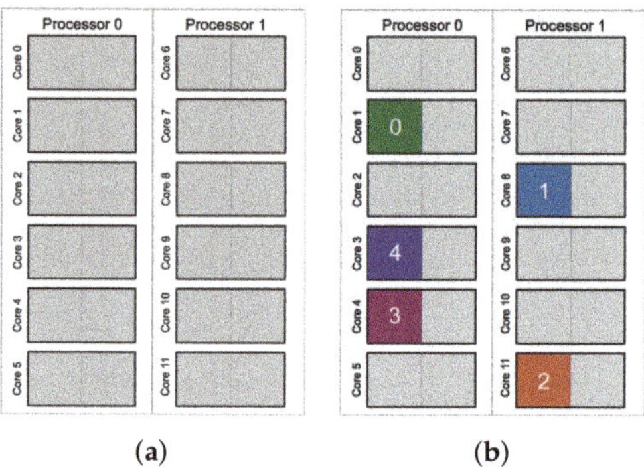

Figure 11. Thread placement when no affinity is used. (**a**) Platform's architecture. (**b**) Example of thread placement without control

For instance, using 20 processes organized into 5 groups of 4 processes, a thread placement option without using affinity features is displayed in Figure 12a. To optimize parallel performance, the optimal thread placement can be forced using OpenMP affinity features as shown in Figure 12b.

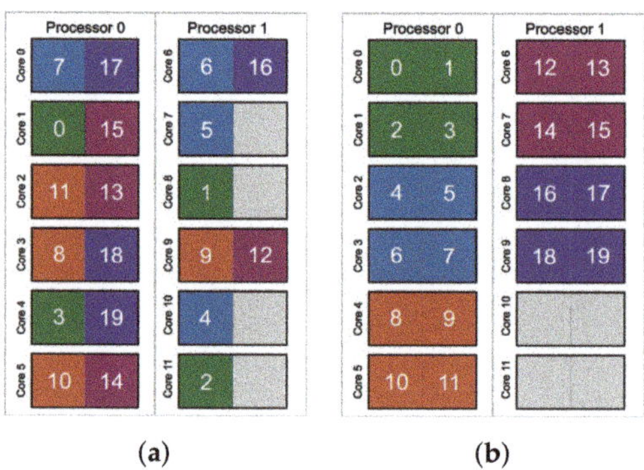

Figure 12. Optimal thread placement. (**a**) Example of thread group placement without control. (**b**) Example of thread group placement with affinity control.

Table 7 shows the parallel speed-up when more than 12 processes are used, i.e., using hyperthreading for the highest computational cost functions. Results manifested in Table 7 have been obtained using 16 and 20 processes by varying the number of groups ($NoCs$) and consequently varying the number of processes per group ($inCs$). Important conclusions can be drawn by analyzing the results of this table: remarkable scalability is obtained through the two-level parallel algorithm, even using logical cores (hyperthreading); although the parallel performance allows setting the $NoCs$ value (i.e., number of groups) according to the desired size of the subpopulations, i.e., according to the optimization performance rather than parallel behavior. All efficiency values are above 72%, except for the Foxholes function (f_{13}), characterized by having only two design variables (see Table 1), which penalizes fine-grained parallelism. Although both fine-grained parallelism and hyperthreading slightly penalize parallel efficiency, a remarkable average greater than 75% parallel efficiency is obtained. The average efficiency barely decreases as the number of processes increases from 16 to 20, resulting in a slight fall of the average efficiency from 75.6% to 74.9%, i.e., the outstanding parallel scalability is maintained.

This outstanding behavior is confirmed by the results shown in Table 8, which are the results conducted on all the available threads (24) when hyperthreading is activated. It is found that the two-level parallel algorithm has remarkable parallel scalability with an average parallel efficiency of 74.4%.

Table 7. Parallel speed-up for the two-level parallel algorithm using groups of processes. Population size = 240.

	16 Processes			20 Processes			
$NoCs; inCs$	8;2	4;4	2;8	10;2	5;4	4;5	2;10
f_1	12.5	12.5	12.1	15.9	15.0	15.2	15.2
f_2	12.5	12.1	11.6	14.4	14.8	15.0	14.7
f_{10}	11.9	11.8	11.8	14.7	14.7	14.7	14.7
f_{13}	10.1	10.0	9.7	12.7	12.4	12.3	11.7
f_{30}	12.2	12.1	12.2	15.3	15.2	15.1	15.1

Table 8. Parallel speed-up for the two-level parallel algorithm using groups of processes. Population size = 240. Number of processes = 24.

	24 Processes			
$NoCs; inCs$	12;2	6;4	4;6	2;12
f_1	19.0	18.3	18.7	17.4
f_2	18.0	17.4	16.9	18.0
f_{10}	17.6	17.6	17.4	17.4
f_{30}	18.2	18.0	18.1	17.8

Tables 9 and 10 show the number of functions evaluations required by the data sharing parallel algorithm to obtain an error of less than 1×10^{-3} (1×10^2 for functions marked with an asterisk), when the total population size is 240 ($popInitSize = 240$) and 60 ($popInitSize = 60$), respectively. These results show that the number of concurrent processes does not modify the optimization behavior. The heuristic nature of the proposed optimization algorithm results in different evaluations for the same function depending on the concurrent processes.

Tables 11 and 12 listed the number of functions evaluations required by the asynchronous parallel algorithm for population sizes 240 ($popInitSize = 240$) and 60 ($popInitSize = 60$), respectively. It is clear that, unlike the sharing data-parallel algorithm, the ratio of convergence depends on the number of concurrent processes used for the asynchronous parallel algorithm. In addition, the convergence ratio slightly worsens as the number of concurrent processes increases, but the outstanding parallel scalability offsets this behavior. Note that this behavior depends on the subpopulation sizes, which depend on the population size.

Table 9. Sharing data parallel algorithm: number of function evaluations for error $<1 \times 10^{-3}$ ($* < 1 \times 10^{2}$). $popInitSize = 240$.

	NoCs			
	1	**2**	**6**	**12**
f_1	75,384	80,657	83,776	76,385
f_2	73,464	70,135	73,034	60,717
f_3	2136	2120	2128	2200
f_4	4152	4889	4005	3507
f_5	840	842	687	312
f_6	9,627,227	9,966,401	9,430,103	9,876,351
f_7 *	960	762	722	583
f_8 *	317,376	374,307	255,284	324,357
f_9	16,848	16,516	17,853	17,829
f_{10}	739,296	854,471	780,928	743,569
f_{11} *	78,720	65,643	75,129	76,902
f_{12}	6,186,240	7,359,497	8,535,793	5,457,901
f_{13}	14,088	9603	7294	11,392
f_{14}	1920	2042	3831	2259
f_{15}	2352	2144	2453	1722
f_{16}	3120	3342	3328	4471
f_{17}	3888	3517	3275	2470
f_{18} *	480	456	453	373
f_{19}	2448	2259	2192	1990
f_{20}	2112	2262	2031	1892
f_{21}	2040	1732	1601	974
f_{22}	3,966,264	3,298,233	5,086,208	7,588,192
f_{23}	57,090	3134	4069	3396
f_{24}	30,408	28,982	30,281	30,248
f_{25} *	46,032	56,975	35,157	41,468
f_{26}	9672	15,605	13,573	10,713
f_{27}	6,840,528	10,618,500	3,333,940	8,731,516
f_{28}	480	440	462	164
f_{29} *	24,168	30,604	25,408	26,676
f_{30} *	9,787,467	8,810,661	8,232,263	10,546,564

Table 10. Sharing data parallel algorithm: number of function evaluations for error $<1 \times 10^{-3}$ ($* < 1 \times 10^{2}$). *popInitSize* = 60.

	NoCs			
	1	**2**	**6**	**12**
f_1	28,074	32,624	28,419	28,128
f_2	24,888	24,323	25,030	22,209
f_3	2082	2361	1867	1670
f_4	2088	2319	1762	2220
f_5	312	314	250	173
f_6	2,654,280	1,896,252	2,914,210	1,951,487
f_7 *	354	325	430	262
f_8 *	196,337	279,480	347,191	238,579
f_9	6420	7143	6844	7113
f_{10}	311,640	262,261	308,376	310,209
f_{11} *	32,424	28,081	28,738	32,903
f_{12}	2,624,640	2,353,703	2,680,174	2,202,838
f_{13}	36,126	18,554	6345	34,818
f_{14}	1998	1944	1917	2689
f_{15}	762	820	753	520
f_{16}	2400	2503	3393	2781
f_{17}	2910	1271	2745	2865
f_{18} *	120	114	105	63
f_{19}	882	762	879	604
f_{20}	834	807	812	629
f_{21}	912	691	743	543
f_{22}	1,907,900	1,110,490	1,874,520	2,209,849
f_{23}	27,760	3956	2633	3782
f_{24}	8304	9840	10,120	9041
f_{25} *	17,022	21,353	28,550	17,609
f_{26}	2208	2626	9685	1920
f_{27}	2,930,112	2,842,249	2,925,193	2,806,709
f_{28}	120	113	113	37
f_{29} *	26,940	12,415	17,103	17,228
f_{30} *	2,939,910	2,650,149	2,863,317	2,815,149

Table 11. Asynchronous parallel algorithm: number of function evaluations for error $<1 \times 10^{-3}$ ($* < 1 \times 10^{2}$). *popInitSize* $= 240$.

	NoCs			
	1	**2**	**6**	**12**
f_1	80,136	83,277	90,626	84,218
f_2	73,824	72,792	73,927	74,209
f_3	2568	2578	2989	3313
f_4	3264	5795	5436	5963
f_5	816	633	750	410
f_6	8,974,650	10,097,595	10,025,662	11,314,284
f_7 *	1032	759	897	481
f_8 *	253,920	34,7465	61,2221	86,7391
f_9	17,184	16,644	23,193	25,164
f_{10}	71,4336	85,8736	1,078,127	1,252,559
f_{11} *	64,656	77,582	88,175	10,3109
f_{12}	8,937,680	7,699,045	10,565,357	11,289,390
f_{13}	46,872	10,347	17,452	20,750
f_{14}	1992	3554	4702	5277
f_{15}	2256	2277	2656	1922
f_{16}	4896	4869	6881	6861
f_{17}	3264	3640	4408	5952
f_{18} *	480	411	413	306
f_{19}	2256	2204	2452	2353
f_{20}	2304	2441	2661	1826
f_{21}	1896	1590	2347	2041
f_{22}	4,256,610	7,094,932	7,742,422	6,726,769
f_{23}	10,1640	49,884	14,750	30,406
f_{24}	31,152	32,214	32,039	33,949
f_{25} *	37,752	47,883	46,427	41,350
f_{26}	8592	3266	2386	3399
f_{27}	8,689,680	9,215,379	10,203,567	10,787,817
f_{28}	528	456	444	199
f_{29} *	21,624	29,572	58,600	46,029
f_{30} *	10,729,470	10,103,632	10,519,594	9,802,382

As earlier recorded, the parallel asynchronous algorithm allows each thread to have its population size without sacrificing parallel performance and thus exploring populations of different characteristics, which could improve the optimization's performance. Table 13 compares the number of function evaluations (# FEs) for functions f_6, f_{22}, and f_{27} when using homogeneous and heterogeneous subpopulation sizes. The latter improving the optimization performance. Moreover, not reaching a good solution due to small populations can be avoided by increasing the number of processes. For instance, 12 processes are used for f_6 and f_{27} (see Table 12).

Table 12. Asynchronous parallel algorithm: number of function evaluations for error $<1 \times 10^{-3}$ ($* < 1 \times 10^{2}$). $popInitSize = 60$.

	NoCs			
	1	**2**	**6**	**12**
f_1	28,308	25,754	37,605	33,407
f_2	24,684	25,346	26,985	26,891
f_3	1644	1163	1876	3188
f_4	2778	2239	4294	4816
f_5	378	307	228	308
f_6	2,742,830	2,918,138	2,936,681	
f_7 *	402	415	376	156
f_8 *	216,387	314,711	495,643	602,778
f_9	6822	6827	9873	11,370
f_{10}	314,562	316,765	415,446	413,730
f_{11} *	28,338	26,143	29,280	35,931
f_{12}	2,444,835	2,056,447	2,802,878	2,993,886
f_{13}	52,848	35,565	30,404	63,877
f_{14}	1680	2625	7972	5786
f_{15}	732	926	851	583
f_{16}	2736	6429	4814	8359
f_{17}	2790	3023	5588	9495
f_{18} *	120	101	105	46
f_{19}	630	846	877	563
f_{20}	792	736	878	732
f_{21}	858	914	930	1437
f_{22}	1,089,000	1,850,238	2,757,559	
f_{23}	1980	11,291	20,030	21,210
f_{24}	8940	9557	8676	11,540
f_{25} *	17,652	21,631	17,447	17,945
f_{26}	3342	1151	2377	3869
f_{27}	2,918,580	2,865,679	2,970,869	2,779,579
f_{28}	120	116	105	30
f_{29} *	23,586	22,921	42,961	59,068
f_{30} *	2,782,130	2,885,935	2,631,535	2,801,275

It is settled that the proposed parallel algorithms achieve a remarkable parallel performance without disordering the optimization behavior. Figures 13 and 14 point the significant improvement in the convergence speed of the proposed ESCA algorithm compared to the SCA algorithm.

Table 13. Asynchronous parallel algorithm: number of function evaluations for error $<1 \times 10^{-3}$, 6 processes and homogeneous and heterogeneous subpopulation sizes. *popInitSize* $= 240$.

	Thread Id.						
	0	**1**	**2**	**3**	**4**	**5**	
	Subpopulation Sizes						# FEs
f_6	40	40	40	40	40	40	10,025,662
	80	60	40	30	20	10	8,365,248
f_{22}	40	40	40	40	40	40	7,742,422
	80	60	40	30	20	10	6,341,866
f_{27}	40	40	40	40	40	40	10,203,567
	80	60	40	30	20	10	9,941,450

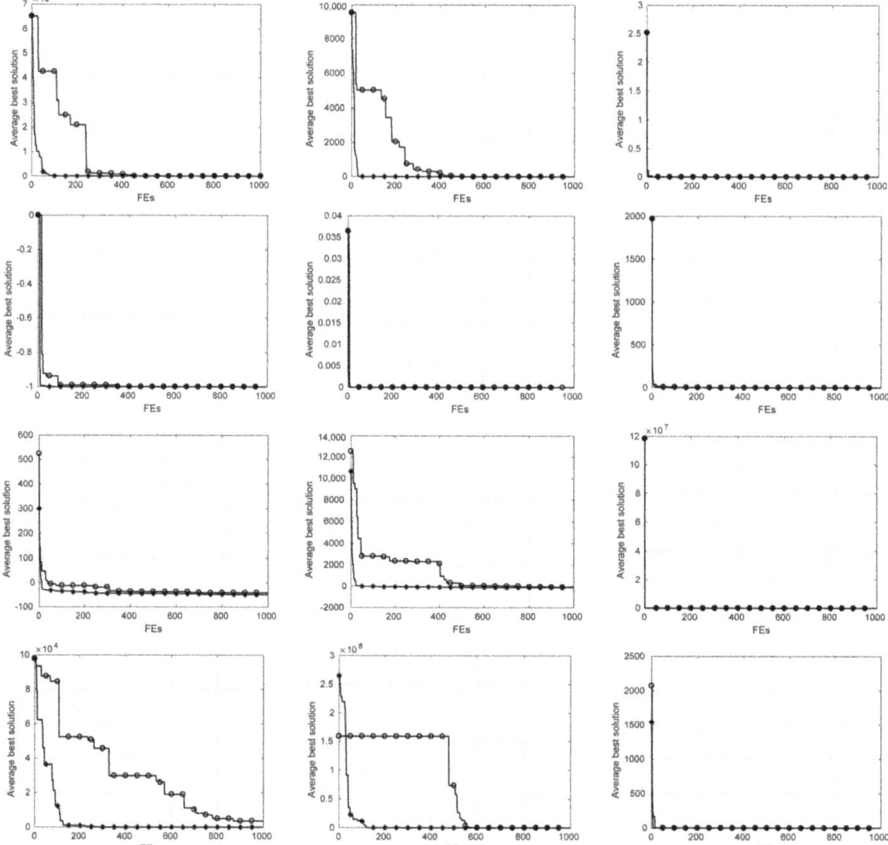

Figure 13. *Cont.*

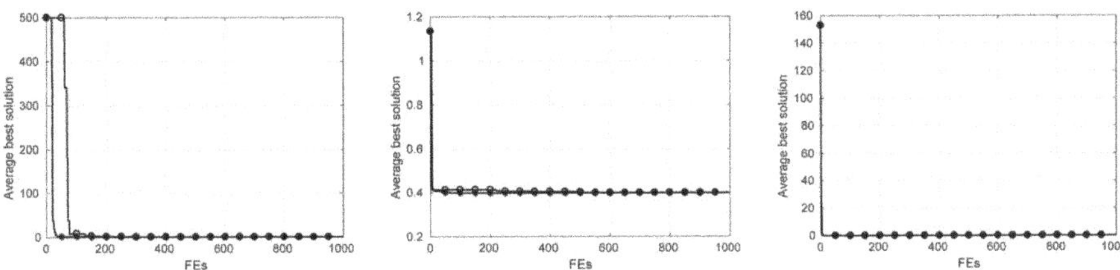

Figure 13. Convergence curves for the benchmark functions $f_1 - f_{15}$ in row-major order. Optimization algorithms are SCA [○], ESCA [∗].

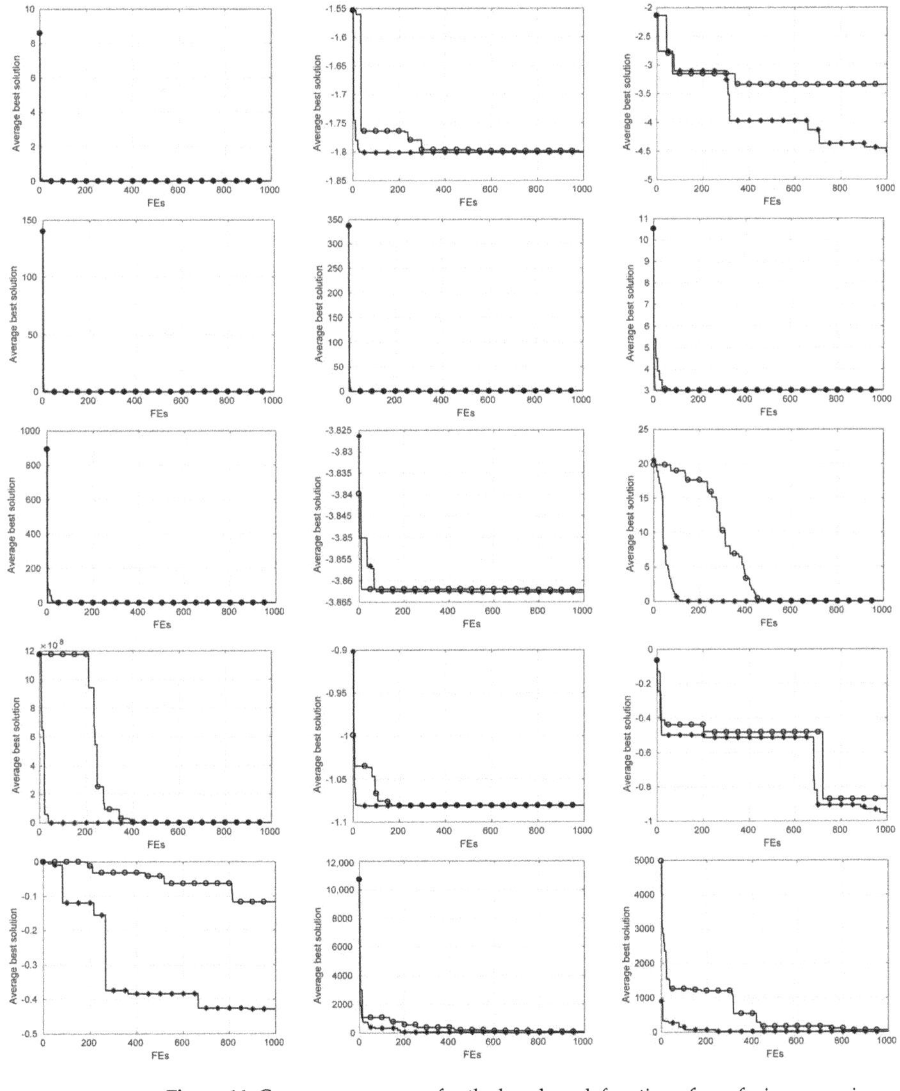

Figure 14. Convergence curves for the benchmark functions $f_{16} - f_{30}$ in row-major order. Optimization algorithms are SCA [○], ESCA [∗].

The last analysis discusses the optimization's behavior when solving the engineering design problems described in Section 4.2. Table 14 compares the convergence ratio of the SCA and ESCA methods when only 10,000 and 20,000 generations are processed. As can be observed from this table, the ESCA outperforms the SCA algorithm in terms of convergence ratio. Similar results are obtained when optimizing the 30 benchmark functions. This behavior confirms that our proposal significantly boosts the SCA algorithm.

Table 14. Convergence ratio for ESCA and SCA algorithms with different population sizes.

	Population Size		
	60	**120**	**240**
	Pressure Vessel Problem		
ESCA-10000	6060.2070	6060.7420	6059.9340
SCA-10000	6079.0610	6091.4340	6068.5540
ESCA-20000	6060.0950	6059.8290	6059.8000
SCA-20000	6065.7460	6066.9530	6069.2260
	Welded beam problem		
ESCA-10000	1.728844	1.726625	1.726300
SCA-10000	1.748143	1.749394	1.747236
ESCA-20000	1.726585	1.726704	1.725514
SCA-20000	1.751480	1.747207	1.738482
	Rolling element bearing problem		
ESCA-10000	81,706.17	81,798.38	81,832.05
SCA-10000	80,673.58	81,333.65	80,318.50
ESCA-20000	81,803.87	81,774.60	81,836.77
SCA-20000	80,224.49	80,335.60	81,086.44

As for solution accuracy, the results on benchmark functions and challenging engineering problems are listed in Table 15. These results are acquired from 30 independent runs on each function, 10,000 iterations, and three population sizes, i.e., 60, 120, and 240. As can be observed from this table, the ESCA algorithm performs better than SCA in almost all functions. These outcomes are statistically compared in Table 16. Indeed, to measure the overall performance of the ESCA algorithm respect to its original counterpart SCA, the non-parametric statistical tests of Friedman, Friedman aligned, and Quade test are employed. The Friedman test or Friedman rank test is a non-parametric test developed by Milton Friedman [48] consisting of arranging the data by blocks, replacing them by their respective order, considering the existence of identical data. Therefore, in the Friedman test the performance of the analyzed algorithms are ranked separately for each data set. This ranking scheme only allows comparisons between sets, since comparisons between sets are meaningless. When the number of algorithms to be compared is small, this can be a disadvantage, in this case inter-dataset comparison may be desirable and we can employ the Friedman aligned or Friedman aligned rank method [49]. The Quade or Quade rank test [50] is also a non-parametric test, which shows its robustness for small data sets. Regardless of the population size, the ESCA is ranked first under all tests.

Table 15. Average values for unconstrained and constrained problems obtained by ESCA and SCA.

	Population Size					
	60		120		240	
	SCA	ESCA	SCA	ESCA	SCA	ESCA
f_1	2.757179×10^{-64}	0.000000	1.712496×10^{-79}	0.000000	4.457065×10^{-94}	0.000000
f_2	8.616185×10^{-65}	0.000000	1.046044×10^{-80}	0.000000	1.112510×10^{-92}	0.000000
f_3	6.811942×10^{-6}	5.491076×10^{-9}	4.783583×10^{-6}	3.114413×10^{-9}	1.401307×10^{-6}	7.104723×10^{-10}
f_4	-9.999516×10^{-1}	-1.000000	-9.999736×10^{-1}	-1.000000	-9.999892×10^{-1}	-1.000000
f_5	0.000000	0.000000	0.000000	0.000000	0.000000	0.000000
f_6	9.900274×10^{-2}	8.185273×10^{-3}	9.765063×10^{-2}	2.929811×10^{-3}	6.404594×10^{-2}	2.334942×10^{-3}
f_7	-4.845251×10^{1}	-4.990339×10^{1}	-4.877389×10^{1}	-4.995156×10^{1}	-4.896195×10^{1}	-4.996917×10^{1}
f_8	-1.262160×10^{2}	-1.539732×10^{2}	-1.339290×10^{2}	-1.787089×10^{2}	-1.501428×10^{2}	-1.862011×10^{2}
f_9	$8.721680 \times 10^{-202}$	0.000000	$3.156447 \times 10^{-257}$	0.000000	$2.251127 \times 10^{-315}$	0.000000
f_{10}	8.175285×10^{-1}	0.000000	1.361425×10^{-3}	0.000000	2.200964×10^{-8}	0.000000
f_{11}	2.701419×10^{1}	2.643757×10^{1}	2.699064×10^{1}	2.614943×10^{1}	2.663097×10^{1}	2.585579×10^{1}
f_{12}	3.584155×10^{-1}	5.114134×10^{-1}	3.100522×10^{-1}	4.890716×10^{-1}	2.815470×10^{-1}	4.889729×10^{-1}
f_{13}	1.064141	1.196414	9.980039×10^{-1}	1.064141	9.980038×10^{-1}	9.980038×10^{-1}
f_{14}	3.979373×10^{-1}	3.978874×10^{-1}	3.979186×10^{-1}	3.978874×10^{-1}	3.979079×10^{-1}	3.978874×10^{-1}
f_{15}	0.000000	0.000000	0.000000	0.000000	0.000000	0.000000
f_{16}	2.880073×10^{-5}	3.813791×10^{-9}	1.142770×10^{-5}	7.944414×10^{-10}	6.238591×10^{-6}	1.858303×10^{-10}
f_{17}	-1.774460	-1.801303	-1.801248	-1.801303	-1.801272	-1.801303
f_{18}	-3.187932	-3.700737	-3.375650	-4.044260	-3.610325	-4.071782
f_{19}	0.000000	0.000000	0.000000	0.000000	0.000000	0.000000
f_{20}	0.000000	0.000000	0.000000	0.000000	0.000000	0.000000
f_{21}	3.000000	3.000000	3.000000	3.000000	3.000000	3.000000
f_{22}	3.214731×10^{-2}	6.673788×10^{-3}	1.718025×10^{-2}	3.599778×10^{-3}	1.316666×10^{-2}	2.510527×10^{-3}
f_{23}	-3.855633	-3.858840	-3.855658	-3.859628	-3.857749	-3.860941
f_{24}	4.588922×10^{-15}	3.996803×10^{-15}	4.233650×10^{-15}	3.878379×10^{-15}	4.115227×10^{-15}	3.996803×10^{-15}
f_{25}	1.888945	1.585867	1.787761	1.515388	1.698389	1.389630
f_{26}	-1.069455	-1.080938	-1.080930	-1.080938	-1.080936	-1.080938
f_{27}	-5.685987×10^{-1}	-6.957021×10^{-1}	-6.135416×10^{-1}	-8.465688×10^{-1}	-7.018316×10^{-1}	-8.571367×10^{-1}
f_{28}	-3.945496×10^{-2}	-1.893884×10^{-1}	-8.203238×10^{-2}	-2.315995×10^{-1}	-1.025653×10^{-1}	-2.631384×10^{-1}
f_{29}	3.258800×10^{1}	3.224237	1.912760×10^{1}	1.364237	1.720010×10^{1}	7.855821×10^{-1}
f_{30}	3.745441×10^{1}	2.581768	2.071528×10^{1}	1.359673	1.765572×10^{1}	9.444320×10^{-1}
Vessel	6.213857×10^{3}	6.097895×10^{3}	6.176765×10^{3}	6.067191×10^{3}	6.150466×10^{3}	6.062122×10^{3}
Beam	1.792532	1.733833	1.783172	1.731625	1.770235	1.729274
Bearing	7.303758×10^{4}	8.116530×10^{4}	7.449770×10^{4}	8.147987×10^{4}	7.689757×10^{4}	8.162418×10^{4}

Table 16. Comparison of solution accuracy for ESCA and SCA algorithms. The average ranking results by Friedman, Friedman aligned, and Quade tests.

	Population Size								
	60			120			240		
	Ranking								
	Friedman	F. aligned	Quad	Friedman	F. aligned	Quad	Friedman	F. aligned	Quad
ESCA	1.1970	22.7121	1.1738	1.2273	23.3485	1.1934	1.2273	22.8333	1.1783
SCA	1.8030	44.2879	1.8262	1.7727	43.6515	1.8066	1.7727	44.1667	1.8217

5.2. Further Comparison with Numerous State-of-the-Art Algorithms

In this section, we compare the sequential version of the ESCA algorithms to several well-known algorithms. Firstly, the comparison algorithms are benchmarked on a set of 30 unconstrained problems. Then, we test these algorithms in solving three challenging engineering problems with constrained and unknown search spaces.

5.2.1. Benchmarking of the Comparison Algorithms

The ESCA algorithm is benchmarked on 30 unconstrained functions that are listed in Tables 1 and 2. The ESCA algorithm runs on each benchmark function 30 times. A comparison to grey wolf algorithm (GWO) [51], whale optimization algorithm (WOA) [52] and Harris hawk optimization algorithm (HHO) [53] is provided as well. To ensure a fair comparison, the individuals are replaced only if there is an improvement of the objective function over the course of iterations of each algorithm, i.e the selection operator used in ESCA was "rank selection" also used by GWO, WOA and HHO. Table 17, compares the convergence speed in terms of the number of functions evaluations (# FEs) required to obtain an error of less than 1×10^{-3} and (1×10^2 for functions marked with an asterisk), for a population size of 120. As can be observed from this table, the ESCA algorithm exhibits the lowest # FEs values for almost all functions. Accordingly, the ESCA algorithm can early converge to a feasible solution for almost all benchmark functions.

Table 17. Number of function evaluations for error $<1 \times 10^{-3}$ ($* <1 \times 10^2$).

	ESCA	GWO	HHO	WOA
f_1	14,502	6920	2635	7567
f_2	12,399	6261	2024	5877
f_3	1051	1461	535	829
f_4	1582	6352	3020	2123
f_5	282	400	307	346
f_6	1,121,028	1,019,746	1,404,298	1,173,481
f_7 *	307	281	214	268
f_8 *	17,545	3946	1329	1165
f_9	7543	3501	2219	149,835
f_{10}	77,362	22,158	6625	1,058,019
f_{11} *	10,727	4732	1054	4077
f_{12}	853,399	1,088,983	11,311	7280
f_{13}	204,227	563,262	46,084	22,772
f_{14}	2043	7718	4079	2882
f_{15}	856	1012	1286	1701

Table 17. *Cont.*

	ESCA	GWO	HHO	WOA
f_{16}	1330	2758	5191	6688
f_{17}	1142	28,240	3073	1279
f_{18}	799,483	1,198,708	1,385,561	1,015,650
f_{19}	880	1046	1227	3460
f_{20}	866	1029	1606	8337
f_{21}	1214	1961	1597	1634
f_{22}	1,058,023	1,142,825	1,481,119	228
f_{23}	1415	96,594	20,375	83,281
f_{24}	15,322	8510	4928	11,575
f_{25} *	7537	2275	674	1708
f_{26}	10,031	11,503	421,650	321,426
f_{27}	822,131	1,128,604	583,872	927,182
f_{28}	962,612	968,637	1,171,682	942,952
f_{29} *	6955	11,138	131,846	33,285
f_{30} *	7654	55,073	158,831	59,033

The statistical data (best cost function, and corresponding average, worst, and standard deviation) are summarized in Table 18. These results are derived from 30 independent runs on each function, a population size of 120 individuals, and 10,000 iterations. It can be seen from this table that the ESCA algorithm holds a competitive performance in terms of solution accuracy as opposed to the comparison algorithms.

Table 18. Statistical data for 30 runs with a population of 120 and 10,000 iterations for f_1 to f_{30}.

		ESCA	GWO	HHO	WOA
f_1	Best	0.000000	0.000000	0.000000	0.000000
	Avg.	0.000000	0.000000	0.000000	0.000000
	Worst	0.000000	0.000000	0.000000	0.000000
	SD	0.000000	0.000000	0.000000	0.000000
f_2	Best	0.000000	0.000000	0.000000	0.000000
	Avg.	0.000000	0.000000	0.000000	0.000000
	Worst	0.000000	0.000000	0.000000	0.000000
	SD	0.000000	0.000000	0.000000	0.000000
f_3	Best	3.262152×10^{-18}	5.547644×10^{-13}	0.000000	1.203238×10^{-19}
	Avg.	8.895248×10^{-11}	1.170037×10^{-10}	0.000000	6.586553×10^{-16}
	Worst	6.298503×10^{-10}	3.953307×10^{-10}	0.000000	1.233480×10^{-14}
	SD	1.412547×10^{-10}	9.315382×10^{-11}	0.000000	2.205548×10^{-15}
f_4	Best	-1.000000	-1.000000	-1.000000	-1.000000
	Avg.	-1.000000	-1.000000	-1.000000	-1.000000
	Worst	-1.000000	-1.000000	-1.000000	-1.000000
	SD	6.943355×10^{-13}	4.337546×10^{-10}	8.599751×10^{-17}	9.634141×10^{-13}
f_5	Best	0.000000	0.000000	0.000000	0.000000
	Avg.	0.000000	0.000000	0.000000	0.000000
	Worst	0.000000	0.000000	0.000000	0.000000
	SD	0.000000	0.000000	0.000000	0.000000

Table 18. *Cont.*

		ESCA	GWO	HHO	WOA
f_6	Best	1.807347×10^{-6}	4.686073×10^{-8}	4.023087×10^{-5}	8.462644×10^{-4}
	Avg.	6.913877×10^{-4}	4.435400×10^{-2}	3.457026×10^{-3}	1.179321×10^{-2}
	Worst	2.241546×10^{-3}	1.330605	6.863861×10^{-3}	2.110537×10^{-2}
	SD	6.032768×10^{-4}	2.388509×10^{-1}	1.911670×10^{-3}	5.070219×10^{-3}
f_7	Best	-5.000000×10^{1}	-5.000000×10^{1}	-5.000000×10^{1}	-5.000000×10^{1}
	Avg.	-4.999999×10^{1}	-5.000000×10^{1}	-5.000000×10^{1}	-5.000000×10^{1}
	Worst	-4.999997×10^{1}	-5.000000×10^{1}	-5.000000×10^{1}	-5.000000×10^{1}
	SD	7.486059×10^{-6}	9.662948×10^{-8}	5.492594×10^{-11}	2.403252×10^{-10}
f_8	Best	-2.099980×10^{2}	-2.100000×10^{2}	-2.100000×10^{2}	-2.100000×10^{2}
	Avg.	-2.099872×10^{2}	-2.063305×10^{2}	-2.100000×10^{2}	-2.100000×10^{2}
	Worst	-2.099745×10^{2}	-1.549028×10^{2}	-2.100000×10^{2}	-2.100000×10^{2}
	SD	7.246266×10^{-3}	1.372988×10^{1}	5.265073×10^{-8}	2.197536×10^{-7}
f_9	Best	0.000000	0.000000	0.000000	$5.909506 \times 10^{-178}$
	Avg.	0.000000	0.000000	0.000000	4.294324×10^{-82}
	Worst	0.000000	0.000000	0.000000	6.977677×10^{-81}
	SD	0.000000	0.000000	0.000000	1.580556×10^{-81}
f_{10}	Best	0.000000	$2.470328 \times 10^{-323}$	0.000000	3.725891×10^{-8}
	Avg.	0.000000	$7.905050 \times 10^{-323}$	0.000000	1.000874×10^{-2}
	Worst	0.000000	$1.729230 \times 10^{-322}$	0.000000	2.032146×10^{-1}
	SD	0.000000	0.000000	0.000000	3.734209×10^{-2}
f_{11}	Best	2.481895×10^{1}	2.522460×10^{1}	2.489752×10^{1}	2.486321×10^{1}
	Avg.	4.935104×10^{3}	2.685818×10^{1}	4.932600×10^{3}	2.612374×10^{4}
	Worst	1.003584×10^{4}	2.889938×10^{1}	1.002894×10^{4}	9.002408×10^{4}
	SD	4.931226×10^{3}	7.683004×10^{-1}	4.928556×10^{3}	3.000263×10^{4}
f_{12}	Best	1.019230×10^{-8}	4.395919×10^{-9}	4.827285×10^{-17}	6.442491×10^{-13}
	Avg.	3.333334×10^{-1}	4.000000×10^{-1}	2.551869×10^{-12}	3.430491×10^{-10}
	Worst	6.666667×10^{-1}	6.666667×10^{-1}	2.321049×10^{-11}	2.552220×10^{-9}
	SD	3.333332×10^{-1}	3.265986×10^{-1}	5.095444×10^{-12}	7.470906×10^{-10}
f_{13}	Best	9.980038×10^{-1}	9.980038×10^{-1}	9.980038×10^{-1}	9.980038×10^{-1}
	Avg.	1.588057	1.923918	9.980038×10^{-1}	9.980038×10^{-1}
	Worst	1.076318×10^{1}	2.982105	9.980038×10^{-1}	9.980038×10^{-1}
	SD	1.831761	9.898436×10^{-1}	4.309420×10^{-16}	6.214605×10^{-16}
f_{14}	Best	3.978874×10^{-1}	3.978874×10^{-1}	3.978874×10^{-1}	3.978874×10^{-1}
	Avg.	3.978874×10^{-1}	3.978878×10^{-1}	3.978874×10^{-1}	3.978874×10^{-1}
	Worst	3.978874×10^{-1}	3.978987×10^{-1}	3.978874×10^{-1}	3.978874×10^{-1}
	SD	2.664066×10^{-10}	2.044411×10^{-6}	3.707297×10^{-15}	7.625589×10^{-12}
f_{15}	Best	0.000000	0.000000	0.000000	0.000000
	Avg.	0.000000	0.000000	0.000000	0.000000
	Worst	0.000000	0.000000	0.000000	0.000000
	SD	0.000000	0.000000	0.000000	0.000000
f_{16}	Best	7.032691×10^{-16}	4.176314×10^{-12}	1.053336×10^{-17}	1.518097×10^{-10}
	Avg.	8.501715×10^{-11}	2.991300×10^{-10}	9.229996×10^{-16}	1.061443×10^{-9}
	Worst	6.188203×10^{-10}	1.038909×10^{-9}	1.010500×10^{-14}	4.095840×10^{-9}
	SD	1.224357×10^{-10}	2.805821×10^{-10}	2.016184×10^{-15}	7.755945×10^{-10}
f_{17}	Best	-1.801303	-1.801303	-1.801303	-1.801303
	Avg.	-1.801303	-1.801303	-1.801303	-1.801303
	Worst	-1.801303	-1.801303	-1.801303	-1.801303
	SD	3.984603×10^{-12}	3.073081×10^{-9}	1.314259×10^{-15}	1.115984×10^{-12}

Table 18. *Cont.*

		ESCA	GWO	HHO	WOA
f_{18}	Best	-4.687657	-4.687658	-4.687658	-4.687658
	Avg.	-4.687651	-4.567539	-4.599323	-4.359473
	Worst	-4.687640	-3.749195	-4.332021	-3.573593
	SD	3.945663×10^{-6}	1.662246×10^{-1}	7.870435×10^{-2}	3.986633×10^{-1}
f_{19}	Best	0.000000	0.000000	0.000000	0.000000
	Avg.	0.000000	0.000000	0.000000	0.000000
	Worst	0.000000	0.000000	0.000000	0.000000
	SD	0.000000	0.000000	0.000000	0.000000
f_{20}	Best	0.000000	0.000000	0.000000	0.000000
	Avg.	0.000000	0.000000	0.000000	0.000000
	Worst	0.000000	0.000000	0.000000	0.000000
	SD	0.000000	0.000000	0.000000	0.000000
f_{21}	Best	3.000000	3.000000	3.000000	3.000000
	Avg.	3.000000	3.000000	3.000000	3.000000
	Worst	3.000000	3.000000	3.000000	3.000000
	SD	3.827852×10^{-13}	5.764236×10^{-9}	1.924979×10^{-14}	9.407358×10^{-11}
f_{22}	Best	2.100529×10^{-5}	6.233180×10^{-7}	2.762363×10^{-4}	2.471215×10^{-3}
	Avg.	1.123810×10^{-3}	1.300996×10^{-1}	6.401639×10^{-3}	6.126825×10^{-2}
	Worst	2.974472×10^{-3}	1.035930	3.782117×10^{-2}	3.768746×10^{-1}
	SD	1.050413×10^{-3}	3.277276×10^{-1}	9.226831×10^{-3}	7.362338×10^{-2}
f_{23}	Best	-3.862780	-3.862780	-3.862780	-3.862780
	Avg.	-3.862780	-3.862255	-3.862780	-3.862254
	Worst	-3.862780	-3.854902	-3.862780	-3.854902
	SD	1.061189×10^{-10}	1.965115×10^{-3}	5.382464×10^{-15}	1.965074×10^{-3}
f_{24}	Best	3.996803×10^{-15}	3.996803×10^{-15}	4.440892×10^{-16}	4.440892×10^{-16}
	Avg.	3.996803×10^{-15}	7.312669×10^{-15}	4.440892×10^{-16}	2.575717×10^{-15}
	Worst	3.996803×10^{-15}	7.549517×10^{-15}	4.440892×10^{-16}	7.549517×10^{-15}
	SD	0.000000	8.862025×10^{-16}	0.000000	1.967404×10^{-15}
f_{25}	Best	1.099003×10^{-3}	4.167573×10^{-8}	2.110681×10^{-7}	1.097794×10^{-7}
	Avg.	9.811139×10^{-2}	9.347600×10^{-2}	4.397173×10^{-3}	3.273148×10^{-7}
	Worst	3.014981×10^{-1}	3.999622×10^{-1}	1.098999×10^{-2}	1.083257×10^{-6}
	SD	1.046370×10^{-1}	9.982078×10^{-2}	5.381619×10^{-3}	2.209943×10^{-7}
f_{26}	Best	-1.080938	-1.080938	-1.080938	-1.080938
	Avg.	-1.080938	-1.080938	-1.075192	-1.075192
	Worst	-1.080938	-1.080938	-1.056311	-1.056311
	SD	1.216749×10^{-10}	4.717320×10^{-10}	1.041639×10^{-2}	1.041639×10^{-2}
f_{27}	Best	-9.649998×10^{-1}	-9.649999×10^{-1}	-9.649999×10^{-1}	-9.649999×10^{-1}
	Avg.	-9.426906×10^{-1}	-9.350842×10^{-1}	-9.355537×10^{-1}	-7.696397×10^{-1}
	Worst	-9.079998×10^{-1}	-7.367849×10^{-1}	-7.035660×10^{-1}	-4.828707×10^{-1}
	SD	2.065763×10^{-2}	4.201816×10^{-2}	6.553091×10^{-2}	1.953920×10^{-1}
f_{28}	Best	-9.649623×10^{-1}	-9.649673×10^{-1}	-5.170000×10^{-1}	-9.079987×10^{-1}
	Avg.	-5.700238×10^{-1}	-4.854299×10^{-1}	-3.504035×10^{-1}	-3.186518×10^{-1}
	Worst	-5.317959×10^{-2}	-5.317959×10^{-2}	-5.317959×10^{-2}	-2.813614×10^{-2}
	SD	2.867891×10^{-1}	2.743351×10^{-1}	1.736198×10^{-1}	2.090066×10^{-1}
f_{29}	Best	2.498726×10^{-4}	1.093726×10^{-5}	9.178611×10^{-13}	4.883815×10^{-8}
	Avg.	5.554048×10^{-3}	1.419868×10^{-1}	2.459967×10^{1}	1.769584×10^{-1}
	Worst	5.564318×10^{-2}	3.434501	3.684844×10^{2}	3.925457
	SD	9.949693×10^{-3}	6.288349×10^{-1}	9.190723×10^{1}	7.128277×10^{-1}
f_{30}	Best	1.696582×10^{-4}	9.049588×10^{-6}	5.440372×10^{-11}	4.601300×10^{-8}
	Avg.	4.315053×10^{-3}	1.263696×10^{1}	3.685149×10^{1}	2.656585×10^{1}
	Worst	2.657023×10^{-2}	3.684844×10^{2}	3.684844×10^{2}	7.966935×10^{2}
	SD	5.023689×10^{-3}	6.609445×10^{1}	1.105443×10^{2}	1.430091×10^{2}

Inferential statistics prove how well a sample of data sustains a particular hypothesis and whether the outcomes can be generalized for other data samples. To evaluate the overall performance of the ESCA algorithm and determine the significance of data in Table 17 (average) and Table 18, non-parametric statistical tests dubbed Friedman, Friedman aligned, and Quade test are employed [54]. Tables 19 and 20 statistically compare the assessed algorithms in terms of convergence speed and solution accuracy, respectively. Tables 21 and 22 estimate the contrast between medians of data in Table 17 (average) and Table 18, respectively, while considering all pairwise comparisons [54]. As can be observed from Table 19, the ESCA algorithm is ranked first under all statistical tests in terms of convergence speed. Similar results are obtained in Table 21 in which the ESCA algorithm always obtain a positive difference value with respect to the comparison algorithms. That is, the ESCA algorithm performs better than others. As for the solution accuracy, the ESCA and HHO algorithms are ranked first with a competitive performance, as shown in Table 20. However, according to the outcomes in Table 22, the proposed algorithm is slightly better than the HHO algorithm. Unlike this latter, the ESCA algorithm always has a positive contrast compared to the other tested algorithms.

The effectiveness of the proposed ESCA algorithm in solving high-dimensional problems is validated in Table 23. The outcomes show that the proposed algorithm exhibits promising and competitive performance compared to the state-of-the-art algorithms.

Table 19. Comparison of convergence speed for the assessed algorithms. The average ranking outcomes through Friedman, Friedman aligned, and Quade tests.

Algorithm	Ranking		
	Friedman	Friedman Aligned	Quade
ESCA	2.1667	54.4667	2.2000
GWO	2.9000	65.8000	2.8387
HHO	2.3667	60.7667	2.5376
WOA	2.5667	60.9667	2.4237

Table 20. Comparison of solution accuracy for the assessed algorithms. The average ranking outcomes through Friedman, Friedman aligned, and Quade tests.

Algorithm	Ranking		
	Friedman	Friedman Aligned	Quade
ESCA	2.2000	53.7000	2.0613
GWO	2.8333	66.2000	2.8828
HHO	2.2000	53.0000	2.2065
WOA	2.7667	69.1000	2.8495

Table 21. Comparison of convergence speed for the assessed algorithms. Contrast Estimation based on medians.

	ESCA	GWO	HHO	WOA
ESCA	0	865.5	159.6	398.9
GWO	−865.5	0	−705.9	−466.6
HHO	−159.6	705.9	0	239.3
WOA	−398.9	466.6	−239.3	0

Table 22. Comparison of solution accuracy for the assessed algorithms. Contrast Estimation based on medians.

	ESCA	GWO	HHO	WOA
ESCA	0	8.290×10^{-16}	4.145×10^{-16}	4.145×10^{-16}
GWO	-8.290×10^{-16}	0	-4.145×10^{-16}	-4.145×10^{-16}
HHO	-4.145×10^{-16}	4.145×10^{-16}	0	0
WOA	-4.145×10^{-16}	4.145×10^{-16}	0	0

Table 23. Statistical data for 30 runs with a population of 120 and 10,000 iterations for high-dimensional functions.

	# N. var.		ESCA	GWO	HHO	WOA
f_1	100	Best	0.000000	0.000000	0.000000	0.000000
		Avg.	0.000000	0.000000	0.000000	0.000000
		Worst	0.000000	0.000000	0.000000	0.000000
		SD	0.000000	0.000000	0.000000	0.000000
	300	Best	0.000000	$1.472678 \times 10^{-269}$	0.000000	0.000000
		Avg.	0.000000	$1.195492 \times 10^{-267}$	0.000000	0.000000
		Worst	0.000000	$9.890031 \times 10^{-267}$	0.000000	0.000000
		SD	0.000000	0.000000	0.000000	0.000000
	500	Best	0.000000	$6.492796 \times 10^{-216}$	0.000000	0.000000
		Avg.	0.000000	$2.937464 \times 10^{-214}$	0.000000	0.000000
		Worst	0.000000	$5.611286 \times 10^{-213}$	0.000000	0.000000
		SD	0.000000	0.000000	0.000000	0.000000
f_2	100	Best	0.000000	0.000000	0.000000	0.000000
		Avg.	0.000000	0.000000	0.000000	0.000000
		Worst	0.000000	0.000000	0.000000	0.000000
		SD	0.000000	0.000000	0.000000	0.000000
	300	Best	0.000000	$2.664037 \times 10^{-269}$	0.000000	0.000000
		Avg.	0.000000	$1.078682 \times 10^{-267}$	0.000000	0.000000
		Worst	0.000000	$1.117063 \times 10^{-266}$	0.000000	0.000000
		SD	0.000000	0.000000	0.000000	0.000000
	500	Best	0.000000	$4.427418 \times 10^{-216}$	0.000000	0.000000
		Avg.	0.000000	$3.940111 \times 10^{-214}$	0.000000	0.000000
		Worst	0.000000	$2.032876 \times 10^{-213}$	0.000000	0.000000
		SD	0.000000	0.000000	0.000000	0.000000
f_{10}	100	Best	$8.324341 \times 10^{-149}$	$1.977371 \times 10^{-107}$	0.000000	2.360440×10^{2}
		Avg.	$7.974122 \times 10^{-116}$	5.051906×10^{-92}	0.000000	7.941213×10^{3}
		Worst	$2.391764 \times 10^{-114}$	1.513262×10^{-90}	0.000000	3.263596×10^{4}
		SD	$4.293318 \times 10^{-115}$	2.716247×10^{-91}	0.000000	7.423773×10^{3}
	300	Best	1.527654×10^{-82}	9.975385×10^{-35}	0.000000	5.113928×10^{5}
		Avg.	1.566300×10^{-55}	4.698595×10^{-7}	0.000000	2.324814×10^{6}
		Worst	4.212831×10^{-54}	1.409572×10^{-5}	0.000000	3.178554×10^{6}
		SD	7.582484×10^{-55}	2.530258×10^{-6}	0.000000	5.939182×10^{5}
	500	Best	1.617417×10^{-70}	4.140184×10^{-14}	0.000000	8.236708×10^{6}
		Avg.	2.137194×10^{-27}	1.223006×10^{-2}	0.000000	1.223004×10^{7}
		Worst	6.411583×10^{-26}	3.383713×10^{-1}	0.000000	1.470168×10^{7}
		SD	1.150914×10^{-26}	6.061667×10^{-2}	0.000000	1.446876×10^{6}

Table 23. *Cont.*

	# N. var.		ESCA	GWO	HHO	WOA
f_{11}	100	Best	9.417182×10^1	9.409247×10^1	9.460401×10^1	9.267136×10^1
		Avg.	9.690864×10^1	9.618143×10^1	9.501840×10^1	9.309575×10^1
		Worst	9.839476×10^1	9.827330×10^1	9.538590×10^1	9.337289×10^1
		SD	1.214620	8.735442×10^{-1}	1.739057×10^{-1}	1.907979×10^{-1}
	300	Best	2.958073×10^2	2.957236×10^2	2.951796×10^2	5.714967
		Avg.	2.976425×10^2	2.970865×10^2	2.957244×10^2	2.828332×10^2
		Worst	2.981833×10^2	2.978485×10^2	2.959295×10^2	2.928548×10^2
		SD	6.213829×10^{-1}	7.024913×10^{-1}	1.326191×10^{-1}	5.145973×10^1
	500	Best	4.973285×10^2	4.950355×10^2	4.935614×10^2	4.904825×10^2
		Avg.	4.978877×10^2	4.969489×10^2	4.939061×10^2	4.910578×10^2
		Worst	4.981244×10^2	4.976162×10^2	4.939489×10^2	4.913822×10^2
		SD	2.206418×10^{-1}	6.608382×10^{-1}	9.846602×10^{-2}	2.614029×10^{-1}
f_{24}	100	Best	3.996803×10^{-15}	1.110223×10^{-14}	4.440892×10^{-16}	4.440892×10^{-16}
		Avg.	3.996803×10^{-15}	1.453652×10^{-14}	4.440892×10^{-16}	2.338870×10^{-15}
		Worst	3.996803×10^{-15}	1.820766×10^{-14}	4.440892×10^{-16}	7.549517×10^{-15}
		SD	0.000000	1.117208×10^{-15}	0.000000	1.995713×10^{-15}
	300	Best	3.996803×10^{-15}	2.176037×10^{-14}	4.440892×10^{-16}	4.440892×10^{-16}
		Avg.	3.996803×10^{-15}	2.614205×10^{-14}	4.440892×10^{-16}	2.457294×10^{-15}
		Worst	3.996803×10^{-15}	2.886580×10^{-14}	4.440892×10^{-16}	7.549517×10^{-15}
		SD	0.000000	3.135436×10^{-15}	0.000000	2.186836×10^{-15}
	500	Best	3.996803×10^{-15}	2.886580×10^{-14}	4.440892×10^{-16}	4.440892×10^{-16}
		Avg.	4.825769×10^{-15}	3.158955×10^{-14}	4.440892×10^{-16}	2.457294×10^{-15}
		Worst	7.549517×10^{-15}	3.597123×10^{-14}	4.440892×10^{-16}	7.549517×10^{-15}
		SD	1.502629×10^{-15}	1.985144×10^{-15}	0.000000	2.371435×10^{-15}
f_{25}	100	Best	4.955085	2.624674	4.766665×10^{-5}	3.631265×10^{-5}
		Avg.	6.169643	4.123167	4.122016×10^{-3}	1.910465×10^{-3}
		Worst	7.315961	5.184571	2.124806×10^{-2}	1.105270×10^{-2}
		SD	5.111456×10^{-1}	5.124575×10^{-1}	5.953081×10^{-3}	4.083615×10^{-3}
	300	Best	2.643887×10^1	2.282209×10^1	2.212745×10^{-3}	3.726640×10^{-3}
		Avg.	2.697167×10^1	2.376806×10^1	8.795785×10^{-3}	8.291253×10^{-3}
		Worst	2.757999×10^1	2.474629×10^1	1.782640×10^{-2}	2.452580×10^{-2}
		SD	3.359007×10^{-1}	4.596044×10^{-1}	4.937780×10^{-3}	5.300295×10^{-3}
	500	Best	4.658280×10^1	4.243303×10^1	8.669046×10^{-3}	2.174499×10^{-2}
		Avg.	4.724426×10^1	4.391827×10^1	2.089506×10^{-2}	3.314949×10^{-2}
		Worst	4.807043×10^1	4.471744×10^1	2.799315×10^{-2}	5.016311×10^{-2}
		SD	3.648753×10^{-1}	5.330605×10^{-1}	4.269011×10^{-3}	7.169631×10^{-3}

5.2.2. Optimization Outcomes for Classical Engineering Problems

The results for the pressure vessel design problem are compared in Tables 24 and 25. The multi-strategy enhanced SCA (MSCA) was presented in [55], which also provides numerical results. The numerical results for the improved harmony search algorithm (IHS) [56], gravitational search algorithm (GSA) [57], DE [10], and HSA [14] were provided in [55]. Moreover, results for PSO [5] were taken from [58]. Results for GA [9] are provided in [59–61] for GA_1, GA_2 and GA_3 respectively. In [62], results for evolutionary strategy ES were provided, while those of the ACO algorithm were reported in [63]. GWO, WOA, WOA [52], and HHO [53] algorithms are included in the comparative study of classical engineering problems, i.e., pressure vessel problem, welded beam design problem, and rolling element bearing design.

The comparison for the pressure vessel problem is exhibited in Tables 24 and 25. The former shows both the variables and the cost function's optimal value, while the latter provides the constraints' value. The proposed ESCA algorithm and the DE algorithm achieve the best feasible results. It should be noted that the solution provided by MSCA and HHO methods are not feasible since both variables d_s and d_h have been considered as continuous variables, which is not correct as they are actually discrete variables. In particular, they must be multiples of 0.0625 inches. The IHS and ACO methods are not feasible because they do not meet the g_3 and g_1 constraints, respectively, as shown in Table 25.

The results of the welded beam design problem are reported in Tables 26 and 27. Table 26 exhibits the optimal cost of the function and its variables for several state-of-the-art algorithms, including; GSA algorithm [57], the ray optimization (RO) algorithm [64], IHS algorithm [56], genetic algorithm (GA_3) [61], the GWO algorithm, the WOA algorithm, and the HHO algorithm. Outcomes reveal that the ESCA algorithm outperforms the state-of-the-art algorithms in solving the welded beam design problem. The constraints of the leading solutions are listed in Table 27. It worth mentioning that the solution provided by HHO algorithm is not feasible as it does not meet the g_2 constraint.

Table 24. Design variables and comparison of the best solutions obtained for pressure vessel problem.

Algorithm	Variables				Function Cost
	ds	dh	R	L	
ESCA	0.8125	0.4375	42.0983	176.6385	6059.7344
SCA	0.8125	0.4375	42.0799	177.0465	6066.1710
MSCA	0.7793	0.3996	40.3255	199.9213	5935.7161
IHS	1.1250	0.6250	58.2902	43.6927	7197.7300
GSA	1.1250	0.6250	55.9887	84.4542	8538.8359
PSO	0.8125	0.4375	42.0913	176.7465	6061.0777
GA_1	0.8125	0.4345	40.3239	200.0000	6288.7445
GA_2	0.8125	0.4375	42.0974	176.6541	6059.9463
GA_3	0.9375	0.5000	48.3290	112.6790	6410.3811
ES	0.8125	0.4375	42.0981	176.6405	6059.7456
DE	0.8125	0.4375	42.0984	176.6377	6059.7340
ACO	0.8125	0.4375	42.1036	176.5727	6059.0888
GWO	0.8125	0.4375	42.0892	176.7587	6061.0135
HHO	0.8176	0.4073	42.0917	176.7196	6000.4626
WOA	0.8125	0.4375	42.0983	176.6390	6059.7410

The results for the rolling element bearing design problem are compared in Table 28. In addition to the SCA algorithm, the proposed ESCA algorithm is compared to the genetic algorithm (GA_4) [65], the TLBO algorithm [66], the mine blasting algorithm (MBA) [67], the supply demand-based optimization algorithm (SDO) [68], and the HHO algorithm. Note that, as shown in Table 29, neither TLBO nor MBA, nor SDO, nor HHO obtain feasible solutions. Indeed, the TLBO violates the g_7 constraint, while MBA, SDO and HHO violate the g_4 constraint. As shown in these tables, ESCA also carries the best feasible result on this constrained maximization problem.

Concisely, the outcomes on the assessed engineering problems prove that ESCA is high-performing in solving challenging problems as opposed to the comparison algorithms.

Table 25. Constraints of the best solutions obtained for the pressure vessel problem.

Algorithm	Constraints			
	g_1	g_2	g_3	g_4
ESCA	-2.81×10^{-6}	-3.59×10^{-2}	-5.57×10^{-1}	-6.34×10^{1}
SCA	-3.59×10^{-4}	-3.61×10^{-2}	-9.97×10^{2}	-6.30×10^{1}
MSCA	-9.75×10^{-4}	-1.49×10^{-2}	-1.26×10^{1}	-4.01×10^{1}
IHS	-1.05×10^{-7}	-6.89×10^{-2}	6.57×10^{-2}	-1.96×10^{2}
GSA	-4.44×10^{-2}	-9.09×10^{-2}	-2.71×10^{5}	-1.56×10^{2}
PSO	-1.39×10^{-4}	-3.59×10^{-2}	-1.16×10^{2}	-6.33×10^{1}
GA_1	-3.42×10^{-2}	-4.98×10^{-2}	-3.04×10^{2}	-4.00×10^{1}
GA_2	-2.02×10^{-5}	-3.59×10^{-2}	-2.49×10^{1}	-6.33×10^{1}
GA_3	-4.75×10^{-3}	-3.89×10^{-2}	-3.65×10^{3}	-1.27×10^{2}
ES	-6.92×10^{-6}	-3.59×10^{-2}	2.90	-6.34×10^{1}
DE	-6.68×10^{-7}	-3.59×10^{-2}	-3.71	-6.34×10^{1}
ACO	9.99×10^{-5}	-3.58×10^{-2}	-1.22	-6.34×10^{1}
GWO	-1.79×10^{-4}	-3.60×10^{-2}	-4.06×10^{1}	-6.32×10^{1}
HHO	-5.21×10^{-3}	-5.74×10^{-3}	-6.57×10^{-6}	-6.33×10^{1}
WOA	-3.39×10^{-6}	-3.59×10^{-2}	-1.25	-6.34×10^{1}

Table 26. Welded beam problem. Function cost and variables.

Algorithm	Variables				Function Cost
	h	l	t	b	
ESCA	0.205727	3.470570	9.036625	0.205730	1.724862
SCA	0.205661	3.471731	9.037817	0.205742	1.725213
GSA	0.182129	3.856979	10.000000	0.202376	1.879952
RO	0.203687	3.528467	9.004233	0.207241	1.735344
IHS	0.203687	3.528467	9.004233	0.207241	1.735344
GA_3	0.248900	6.173000	8.178900	0.253300	2.433100
GWO	0.205676	3.478377	9.03681	0.205778	1.726240
HHO	0.204039	3.531061	9.027463	0.206147	1.731990
WOA	0.205396	3.484293	9.037426	0.206276	1.730499

Table 27. Welded Beam problem. Constraints.

Algorithm	Constraints						
	g_1	g_2	g_3	g_4	g_5	g_6	g_7
ESCA	-7.80×10^{-2}	-5.98×10^{-2}	-3.00×10^{-6}	-3.43	-8.07×10^{-2}	-2.36×10^{-1}	-3.20×10^{-2}
SCA	-0.699753	-9.721939	-0.000081	-3.432575	-0.080661	-0.235547	-1.602377
GSA	-5.35×10^2	-5.10×10^3	-2.02×10^{-2}	-3.26	-5.71×10^{-2}	-2.39×10^{-1}	-1.33×10^4
RO	-2.24	-4.13	-3.55×10^{-3}	-3.42	-7.87×10^{-2}	-2.35×10^{-1}	-1.24×10^4
IHS	-2.24	-4.13	-3.55×10^{-3}	-3.42	-7.87×10^{-2}	-2.35×10^{-1}	-1.24×10^4
GA_3	-5.76×10^3	-2.56×10^2	-4.40×10^{-3}	-2.98	-1.24×10^{-1}	-2.34×10^{-1}	-2.39×10^4
GWO	-2.12×10^1	-8.29	-1.02×10^{-4}	-3.43	-8.07×10^{-2}	-2.36×10^{-1}	-4.31
HHO	-6.21×10^1	5.72×10^{-2}	-2.11×10^{-3}	-3.43	-7.90×10^{-2}	-2.36×10^{-1}	-3.26×10^1
WOA	-2.15×10^1	-8.48×10^1	-8.80×10^{-4}	-3.43	-8.04×10^{-2}	-2.36×10^{-1}	-4.83×10^1

Table 28. Design variables and comparison of the best solutions obtained for the rolling element bearing design problem.

Design Variables	Algorithm						
	SCA	GA_4	TLBO	MBA	SDO	HHO	ESCA
D_m	125.719015	125.717100	125.719100	125.715300	125.700000	125.000000	125.718960
D_b	21.425557	21.423000	21.425590	21.423300	21.424905	21.000000	21.425563
Z	11.000000	11.000000	11.000000	11.000000	11.000000	11.090000	11.000000
f_i	0.515000	0.515000	0.515000	0.515000	0.515002	0.515000	0.515000
f_o	0.515000	0.515000	0.515000	0.515000	0.515930	0.515000	0.515000
K_{Dmin}	0.490213	0.415900	0.424266	0.488805	0.487755	0.400000	0.465124
K_{Dmax}	0.672451	0.651000	0.633948	0.627829	0.629992	0.600000	0.653542
ϵ	0.300000	0.300043	0.300000	0.300149	0.300039	0.300000	0.300000
e	0.070763	0.022300	0.068858	0.097305	0.053510	0.050474	0.020149
ψ	0.760058	0.751000	0.799498	0.646095	0.665982	0.600000	0.736634
Function cost	81,859.508	81,841.511	81,859.738	81,843.686	81,575.185	83,011.883	81,859.552

Table 29. Constraints of the best solutions obtained for the rolling element bearing design problem.

Constraints	Algorithm						
	SCA	GA_4	TLBO	MBA	SDO	HHO	ESCA
g_1	0.000009	0.000822	0.000004	0.000564	-0.001272	0.013477	0.000003
g_2	8.536204	13.733000	13.152560	8.630250	8.706960	14.000000	10.292446
g_3	4.220456	2.724000	1.525180	1.101430	1.249630	0.000000	2.896814
g_4	1.376183	1.107000	2.559350	-2.040450	-1.445445	-3.000000	0.673457
g_5	0.719015	0.717100	0.719100	0.715300	0.700000	0.000000	0.718960
g_6	16.971735	4.857900	16.495400	23.610950	12.677500	12.618500	4.318290
g_7	0.000047	0.002129	-0.000022	0.000518	0.009240	0.700000	0.000070
g_8	0.000000	0.000000	0.000000	0.000000	0.000002	0.000000	0.000000
g_9	0.000000	0.000000	0.000000	0.000000	0.000930	0.000000	0.000000

6. Conclusions

This paper proposed an enhanced SCA algorithm dubbed the ESCA algorithm in which the diversification behavior of the SCA algorithm is reduced at the end of the optimization course. Indeed, the SCA algorithm's exploitation abilities are strengthened with a best-guided strategy that refines the current solution and leads the algorithm to converge swiftly toward the optimum. Experimental tests on benchmark functions and challenging engineering problems prove the supremacy of the proposed algorithm in overall performance, i.e., solution accuracy and convergence speed, compared to a set of state-of-the-art algorithms. This domination is confirmed through statistical tests. The proposed ESCA algorithms are ranked first according to Friedman, Friedman aligned, and Quade tests in terms of convergence speed and solution accuracy. Furthermore, one-level parallel ESCA algorithms that work synchronously and asynchronously are designed as well. They efficiently utilize multicore architectures by joining coarse-grained and fine-grained parallel techniques. The parallel scalability of these algorithms yields an efficient use of the physical and logical cores when hyperthreading is enabled, which increases the total number of threads that are efficiently used when the two-level parallel algorithm is executed. It was identified that the one-level parallel ESCA algorithms diminish the computing time, on average, by 87.4% and 90.8%, respectively, using 12 processing cores. Moreover, it has been shown that parallel performance can be improved by affinity techniques that permit mapping processes over the cores of multicore processors. In fact, the two-level parallel algorithms provide extra reductions of the computing time by 91.4%, 93.1%, and 94.5% with 16, 20, and 24 processing cores. Considering its outstanding optimization performance and computational behavior capability of extracting the maximum performance from the available computational resources, the proposed algorithm is particularly fitting for high computational complexity problems.

Author Contributions: H.M. and A.B. conceived the optimization algorithms; H.M., J.-L.S.-R., A.J.-M., D.G.-S and J.G.-G. conceived the parallel algorithms; H.M., J.G.-G. and D.G.-S. codified the parallel algorithms; A.B., H.M., J.G.-G., J.-L.S.-R. and A.J.-M. performed numerical experiments; H.M., A.B. and J.G.-G. analyzed the data; H.M. wrote the original draft. A.B., J.-L.S.-R. and A.J.-M. reviewed and edited the manuscript. All authors have read and agreed to the published version of the manuscript.

Funding: This research was supported by the Spanish Ministry of Science, Innovation and Universities and the Research State Agency under Grant RTI2018-098156-B-C54 cofinanced by FEDER funds and the Ministry of Science and Innovation and the Research State Agency under Grant PID2020-120213RB-I00 cofinanced by FEDER funds.

Institutional Review Board Statement: Not applicable.

Informed Consent Statement: Not applicable.

Data Availability Statement: Not applicable.

Conflicts of Interest: The authors declare no conflict of interest.

References

1. Dorigo, M.; Di Caro, G. The Ant Colony Optimization Meta-heuristic. In *New Ideas in Optimization*; McGraw-Hill Ltd.: Maidenhead, UK, 1999; pp. 11–32.
2. Schwefel, H.P. Evolutionsstrategie Und Numerische Optimierung. Ph.D. Thesis, Department of Process Engineering, Technical University of Berlin, Berlin, Germany, 1975.
3. Bäck, T.; Rudolph, G.; Schwefel, H.P. Evolutionary Programming and Evolution Strategies: Similarities and Differences. In Proceedings of the Second Annual Conference on Evolutionary Programming, La Jolla, CA, USA, 25–26 February 1993; pp. 11–22.
4. Koza, J.R. *Genetic Programming: A Paradigm for Genetically Breeding Populations of Computer Programs to Solve Problems*; Technical Report; Stanford University: Stanford, CA, USA, 1990.
5. Poli, R.; Kennedy, J.; Blackwell, T. Particle swarm optimization. *Swarm Intell.* **2007**, *1*, 33–57. [CrossRef]
6. Eusuff, M.; Lansey, K.; Pasha, F. Shuffled frog-leaping algorithm: A memetic meta-heuristic for discrete optimization. *Eng. Optim.* **2006**, *38*, 129–154. [CrossRef]

7. Karaboga, D.; Basturk, B. On the Performance of Artificial Bee Colony (ABC) Algorithm. *Appl. Soft Comput.* **2008**, *8*, 687–697. [CrossRef]
8. Ingber, L. Simulated annealing: Practice versus theory. *Math. Comput. Model.* **1993**, *18*, 29–57. [CrossRef]
9. Holland, J.H. *Adaptation in Natural and Artificial Systems: An Introductory Analysis with Applications to Biology, Control, and Artificial Intelligence*; MIT Press: Cambridge, MA, USA, 1992.
10. Price, K.V. An Introduction to Differential Evolution. In *New Ideas in Optimization*; McGraw-Hill Ltd.: Maidenhead, UK, 1999; pp. 79–108.
11. Storn, R. On the usage of differential evolution for function optimization. In Proceedings of the North American Fuzzy Information Processing, Berkeley, CA, USA, 19–22 June 1996; pp. 519–523.
12. Bilal; Pant, M.; Zaheer, H.; Garcia-Hernandez, L.; Abraham, A. Differential evolution: A review of more than two decades of research. *Eng. Appl. Artif. Intell.* **2020**, *90*, 103479.
13. Farmer, J.D.; Packard, N.H.; Perelson, A.S. The Immune System, Adaptation, and Machine Learning. *Phys. D* **1986**, *2*, 187–204. [CrossRef]
14. Kim, J.H. Harmony Search Algorithm: A Unique Music-inspired Algorithm. *Procedia Eng.* **2016**, *154*, 1401–1405. [CrossRef]
15. Mirjalili, S. SCA: A Sine Cosine Algorithm for solving optimization problems. *Knowl.-Based Syst.* **2016**, *96*, 120–133. [CrossRef]
16. Kumar-Majhi, S. An Efficient Feed Foreword Network Model with Sine Cosine Algorithm for Breast Cancer Classification. *Int. J. Syst. Dyn. Appl. (IJSDA)* **2018**, *7*, 202397. [CrossRef]
17. Rajesh, K.; Dash, S. Load frequency control of autonomous power system using adaptive fuzzy based PID controller optimized on improved sine cosine algorithm. *J. Ambient. Intell. Humaniz. Comput.* **2019**, *10*, 2361–2373. [CrossRef]
18. Khezri, R.; Oshnoei, A.; Tarafdar Hagh, M.; Muyeen, S. Coordination of Heat Pumps, Electric Vehicles and AGC for Efficient LFC in a Smart Hybrid Power System via SCA-Based Optimized FOPID Controllers. *Energies* **2018**, *11*, 420. [CrossRef]
19. Ramanaiah, M.L.; Reddy, M.D. Sine cosine algorithm for loss reduction in distribution system with unified power quality conditioner. *i-Manag. J. Power Syst. Eng.* **2017**, *5*, 10.
20. Dhundhara, S.; Verma, Y.P. Capacitive energy storage with optimized controller for frequency regulation in realistic multisource deregulated power system. *Energy* **2018**, *147*, 1108–1128. [CrossRef]
21. Singh, V.P. Sine cosine algorithm based reduction of higher order continuous systems. In Proceedings of the 2017 International Conference on Intelligent Sustainable Systems (ICISS), Palladam, India, 7–8 December 2017; pp. 649–653. [CrossRef]
22. Das, S.; Bhattacharya, A.; Chakraborty, A.K. Solution of short-term hydrothermal scheduling using sine cosine algorithm. *Soft Comput.* **2018**, *22*, 6409–6427. [CrossRef]
23. Kumar, V.; Kumar, D. *Handbook of Research on Machine Learning Innovations and Trends*; IGI Global: Hershey, PA, USA, 2017; pp. 715–726. [CrossRef]
24. Yıldız, B.S.; Yıldız, A.R. Comparison of grey wolf, whale, water cycle, ant lion and sine-cosine algorithms for the optimization of a vehicle engine connecting rod. *Mater. Test.* **2018**, *60*, 311–315. [CrossRef]
25. Elfattah, M.A.; Abuelenin, S.; Hassanien, A.E.; Pan, J.S. Handwritten Arabic Manuscript Image Binarization Using Sine Cosine Optimization Algorithm. In Proceedings of the International Conference on Genetic and Evolutionary Computing, Fuzhou, Fujian, China, 7–9 November 2016; Volume 536, pp. 273–280.
26. Mirjalili, S.M.; Mirjalili, S.Z.; Saremi, S.; Mirjalili, S. *Studies in Computational Intelligence*; Springer: Berlin, Germany, 2020; Volume 811, pp. 201–217. [CrossRef]
27. Ewees, A.A.; Abd Elaziz, M.; Al-Qaness, M.A.A.; Khalil, H.A.; Kim, S. Improved Artificial Bee Colony Using Sine-Cosine Algorithm for Multi-Level Thresholding Image Segmentation. *IEEE Access* **2020**, *8*, 26304–26315. [CrossRef]
28. Gupta, S.; Deep, K.; Mirjalili, S.; Kim, J.H. A modified sine cosine algorithm with novel transition parameter and mutation operator for global optimization. *Expert Syst. Appl.* **2020**, *154*, 113395. [CrossRef]
29. Gupta, S.; Deep, K. A novel hybrid sine cosine algorithm for global optimization and its application to train multilayer perceptrons. *Appl. Intell.* **2020**, *50*, 993–1026. [CrossRef]
30. Rizk-Allah, R.M. An improved sine–cosine algorithm based on orthogonal parallel information for global optimization. *Soft Comput.* **2019**, *23*, 7135–7161. [CrossRef]
31. Belazzoug, M.; Touahria, M.; Nouioua, F.; Brahimi, M. An improved sine cosine algorithm to select features for text categorization. *J. King Saud-Univ.-Comput. Inf. Sci.* **2020**, *32*, 454–464. [CrossRef]
32. Gupta, S.; Deep, K. Improved sine cosine algorithm with crossover scheme for global optimization. *Knowl.-Based Syst.* **2019**, *165*, 374–406. [CrossRef]
33. Qu, C.; Zeng, Z.; Dai, J.; Yi, Z.; He, W. A modified sine-cosine algorithm based on neighborhood search and greedy levy mutation. *Comput. Intell. Neurosci.* **2018**, *2018*. [CrossRef] [PubMed]
34. Rosli, S.J.; Rahim, H.A.; Abdul Rani, K.N.; Ngadiran, R.; Ahmad, R.B.; Yahaya, N.Z.; Abdulmalek, M.; Jusoh, M.; Yasin, M.N.M.; Sabapathy, T.; et al. A Hybrid Modified Method of the Sine Cosine Algorithm Using Latin Hypercube Sampling with the Cuckoo Search Algorithm for Optimization Problems. *Electronics* **2020**, *9*, 1786. [CrossRef]
35. Abd Elaziz, M.; Oliva, D.; Xiong, S. An improved opposition-based sine cosine algorithm for global optimization. *Expert Syst. Appl.* **2017**, *90*, 484–500. [CrossRef]
36. Sindhu, R.; Ngadiran, R.; Yacob, Y.M.; Zahri, N.A.H.; Hariharan, M. Sine–cosine algorithm for feature selection with elitism strategy and new updating mechanism. *Neural Comput. Appl.* **2017**, *28*, 2947–2958. [CrossRef]

37. Long, W.; Wu, T.; Liang, X.; Xu, S. Solving high-dimensional global optimization problems using an improved sine cosine algorithm. *Expert Syst. Appl.* **2019**, *123*, 108–126. [CrossRef]
38. Issa, M.; Hassanien, A.E.; Oliva, D.; Helmi, A.; Ziedan, I.; Alzohairy, A. ASCA-PSO: Adaptive sine cosine optimization algorithm integrated with particle swarm for pairwise local sequence alignment. *Expert Syst. Appl.* **2018**, *99*, 56–70. [CrossRef]
39. Chegini, S.N.; Bagheri, A.; Najafi, F. PSOSCALF: A new hybrid PSO based on Sine Cosine Algorithm and Levy flight for solving optimization problems. *Appl. Soft Comput.* **2018**, *73*, 697–726. [CrossRef]
40. Nenavath, H.; Jatoth, R.K.; Das, S. A synergy of the sine-cosine algorithm and particle swarm optimizer for improved global optimization and object tracking. *Swarm Evol. Comput.* **2018**, *43*, 1–30. [CrossRef]
41. Singh, N.; Singh, S. A novel hybrid GWO-SCA approach for optimization problems. *Eng. Sci. Technol. Int. J.* **2017**, *20*, 1586–1601. [CrossRef]
42. Nenavath, H.; Jatoth, R.K. Hybridizing sine cosine algorithm with differential evolution for global optimization and object tracking. *Appl. Soft Comput.* **2018**, *62*, 1019–1043. [CrossRef]
43. Migallón, H.; Jimeno-Morenilla, A.; Sánchez-Romero, J.L.; Rico, H.; Rao, R.V. Multipopulation-based multi-level parallel enhanced Jaya algorithms. *J. Supercomput.* **2019**, *75*, 1697–1716. [CrossRef]
44. García-Monzó, A.; Migallón, H.; Jimeno-Morenilla, A.; Sánchez-Romero, J.L.; Rico, H.; Rao, R.V. Efficient Subpopulation Based Parallel TLBO Optimization Algorithms. *Electronics* **2018**, *8*, 19. [CrossRef]
45. Free Software Foundation, Inc. GCC, the GNU Compiler Collection. Available online: https://www.gnu.org/software/gcc/index.html (accessed on 15 October 2021).
46. OpenMP Architecture Review Board. OpenMP Application Program Interface, Version 3.1. 2011. Available online: http://www.openmp.org (accessed on 15 October 2021).
47. Dimakopoulos, V.V.; Hadjidoukas, P.E.; Philos, G.C. A Microbenchmark Study of OpenMP Overheads under Nested Parallelism. In *OpenMP in a New Era of Parallelism*; Eigenmann, R., de Supinski, B.R., Eds.; Springer: Berlin/Heidelberg, Germany, 2008; pp. 1–12. [CrossRef]
48. Friedman, M. The use of ranks to avoid the assumption of normality implicit in the analysis of variance. *J. Am. Stat. Assoc.* **1937**, *32*, 675–701. [CrossRef]
49. Hodges, J.; Lehmann, E.L. Rank methods for combination of independent experiments in analysis of variance. In *Selected Works of EL Lehmann*; Springer: Berlin/Heidelberg, Germany, 2012; pp. 403–418.
50. Quade, D. On analysis of variance for the k-sample problem. *Ann. Math. Stat.* **1966**, *37*, 1747–1758. [CrossRef]
51. Mirjalili, S.; Mirjalili, S.M.; Lewis, A. Grey wolf optimizer. *Adv. Eng. Softw.* **2014**, *69*, 46–61. [CrossRef]
52. Mirjalili, S.; Lewis, A. The whale optimization algorithm. *Adv. Eng. Softw.* **2016**, *95*, 51–67. [CrossRef]
53. Heidari, A.A.; Mirjalili, S.; Faris, H.; Aljarah, I.; Mafarja, M.; Chen, H. Harris hawks optimization: Algorithm and applications. *Future Gener. Comput. Syst.* **2019**, *97*, 849–872. [CrossRef]
54. García, S.; Fernández, A.; Luengo, J.; Herrera, F. Advanced nonparametric tests for multiple comparisons in the design of experiments in computational intelligence and data mining: Experimental analysis of power. *Inf. Sci.* **2010**, *180*, 2044–2064. [CrossRef]
55. Chen, H.; Wang, M.; Zhao, X. A multi-strategy enhanced sine cosine algorithm for global optimization and constrained practical engineering problems. *Appl. Math. Comput.* **2020**, *369*, 124872. [CrossRef]
56. Mahdavi, M.; Fesanghary, M.; Damangir, E. An improved harmony search algorithm for solving optimization problems. *Appl. Math. Comput.* **2007**, *188*, 1567–1579. [CrossRef]
57. Rashedi, E.; Nezamabadi-pour, H.; Saryazdi, S. GSA: A Gravitational Search Algorithm. *Inf. Sci.* **2009**, *179*, 2232–2248. [CrossRef]
58. He, Q.; Wang, L. An effective co-evolutionary particle swarm optimization for constrained engineering design problems. *Eng. Appl. Artif. Intell.* **2007**, *20*, 89–99. [CrossRef]
59. Coello, C.A.C. Theoretical and numerical constraint-handling techniques used with evolutionary algorithms: A survey of the state of the art. *Comput. Methods Appl. Mech. Eng.* **2002**, *191*, 1245–1287. [CrossRef]
60. Coello, C.A.C.; Montes, E.M. Constraint-handling in genetic algorithms through the use of dominance-based tournament selection. *Adv. Eng. Inform.* **2002**, *16*, 193–203. [CrossRef]
61. Deb, K. GeneAS: A robust optimal design technique for mechanical component design. In *Evolutionary Algorithms in Engineering Applications*; Springer: Berlin/Heidelberg, Germany, 1997; pp. 497–514.
62. Mezura-Montes, E.; Coello, C.A.C. An empirical study about the usefulness of evolution strategies to solve constrained optimization problems. *Int. J. Gen. Syst.* **2008**, *37*, 443–473. [CrossRef]
63. Kaveh, A.; Talatahari, S. An improved ant colony optimization for constrained engineering design problems. *Eng. Comput.* **2010**, *27*, 155–182. [CrossRef]
64. Kaveh, A.; Khayatazad, M. A new meta-heuristic method: Ray Optimization. *Comput. Struct.* **2012**, *112–113*, 283–294. [CrossRef]
65. Rajeswara Rao, B.; Tiwari, R. Optimum design of rolling element bearings using genetic algorithms. *Mech. Mach. Theory* **2007**, *42*, 233–250. [CrossRef]
66. Rao, R.V.; Savsani, V.; Vakharia, D. Teaching-learning-based optimization: A novel method for constrained mechanical design optimization problems. *Comput.-Aided Des.* **2011**, *43*, 303–315. [CrossRef]

67. Sadollah, A.; Bahreininejad, A.; Eskandar, H.; Hamdi, M. Mine blast algorithm: A new population based algorithm for solving constrained engineering optimization problems. *Appl. Soft Comput.* **2013**, *13*, 2592–2612. [CrossRef]
68. Zhao, W.; Wang, L.; Zhang, Z. Supply-Demand-Based Optimization: A Novel Economics-Inspired Algorithm for Global Optimization. *IEEE Access* **2019**, *7*, 73182–73206. [CrossRef]

Article

ReRec: A Divide-and-Conquer Approach to Recommendation Based on Repeat Purchase Behaviors of Users in Community E-Commerce

Jun Wu [1,2], Yuanyuan Li [1], Li Shi [2], Liping Yang [1], Xiaxia Niu [1] and Wen Zhang [3,*]

[1] School of Economics and Management, Beijing University of Chemical Technology, Beijing 100029, China; wujun@mail.buct.edu.cn (J.W.); lyyletter@126.com (Y.L.); m17856381770@163.com (L.Y.); 15136212624@163.com (X.N.)
[2] College of Information Science and Technology, Beijing University of Chemical Technology, Beijing 100029, China; simon_shl@126.com
[3] College of Economics and Management, Beijing University of Technology, Beijing 100124, China
* Correspondence: zhangwen@bjut.edu.cn

Abstract: Existing studies have made a great endeavor in predicting users' potential interests in items by modeling user preferences and item characteristics. As an important indicator of users' satisfaction and loyalty, repeat purchase behavior is a promising perspective to extract insightful information for community e-commerce. However, the repeated purchase behaviors of users have not yet been thoroughly studied. To fill in this research gap from the perspective of repeated purchase behavior and improve the process of generation of candidate recommended items this research proposed a novel approach called ReRec (Repeat purchase Recommender) for real-life applications. Specifically, the proposed ReRec approach comprises two components: the first is to model the repeat purchase behaviors of different types of users and the second is to recommend items to users based on their repeat purchase behaviors of different types. The extensive experiments are conducted on a real dataset collected from a community e-commerce platform, and the performance of our model has improved at least about 13.6% compared with the state-of-the-art techniques in recommending online items (measured by F-measure). Specifically, for active users, with $w = 1$ and $N_{(U_A)} \in [5, 25]$, the results of ReRec show a significant improvement (at least 50%) in recommendation. With α and σ as 0.75 and 0.2284, respectively, the proposed ReRec for unactive users is also superior to (at least 13.6%) the evaluation indicators of traditional Item CF when $N_{(U_B)} \in [6, 25]$. To the best of our knowledge, this paper is the first to study recommendations in community e-commerce.

Keywords: ReRec; community e-commerce; repeat purchase; user behavior modeling; recommendation system

Citation: Wu, J.; Li, Y.; Shi, L.; Yang, L.; Niu, X.; Zhang, W. ReRec: A Divide-and-Conquer Approach to Recommendation Based on Repeat Purchase Behaviors of Users in Community E-Commerce. *Mathematics* **2022**, *10*, 208. https://doi.org/10.3390/math10020208

Academic Editors: Antonin Ponsich, Mariona Vila Bonilla and Bruno Domenech

Received: 7 December 2021
Accepted: 4 January 2022
Published: 10 January 2022

1. Introduction

Community e-commerce, which combines the features of traditional e-commerce and mobile commerce, is a representative of community economy [1] and marks the rise of a new commercial ideology. Generally speaking, community e-commerce refers to a novel business model that takes communities as service units and provides a more convenient manner in online shopping than traditional e-commerce for community residents [2,3]. On the one hand, unlike traditional e-commerce that provides products and services all over the world or a country, community e-commerce focuses on a relatively stable group of consumers in a local area as a compatible complement for B2B, B2C and C2C models. On the other hand, like traditional e-commerce, the huge amount of online information and items brings about a heavy burden for online consumers, the users of community e-commerce also suffer from the endless choices and decisions in online shopping and the merchants in community e-commerce are still struggling to predict the interests of users in online items beforehand, in order to manage their inventories. For this reason, it is urgent to develop a recommendation system for community e-commerce platforms to predict the items that a user may possibly purchase in the near future based on the user's purchase history [4].

In community e-commerce, it is a usual case that a user would purchase the same item repeatedly and periodically. In the scenario of traditional e-commerce, these items will not be recommended to the user repeatedly in the future. However, with the focus on limited number of users in a local area, the recommendation for repeat purchase is crucial for the success of community e-commerce. For instance, by observing user behaviors on the community e-commerce platform T-app (see Section 5.1), we find that from 1 January 2018 to 1 April 2019, among 955 users who have made purchases on T-app, 58.74% have repeat purchases. For these users with repeat purchase, their average repurchase is 3.61 times, and 10.33% of them repurchase the same item six times. In an extreme case, we find that one user has repurchased the same item up to 43 times during the investigated time duration. Among all the 105 types of items, 82 (78.10%) have been repurchased by users. Therefore, it can be seen that repeated purchase behavior is an essential user characteristic that should be paid enough attention to when community e-commerce platforms make recommendation plans.

Existing studies have proposed many recommendation algorithms to predict users' potential interests in items by characterizing user preferences and item characteristics, e.g., the nearest neighborhood based recommendation algorithm [5–7], the matrix factorization based recommendation algorithm [8,9] and the context aware recommendation algorithm [10,11]. Clearly, the basic idea of these algorithms is straightforward—that if a user purchased an item in the past, he or she will also purchase similar items, or items purchased similar users at that time, in the future. However, if an item has already been purchased by a user, then the item will not be recommended by theses algorithms to the user. That is to say, the repeated purchase behavior of users has not yet been thoroughly studied. To fill in this research gap, this paper proposes a novel approach called ReRec (Repeat purchase Recommender) for recommending items to users in community e-commerce. To the best of our knowledge, this paper is the first to conduct item recommendation in community e-commerce. For industrial applications, the proposed method can help manage and identify loyal users and segment users and to improve customer relationship management (CRM) processes. In addition, for managers, this method can also help them formulate precision marketing strategies, recognize the market, and advance the sustainable development of products.

Specifically, ReRec comprises two components. The first component is to model the repeat purchase behaviors of different types of user. This research models the repeat purchase behaviors of the users in community e-commerce. based on their activity in the community and the stability of their interests in items, in a divide-and-conquer manner, using these categories: active users with stable interest (ASI), active users with unstable interest (AUSI), inactive users with stable interest (IASI) and inactive users with unstable interest (IAUSI). The second component is to recommend items to users based on repeat purchase behaviors. This research proposes the ReRec approach in four variants to deal with different types of users and interests, i.e., recommendation for active users with stable interest (ReRec-ASI), recommendation for active users with unstable interest (ReRec-AUSI), recommendation for inactive users with stable interest (ReRec-IASI) and recommendation for inactive users with unstable interest (ReRec-IAUSI). Finally, extensive experiments based on a real community e-commerce platform are conducted and the experimental results demonstrate that the proposed ReRec approach outperforms state-of-the-art techniques significantly.

The rest of this paper is organized as follows. Section 2 states the problem. Section 3 presents related works. Section 4 proposes the ReRec approach. Section 5 conducts the experiments. Section 6 concludes the paper and indicates future work.

2. Problem Statement

The problem studied in this paper is one of recommendation for repeat purchase in community e-commerce, which is different from that of traditional recommendation, such as collaborative filtering [5,6,12]. Essentially, this research can formulate the problem as follows. Assume that there are a set of users as $U = \{u_k | 1 \leq k \leq m\}$, and a set of items as $I = \{i_s | 1 \leq s \leq n\}$ in community e-commerce. The historical sales data until time t is recorded as a matrix $R_{UI}^t = \left\{ r_{u_k i_s}^t \middle| 1 \leq k \leq m, 1 \leq s \leq n \right\}$, where $r_{u_k i_s}^t$ is the number of cumulative purchases of the user u_k of the item i_s at t. Note that the user u_k has purchased the item i_s repeatedly and periodically. Let $\widetilde{R}_{u_k i_s}^{t+1}$ be the possibility that the user u_k purchases the item i_s on $t+1$. We need to speculate the possibilities of user u_k purchasing all the possible items i_s $(1 \leq s \leq n)$ on $t+1$, i.e., $\widetilde{R}_{u_k i_s}^{t+1}$ for all the items i_s on $t+1$. After deriving the $\widetilde{R}_{u_k i_s}^{t+1}$, it sorts all the possibilities in descending order for user u_k, and uses the top N items as the recommendation list to him or her.

3. Related Works

3.1. Nearest Neighborhood Based Recommendation

On the aspect of nearest neighborhood recommendation, the user-based nearest neighbor method and item-based nearest neighbor method are usually adopted. Resnick et al. [13] propose user-based collaborative filtering to recommend internet news to readers according to readers' rating scores of the internet news. This algorithm firstly calculates the similarity between users, and then for a given user it recommends items that are of interest to similar users to him or her. Considering the large number of items in a recommender system, Sarwar et al. [14] propose item-based recommendation to compute and store items' similarities beforehand in the system and use these similarities in real time when needed to produce a recommendation list for a user. The basic idea of the item-based algorithm is to assume that people will like items that are similar to those items they have purchased before. Since a user has purchased an item in history, he or she would also purchase similar items in the future. The item-based algorithm is very similar to the user-based algorithm. More details about user-based collaborative filtering and item-based collaborative filtering approaches can be found in the available literature [5–7,12,15]. The advantage of the nearest neighborhood algorithms is that they are easy to implement in real practice because of their simple mathematical form and consolidated intuitiveness. However, due to the sparse nature of the historical purchasing data, it is difficult to measure similarities between users and items [8]. Moreover, because the users' interests in items can change very frequently, it makes the computation complexity of real-time recommendation intractable [4].

3.2. Matrix Factorization Based Recommendation

On the aspect of matrix factorization based recommendation, SVD (Single Value Decomposition), SVD++ and NMF (non-negative Matrix Factorization) are the most representative techniques. SVD is a basic matrix decomposition method used in the recommender systems proposed by Chen et al. [9] and Brand [16]. It decomposes the original matrix R with higher dimensions into three matrix multiplication forms with lower dimensions, which brings convenience to matrix calculation and storage. Specifically, SVD decomposes the rating matrix $R_{m \times n}$ into three matrices: left singular vector $P_{m \times n}$, right singular vector $Q_{m \times n}$ and singular value diagonal matrix $S_{m \times n}$ as in Equation (1). Both P and Q matrices are orthogonal and matrix S is a diagonal matrix composed of singular values where all the singular values are aligned in descending order from the largest to the smallest. For all the singular values $S_{ii} \geq 0$, the rank of the rating matrix R is a, and the number of ranks that can be taken is $\{a_h | 1 \leq h \leq \min(m, n)\}$.

$$R = PSQ^T \tag{1}$$

SVD++ is an extension of traditional SVD that takes into account both explicit and implicit information for recommendation [17]. Here, explicit information refers to the users' rating of an item, and implicit information refers to the users' implicit feedback, such as browsing, buying, and clicking history [8]. The prediction rating $\hat{\gamma}_{ui}$ of SVD++ is defined in Equation (2).

$$\hat{\gamma}_{ui} = b_{ui} + q_i^T(p_u) = \mu + b_u + b_i + q_i^T(p_u) \tag{2}$$

The prediction rating $\hat{\gamma}_{ui}$ is composed of two parts: one is the deviation of different users to different products b_{ui}, the other is the product of the user preference vector p_u and the product feature vector q_i, where μ denotes the benchmark value in the score, b_u is the deviation value of user rating, and b_i is the score deviation of the product. These parameters need to be trained to obtain specific values.

As for NMF, the rating matrix R is approximated by the product of two low-dimensional matrices P and Q, as shown in Equation (3). The NMF problem is non-convex and is usually solved by the gradient descent method [18].

$$R = P^T Q \tag{3}$$

The advantage of the matrix decomposition method is that the users' preference in the item is regarded as the product of two components, i.e., as the users' latent vector representing the user preference and the item's latent vector representing the item's characteristics. Both the user's latent vector and the item's latent vector can be stored in the memory of the recommender system in advance, so it is convenient to compute and predict the user's preference in the item in real time. However, the matrix factorization method also has some defects. Because most view the user item rating matrix from a global perspective and perform matrix decomposition, their performance will be affected due to the large scale of the original user project scoring matrix and the sparse data.

3.3. Context-Aware Recommendation

The collaborative filtering algorithm for recommendation only considers the interactive information between users and items, such as the users' rating matrix for items. Meanwhile, other information, such as contextual situation information during interactive behavior, is generally not considered. A context-aware recommender system (CARS) is used to recommend items to users based on relevant contextual information such as time, weather and location. Contextual information can improve the performance of recommendation and user satisfaction when it is combined with the recommendation algorithm. Gorgoglione et al. [19] report that the context-aware recommendation system can achieve more accurate recommendation by adding contextual information in the experiments, and this recommendation system can significantly increase the platform profit and users' stickability. Time information can consist of the time when users purchase, comment, search or perform other behaviors, or the time of the season or holiday. For instance, around the time of the Dragon Boat Festival in China, users may have a higher preference for rice dumplings than usual.

There are also some studies showing that reasonable use of time information can improve the performance of the recommendation algorithm. Zimdars et al. [20] make use of time series forecasting in collaborative filtering for recommendation. Campos et al. [21] find that there is a time-dependent characteristic of user behaviors in online shopping. For instance, the same user may have different preference patterns on different dates, months and seasons. Liang et al. [22] propose the Time SVD algorithm to integrate four kinds of time-affected factors into time functions and they find that the performance of the Time SVD algorithm is significantly better than that of the traditional SVD algorithm. Qin et al. [23] claim that users of different professions have obvious differences in understanding items, and there is an important relationship between user hierarchy classification and user interest. Traditional collaborative filtering algorithms do not consider change in users' interests. However, in real practice, users' interests are constantly changing with time and the influence of the environment. Therefore, some studies introduce the concept of user interest drift [24,25]. Chen et al. [26] provide a matrix decomposition optimization model that is constructed to think about the score matrix and combines time information and the original score matrix to improve the recommendation efficiency. Wu et al. [27] include the time factor in order optimize the weights of users' ratings based on time and user similarities.

4. The Proposed Approach

4.1. The Overview of the ReRec Approach

The overall structure of the proposed ReRec approach is shown in Figure 1. As can be seen, the proposed ReRec approach is composed of two components, i.e., repeat purchase behavior analysis and item recommendation. Before the analysis, this research collects the user-item purchase records as a basic data matrix, i.e., the original user-item interaction matrix. In the original user-item interaction matrix, the row label is user ID, the column label is item ID and the element is the cumulative purchase quantity of an item by the corresponding user at time t. Then, users are classified according to activeness and user-items are classified according to stableness. As shown in the yellow area of Figure 1, users are partitioned by mathematical modeling as the active and inactive users, and the items are partitioned as stable and unstable interest. The nodes with black circles denote the users. The nodes with blue circles denote the items. The nodes with red circles denote user–item interaction. The nodes with dotted circles denote the immediate process. In addition, the partition process can be visualized as the user partition matrix and the item partition matrix derived from the original user–item interaction. As for the user partition matrix, a user ID with yellow indicates an active user and a user ID with green indicates an inactive user. As for the item partition matrix, an item ID with red indicates the stable interest of its user and an item ID with blue indicates the unstable interest of its user. Results of the combined user and item classification can be seen in the joint user-item partition matrix.

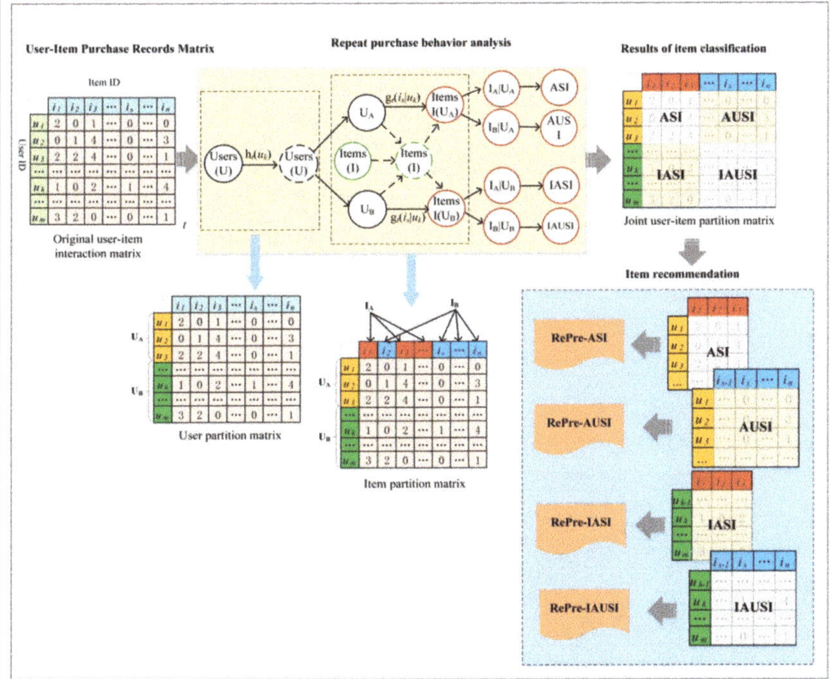

Figure 1. The overall structure of the proposed ReRec approach.

With the above joint user-item partition matrix at hand, this research conducts the item recommendation by using a divide-and-conquer approach. That is, it partitions the repeat purchase behaviors of users into four types: active users with stable interest (ASI), inactive users with stable interest (IASI), active users with unstable interest (AUSI), and inactive users with unstable interest (IAUSI). Furthermore, this research proposes the ReRec recommendation algorithm with its four variants to deal with the repeat purchase behaviors of the four types one by one: the ReRec-ASI approach, the ReRec-IASI approach, the ReRec-AUSI approach and the ReRec-IAUSI approach, which are shown in the blue area of Figure 1.

4.2. Repeat Purchase Behavior Modeling

As community e-commerce focuses on the residents in the local community, the characteristics of user purchase data are different from that of the large-scale e-commerce platform, such as Alibaba, JD and Amazon. Firstly, the consumer group for community e-commerce is relatively stable. That is, the users of community e-commerce are local residents in a limited area such as a residential area, an office area or a campus. Secondly, the number of item types in a community e-commerce is relatively small. Therefore, it can study the characteristics of user-item interactions in a finer granularity than that of the traditional recommendation algorithms and this research holds that the study of fine-grained interactions between users and items is beneficial for improvement of the recommendation algorithm. For this purpose, this research classifies and studies the repeat purchase behaviors of users based on their historical purchase data.

4.2.1. The Classification Models

As the activeness of users is related to the transaction volume of the users' base over time [28], this research adopts a mathematical modeling method to model the behaviors of users along with user-item by purchase volume and the length of time in using community e-commerce.

The mathematical models of user classification are shown in Equations (4)–(6), where $h_t(u_k)$ is the user activeness. $h_t(u_k)$ is positively related to the number of item types purchased by users u_k at time t and the number of days of user u_k when using community e-commerce. This research standardizes these factors to eliminate inconsistent dimensions. Equations (4)–(6) can divide all the users of the e-commerce platform into two types, as active users and inactive users.

$$h_t(u_k) = \frac{h'_t(u_k) - min\{h'_t(u_k)\}}{max\{h'_t(u_k)\} - min\{h'_t(u_k)\}}, u_k \in U \tag{4}$$

$$h'_t(u_k) = \frac{N^t_{type}(u_k) - min\{N^t_{type}(u_k)\}}{max\{N^t_{type}(u_k)\} - min\{N^t_{type}(u_k)\}} * \frac{\Delta t(u_k) - min\{\Delta t(u_k)\}}{max\{\Delta t(u_k)\} - min\{\Delta t(u_k)\}} \tag{5}$$

$$\Delta t(u_k) = t^{u_k}_{last} - t^{u_k}_{start} \tag{6}$$

Meanwhile, the mathematical models of item classification are shown in Equations (7)–(9), where $g_t(i_s|u_k)$ is the interest stableness. $g_t(i_s|u_k)$ is positively related to the total number of item i_s purchased by user u_k before time t and the time interval between the last purchase of user u_k as well as the earliest purchase of item i_s. This research also standardized these factors to eliminate inconsistent dimensions. Equations (7)–(9) can divide users' interests in items into stable interest and unstable interest.

The symbolic definitions of the classification models are shown in Table 1. This research defines the four types of user repeat purchase behaviors based on user activeness and item stableness.

$$g_t(i_s|u_k) = \frac{g'_t(i_s|u_k) - min\{g'_t(i_s|u_k)\}}{max\{g'_t(i_s|u_k)\} - min\{g'_t(i_s|u_k)\}}, u_k \in U, i_s \in I \tag{7}$$

$$g'_t(i_s|u_k) = \frac{N^t_{num}(i_s|u_k) - min\{N^t_{num}(i_s|u_k)\}}{max\{N^t_{num}(i_s|u_k)\} - min\{N^t_{num}(i_s|u_k)\}} * \frac{\Delta t(i_s|u_k) - min\{\Delta t(i_s|u_k)\}}{max\{\Delta t(i_s|u_k)\} - min\{\Delta t(i_s|u_k)\}} \tag{8}$$

$$\Delta t(i_s|u_k) = t^{u_k i_s}_{last} - t^{u_k i_s}_{start} \tag{9}$$

Table 1. Symbolic definition.

Index	Symbols	Definition Description
1	u_k	User u_k , $u_k \in U$, $U = \{u_1, u_2, \cdots, u_k, \cdots, u_m\}$
2	i_s	Item i_s , $i_s \in I$, $I = \{i_1, i_2, \cdots, i_s, \cdots, i_n\}$
3	$h'_t(u_k)$	The activeness of user u_k (in using community e-commerce) at time t
4	$h_t(u_k)$	The activeness of user u_k at time t after standardization
5	$N^t_{type}(u_k)$	The number of item types purchased by users u_k at time t
6	$\Delta t(u_k)$	The number of days of user u_k in using community e-commerce
7	$t^{u_k}_{last}$	The last time that user u_k purchased an item in using community e-commerce
8	$t^{u_k}_{strat}$	The first time that user u_k purchased an item in using community e-commerce
9	$g'_t(i_s|u_k)$	The stability of user u_k purchasing item i_s after standardization
10	$g_t(i_s|u_k)$	The stability of user u_k purchasing item i_s
11	$N^t_{num}(i_s|u_k)$	The total number of item i_s purchased by user u_k before time t
12	$\Delta t(i_s|u_k)$	The time interval between the last purchase of user u_k and the earliest purchase of item i_s
13	$t^{u_k i_s}_{last}$	The last time user u_k purchasing item i_s
14	$t^{u_k i_s}_{start}$	The first time user u_k purchasing item i_s
15	n_{U_A}	The number of users in U_A
16	n_{U_B}	The number of users in U_B

4.2.2. User–Item Interaction

The user interacts with the item when a purchase record occurs. This section defines the user–item interactions theoretically by using mathematical modelling. It calculates the activeness of a user by mathematical model $h_t(u_k)$, and the stableness of user-item by $g_t(i_s|u_k)$. The following shows the definitions of active user and inactive user, and the definitions of stableness interest and unstableness interest.

Definition 1. *Assume that U_A denotes a set of active users. If a user u_k from U_A uses the community e-commence software for a relatively long time and purchases a variety of items, the user is an active user, where if $h_t(u_k) \geq \delta$, user $u_k \in U_A$; δ is the threshold of user activeness, $\delta \in (0,1)$, and δ is decided by the cumulative distribution of $h_t(u_k)$ of all users.*

Definition 2. *Assume that U_B denotes a set of inactive users. If a user u_k uses the threshold of user activeness for a relatively short time or purchases fewer types of item, the user is an inactive user, where if $h_t(u_k) < \delta$, user $u_k \in U_B$. So the set of all users U consists of U_A and U_B, i.e., $U = U_A \cup U_B$.*

Definition 3. *Let $I_A(u_k)$ be a set of stable interests of user u_k. If the number of item i_s purchased by user u_k is relatively large and the time span of the purchase behavior is long, item i_s is users with stable interest, where $g_t(i_s|u_k) \geq \theta$, $i_s(u_k) \in I_A(u_k)$. θ is the threshold of user-item stableness, $\theta \in (0,1)$, and the value of θ is decided by the cumulative distribution of $g_t(i_s|u_k)$ of all items.*

Definition 4. *Let $I_B(u_k)$ be a set of unstable interests of user u_k. If the number of items i_s purchased by user u_k is relatively small or the time span of the purchase behavior is short, item i_s is users with unstable interest, where $g_t(i_s|u_k) < \theta$ and $i_s(u_k) \in I_B(u_k)$. Thus, the set of all items for user u_k i.e., $I(u_k)$ consists of $I_A(u_k)$ and $I_B(u_k)$, $I(u_k) = I_A(u_k) \cup I_B(u_k)$.*

With the above definitions, the user–item interactions can be divided into four categories. The notation $(I_A|U_A)$ denotes the active users with stable interest (ASI). The notation $(I_B|U_A)$ denotes the active users with unstable interest (AUSI). The notation $(I_A|U_B)$ denotes the inactive users with stable interest (IASI). The notation $(I_B|U_B)$ denotes the inactive users with unstable interest (IAUSI). Figure 2 shows the classification process as a whole. Mathematical functions of the classification are shown as Equation (10).

$$i_s(u_k) \in \begin{cases} I_A|U_A & , if\, h_t(u_k) \geq \delta \text{ and } g_t(i_s|u_k) \geq \theta \\ I_B|U_A & , if\, h_t(u_k) \geq \delta \text{ and } g_t(i_s|u_k) < \theta \\ I_A|U_B & , if\, h_t(u_k) < \delta \text{ and } g_t(i_s|u_k) \geq \theta \\ I_B|U_B & , if\, h_t(u_k) < \delta \text{ and } g_t(i_s|u_k) < \theta \end{cases} \tag{10}$$

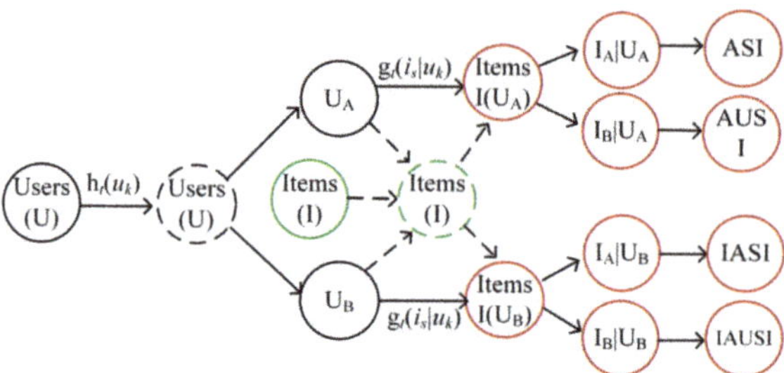

Figure 2. The classification structure of user–item interactions.

4.3. Item Recommendation

4.3.1. Model of ReRec-ASI

The overall interests of ASI users remain active and they have stable interests in items in $I_A|U_A$. This improves the algorithm upon the repurchase cycle of items. Generally, when a user has just purchased an item, the possibility of repeating the purchase immediately is very low. However, as time goes on, with the user running out of the item, he/she is more likely to make repeated purchase. For this reason, this research could prioritize the recommendation of the item to the user. This research develops a time incentive factor $w_\alpha(t^{u_k i_s})$ based on relationship of the last purchase time and the repurchase cycle to improve the user–KNN recommendation algorithm. Due to users in ASI having stable purchase interests, it assumes that their stable interests do not change over time, and the time incentive factor $w_\alpha(t^{u_k i_s})$ is a periodic piecewise constant function. The model of the time incentive function is as shown in Equation (11).

$$w_\alpha\left(t^{u_k i_s}\right) = \begin{cases} -w, & t^{u_k i_s}_{last} \leq t^{u_k i_s} < t^{u_k i_s}_{last} + \alpha T^{i_s} \\ w, & t^{u_k i_s}_{last} + \alpha T^{i_s} \leq t^{u_k i_s} < t^{u_k i_s}_{last+1} \end{cases} \tag{11}$$

Here, $t^{u_k i_s}_{last+1} = t^{u_k i_s}_{last} + T^{i_s}$, T^{i_s} is the repurchase cycle of item i_s, αT^{i_s} is the best time to recommend from time $t^{u_k i_s}_{last}$ to the next purchase time $t^{u_k i_s}_{last+1}$, and α is a lead-time factor and $\alpha \in (0,1)$.

To be specific, as users in ASI have stable purchasing interest and obvious repeat purchase behavior, it considers a periodic time incentive factor for item recommendation in ASI. That is, the time incentive factor changes with the repurchase cycle. In particular, if the last time user u_k purchases item i_s is time $t_{last}^{u_k i_s}$, he/she will purchase item i_s repeatedly at time $t_{last}^{u_k i_s} + T^{i_s}$, i.e., $t_{last+1}^{u_k i_s}$. When the recommendation time $t^{u_k i_s} \in \left[t_{last}^{u_k i_s}, t_{last}^{u_k i_s} + \alpha T^{i_s} \right)$, it is very unlikely for user u_k to make a repeat purchase. Thus, a negative time incentive factor $-w$ should be combined with the recommendation algorithm. However, when the recommendation time $t^{u_k i_s}$ is close to the next time of repeat purchase $t_{last+1}^{u_k i_s}$, and $t^{u_k i_s} \in \left[t_{last}^{u_k i_s} + \alpha T^{i_s}, t_{last+1}^{u_k i_s} \right)$, it is very likely for user u_k to repeat purchase item i_s. Thus, a positive time incentive factor w should be combined with the recommendation algorithm. This time incentive process is carried out periodically with repeated purchase.

Next, this research employs cosine similarity to calculate the similarity between users. The similarity between user u_k and user $u_{k'}$ at time t is shown in Equation (12):

$$\text{sim}(u_k, u_{k'})_t = \frac{u_k * u_{k'}}{\|u_k\| * \|u_{k'}\|} \tag{12}$$

where u_k, $u_{k'}$ are the vectors of historical purchase records of user u_k and user $u_{k'}$ before time t, respectively. The $\widetilde{R}_{u_k i_s}^{t+1}$ function of this kind of items is established as Equation (13).

$$\widetilde{R}_{u_k i_s}^{t+1} = \sum_{\substack{u_{k'} \in U_A \\ i_s \in I_A | U_A}} q_{u_{k'} i_s}^t * \text{sim}(u_k, u_{k'})_t + x_{u_k i_s} w_\alpha \left(t^{u_k i_s} \right) + \left(1 - x_{u_k i_s} \right) \overline{w_\alpha \left(t^{u_{k'} i_s} \right)} \tag{13}$$

Here, $q_{u_{k'} i_s}^t$ is the cumulative purchase of item i_s by user $u_{k'}$ at time t. $\text{sim}(u_k, u_{k'})_t$ is the similarity between user u_k and user $u_{k'}$ at time t. $w_\alpha \left(t^{u_k i_s} \right)$ is the time incentive factor if user u_k purchased item i_s at time t. If user u_k did not purchase item i_s before time t, it uses $\overline{w_\alpha \left(t^{u_{k'} i_s} \right)}$ to incentive the recommendation process, where $w_\alpha \left(t^{u_{k'} i_s} \right)$ is the time incentive factor by users $u_{k'}$ in U_A except user u_k and $\overline{w_\alpha \left(t^{u_{k'} i_s} \right)}$ is the average time incentive factor by all other users $u_{k'}$. $\overline{w_\alpha \left(t^{u_{k'} i_s} \right)}$ is established as Equation (14). $x_{u_k i_s}$ is a 0–1 variable.

$$\overline{w_\alpha \left(t^{u_{k'} i_s} \right)} = \frac{\sum_{\substack{k' \neq k \\ u_{k'} \in U_A}} x_{u_{k'} i_s} w_\alpha \left(t^{u_{k'} i_s} \right)}{\sum_{\substack{k' \neq k \\ u_{k'} \in U_A}} x_{u_{k'} i_s}} \tag{14}$$

Here, it regulates $x_{u_k i_s} = \begin{cases} 1 & \textit{if user } u_k \textit{ ever purchased item } i_s, \\ 0 & \textit{else} \end{cases}$.

4.3.2. Model of ReRec-AUSI

The overall interests of the AUSI users remain active, but they purchase items in $I_B | U_A$ of their random interest, where the activeness of users is more than threshold δ but the stableness of user-item interest is less than threshold θ. The repeat purchase behavior of users is not significant. Hence, the proposed ReRec-ASI based on the repeat purchase cycle of items will be invalid for item recommendation in AUSI. For this reason, this research considers the recommendation algorithm for the AUSI users by combining the user-KNN algorithm and the one-time hot-sale index, assuming that items with higher one-time hot-sale index in AUSI may be preferred by users. In particular, one-time hot-sale index of item i_s, denoted by $\tau_{i_s}^t$, refers to an index that is the largest single sales quantity before time t of item i_s, after the range standardized calculation. The calculation of $\tau_{i_s}^t$ is as Equation (15), where $c_{max}^{i_s t}$ is the largest one-time sales at time t of item i_s, $max \sum_{i_{s'} \in I_B | U_A} c_{max}^{i_{s'} t}$ is the largest

$c_{max}^{i_{s'}t}$ among all the $c_{max}^{i_{s'}t}$ of items in $I_B|U_A$, and $min\sum_{i_{s'}\in I_B|U_A} c_{max}^{i_{s'}t}$ is the smallest $c_{max}^{i_{s'}t}$ among all the $c_{max}^{i_{s'}t}$ of items in $I_B|U_A$. The bigger the largest single sales quantity, the greater the one-time hot-sale index. $\tau_{i_s}^t$ is a decimal between 0 and 1.

$$\tau_{i_s}^t = \frac{c_{max}^{i_s t} - min\sum_{i_{s'}\in I_B|U_A} c_{max}^{i_{s'}t}}{max\sum_{i_{s'}\in I_B|U_A} c_{max}^{i_{s'}t} - min\sum_{i_{s'}\in I_B|U_A} c_{max}^{i_{s'}t}} \tag{15}$$

It is similar to the ReRec-ASI approach that this research considers the ReRec-AUSI method by adding one-time hot-sale index to the user-KNN recommendation algorithm. However, as users in AUSI have unstable interest, this research recognizes the similarity by reversing it from 1, and then multiplying by the cumulative purchase amount of other users for item i_s. The improved similarity can pledge that not only the recommended items were purchased by similar users, but also are not always recommended. This is in line with the characteristics of unstable purchase interest of users in AUSI. Moreover, combined with the one-time hot-sale index, the improved similarity will further better the hit rate of recommended items. The $\widetilde{R}_{u_k i_s}^{t+1}$ function of ReRec-AUSI is established as Equation (16), where $q_{u_{k'} i_s}^t$ is the cumulative purchase quantity of item i_s by user $u_{k'}$ at time t. $sim(u_k, u_{k'})_t$ is the similarity between user u_k and user $u_{k'}$ at time t based on KNN algorithm. $\tau_{i_s}^t$ is the one-time hot-sale index at time t of item i_s.

$$\widetilde{R}_{u_k i_s}^t = \sum_{\substack{u_{k'}\in U_A \\ i_s \in I_B|U_A}} q_{u_{k'} i_s}^t * (1 - sim(u_k, u_{k'})_t) + \tau_{i_s}^t \tag{16}$$

4.3.3. Model of ReRec-IASI

The overall interests of users in IASI remain inactive, but they purchase items in $I_A|U_B$ of their stable interest. This research improves the algorithm upon repurchase cycle of items. Especially, it is similar to the behavior of users in ASI in that when a user has just purchased an item the possibility of repeating the purchase immediately is very low, but, as time goes on, with the user running out of the item, he/she is more likely to make repeated purchase. However, as the users in IASI remain inactive, the proposed ReRec-ASI for active users will be invalid for item recommendation in IASI, and the similarity based on users is unreliable. For this reason, this research prioritizes the item-KNN recommendation algorithm by adding a time incentive factor. Considering the characteristic of users in IASI, it assumes that the trajectory of their purchasing interest conforms to the Eibinghaus forgetting curve [29] and the interest declines over time. So, similar but different from the time incentive function in ReRec-ASI is that the principal of function segmentation of time incentive factor $w_b(t^{u_k i_s})$ of ReRec-IASI is the same, but is improved by the Eibinghaus forgetting curve, and is a periodic piecewise exponential function. The model of the time incentive function is as shown in Equation (17).

$$w_b\left(t^{u_k i_s}\right) = \begin{cases} -e^{-\frac{t^{u_k i_s} - t_{last}^{u_k i_s}}{\sigma}} & , t_{last}^{u_k i_s} \le t^{u_k i_s} < t_{last}^{u_k i_s} + \alpha T^{i_s} \\ 1 - e^{-\frac{t^{u_k i_s} - t_{last}^{u_k i_s}}{\sigma}} & , t_{last}^{u_k i_s} + \alpha T^{i_s} \le t^{u_k i_s} < t_{last+1}^{u_k i_s} \end{cases} \tag{17}$$

Here, $t_{last+1}^{u_k i_s} = t_{last}^{u_k i_s} + T^{i_s}$, T^{i_s} is the repurchase cycle of item i_s. αT^{i_s} is the best time to recommend from time $t_{last}^{u_k i_s}$ to the next purchase time $t_{last+1}^{u_k i_s}$, where α is a lead-time factor and $\alpha \in (0, 1)$. σ is the forgetting rate, and $\sigma \in (0, 1)$.

To be specific, as users in IASI have stable purchasing interest in items and obvious repeat purchase behavior, this research considers a periodic time incentive factor to item recommendation in IASI, i.e., the time incentive factor according to improved Eibinghaus forgetting curve changes with the repurchase cycle. In particular, if the last time user u_k purchases item i_s is time $t_{last}^{u_k i_s}$, generally he/she will purchase item i_s repeatedly at time $t_{last}^{u_k i_s} + T^{i_s}$, i.e., $t_{last+1}^{u_k i_s}$. When the recommendation time is $t^{u_k i_s} \in \left[t_{last}^{u_k i_s}, t_{last}^{u_k i_s} + \alpha T^{i_s} \right)$, it is very unlikely for user u_k to make a repeat purchase. For this reason, a negative time incentive factor $-e^{-\frac{t^{u_k i_s} - t_{last}^{u_k i_s}}{\sigma}}$ should be considered in the recommendation algorithm. However, when the recommendation time $t^{u_k i_s}$ is close to the next time of repeat purchase $t_{last+1}^{u_k i_s}$, where $t^{u_k i_s} \in \left[t_{last}^{u_k i_s} + \alpha T^{i_s}, t_{last+1}^{u_k i_s} \right)$, it is very likely for user u_k to repeat purchase item i_s, so a positive time incentive factor $1 - e^{-\frac{t^{u_k i_s} - t_{last}^{u_k i_s}}{\sigma}}$ should be considered in the recommendation algorithm. This time incentive process is also carried out periodically with repeated purchase.

Next, it uses cosine similarity to calculate the similarity between items. The similarity between item i_s and item $i_{s'}$ at time t is shown in Equation (18).

$$\text{sim}(i_s, i_{s'})_t = \frac{i_s * i_{s'}}{\| i_s \| * \| i_{s'} \|} \tag{18}$$

where i_s, $i_{s'}$ are the vectors of historical purchase records of item i_s and item $i_{s'}$ before time t, respectively. So, the $\widetilde{R}_{u_k i_s}^{t+1}$ function of this kind of items is established as Equation (19).

$$\widetilde{R}_{u_k i_s}^{t+1} = \sum_{i_{s'} \in I_A | U_B} q_{u_k i_{s'}}^t * \text{sim}(i_s, i_{s'})_t + x_{u_k i_s} w_b \left(t^{u_k i_s} \right) + \left(1 - x_{u_k i_s} \right) \overline{w_b \left(t^{u_{k'} i_s} \right)} \tag{19}$$

Here, $q_{u_k i_{s'}}^t$ is the cumulative purchase of item $i_{s'}$ by user u_k at time t. $\text{sim}(i_s, i_{s'})_t$ is the similarity between item i_s and item $i_{s'}$ at time t. $w_b \left(t^{u_k i_s} \right)$ is the time incentive factor when user u_k purchases item i_s at time t. If user u_k did not purchase item i_s before time t, this research uses $\overline{w_b \left(t^{u_{k'} i_s} \right)}$ to incentivize the recommendation process, where $w_b \left(t^{u_{k'} i_s} \right)$ is the time incentive factor by users $u_{k'}$ in U_B except user u_k. $\overline{w_b \left(t^{u_{k'} i_s} \right)}$ is the average value of the time incentive factor when user $u_{k'}$ who is not user u_k, purchases item i_s at time t. $x_{u_k i_s}$ is a 0–1 variable and it is modeled as Equation (20).

$$x_{u_k i_s} = \begin{cases} 1 & \text{if user } u_k \text{ ever purchased item } i_s, \\ 0 & \text{else} \end{cases} \tag{20}$$

4.3.4. Model of ReRec-IAUSI

The overall interests of the IAUSI users remain inactive and they usually purchase items in $I_B | U_B$ of their random interests, where the activeness of users is less than threshold δ and the stableness of user–item interest is also less than threshold θ. Users do not have declining repeat purchase behavior. Hence, the proposed ReRec-IASI based on declining repeat purchase cycle of items will be invalid for item recommendation in IAUSI. For this reason, this research considers the recommendation algorithm for the IAUSI users by combining the item–KNN algorithm and total hot-sale index, where it assumes that items with higher total hot-sale index in IAUSI may be preferred by users. In particular, the total hot-sale index of item i_s, denoted by $\varphi_{i_s}^t$, refers to an index that is the largest total sales quantity before time t of item i_s, after the range standardized calculation. The calculation of $\varphi_{i_s}^t$ is as in Equation (21), where $C^{i_s t}$ is the largest total sales before time t of item i_s, $max \sum_{i_{s'} \in I} C^{i_{s'} t}$ is the largest $C^{i_s t}$ among all the $C^{i_s t}$ of items in $I_B | U_B$, and $min \sum_{i_{s'} \in I} C^{i_{s'} t}$ is the smallest $C^{i_s t}$ among all the $C^{i_s t}$ of items in $I_B | U_B$. The bigger the largest total sales quantity, the greater the total hot-sale index. $\varphi_{i_s}^t$ is a decimal between 0 and 1.

$$\varphi_{i_s}^t = \frac{C^{i_s t} - min \sum_{i_{s'} \in I_B | U_B} C^{i_{s'} t}}{max \sum_{i_{s'} \in I_B | U_B} C^{i_{s'} t} - min \sum_{i_{s'} \in I_B | U_B} C^{i_{s'} t}} \tag{21}$$

Similar to the ReRec-IASI approach, this research considers the ReRec-IASUI method by adding an incentive factor which is a hot-sale index to the item–KNN recommendation algorithm. However, as users in IAUSI have unstable interest, the research recognizes the similarity by reversing it from 1, and then multiplying by the cumulative purchase amount of other users for item i_s. The improved similarity can show not only that the recommended items were purchased by similar users, but also that the recommended items are diverse. This is in line with the characteristics of unstable purchase interest of users in IAUSI. Moreover, combined with the total hot-sale index, the improved similarity will further increase the hit rate of recommended items. The $\widetilde{R}^{t+1}_{u_k i_s}$ function of ReRec-IAUSI can be formed as Equation (22).

$$\widetilde{R}^{t+1}_{u_k i_s} = \sum_{i_{s'} \in I_B | U_B} q^t_{u_k i_{s'}} * (1 - \text{sim}(i_s, i_{s'})_t) + \varphi^t_{i_s} \tag{22}$$

where $q^t_{u_k i_{s'}}$ is the cumulative purchase of item $i_{s'}$ by user u_k at time t. $\text{sim}(i_s, i_{s'})_t$ is the similarity between item i_s and item $i_{s'}$ at time t. $\varphi^t_{i_s}$ is the hot-sale index at time t of item i_s.

5. Experiments

5.1. The Dataset

The dataset used in this paper comes from a community e-commerce platform T-app, with 11,350 purchase records from June 2017 to August 2019. It contains 1064 users and 137 kinds of items. The characteristics of each record include user ID, item ID, purchase time, purchase quantity, price, payment method and other attributes. Specifically, the data from June 2017 to April 2019 (10,343 records) are used as the training set, and the data from April 2019 to August 2019 (1007 records) are used as the test set. The user–item recommendation models are trained on the training set, and are tested on the test set.

The purchase behavior of users on the T-app platform has obvious characteristics of repurchase. For instance, by analyzing the data of a time phase, it is found that among 955 users who have made purchases, 58.74% have repeat purchases. In Figure 3, it can be seen that the total repurchases of 23% of repurchase users is larger than 15. The average repurchase time of repurchase users is 3.61. Among the repurchase users, 10.33% repurchase the same item more than six times. In an extreme case, it is found that one user has repurchased the same item up to 43 times under the investigated time duration. In Figure 4, it can be seen that, among all the types of item (105 types), 78.10% (82 types) have been repurchased by users, and in 17% of the repurchased items, the total number of times repurchased by users is more than 120.

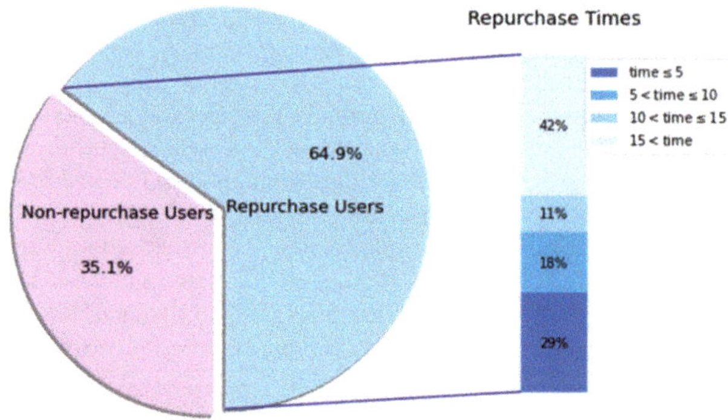

Figure 3. Proportions of repurchase users.

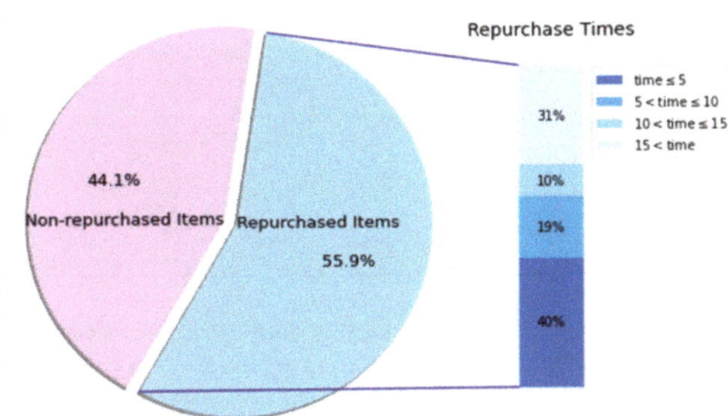

Figure 4. Proportions of repurchased items.

5.2. Experimental Setup

In the traditional collaborative filtering recommendation, the user–item score matrix is usually used as the original data for the recommendation calculation. This paper adopts offline experiments for verification, and the user's cumulative purchase is used as the score. First, according to the user classification model, the user-item is classified into four categories: active users with stable interest, active users with unstable interest, inactive users with stable interest and inactive users with unstable interest. Then, the recommendation calculation is carried out for each category, and the improved recommendation algorithms for active users and inactive users are evaluated respectively. The results are then compared with that of the traditional CF, SVD, SVD++, and NMF algorithms.

The repurchase cycle refers to the time interval between the nth and the $(n + 1)$th purchase of item i_s by user u_k. For an item, the repurchase cycle of different users at the same time period is different, and that of the same user at different time periods is also different. So, if the items' repurchase cycle is calculated by each user by time, it could be highly random and prone to overfitting. Therefore, for the active users' stable purchase behavior, the average repurchase cycle of the top three users in purchase quantity of a certain item is used as the repurchase cycle. For the items included in $I_A|U_B$, as the overall interest of users is inactive, the repurchase cycle of the user who purchases the largest quantity of an item is regarded as the repurchase cycle of this item. Examples of repurchase cycle for some items included in $I_A|U_A$ are shown in Table 2 and for some items included in $I_A|U_B$ in Table 3.

Table 2. Repurchase cycle of typical items included in $I_A|U_A$.

Item ID	Repurchase Cycle (Days)	Name
2	14.07	ZY
38	24.65	TB-Mo
61	21.25	TB-Th
68	14.84	ZQB-F
69	20.37	ZYB-We
73	15.51	HB-We

Table 3. Repurchase cycle of typical items included in $I_A|U_B$.

Item ID	Repurchase Cycle (Days)	Name
2	10	ZY
38	16	TB-Mo
61	24	TB-Th
68	12	ZQB-Fr
69	14	ZYB-We
73	11	HB-We

In the experiments, each type of user behavior model can produce a corresponding item recommendation list. After sorting in descending order according to the purchase possibility, the recommended items can be selected according to the top N method. $N_{(U_A)}$ is the number of active users and $N_{(U_B)}$ is the number of inactive users, and they can be expressed as Equations (23) and (24), respectively.

$$N_{(U_A)} = N_{(I_A|U_A)} + N_{(I_B|U_A)} \tag{23}$$

$$N_{(U_B)} = N_{(I_A|U_B)} + N_{(I_B|U_B)} \tag{24}$$

Here, $N_{(I_A|U_A)}$ is the recommended item quantity from items included in $I_A|U_A$. $N_{(I_B|U_A)}$, $N_{(I_A|U_B)}$, $N_{(I_B|U_B)}$ are similar in meaning to $N_{(I_A|U_A)}$. So, it is easy to discover that the recommendation list of active users is composed of $N_{(I_A|U_A)}$ stable interests and $N_{(I_B|U_A)}$ unstable interests. Similarly, the recommendation list of inactive users is composed of $N_{(I_A|U_B)}$ stable interests and $N_{(I_B|U_B)}$ unstable interests.

Considering the actual situation of T-app, its operators should select the best combination of items in different user–item classifications for recommendation. Hence, here this research uses the grid search method to test the models. Firstly, let the total number of recommendation items be less than the number of all items, N_{max}, for each type of user. Both the number of stable items and unstable items should be less than N_{max}. That is to say, it has constraints (25) and (26). In the test experiment, N_{max} is set as 25. Secondly, with constraints (25), $N_{(U_A)}$ has multiple combinations of $N_{(I_A|U_A)}$ and $N_{(I_B|U_A)}$, and it is the same as $N_{(U_B)}$. For instance, when the total number of recommended items N_{max} is 5, $(N_{(I_A|U_A)}, N_{(I_B|U_A)})$ can be able to (0,5), (1,4), (2,3), (3,2), (4,1), (5,0). It can select the optimal combination among the six combinations as the recommended combination when $N_{(U_A)} = 5$.

$$\begin{cases} 0 \leq N_{(I_A|U_A)} \leq N_{max} \\ 0 \leq N_{(I_B|U_A)} \leq N_{max} \\ 0 \leq N_{(I_A|U_A)} + N_{(I_B|U_A)} \leq N_{max} \end{cases} \tag{25}$$

$$\begin{cases} 0 \leq N_{(I_A|U_B)} \leq N_{max} \\ 0 \leq N_{(I_B|U_B)} \leq N_{max} \\ 0 \leq N_{(I_A|U_B)} + N_{(I_B|U_B)} \leq N_{max} \end{cases} \tag{26}$$

5.3. Evaluation Metrics

Three evaluating indicators are used to gauge the algorithm performance, precision (Pre), recall (Rec) and F-measure, defined in Equations (27)–(29). Precision is defined as the ratio of items that users like to all recommended items in the recommended list. Recall is defined as the ratio of the items that users like in the recommended list to all the items that users like in the system. Generally, precision and recall must be used at the same time to fully evaluate the quality of the algorithm. Some researchers have proposed an indicator called F-measure that comprehensively integrates the precision and the recall. Therefore, the evaluation indicators used in this paper are precision, recall, and F-measure to measure the precision of item recommendation. The three expressions are shown as (27)–(29).

$$Pre = \frac{TP}{TP + FP} \tag{27}$$

$$Rec = \frac{TP}{TP + FN} \tag{28}$$

$$F - measure = \frac{2 \times Pre \times Rec}{Pre + Rec} \tag{29}$$

Here, TP is the number of items that have been recommended and purchased; FP is the number of items that are recommended but not purchased; and FN is the number of items that have not been recommended but purchased.

5.4. Experimental Results

Figure 5 shows the comparison results of the proposed ReRec algorithm on active users (i.e., the combination of ReRec-ASI and ReRec-AUSI) compared with four baseline methods, traditional User CF, SVD, SVD++ and NMF algorithms. It sets $w = 1$, and $N_{(U_A)} \in [5, 25]$. It can be seen from Figure 5 that, on the purchase prediction of active users, the proposed ReRec algorithm performs better than the traditional User CF, SVD, SVD++ and NMF algorithms in terms of the three evaluation indicators, precision, recall and F-measure. This indicates that the proposed ReRec algorithm for active users in this paper improves the hit rate of item recommendation and ensures the precision of recommendation results.

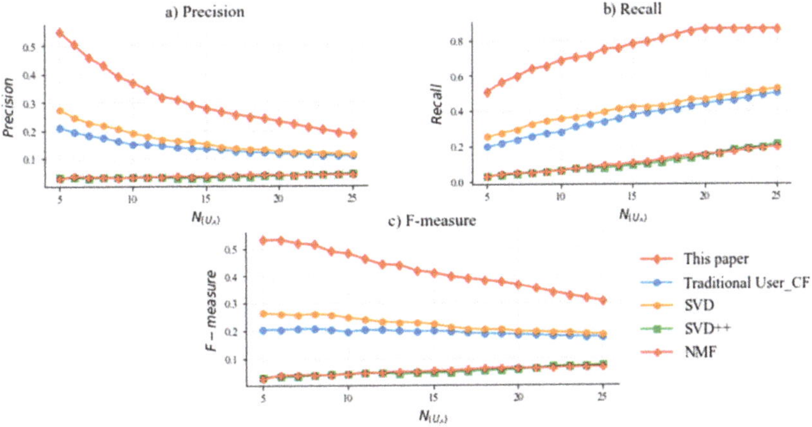

Figure 5. The comparison results of the proposed ReRec algorithm (the combination of ReRec-ASI and ReRec-AUSI) and four baselines on active users.

Figure 6 shows the comparison results of the proposed ReRec approach with the baselines on inactive users (the combination of ReRec-IASI and ReRec-IAUSI). It sets the parameters α and σ as 0.75 and 0.2284, respectively. The total number of recommendation items of $N_{(U_B)}$ is the same as $N_{(U_A)}$. It can be seen that, in the purchase prediction of inactive users, when $N_{(U_B)} \in [6, 25]$, the improved Item CF algorithm proposed in this paper is superior to the evaluation indicators of traditional Item CF, SVD, SVD++ and NMF algorithms in terms of precision, recall and F-measure. Because the number of item type in the test data is relatively smaller than the number of users, the purchase prediction performance for inactive users is not as good as that for active users. However, the purchase prediction of inactive users based on the improved Item CF algorithm still improves the hit rate of item recommendation within a certain range, and also ensures a higher precision of recommendation results.

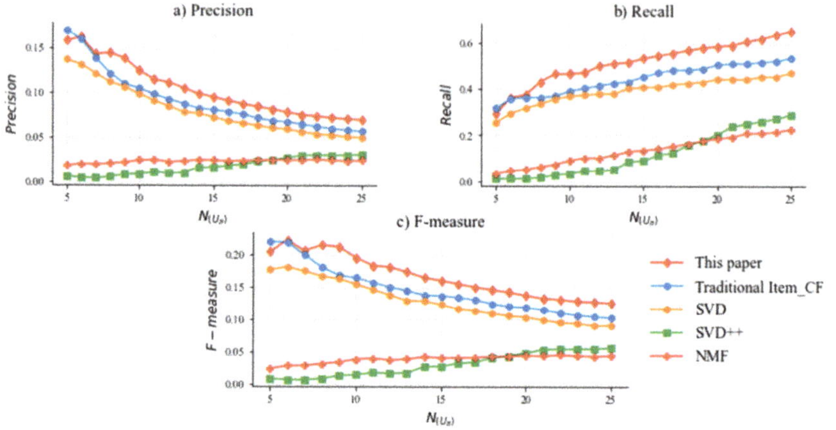

Figure 6. The comparison results of the proposed ReRec approach (the combination of ReRec-IASI and ReRec-IAUSI) and four baselines on inactive users.

The poor performance of the baselines can be explained because all ratings in the user item rating matrix are regarded as equal, ignoring the heterogeneity of users' interests, i.e., user's personalized interest and users' public interest. The SVD method, which is derived from linear algebra, has a solid mathematical foundation in matrix approximation. However, it lacks a user's preference model and an item's preference model of the user's interest in the item. In the SVD++ method, a bias model and the latent vectors of the user and the item are used to model the user's interest in the item. Using stochastic gradient descent to update the bias vector and latent vector of each observed rating in the user item rating matrix can result in a large amount of computation. The advantage of the NFM model is that the elements of latent users and item vectors can be non-negative, while its disadvantage is that the precision of rating prediction is reduced.

In summary, none of the baselines improve the recommendation algorithms according to different types of user behavior on the temporal horizon. Although some scholars have added the user's personalized behavior into the item recommendation algorithm, they more often than not ignore user loyalty in recommendations that may drive the users' repeat purchase. It holds that the users' loyalty to the shopping platform and items has a non-negligible impact on the successful recommendation of items. Following this line of thought, this research proposes the ReRec algorithm based on user behavior classification and item repurchase cycle. The proposed ReRec algorithm can predict the possibility of repeat purchase in order to recommend the top N items to users and improve the user experience of the recommendation system.

5.5. Sensitivity Analysis of Parameter w

In the proposed ReRec approach for active users, incentive factor w is an important parameter. In order to analyze the influence of w on the recommendation process, it conducts sensitivity analysis on the parameter w. Figure 7 illustrates the F-measures with $N_{(I_A|u_A)}$ and $N_{(I_B|u_A)}$, when other conditions are fixed and w varies. The following conclusions can be drawn from Figure 7. When $w \in [0,5]$, for $N_{(I_A|u_A)} \leq 10$ and $N_{(I_B|u_A)} \leq 15$, the F-measures with various combination of $N_{(I_A|u_A)}$ and $N_{(I_B|u_A)}$ are better than that of other conditions.

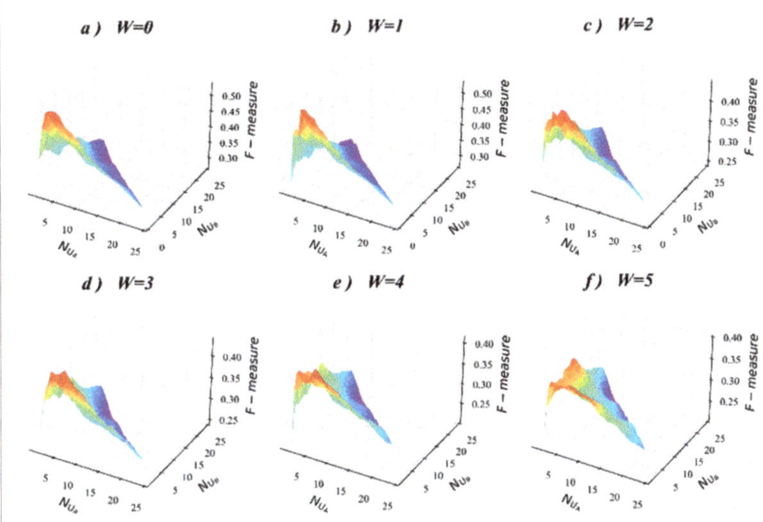

Figure 7. The F-measures of different values of w.

Figure 8 illustrates the F-measures with incentive factor w given $N_{(u_A)} = 3$. It can be seen that, with the value of w increasing in [0,5], the value of F-measure first increases and then decreases. When $w > 5$, the values of F-measure are kept stable. Therefore, the research further analyzes the evaluation indicators with $w \in [0,5]$.

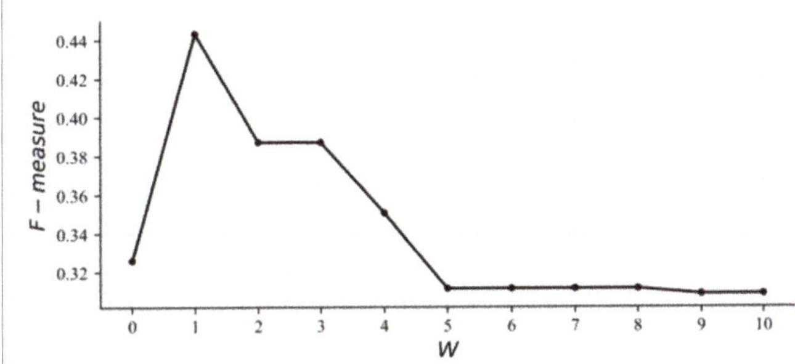

Figure 8. The F-measures for incentive factor w.

Figure 9 illustrates the evaluating indicators (precision, recall and F-measure) with the total recommended quantity $N_{(u_A)}$ when $w \in [0,5]$. It can be seen that when $N_{(u_A)}$ increases in the range [0,5], the variation trend of precision is relatively unstable. In comparison, the recall and F-measure go up firstly and then go down. While $N_{(u_A)}$ increases in the range [5,25], the precisions gradually decrease, while the recall increases. As a result, the F-measures decrease. It is evident that when $N_{(u_A)} > 7$, the performances of three evaluating indicators at $w = 1$ are better than that at other values of w. Therefore, the ReRec algorithm should be used with the setting as $w = 1$.

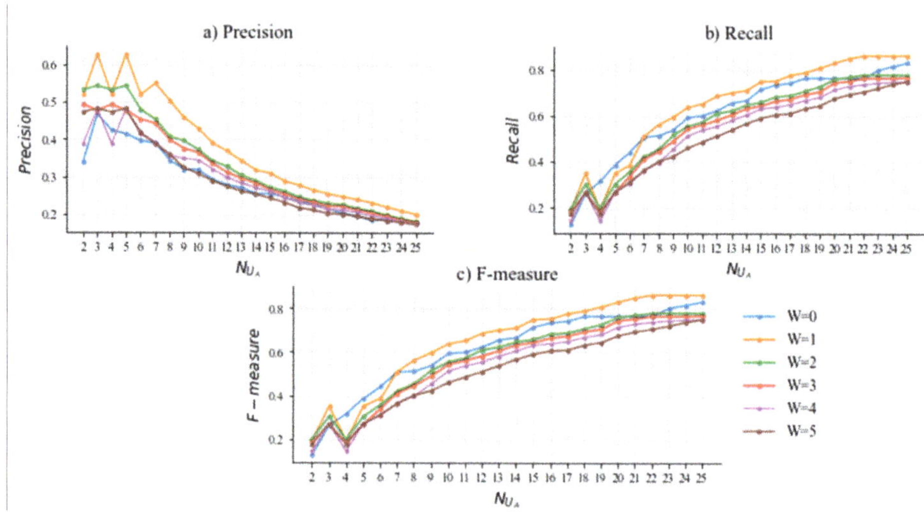

Figure 9. The change of three evaluating indicators at given w.

In the proposed ReRec approach for inactive users, the grid search method is adopted to carry out ReRec-IASI and ReRec-IAUSI. The precision, recall and F-measures with $N_{(I_A|U_B)}$ and $N_{(I_B|U_B)}$ are shown in Figure 10. It can be seen that when $N_{(I_A|U_B)}$ is fixed, the precisions of the recommendation results are decreasing along with the increase of $N_{(I_B|U_B)}$. When $N_{(U_B)}$ is small, the precisions and F-measures are large. The recalls of recommendation results are large when $N_{(I_A|U_B)}$ and $N_{(I_B|U_B)}$ and are approximately equal to each other.

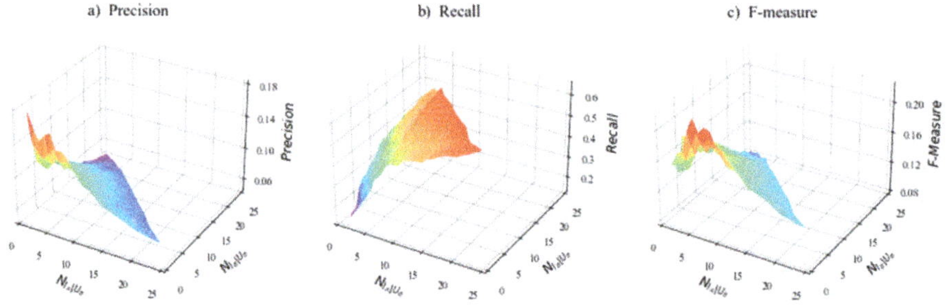

Figure 10. The changes of three evaluation indicators with the combination of inactive users.

5.6. Discussion of Important Results

The proposed methods are trained in the training set and evaluated in the test set. Three evaluating indicators are used to gauge the algorithm performance as precision, recall and F-measure (Equations (27)–(29)). We conduct our experiments on a real-life community e-commerce platform. Results show that the proposed ReRec method provides better performance compared to the existing methods (namely traditional CF, SVD, SVD++, NMF). The discussion of the important results of the proposed methods is analyzed as follows.

Four types of user-item interactions are obtained before applying ReRec: active users with stable interest (ASI), inactive users with stable interest (IASI), active users with unstable interest (AUSI), and inactive users with unstable interest (IAUSI). For active users, the hit rate of item recommendation shows a marked improvement, while for inactive users, the hit ratio increased slightly. Compared with inactive users, active users use the platform more frequently, so it is easier to detect their buying interest. The reason for the poor performance of the baselines may be that none of the baselines improves the recommendation algorithms according to different types of user behavior on the temporal horizon.

The performance of ReRec is analyzed based on varying the values of incentive factor w. With the value of w increasing in [0,5], the value of F-measure first increases and then decreases. When $w > 5$, the values of the F-measure are kept stable and low. When $w = 1$, the ReRec algorithm shows the highest precision. It is evident that when $N_{(U_A)} > 7$, the performances of three evaluating indicators at $w = 1$ are better than that at other values of w. Therefore, the ReRec algorithm should be used with the setting as $w = 1$.

Finally, the practical contribution is summarized. The result of recommendation is stable, which can provide support for business management decision-making in enterprises, and the effectiveness of the algorithm is verified. For instance, precise marketing strategies based on customer heterogeneity can be implemented, thus reducing the operating costs of community e-commerce platforms.

6. Concluding Remarks

To fill in the research gap from the perspective of repeated purchase behavior and improve the process of the generation of a recommendation list, this research proposed a novel approach called ReRec (Repeat purchase Recommender) to recommending items to users in a divide and conquer manner. The proposed method includes ReRec-ASI, ReRec-AUSI, ReRec-IASI and ReRec-IAUSI. Experiments are conducted on a real dataset collected from a community e-commerce platform. Compared with well-known existing methods (e.g., SVD, SVD++) the ReRec method improves the recommend performance by at least 13.6% (measured by F-measure). Specifically, for active users, with $w = 1$ and $N_{(U_A)} \in [5, 25]$, the ReRec-ASI, ReRec-AUSI shows a significant improvement (at least 50%) in recommendation. With α and σ as 0.75 and 0.2284, respectively, the proposed ReRec-IASI and ReRec-IAUSI are also superior to (by at least 13.6%) the evaluation indicators of traditional Item CF when $N_{(U_B)} \in [6, 25]$.

Although the proposed ReRec approach performs well in this study, there are still some gaps to be explored in the future:

Firstly, the size of test set needs to be expanded, because this paper only uses four months of consumption data to test the algorithm at present. The amount of data that can be used now on T-app is limited. In the future, there will be more consumption data available this large-scale data can be used to verify the algorithm.

Secondly, when it has consumption data over a long time, such as consumption data for several years, it would attempt to improve the recommendation algorithms with centralized consumption behaviors such as seasonal consumption and holiday consumption, which is an interesting problem in recommendation.

Author Contributions: Conceptualization, J.W. and W.Z.; methodology, J.W., Y.L. and W.Z.; software, Y.L. and L.Y.; validation, Y.L., L.Y. and X.N.; formal analysis, J.W., Y.L., L.S. and W.Z.; investigation, Y.L. and L.S.; resources, J.W. and L.S.; data curation, Y.L., L.Y. and X.N.; writing—original draft preparation, Y.L., L.Y. and X.N.; writing—review and editing, J.W. and W.Z.; visualization, Y.L. and L.Y.; supervision, J.W. and W.Z.; project administration, J.W., L.S. and W.Z.; funding acquisition, J.W. and W.Z. All authors have read and agreed to the published version of the manuscript.

Funding: This research is funded by National Natural Science Foundation of China under Grant Nos. 72174018 and 71932002; Beijing Youth Talent Fund under Grant No. Q0011019202001; Beijing Natural Science Fund under Grant No. 9222001; Beijing University of Chemical Technology First-Class Discipline Construction (XK1802-5), and Beijing University of Chemical Technology (GJD202002).

Institutional Review Board Statement: Not applicable.

Informed Consent Statement: Not applicable.

Data Availability Statement: Not applicable.

Conflicts of Interest: The authors declare no conflict of interest.

References

1. Zsolnai, L. Green business or community economy? *Int. J. Soc. Econ.* **2002**, *29*, 652–662. [CrossRef]
2. Lao, J.; Zhong, Y.; Tan, Z. Study of community E-commerce model based on intelligent building. *J. Intell.* **2007**, *26*, 39–41.
3. Kim, H.K.; Oh, H.Y.; Gu, J.C.; Kim, J.K. Commenders: A recommendation procedure for online book communities. *Electron. Commer. Res. Appl.* **2011**, *10*, 501–509. [CrossRef]
4. Zhang, W.; Du, Y.; Yang, Y.; Yoshida, T. DeRec: A data-driven approach to accurate recommendation with deep learning and weighted loss function. *Electron. Commer. Res. Appl.* **2018**, *31*, 12–23. [CrossRef]
5. Iwanaga, J.; Nishimura, N.; Sukegawa, N.; Takano, Y. Improving collaborative filtering recommendations by estimating user preferences from clickstream data. *Electron. Commer. Res. Appl.* **2019**, *37*, 100877. [CrossRef]
6. Ghasemi, N.; Momtazi, S. Neural text similarity of user reviews for improving collaborative filtering recommender systems. *Electron. Commer. Res. Appl.* **2020**, *45*, 101019. [CrossRef]
7. Riyahi, M.; Sohrabi, M.K. Providing effective recommendations in discussion groups using a new hybrid recommender system based on implicit ratings and semantic similarity. *Electron. Commer. Res. Appl.* **2020**, *40*, 100938. [CrossRef]
8. Verstrepen, K.; Bhaduriy, K.; Cule, B.; Goethals, B. Collaborative filtering for binary, positive-only data. In Proceedings of the 23rd ACM SIGKDD Conference, Halifax, NS, Canada, 13–17 August 2017; pp. 1–21.
9. Chen, J.; Wei, L.; Zhang, L. Dynamic evolutionary clustering approach based on time weight and latent attributes for collaborative filtering recommendation. *Chaos Solitons Fractals* **2018**, *114*, 8–18. [CrossRef]
10. Verbert, K.; Manouselis, N.; Ochoa, X.; Wolpers, M.; Drachsler, H.; Bosnic, I.; Duval, E. Context-aware recommender sys-tems for learning: A survey and future challenges. *IEEE Trans. Learn. Technol.* **2012**, *5*, 318–335. [CrossRef]
11. Mezni, H.; Benslimane, D.; Bellatreche, L. Context-aware service recommendation based on knowledge graph em-bedding. *IEEE Trans. Knowl. Data Eng.* **2021**, *99*, 1–14. [CrossRef]
12. Nguyen, V.-D.; Sriboonchitta, S.; Huynh, V.-N. Using community preference for overcoming sparsity and cold-start problems in collaborative filtering system offering soft ratings. *Electron. Commer. Res. Appl.* **2017**, *26*, 101–108. [CrossRef]
13. Resnick, P.; Iacovou, N.; Suchak, M.; Bergstrom, P. Group Lens: An open architecture for collaborative filtering of net news. In Proceedings of the ACM 1994Conference on Computer Supported Cooperative Work, Chapel Hill, NC, USA, 22–26 October 1994; pp. 175–186.
14. Sarwar, B.; Karypis, G.; Konstan, J.; Riedl, J. Item-based collaborative filtering recommendation algorithms. In Proceedings of the 10th International Conference on World Wide Web, Hong Kong, China, 1–5 May 2001; pp. 285–295.
15. Wang, C.; Zheng, Y.; Jiang, J.; Ren, K. Toward Privacy-Preserving Personalized Recommendation Services. *Engineering* **2018**, *4*, 21–28. [CrossRef]
16. Brand, M. Fast online SVD revisions for lightweight recommender systems. In Proceedings of the Third SIAM International Conference on Data Mining, San Francisco, CA, USA, 1–3 May 2003; pp. 37–46.
17. Koren, Y. Factorization meets the neighborhood: A multifaceted collaborative filtering model. In Proceedings of the 14th ACM SIGKDD International Conference on Knowledge Discovery and Data Mining, Las Vegas, NV, USA, 24–27 August 2008; pp. 426–434.
18. Kim, J.; Park, H. Toward Faster Nonnegative Matrix Factorization: A New Algorithm and Comparisons. In Proceedings of the 2008 Eighth IEEE International Conference on Data Mining, Pisa, Italy, 15–19 December 2008; pp. 353–362. [CrossRef]
19. Gorgoglione, M.; Panniello, U.; Tuzhilin, A. The effect of context-aware recommendations on customer purchasing behavior and trust. In Proceedings of the Fifth ACM Conference on Recommender Systems, Chicago, IL, USA, 23–27 October 2011; pp. 85–92. [CrossRef]
20. Zimdars, A.; Chickering, D.M.; Meek, C. Using temporal data for making recommendations. In Proceedings of the Seventeenth Conference on Uncertainty in Artificial Intelligence, Seattle, WA, USA, 2–5 August 2001; pp. 580–588.
21. Campos, P.G.; Díez, F.; Bellogín, A. Temporal rating habits: A valuable tool for rating discrimination. In Proceedings of the 2nd Challenge on Context-Aware Movie Recommendation, Chicago, IL, USA, 27 October 2011; pp. 29–35.
22. Liang, X.; Yang, Q. Time-dependent models in collaborative filtering based recommender system. In Proceedings of the 2009 IEEE/WIC/ACM International Joint Conference on Web Intelligence and Intelligent Agent Technology, Milano, Italy, 15–18 September 2009; pp. 450–457.
23. Qin, G.; Du, X. An efficient collaborative filtering algorithm with user hierarchy. *Comput. Sci.* **2004**, *10*, 138–140.
24. Xing, C.; Gao, F.; Zhan, S.; Zhou, L. A collaborative filtering recommendation algorithm incorporated with user interest change. *J. Comput. Res. Dev.* **2007**, *02*, 296–301. [CrossRef]
25. Zhang, Y.; Liu, Y. A Collaborative Filtering Algorithm Based on Time Period Partition. In Proceedings of the Third International Symposium on Intelligent Information Technology & Security Informatics, Washington, DC, USA, 2–4 April 2010; pp. 777–780.
26. Chen, J.; Lu, Y.; Shang, F.; Zhu, T. A novel recommendation scheme with multifactorial weighted matrix decomposition strategies via forgetting rule. *Eng. Appl. Artif. Intell.* **2021**, *101*, 104191. [CrossRef]
27. Wu, F.; Yu, L.; Feng, M. A collaborative filtering algorithm based on time effect. *Comput. Eng. Sci.* **2017**, *39*, 2095–2101.
28. Fader, P.S.; Hardie, B.G.; Lee, K.L. RFM and CLV: Using iso-value curves for customer base analysis. *J. Mark. Res.* **2005**, *42*, 415–430. [CrossRef]
29. Hermann, E. Memory: A Contribution to Experimental Psychology. *Ann. Neurosci.* **2013**, *20*, 155–156.

Article

Joint Optimization of Ticket Pricing Strategy and Train Stop Plan for High-Speed Railway: A Case Study

Jin Qin [1,2,*], Xiqiong Li [1], Kang Yang [1] and Guangming Xu [1]

[1] School of Traffic & Transportation Engineering, Central South University, Changsha 410075, China; 194211018@csu.edu.cn (X.L.); 194211007@csu.edu.cn (K.Y.); 203070@csu.edu.cn (G.X.)
[2] Rail Data Research and Application Key Laboratory of Hunan Province, Changsha 410075, China
* Correspondence: qinjin@csu.edu.cn

Abstract: In this study, we examined ticket pricing and train stop planning for the high-speed railway (HSR), which integrates two key aspects of railway operation and organization. We considered that passenger demand is sensitive to the generalized travel cost (depending on the ticket price and the travel time) and that the train stop plan can affect the travel time and passenger distribution. Then, a mixed-integer non-linear optimization model was proposed for the joint problem of ticket pricing and train stop planning to maximize HSR's transport revenue and minimize passengers' travel time. Based on the high similarity between combinatorial optimization problems and the solid annealing principle, we designed a combined simulated annealing (CSA) algorithm to solve practical problems. The results of a numerical example in the real HSR network showed that the proposed method can improve transport revenue by 5.1% and reduce passengers' travel time loss by 11.15% without increasing transport capacity.

Keywords: high-speed railway; ticket pricing; train stop plan; seat allocation; joint optimization; combined simulated annealing algorithm

MSC: 90B10

Citation: Qin, J.; Li, X.; Yang, K.; Xu, G. Joint Optimization of Ticket Pricing Strategy and Train Stop Plan for High-Speed Railway: A Case Study. *Mathematics* **2022**, *10*, 1679. https://doi.org/10.3390/math10101679

Academic Editors: Antonin Ponsich, Mariona Vila Bonilla and Bruno Domenech

Received: 31 March 2022
Accepted: 11 May 2022
Published: 13 May 2022

Publisher's Note: MDPI stays neutral with regard to jurisdictional claims in published maps and institutional affiliations.

1. Introduction

High-speed railway (HSR) is the preferred transport mode for medium-to-long-distance passengers across the world, especially in China. Up to now, China has constructed an HSR network with an operational mileage of 40,000 km, which has successfully transported more than 18 billion passengers. HSR has enabled most of the railway passenger transport in China, and is playing an increasingly important role in the entire passenger transport market.

However, compared with other competitive transport modes, such as air and highway transportation, the market-oriented operation of the HSR is relatively backward in China. This is mainly manifested in its fixed pricing strategy. The ticket price of HSR in China is calculated by multiplying the transport mileage by the price rate depending on seat class and has been strictly controlled by the government for the past few years. This price mechanism ignores the effect of passenger demand on price adjustment, which is not conducive to the sustainable development of the HSR.

Realizing the drawbacks of the existing pricing strategy, the National Development and Reform Commission has allowed railway enterprises to set HSR ticket prices independently since 2016. Several railway enterprises have implemented price reforms for some high-speed trains, and the practical results have demonstrated that dynamic pricing for HSR can indeed help railway operators improve ticket revenue, which means realizing revenue management (RM) for HSR.

Since originating in the 1970s, RM has been a long-standing issue in many service industries, such as airlines and hospitality [1–7]. The successful application of RM in airlines has encouraged many researchers to conduct RM studies for railways [8–12]. Compared with conventional railways, the policies in pricing and operation modes for HSR are more flexible, which provides more RM applicability for HSR. In Japan and some European countries, railway enterprises implemented RM for HSR earlier than for conventional railways [9]. China Railway has obtained pricing rights for HSR, and has an opportunity to further realize RM.

Ticket pricing and seat control are two significant strategies used by railway enterprises to realize RM. They are interrelated and complementary to each other, and thus should be comprehensively considered in an RM system. Since the train stop plan has a significant impact on seat allocation, it is also mutually interrelated with ticket pricing. The ticket pricing and seat allocation problem has generated numerous studies, but often the two issues have been treated separately. Moreover, such papers have assumed that the train stop plan is fixed to compute the optimal allocation of resources.

This paper aims at filling this gap by establishing a mathematical model for the joint problem of ticket pricing, seat allocation, and train stop planning for HSR. The goal is to balance transport supply and passenger demand, improve the revenue of railway enterprises, and reduce the total travel time of passengers.

Our contributions can be summarized as follows:

(1) Few studies consider the joint problem of HSR pricing, seat allocation, and train stop planning. This is one of the limited number of papers that jointly optimize pricing, seat allocation, and stop planning for HSR.
(2) Considering the impact of the stop plan on the travel time, an elastic passenger demand function related to the ticket price and travel time is constructed.
(3) Based on the simulated annealing theory, an efficient solution algorithm is designed for the combinatorial optimization problem of ticket pricing, seat allocation, and train stop planning for HSR.

The remainder of this paper is organized as follows. Section 2 presents the research on ticket pricing, seat allocation, and train stop planning. Section 3 describes the elastic passenger demand and choice behaviors in a mathematical way, and formulates a collaborative ticket pricing and stop planning model. Section 4 elaborates on the design of the solution algorithm and the specific implementation steps. Section 5 provides an empirical analysis. Section 6 concludes the research and gives future research directions.

2. Literature Review

Dynamic pricing is a classic strategy used by enterprises to improve revenue. It involves selling the same product to different consumers at the right time at different prices. The basics of dynamic pricing comprise the supply capacity of enterprises and the market demand. A suitable dynamic pricing strategy can also regulate and guide the market demand, which is beneficial to the operation of enterprises. Many papers have examined dynamic pricing problems for railways. Vuuren [13] and Jarocka and Ryciuk [14] focused on dynamic pricing for the peak and off-peak periods. In their studies, social welfare and enterprise profits were the main considerations for peak and off-peak pricing, respectively. In a number of studies, division of the ticket pre-sale horizon was the first stage of the dynamic pricing process. Mutations in passenger ticketing demand were usually regarded as signals from which to adjust ticket prices [15]. Based on the number of research efforts on passenger choice behavior [16,17], railway RM models for dynamic pricing of competing routes have been proposed [18,19]. They assumed that passengers can choose other transport modes providing different alternative timetables. In this context, Chen and Gao [19] developed a new method to compute the generalized travel cost, and then use the logit model to allocate passenger flows to different routes. Numerical experiments suggested that their RM models can lead to significant revenue gains.

In terms of seat allocation, most existing research has assumed that passenger demand is fixed [20]. Nevertheless, railway passenger demand varies dynamically. Thus, Jiang et al. [21] studied an approach for HSR seat allocation with dynamic adjustments.

They integrated dynamic seat allocation and short-term demand forecasting to improve the utilization of seats. Yan et al. [22] developed a seat allocation model for multiple HSR trains with flexible train formation. The authors allowed changes in the formation of each train to gain flexible capability. The study provided decision support for seat allocation and train formation simultaneously.

In recent years, the integration of pricing and seat allocation decisions has received more attention. The joint study was first conducted on a single train. Hetrakul et al. [23] supposed that the daily passenger demand was fixed, and used the multinomial logit and latent class models to obtain the ticket reservation time of passengers. Finally, they proposed a collaborative optimization model for railway dynamic pricing and seat allocation. Later, a limited number of studies focused on the joint pricing and seat allocation problem for multiple HSR trains with different stop patterns [24–27]. However, little research has included flexible train stop planning in the joint pricing and seat allocation problem.

The train stop plan is the most important part of the operation scheme for HSR. It determines whether a train can allocate tickets to an origin-destination pair (OD) and affects the total stopping time of each train. The stop plan is usually formulated according to passenger demand. Trains are generally scheduled to stop at large stations and partial small stations to form a good stop plan. On the one hand, this can improve the service quality of each station; on the other hand, it helps railway enterprises save the stopping cost by arranging resources rationally [28]. The stop plan formulation problem has been traditionally modeled and solved considering fixed train formation and fixed stopping time in order to minimize the total cost or the total travel time of passengers [29–31]. Jin et.al. [32] argued that a fixed stopping time may be insufficient for passengers to get on or off the train. Thus, their optimization model considered flexible train formation and flexible stopping time. In addition, many studies have incorporated timetabling in the stop plan optimization problem [33,34], and Qi et al. [35] further ameliorated the seat allocation during the joint optimization process.

In Table 1, we summarize the studies reviewed in this section and compare them with the study we propose. Comparisons are carried from the presence of a competitor and optimization aspects.

Table 1. Summary of literature review.

Reference	Competitor	Optimization Aspects			
		Pricing	Seat Allocation	Stop Plan	Timetable
[13–15]	×	√	×	×	×
[18,19]	√				
[20–22]	×	×	√	×	×
[23–27,35]	×	√	√	×	×
[28–32]	×	×	×	√	×
[33,34]	×	×	×	√	√
This research	×	√	√	√	×

3. Mathematical Formulation

3.1. Problem Analysis

The ticket price of HSR comprises the fees paid by passengers to purchase the HSR transport service, and should be determined considering national policies, transport costs, capacity supply, and passenger demand. In recent years, China Railway has tried to independently set prices for high-speed trains under flexible supply and demand. Figure 1 depicts the price variation with supply and demand, where Q and G denote passenger demand and capacity supply, respectively; D and S represent the demand curve and supply curve, respectively; and the intersection (B) of the two curves is the equilibrium point of supply and demand.

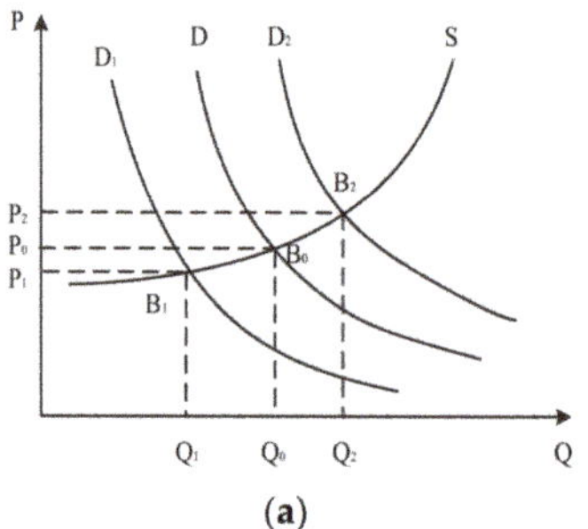 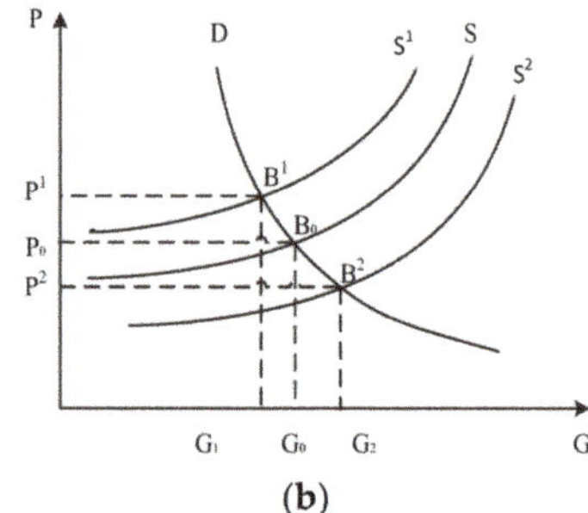

(a) **(b)**

Figure 1. The impact of supply and demand on ticket price. (**a**) The demand curve; (**b**) The supply curve.

The HSR train stop plan refers to the composition of trains' stopping patterns under the given train operating section, class, and number of trains. It determines the stopping sequence of each train on the running path, and plays a crucial role in capacity supply. The stop plan determines the ODs for which each train can provide capacity, which in turn affects the number of tickets that each train allocates to each OD. As shown in Figure 2, Train 1 stops at Station 2 and Station 4 while Train 2 stops at Station 3. Assuming that the train capacity is 1000, the transport supply provided by the stop plan is shown in Table 2.

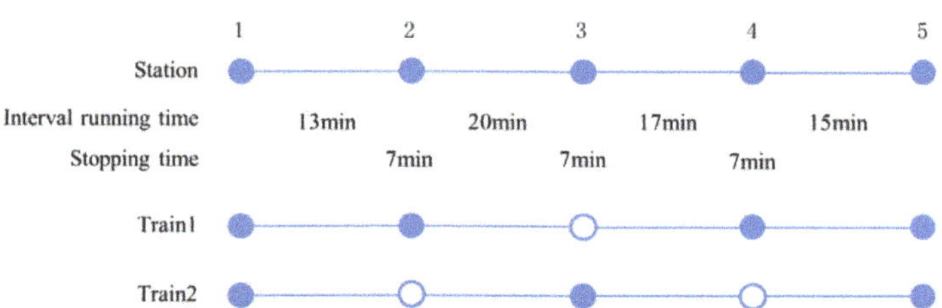

Figure 2. An example of a train stop plan.

Table 2. Transportation supply provided by the sample stop plan.

OD	Availability of Transportation		OD Service Frequency	Possible Maximum Number of Tickets
	Train 1	Train 2		
(1,2)	√	×	1	1000
(1,3)	×	√	1	1000
(1,4)	√	×	1	1000
(1,5)	√	√	2	2000
(2,3)	×	×	0	0
(2,4)	√	×	1	1000
(2,5)	√	×	1	1000
(3,4)	×	×	0	0
(3,5)	×	√	1	1000
(4,5)	√	×	1	1000

From Table 2, we can see that the train stop plan cannot serve OD (2,3) and (3,4). Thus, the corresponding number of tickets is 0. OD (1,5) can be served by two trains, so the number of tickets allocated to it can be up to 2000. This simple example can reflect the impact of the stop plan on the transport supply. By affecting the market supply and demand, the stop plan has an indirect impact on ticket prices.

For HSR trains, "stop depending on passenger demand" is an important principle. HSR is an alternative transportation mode whose passenger demand is affected by the ticket price; that is, the stop plan is indirectly affected by the ticket price.

On an HSR line consisting of N stations and N − 1 sections, there are L trains that run with different stop modes. According to the ticket-buying behavior of HSR passengers, the ticket pre-sale period is divided into multiple periods. Then, based on the price elasticity of passenger demand and the impact of the stop plan on passenger demand, the ticket pricing strategy, seat allocation scheme, and train stop plan should be decided simultaneously.

The variables and parameters used to model the problem are defined in Table 3.

Table 3. Notation.

Notation	Definition	Unit
W	Set of all OD in a certain direction on an HSR line, any OD $(r,s) \in W$ and $1 \leq r \leq s \leq N$.	-
L	Set of all trains that depart on a certain day on the line, any train $l \in L$.	-
β	Set of predetermined price discounts, $\beta = \{\beta_1, \beta_2, \cdots, \beta_M\}$ and $\beta_1 < \beta_2 < \cdots < \beta_M$.	-
C_l	The carrying capacity of train l.	passengers
K	The number of ticket pre-sale periods, $k = 1, 2, \ldots, K$.	-
p_{lrs}	The published ticket price (ceiling price) of train l on (r,s).	
p_{lrs}^k	The ticket price of train l on (r,s) in pre-sale period k.	yuan
c_{lrs}^k	The generalized travel cost of the train l on (r,s) in pre-sale period k.	yuan
t_{lrs}	The travel time of train l on (r,s). $t_{lrs} = +\infty$ when train l cannot provide transport service for (r,s).	minutes
φ_{lrs}^k	The sharing rate of train l for the passenger flow on (r,s) in pre-sale period k.	%
q_{lrs}^k	The elastic passenger flow demand of train l on (r,s) in pre-sale period k.	passengers
π_l^n	The stopping cost of train l at station n.	yuan
u_l^n	The stopping time of train l at station n.	minutes
$r_{lrs}^{\rightarrow k}$	The passenger flow that is rejected in pre-sale period $k-1$ and continues to choose train l in period k.	passengers
	Parameters	
η^k	The elastic demand function coefficient in pre-sale period k.	-
ν	The time value conversion coefficient.	yuan/min
θ	The utility conversion coefficient.	-
	Decision variables	
y_{lrs}^k	The price discount of train l on (r,s) in pre-sale period k. $y_{lrs}^k \in \beta$ and y_{lrs}^k is a discrete variable.	-
x_l^n	$x_l^n = 1$ when train l stops at station n; otherwise $x_l^n = 0$. x_l^n is a binary variable.	-
z_{lrs}^k	The number of seats that train l allocates to (r,s) in pre-sale period k. z_{lrs}^k is an integer variable.	tickets

From Table 3, the price discount β gives the lowest and highest fare levels β_1 and β_M. β_M is the published ticket price set by the government to ensure HSR's social welfare, while β_1 is determined based on the operational cost of HSR. Then, the ticket price can be represented by

$$p_{lrs}^k = y_{lrs}^k \cdot p_{lrs} \tag{1}$$

where p_{lrs}^k is the ticket price of train l on (r,s) in pre-sale period k, y_{lrs}^k is the corresponding price discount, and p_{lrs} is the published ticket price of train l on (r,s).

To simplify the research and formulation of the problem, some assumptions used in our model are as follows:

(1)	All ODs have the same demand elasticity in the same ticket pre-sale period.
(2)	For the same OD and same train, the ticket price will not be reduced as the train departure time approaches.
(3)	Passengers who fail to obtain tickets in a certain period will continue to buy the tickets in the next period.

(4)   Different seat classes, ticket overbooking, and cancellations are not considered.

3.2. Elastic Passenger Demand and Choice Behavior

Generalized travel cost is the most critical influencer of passenger demand. It is described by the ticket price and the travel time (Deng et al. [27]):

$$c_{lrs}^{k} = p_{lrs}^{k} + v \cdot t_{lrs} \tag{2}$$

where t_{lrs} is the travel time of train l on OD (r,s) in pre-sale period k, and v is the time value conversion coefficient.

Then, the average generalized travel cost of (r,s) is

$$c_{rs}^{k} = \frac{1}{\sum_{l \in L} x_l^r \cdot x_l^s} \sum_{l \in L} \sum_{k=1}^{K} c_{lrs}^{k} \cdot x_l^r \cdot x_l^s \tag{3}$$

The passenger flow demand will vary flexibly with the generalized travel cost. Here, the log-linear function is used (Qi et al. [35]) to describe the elastic demand of (r,s) in period k.

$$q_{rs}^{k}\left(c_{rs}^{k}\right) = q_{rs}^{k0}\left(c_{rs}^{k0}\right) \cdot \exp\left[-\eta^{k}\left(\frac{c_{rs}^{k}}{c_{rs}^{k0}} - 1\right)\right] \tag{4}$$

where c_{rs}^{k0} is the average generalized travel cost of (r,s) under the fixed pricing strategy, $q_{rs}^{k0}\left(c_{rs}^{k0}\right)$ is the corresponding passenger demand, and η^{k} is the elastic demand function coefficient in pre-sale period k.

For all trains that can serve (r,s), we use the logit model to describe passengers' choice behavior among them (Qin et al. [26]):

$$\varphi_{lrs}^{k} = \frac{\exp\left(-\theta c_{lrs}^{k}\right)}{\sum_{l' \in L} \exp\left(-\theta c_{l'rs}^{k}\right)} \tag{5}$$

where θ is the utility conversion coefficient.

Then, the elastic passenger demand of train l on (r,s) in pre-sale period k can be obtained by

$$q_{lrs}^{k} = q_{rs}^{k}\left(c_{rs}^{k}\right) \cdot \varphi_{lrs}^{k} = q_{rs}^{k0}\left(c_{rs}^{k0}\right) \cdot \exp\left[-\eta^{k}\left(\frac{c_{rs}^{k}}{c_{rs}^{k0}} - 1\right)\right] \cdot \frac{\exp\left(-\theta c_{lrs}^{k}\right)}{\sum_{l' \in L} \exp\left(-\theta c_{l'rs}^{k}\right)} \tag{6}$$

3.3. Integrated Optimization Model of Ticket Pricing and Stop Planning

3.3.1. Objective Function

In order to improve the revenue of the enterprise, the objective function is constructed based on two aspects: maximizing the transport revenue for HSR and minimizing the total time loss for passengers.

(1)   Transport revenue

The revenue is the difference between the ticket income and the stopping cost of all trains.

$$R = \sum_{k=1}^{K} \sum_{(r,s) \in W} \sum_{l \in L} y_{lrs}^{k} \cdot p_{lrs}^{k} \cdot z_{lrs}^{k} - \sum_{l \in L} \sum_{n=2}^{N-1} x_l^n \cdot \pi_l^n \tag{7}$$

where z_{lrs}^{k} is the number of seats that train l allocates to (r,s) in pre-sale period k, x_l^n is the stopping variable, and π_l^n is the stopping cost of train l at station n.

(2)   Time loss of passengers

The time loss of passengers is determined by trains' stopping time at each intermediate station. We introduce passengers' average unit time value ∂ to describe it.

$$V = v \cdot \sum_{(r,s) \in W} \sum_{l \in L} \left[\sum_{k=1}^{K} z_{lrs}^{k} \cdot \sum_{n=r+1}^{s-1} x_{l}^{n} u_{l}^{n} \right] \tag{8}$$

where u_{l}^{n} is the stopping time of train l at station n.

Finally, use weighting coefficients ω_1 and ω_2 to unify the sub-objective functions as a single objective function

$$\max Z = \omega_1 \cdot R - \omega_2 \cdot V \tag{9}$$

where ω_1 and ω_2 should be determined according to the importance of each sub-objective function. Here, the values of ω_1 and ω_2 are, respectively, taken as 0.6 and 0.4.

3.3.2. Constraints

(1)   Price constraints

At first, the ticket price must be positive.

$$y_{lrs}^{k} > 0 \ (r,s) \in W, l \in L, 1 \leq k \leq K \tag{10}$$

For the railway enterprises to better organize passenger transport, it is necessary to prevent passengers from buying tickets near the departure time. Thus, the closer to the departure time, the higher the ticket price should be.

$$y_{lrs}^{k-1} \leq y_{lrs}^{k} \ (r,s) \in W, l \in L, 2 \leq k \leq K \tag{11}$$

(2)   Ticket (seat) constraints

Firstly, the number of tickets that any train l allocates to (r,s) in period k should be an integer.

$$z_{lrs}^{k} \in N \ (r,s) \in W, l \in L, 1 \leq k \leq K \tag{12}$$

Secondly, a train may allocate tickets to OD (r,s) only when it can serve the OD.

$$(x_{l}^{r} \cdot x_{l}^{s} - 1) \cdot z_{lrs}^{k} = 0 \ (r,s) \in W, l \in L, 1 \leq k \leq K \tag{13}$$

Finally, the number of tickets (seats) that train l allocates to (r,s) in period k should not exceed the passenger demand (Deng et al. [27]).

$$z_{lrs}^{k} \leq q_{lrs}^{k} + r_{lrs}^{\rightarrow k} \ (r,s) \in W, l \in L, 1 \leq k \leq K \tag{14}$$

where $r_{lrs}^{\rightarrow k}$ refers to the passenger demand that is rejected in pre-sale period $k-1$ and continues to choose train l in period k. The setting of this variable makes it possible for the passenger demand rejected in early periods to be satisfied in subsequent periods, so that more passenger demand can be satisfied. According to the assumption (3), passengers who fail to obtain tickets in a certain period will continue to buy tickets in the next period. The transfer passenger demand can be regarded as part of the initial passenger flow in the next period. Therefore, if the rejected passenger demand is denoted as f_{rs}^{k-1}, $r_{lrs}^{\rightarrow k}$ can be obtained according to:

$$r_{lrs}^{\rightarrow k} = f_{rs}^{k-1} \cdot \varphi_{lrs}^{k} = f_{rs}^{k-1} \cdot \exp\left[-\eta^{k} \left(\frac{c_{rs}^{k}}{c_{rs}^{k0}} - 1 \right) \right] \cdot \frac{\exp\left(-\theta c_{lrs}^{k} \right)}{\sum_{l' \in L} \exp\left(-\theta c_{l'rs}^{k} \right)} \tag{15}$$

(3)   Capacity constraints

The number of tickets allocated by a train to any section cannot exceed the train's capacity.

$$\sum_{k=1}^{K}\sum_{r=1}^{j}\sum_{s=j+1}^{N} z_{lrs}^{k} \leq C_l \; l \in \boldsymbol{L}, \forall j \in [1, N-1] \tag{16}$$

where C_l is the carrying capacity of train l.

(4) Reachability constraints

For each OD pair, there must be trains that can serve it.

$$\sum_{l \in L} x_l^r \cdot x_l^s \geq 1 \; (r,s) \in \boldsymbol{W} \tag{17}$$

To sum up the above expressions (1)–(17), Equations (2), (4), (5), and constraint (14) refer to existing studies, while others are newly proposed here.

The objective function (9), Constraint (13), Constraint (15), and Constraint (17) are nonlinear, while other constraints are linear. Constraint (13) can be linearized by introducing a large enough positive number $\boldsymbol{M}$:

$$z_{lrs}^{k} \leq M \cdot x_l^r \; (r,s) \in \boldsymbol{W}, l \in \boldsymbol{L}, 1 \leq k \leq K \tag{18}$$

$$z_{lrs}^{k} \leq M \cdot x_l^s \; (r,s) \in \boldsymbol{W}, l \in \boldsymbol{L}, 1 \leq k \leq K \tag{19}$$

Constraint (17) can be linearized by introducing a new binary variable g_{lrs}:

$$\sum_{l \in L} g_{lrs} \geq 1 \; (r,s) \in \boldsymbol{W} \tag{20}$$

$$g_{lrs} \leq x_l^r \; (r,s) \in \boldsymbol{W}, l \in \boldsymbol{L} \tag{21}$$

$$g_{lrs} \leq x_l^s \; (r,s) \in \boldsymbol{W}, l \in \boldsymbol{L} \tag{22}$$

Finally, the joint optimization model for HSR ticket pricing and train stop planning can be formulated as follows s.t.

$$maxZ = \omega_1 \cdot R - \omega_2 \cdot V \tag{23}$$

Constraints (10)–(12), (14)–(16), and (18)–(22)

Table 4 details the variables and constraints in the formulated model.

Table 4. Characteristics of the formulated model.

Item	Type	Size	Characteristic
y_{lrs}^{k}	Variable	$(|L| \cdot N(N-1) \cdot)2$	Discrete
x_l^{n}	Variable	$|L| \cdot (N-2)$	Binary
z_{lrs}^{k}	Variable	$(|L| \cdot N(N-1) \cdot K)2$	Integer
g_{lrs}	Variable	$(|L| \cdot N(N-1))2$	Binary
Constraint (10)	Constraint	$(|L| \cdot N(N-1) \cdot K)2$	Linear
Constraint (11)	Constraint	$(|L| \cdot N(N-1) \cdot (K-1))2$	Linear
Constraint (12)	Constraint	$(|L| \cdot N(N-1) \cdot K)2$	Linear
Constraint (14)	Constraint	$(|L| \cdot N(N-1) \cdot K)2$	Linear
Constraint (15)	Constraint	$(|L| \cdot N(N-1) \cdot (K-1))2$	Nonlinear
Constraint (16)	Constraint	$|L| \cdot (N-1)$	Linear
Constraint (18)	Constraint	$(|L| \cdot N(N-1) \cdot K)2$	Linear
Constraint (19)	Constraint	$(|L| \cdot N(N-1) \cdot K)2$	Linear
Constraint (20)	Constraint	$(N(N-1))2$	Linear
Constraint (21)	Constraint	$(|L| \cdot N(N-1))2$	Linear
Constraint (22)	Constraint	$(|L| \cdot N(N-1))2$	Linear

4. Solution Method

As an HSR line usually has many intermediate stations and an HSR train has many seats, the model proposed above is a super-large-scale mixed-integer nonlinear programming model. Numerous variables and constraints make it difficult to solve the model with efficient and accurate algorithms. Thus, a heuristic algorithm is chosen for solving and computational analysis. Moreover, the joint problem is a combinatorial optimization problem, which has a strong similarity with the solid annealing principle of SA (as shown in Table 5). Thus, we will design an efficient method based on SA to solve the joint model.

Table 5. Solid annealing vs. combinatorial optimization.

Solid Annealing	Combinatorial Optimization
State	Solution
System energy	The objective function
The lowest energy state	The optimal solution
Heating to melt	Setting the initial temperature
Isothermal process	Generating and accepting (or rejecting) new solutions
Cooling process	Changing the current temperature

The simulated annealing (SA) algorithm is a stochastic algorithm based on Monte-Carlo iteration, which involves asymptotic convergence and allows random movements in the searched neighborhood to escape local minima. It randomly searches for the global optimal solution in the solution collection, and can jump out with a certain probability when falling into the local optimum. In addition, it is easy to implement and less limited by the initial solution. Although proposed more than 40 years ago, it still attracts some attention and is broadly used in many existing solutions for different variants of optimization problems.

HSR's dynamic pricing, seat allocation scheme, and stop plan are mutually influential and restrictive. Once obtained, the optimal stop plan remains unchanged in the pre-sale period. However, the ticket price and seats allocated for the same OD pair are different for each period. Thus, the optimization process can be divided into two layers. The stop plan is first optimized in the outer layer, and then the ticket price and seat allocation scheme under the determined stop plan are optimized in the inner layer level. Thus, the CSA algorithm is proposed for solving the problem. In this way, the algorithm only needs to search for the optimal solution in a smaller solution space in each iteration, so that improves the possibility of searching for the global optimum in the same computing time. If an appropriate termination strategy is adopted, the algorithm can effectively save computing time.

The CSA algorithm needs to start the iterative process based on the initial solution. For our problem, the initial solution includes three aspects: the initial stop plan, ticket prices, and seat allocation scheme. We set them according to the current operational mode of HSR.

4.1. Neighborhood Structure

The neighborhood of the CSA algorithm is divided into inner and outer layers to be constructed separately.

4.1.1. Outer Neighborhood

The outer layer optimizes the stop plan. The stop mode of a train is described by a vector composed of the 0–1 variable, in which 1 indicates stopping at the station and 0 indicates no stopping.

As shown in Figure 3, the neighborhood structure of the outer layer can be constructed in two ways. One involves randomly selecting two intermediate stopping variables of one train for inversion. Another involves randomly selecting two trains, and inverting one intermediate stopping variable of each train. If the newly obtained stop plan does not satisfy the OD pair reachability constraint, we return to the reconstruction.

$$
\left\{
\begin{array}{c}
\{1,1,0,\dots,0,1,1\} \\
\{1,\mathbf{1},1,\dots,0,\mathbf{0},1\} \\
\dots\dots \\
\{1,0,0,\dots,1,1,1\}
\end{array}
\right\}
\quad\rightarrow\quad
\left\{
\begin{array}{c}
\{1,1,0,\dots,0,1,1\} \\
\{1,\mathbf{0},1,\dots,0,\mathbf{1},1\} \\
\dots\dots \\
\{1,0,0,\dots,1,1,1\}
\end{array}
\right\}
$$

or

$$
\left\{
\begin{array}{c}
\{1,\mathbf{1},0,\dots,0,1,1\} \\
\{1,1,1,\dots,0,0,1\} \\
\dots\dots \\
\{1,0,0,\dots,\mathbf{1},1,1\}
\end{array}
\right\}
\quad\rightarrow\quad
\left\{
\begin{array}{c}
\{1,\mathbf{0},0,\dots,0,1,1\} \\
\{1,0,1,\dots,0,1,1\} \\
\dots\dots \\
\{1,0,0,\dots,\mathbf{0},1,1\}
\end{array}
\right\}
$$

The current stop plan   The neighborhood stop plan

Figure 3. Neighborhood structure of train stop plan.

4.1.2. Inner Neighborhood

The inner layer determines the ticket price and seat allocation. We randomly select a coefficient y_{lrs}^{k} from the current price discount solution and β' from the set $\left\{\beta_i \in \beta \,\middle|\, y_{lrs}^{k-1} \leq \beta_i \leq y_{lrs}^{k+1}\right\}$. Then let $y_{lrs}^{k} \leftarrow \beta'$ to construct the neighborhood ticket price solution.

Under the new stop plan and ticket prices, we obtain the passenger demand by Formulas (1)–(6) and round it as the pre-seat allocation scheme. If it satisfies all constraints, the scheme is regarded as a new feasible seat allocation scheme. If any constraints are not satisfied, corresponding adjustments are required. For any train l' and any section $(j', j'+1)$ that exceed the train capacity limit (constraints (16)), we adjust the pre-seat allocation scheme to obtain a feasible scheme.

$$
z_{l'rs}^{k} = \frac{q_{l'rs}^{k}}{\sum_{k=1}^{K}\sum_{r=1}^{j'}\sum_{s=j'+1}^{N} q_{l'rs}^{k}} \times C_{l'} \; l' \in L, r \leq j' < j'+1 \leq s, 1 \leq k \leq K \tag{24}
$$

4.2. Implementation Process

In the CSA, the initial temperature is given by the objective function value of the initial feasible solution, the temperature drops proportionally with a given cooling parameter, and the number of iterations under the same temperature is controlled by an upper limit. When the current temperature is lower than the given end temperature or the current best solution remains unchanged within the specified number of iterations, the algorithm is terminated.

The steps of the CSA algorithm are as follows.

Step 1: Initialization. Set the initial temperature T_0, the cooling parameter α, the end temperature T_{min}, and the maximum iteration numbers I_1 and I_2 for the inner and outer layer algorithms under the same temperature. Determine the initial feasible solution S and calculate the objective function value $Z(S)$. Let the current temperature $t = T_0$, and the current iteration number $i = i' = 1$. The best solution $\overline{S} = S$.

Step 2: Construction of the outer neighborhood solution. Implement the outer neighborhood structure method to obtain a new train stop plan.

Step 3: Perform the inner algorithm.

Step 3.1: Obtain the initial solution of the inner layer. Under the new stop plan, use the inner neighborhood construction method to obtain a new pricing strategy and seat allocation scheme; that is, a new feasible solution S'. Then, calculate the objective function value $Z(S')$.

Step 3.2: Construction of the inner neighborhood solution. Based on the feasible solution S', implement the inner neighborhood method to obtain a new pricing strategy and seat allocation scheme, which constitutes a new feasible solution S''. Then, calculate the objective function value $Z(S'')$.

Step 3.3: Metropolis criterion test for the inner layer. If $Z(S'') \geq Z(S')$, let $S' = S''$. Otherwise, randomly generate a number ρ from the interval (0,1), and if $\rho \leq exp\left(-\frac{(Z(S')-Z(S''))}{t}\right)$, let $S' = S''$, otherwise, refuse the inferior solution and keep the current solution S' unchanged.

Step 3.4: Iteration times test for the inner layer. Let $i' \leftarrow i' + 1$. If $i' \geq I_1$, set $i' = 1$, output the solution S', and go to step 4; otherwise go to step 3.2.

Step 4: Metropolis criterion test for the outer layer. If $Z(S') \geq Z(S)$, let $\overline{S} = S'$, $S = S'$. Otherwise, randomly generate a number ρ from the interval (0,1), and if $\rho \leq exp\left(-\frac{(Z(S)-Z(S'))}{t}\right)$, let $S = S'$, otherwise, refuse the inferior solution and keep the current solution unchanged.

Step 5: Iteration number test for the outer layer. Update the iteration times: $i \leftarrow i + 1$. If $i \geq I_2$, let $t \leftarrow t \cdot \alpha$, $i = 1$, and go to step 6; otherwise go to step 2.

Step 6: Termination check. If $t < T_{min}$ or the current optimal solution $\overline{S}$ remains unchanged in τ iterations, terminate the algorithm and output the optimal solution $\overline{S}$; otherwise, reset $i = 1$ and go to step 2.

It should be noted that the CSA algorithm records the best solution $\overline{S}$. Once a better solution is found, the algorithm will replace $\overline{S}$ with the better one regardless of the Metropolis criterion. Thus, the utilization efficiency of the searched optimal solution can be improved significantly.

5. Empirical Analysis

5.1. Basic Data

The Beijing–Shanghai HSR line is taken as an example to verify the feasibility of the model and algorithm. Among the trains that depart from Beijing South Station and arrive at Shanghai Hongqiao station, four trains G11, G19, G21, and G113 with different stopping patterns are selected for analysis. As Figure 4 shows, these four trains involve 12 stations: Beijing South, Tianjin South, Dezhou East, Jinan West, Taian, Qufu East, Tengzhou East, Xuzhou West, Nanjing South, Zhenjiang South, Suzhou North, and Shanghai Hongqiao. We number them 1–12 sequentially, where 1 and 12 represent Beijing South and Shanghai Hongqiao, respectively. To simplify the problem, we suppose that all HSR transport services adopt the same price discount in the same pre-sale period.

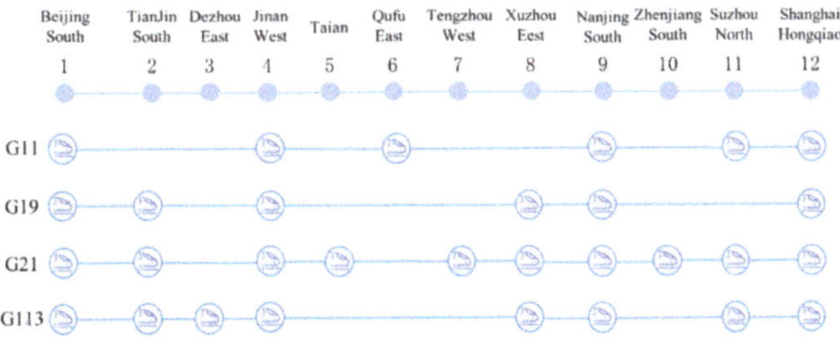

Figure 4. The current train stop plan.

Each train is formed of 16 cars, and its stopping cost and stopping time at each station are 3200 yuan and 7 min, respectively. The seat number of G11, G21, and G113 is 1043, while that of G19 is 1015.

According to the ticket data from 1 August 2016 to 31 July 2017, we can depict passengers' booking rules in Figure 5 and further divide the ticket pre-sale period into four periods: pre-sale days 1–19, 20–25, 26–28, and 29–30. The elastic demand function coefficient η^k reflects the sensitivity of passenger demand to ticket prices, which intensifies as the departure date approaches. We take η^k as 1.8, 1.4, 1.2, and 0.8 for periods 1–4, respectively.

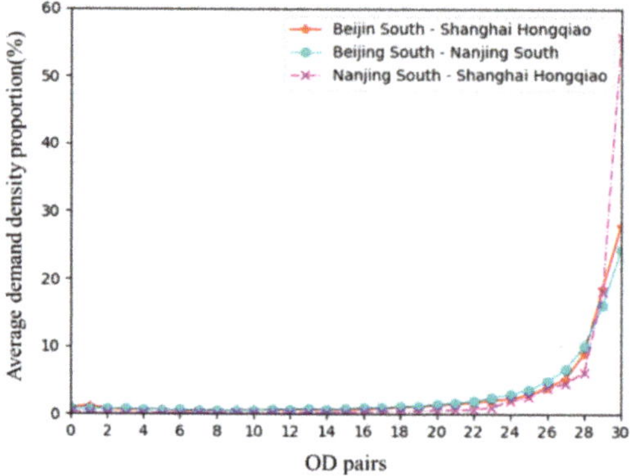

Figure 5. Average demand density proportion of several OD pairs.

The highest (corresponding to the published price) and lowest price discounts are 1 and 0.56, respectively, and we take the interval of 0.04 to define the price discount coefficient set $\beta = \{0.56, 0.60, \ldots, 0.96, 1\}$. According to Qi et al. [35], we have $\nu = 0.6$ yuan/minute and $\theta = 0.012$. For the CSA algorithm, $I_1 = 7920$, $I_2 = 400$, $\alpha = 0.95$, $T_{min} = 0.001$, and $\tau = 100$.

5.2. Results and Analysis

We used the Python programming language to implement the algorithm and solve the model. Figure 6 presents the iteration process of the CSA algorithm. The objective value starts to converge around 130 iterations and remains unchanged thereafter.

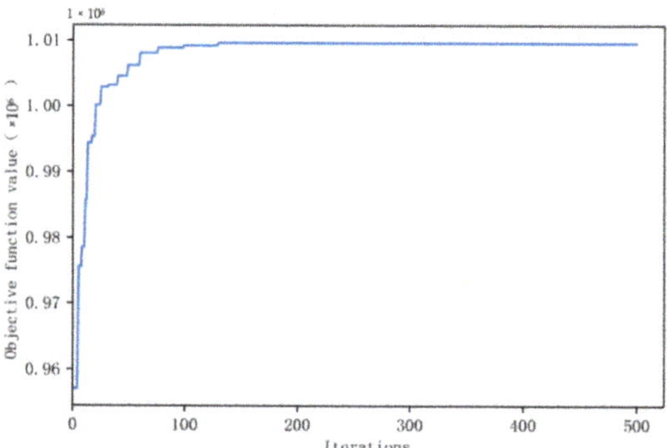

Figure 6. The iteration process.

The results show that the transportation revenue and the total passenger time loss are 1,717,166 yuan and 85,829 min, respectively. Compared with 1,633,840 yuan and 96,600 min under the current fixed operational mode, the revenue increased by 5.10% and the time loss decreased by 11.15%.

Figure 7 shows the optimal stop plan. Compared to the current stop plan shown in Figure 4, the stop frequency is reduced from 22 to 17, and the total stopping cost is reduced by 16,000 yuan (a decrease of 22.73%). G19 and G113 reduced by three and four stops, while G11 and G21 added stops at Station 2 and Station 3, respectively. For some ODs, it indicated that their passenger demand does not necessitate so many trains to serve them. One additional stop for G11 and G21 trains can help meet the demands of these ODs. Moreover, for passengers taking G11 and G21 trains, their stopping time loss only increased by one-stop time, which is less than the reductions in the stopping time loss for passengers on G19 and G113.

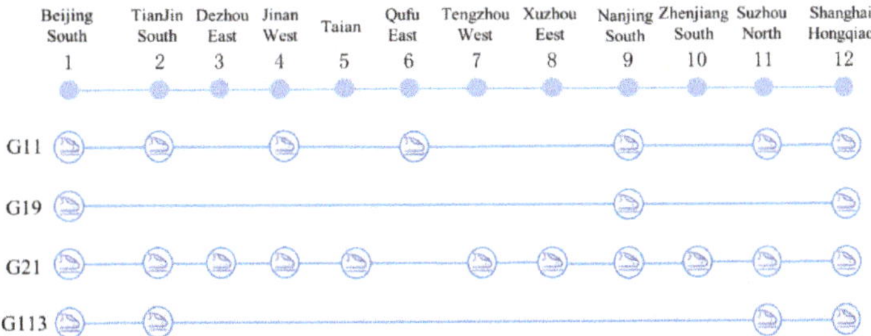

Figure 7. The optimized train stop plan.

Under the fixed operational mode, the passenger flows of G11, G19, G21, and G113 are 1157, 1002, 1113, and 1239, respectively. After optimization, the passenger flows of G11 and G21 increase to 1334 and 1441, while those of G19 and G113 decrease to 705 and 780, respectively. The reason is that G11 and G21 can serve more OD pairs. Conversely, as G19 and G113 reduce stops at several intermediate stations, they no longer serve part of the OD pairs. As a result, the passenger flow of these OD pairs transfers to G11 and G21.

Table 6 gives the optimal dynamic pricing strategy for some ODs (the ticket price is accurate to 0.5 yuan). For any OD, the closer the period is to the departure date, the higher the ticket price. The dynamic price is lower than the fixed price for the first two periods, while it is higher for the last two periods. This is because passengers are generally more sensitive to the ticket price during early ticket pre-sale time, so the passenger demand is rich in elasticity. Discounts on ticket prices in early periods can attract more passengers to buy tickets.

Table 6. Ticket prices for different ticket pre-sale periods.

OD Pairs	Dynamic Prices (Yuan)				Public Prices (Yuan)	Current Prices (Yuan)
	$k = 1$	$k = 2$	$k = 3$	$k = 4$		
1–4	129	175.5	203	221.5	230.5	184.5
1–9	310.5	421.5	488	532	554.5	443.5
1–11	366.5	497.5	576	628	654.5	523.5
1–12	387	525.5	608.5	663.5	691	553.0
8–12	195.5	265	307	335	349	279.0
9–12	94	128	148	161.5	168	134.5

As shown in Figure 8, the passenger flow of most ODs has increased significantly in the first period. By exchanging low ticket prices for more passengers, the revenue can be enhanced. However, passengers booking tickets in the last two periods usually make their travel plans temporarily. Most of them are less sensitive to price and more willing to accept higher ticket prices. Thus, raising ticket prices hardly affects their demand. It can be seen from Figure 8 that the passenger flow of most ODs falls by less than 15% for the last two periods. The contribution of raising the price is more significant than the loss of passenger flow, which can also expand the revenue.

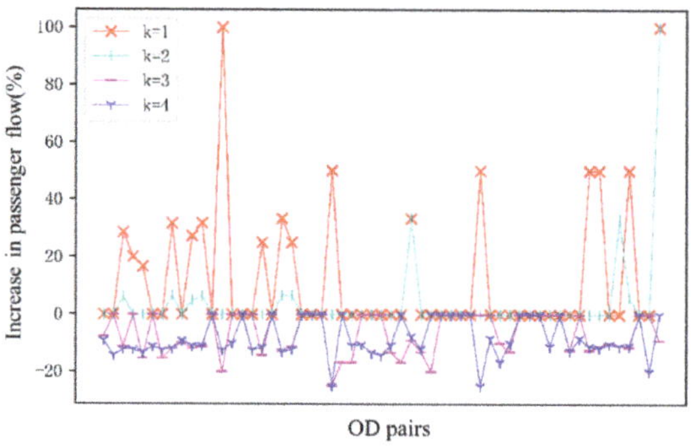

Figure 8. Passenger flow growth of each OD pair.

Figure 9 compares the ticket sales under the fixed price (initial case) and dynamic price (optimal case). The closer the period is to the departure date, the higher the ticket sale volume. For all ODs, the tickets are sold in the latter two periods, which matches the distribution of passenger demand and thus is conducive to obtaining more ticket revenue.

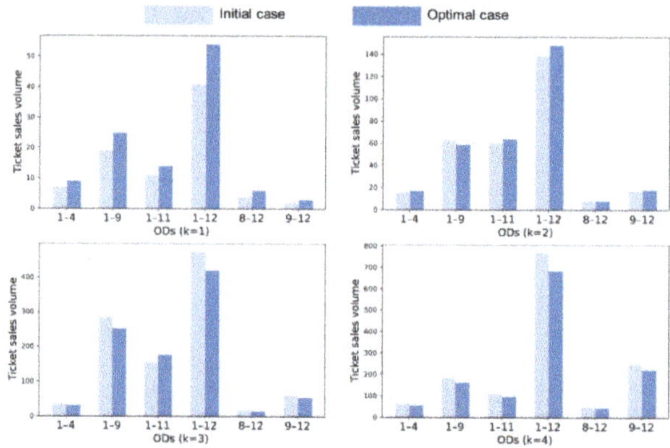

Figure 9. Comparison of ticket sales volume for different ODs in different periods.

Compared to the fixed price, the ticket sales volume of most ODs in the first two periods increased significantly, which indicates that the dynamic pricing strategy can effectively guide passengers to purchase tickets earlier, which is beneficial for railway enterprises to organize passenger transport.

Table 7 gives the passenger flow comparison of each OD before and after optimization. The optimized passenger flow of some ODs increased, but that of more ODs decreased slightly. For most ODs, passengers prefer to buy tickets in the last two periods. The rise in ticket prices will cause some passengers to give up HSR and choose another transport mode. In addition, the passenger demand for the first two periods is relatively weak. Although the ticket price is discounted, only a small number of passengers were attracted. Thus, the passenger flow of most ODs decreased overall.

Table 7. Passenger flow of each OD pair (before/after optimization).

1	2	3	4	5	6	7	8	9	10	11	12
1	57/63	10/9	189/174	79/74	53/55	9/11	58/53	550/521	33/29	359/345	1422/1334
2		5/5	22/20	13/15	-	3/3	12/12	50/46	7/7	86/87	140/137
3			5/5	-	-	-	3/3	0/0	-	10/10	12/11
4				43/38	21/23	16/14	22/19	53/48	11/12	30/28	110/97
5					-	9/10	0/0	5/5	0/0	0/0	11/14
6						-	-	15/10	0/0	16/19	42/45
7							1/1	/0	/0	9/8	/0
8								83/74	14/17	49/45	81/75
9									59/65	139/126	329/299
10										2/2	9/8
11											145/129

Although the total stopping frequency of all trains decreased, the passenger flow of each OD hardly fluctuates. The underlined data in Table 7 denotes the OD whose service frequency is reduced. We can see that the passenger flow fluctuation of these ODs (13.64~1.16%) is not greater than that of other ODs (-33.33~27.27%). This indicates that our model can reasonably adjust the stop plan according to the dynamic passenger demand, which can reduce the stopping cost while maintaining passenger flows and ensuring service quality.

Figure 10 shows the improvement of the objective function value (OFV) under different passenger demands, which indicates that our method can always effectively improve the quality of the optimal combinatorial scheme. The OFV is growing with increases in passenger demand. But after the demand reaches a certain level, the growth rate of the OFV gradually slows as the demand continuously increases. This is because after the total demand has exceeded the transport capacity of the HSR system, the ticket prices for later pre-sale periods will reach the ceiling price. Then, when the demand continues to increase, the highest prices should remain unchanged, which results in the reduction of the OFV's growth rate.

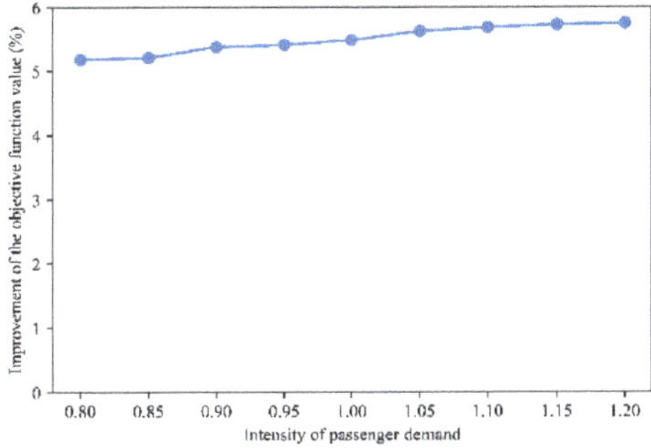

Figure 10. Optimization effect under different initial passenger demand.

6. Conclusions

In this paper, we introduced a mixed-integer non-linear model for jointly optimizing HSR ticket pricing and train stop planning considering multiple trains with multiple stopping patterns, and proposed a CSA algorithm combining the characteristics of the problem. The objective was to maximize the revenue of railway enterprises and minimize the total travel time of passengers. An empirical study based on ticket reservation data was conducted to present the performance of the proposed model and algorithm.

The results obtained illustrate the impacts that the strategy derived from the optimization model has on passenger demand, capacity allocation, and revenue. The solution from the proposed method provides a significant improvement in revenue from the initial 5.1% and causes a marked decline in the total travel time loss of passengers from the initial 11.5%. The dynamic pricing strategy encourages more passengers to buy tickets in earlier pre-sale periods. The optimized train stop plan decreases the total stopping cost of railway enterprises and the total time loss of passengers. In conclusion, this paper has illustrated how railway operators can exploit existing data sources to further realize RM.

The following areas indicate possible directions for future research. Other factors should be considered in the railway RM problem, such as seat classes and ticket cancellations. It would also be interesting to develop a more efficient algorithm that combines exact and metaheuristic methods to solve the proposed model.

Author Contributions: Conceptualization, J.Q.; funding acquisition, J.Q.; data curation, X.L.; software, X.L. and K.Y.; validation, J.Q. and G.X.; writing–original draft, X.L. and J.Q.; writing–review and editing, X.L. and J.Q. All authors contributed equally in writing this article. All authors have read and agreed to the published version of the manuscript.

Funding: This research was funded by the National Natural Science Foundation of China, grant number 72171236 & U2034208; Natural Science Foundation of Hunan Province, grant number 2020JJ5783.

Institutional Review Board Statement: Not applicable.

Informed Consent Statement: Not applicable.

Data Availability Statement: Not applicable.

Acknowledgments: The authors appreciate the Institute of Computing Technology of China Academy of Railway Sciences Corporation Limited that provided the data access.

Conflicts of Interest: The authors declare no conflict of interest.

References

1. Weatherford, L.R.; Kimes, S.E. A comparison of forecasting methods for hotel revenue management. *Int. J. Forecast.* **2003**, *19*, 401–415. [CrossRef]
2. Noone, B.M.; Mattila, A.S. Hotel revenue management and the internet: The effect of price presentation strategies on customers' willingness to book. *Int. J. Hosp. Manag.* **2009**, *28*, 272–279. [CrossRef]
3. Heo, C.Y.; Lee, S. Influences of consumer characteristics on fairness perceptions of revenue management pricing in the hotel industry. *Int. J. Hosp. Manag.* **2011**, *30*, 243–251. [CrossRef]
4. Grauberger, W.; Kimms, A. Airline revenue management games with simultaneous price and quantity competition. *Comput. Oper. Res.* **2016**, *75*, 64–75. [CrossRef]
5. Lawhead, R.J.; Gosavi, A. A Bounded Actor-Critic Reinforcement Learning Algorithm Applied to Airline Revenue Management. *Eng. Appl. Artif. Intell.* **2019**, *82*, 252–262. [CrossRef]
6. Selcuk, A.M.; Avar, Z.M. Dynamic pricing in airline revenue management. *J. Math. Anal. Appl.* **2019**, *478*, 1191–1217. [CrossRef]
7. Kyparisis, G.J.; Koulamas, C. Optimal pricing and seat allocation for a two-cabin airline revenue management problem. *Int. J. Prod. Econ.* **2018**, *201*, 18–25. [CrossRef]
8. Armstrong, A.; Meissner, J. *Railway Revenue Management: Overview and Models*; Lancaster University Management School: Lancaster, UK, 2010.
9. Abe, I. Revenue Management in the Railway Industry in Japan and Portugal: A Stakeholder Approach. Technology and Policy Program. Ph.D. Thesis, Massachusetts Institute of Technology, Cambridge, MA, USA, 2007.
10. Bharill, R.; Rangaraj, N. Revenue management in railway operations: A study of the Rajdhani Express, Indian Railways. *Transp. Res. Part A Policy Pract.* **2008**, *42*, 1195–1207. [CrossRef]
11. Riss, M.; Côté, J.P.; Savard, G. A new revenue optimization tool for high-speed railway: Finding the right equilibrium between revenue growth and commercial objectives. In Proceedings of the 8th World Congress on Railway Research, Seoul, Korea, 18–22 May 2008.
12. Crevier, B.; Cordeau, J.F.; Savard, G. Integrated operations planning and revenue management for rail freight transportation. *Transp. Res. Part B Methodol.* **2012**, *46*, 100–119. [CrossRef]
13. Vuuren, D.V. Optimal pricing in railway passenger transport: Theory and practice in the Netherlands. *Transp. Policy* **2002**, *9*, 95–106. [CrossRef]

14. Jarocka, M.; Ryciuk, U. Pricing in the railway transport. In Proceedings of the 9th International Scientific Conference "Business and Management 2016", Vilnius, Lithuania, 12–13 May 2016.
15. Zhang, J.; Yang, X.; Du, G. Research on High-speed Railway Fare Optimization based on Change of Passenger Flow. In Proceedings of the 3rd International Conference on Mechatronics Engineering and Information Technology (ICMEIT 2019), Dalian, China, 29–30 March 2019.
16. Dargay, J.M.; Clark, S.; Rose, J. The determinants of long distance travel in Great Britain. *Transp. Res. Part A Policy Pract.* **2012**, *46*, 576–587. [CrossRef]
17. Georggi, N.L.; Pendyala, R.M. Analysis of long-distance travel behavior of the elderly and the low-income. *Transp. Res. Rec.* **2012**, *E-C026*, 121–150.
18. Sato, K.; Sawaki, K. Dynamic pricing of high-speed rail with transport competition. *J. Revenue Pricing Manag.* **2011**, *11*, 548–559. [CrossRef]
19. Chen, J.H.; Gao, Z.Y. Optimal Railway Passenger-ticket Pricing. *J. China Railw. Soc.* **2005**, *27*, 16–19.
20. Wang, X.C.; Wang, H.; Zhang, X.N. Stochastic seat allocation models for passenger rail transportation under customer choice. *Transp. Res. Part E* **2016**, *96*, 95–112. [CrossRef]
21. Jiang, X.S.; Chen, X.Q.; Zhang, L.; Zhang, R. Dynamic Demand Forecasting and Ticket Assignment for High-Speed Rail Revenue Management in China. *Transp. Res. Rec.* **2015**, *2475*, 37–45. [CrossRef]
22. Yan, Z.Y.; Li, X.J.; Zhang, Q.; Han, B.M. Seat allocation model for high-speed railway passenger transportation based on flexible train composition. *Comput. Ind. Eng.* **2020**, *142*, 106383. [CrossRef]
23. Hetrakul, P.; Cirillo, C. A latent class choice based model system for railway optimal pricing and seat allocation. *Transp. Res. Part E* **2014**, *61*, 68–83. [CrossRef]
24. Hu, X.L.; Shi, F.; Xu, G.M.; Qin, J. Joint optimization of pricing and seat allocation with multiperiod and discriminatory strategies in high-speed rail networks. *Comput. Ind. Eng.* **2020**, *148*, 106690. [CrossRef]
25. Wu, X.K.; Qin, J.; Qu, W.X.; Zeng, Y.J.; Yang, X. Collaborative Optimization of Dynamic Pricing and Seat Allocation for High-speed Railways: An Empirical Study from China. *IEEE Access* **2019**, *7*, 139409–139419. [CrossRef]
26. Qin, J.; Zeng, Y.J.; Yang, X.; He, Y.X.; Wu, X.K.; Qu, W.X. Time-Dependent Pricing for High-Speed Railway in China Based on Revenue Management. *Sustainability* **2019**, *11*, 4272. [CrossRef]
27. Deng, L.B.; Zeng, N.X.; Chen, Y.X.; Xiao, L.W. A divide-and-conquer optimization method for dynamic pricing of high-speed railway based on seat allocation. *J. Railw. Sci. Eng.* **2019**, *16*, 2407–2413.
28. Vuchic, V.R. *Urban Transit*; Wiley: Hoboken, NJ, USA, 2005.
29. Chang, Y.H.; Ye, C.H.; Shen, C.C. A multi-objective model for passenger train services planning: Application to taiwan's high-speed rail line. *Transp. Res. Part B* **2000**, *34*, 91–106. [CrossRef]
30. Bussieck, M.R.; Lindner, T.; Lübbecke, M.E. A fast algorithm for near cost optimal line plans. *Math. Methods Oper. Res.* **2004**, *59*, 205–220. [CrossRef]
31. Yang, L.X.; Qi, J.G.; Li, S.K.; Gao, Y. Collaborative optimization for train scheduling and train stop planning on high-speed railways. *Omega* **2016**, *64*, 57–76. [CrossRef]
32. Jin, G.W.; He, S.W.; Li, J.B.; Guo, X.L.; Li, Y.B. An Approach for Train Stop Planning with Variable Train Length and Stop Time of High-speed Rail Under Stochastic Demand. *IEEE Access* **2019**, *7*, 129690–129708. [CrossRef]
33. Dong, X.L.; Li, D.W.; Yin, Y.H.; Ding, S.X.; Cao, Z.C. Integrated optimization of train stop planning and timetabling for commuter railways with an extended adaptive large neighborhood search metaheuristic approach. *Transp. Res. Part C Emerg. Technol.* **2020**, *117*, 102681. [CrossRef]
34. Qi, J.G.; Li, S.K.; Gao, Y.; Yang, K.; Liu, P. Joint optimization model for train scheduling and train stop planning with passengers distribution on railway corridors. *J. Oper. Res. Soc.* **2018**, *69*, 556–570. [CrossRef]
35. Xu, G.M.; Zhong, L.H.; Hu, X.L.; Liu, W. Optimal pricing and seat allocation schemes in passenger railway systems. *Transp. Res. Part E Logist. Transp. Rev.* **2022**, *157*, 102580. [CrossRef]

 *mathematics*

Article

Infeasibility Maps: Application to the Optimization of the Design of Pumping Stations in Water Distribution Networks

Jimmy H. Gutiérrez-Bahamondes [1], Daniel Mora-Melia [2,3,*], Bastián Valdivia-Muñoz [1], Fabián Silva-Aravena [4] and Pedro L. Iglesias-Rey [3]

[1]   PhD in Engineering Systems, Faculty of Engineering, Universidad de Talca, Camino Los Niches Km 1, Curicó 3340000, Chile
[2]   Department of Engineering and Construction Management, Faculty of Engineering, Universidad de Talca, Camino Los Niches Km 1, Curicó 3340000, Chile
[3]   Department of Hydraulic Engineering and Environment, Universitat Politècnica de València, Camino de Vera s/n, 46022 Valencia, Spain
[4]   Faculty of Social and Economic Sciences, Universidad Católica del Maule, Avenida San Miguel 3605, Talca 3480094, Chile
*   Correspondence: damora@utalca.cl; Tel.: +56-9-51265044

Abstract: The design of pumping stations in a water distribution network determines the investment costs and affects a large part of the operating costs of the network. In recent years, it was shown that it is possible to use flow distribution to optimize both costs concurrently; however, the methodologies proposed in the literature are not applicable to real-sized networks. In these cases, the space of solutions is huge, a small number of feasible solutions exists, and each evaluation of the objective function implies significant computational effort. To avoid this gap, a new method was proposed to reduce the search space in the problem of pumping station design. This method was based on network preprocessing to determine in advance the maximum and minimum flow that each pump station could provide. According to this purpose, the area of infeasibility is limited by ranges of the decision variable where it is impossible to meet the hydraulic constraints of the model. This area of infeasibility is removed from the search space with which the algorithm works. To demonstrate the benefits of using the new technique, a new real-sized case study was presented, and a pseudo-genetic algorithm (PGA) was implemented to resolve the optimization model. Finally, the results show great improvement in PGA performance, both in terms of the speed of convergence and quality of the solution.

Keywords: pump stations; optimization; energy efficiency; water networks; search space reduction

MSC: 90-08; 68Uxx

Citation: Gutiérrez-Bahamondes, J.H.; Mora-Melia, D.; Valdivia-Muñoz, B.; Silva-Aravena, F.; Iglesias-Rey, P.L. Infeasibility Maps: Application to the Optimization of the Design of Pumping Stations in Water Distribution Networks. *Mathematics* **2023**, *11*, 1582. https://doi.org/10.3390/math11071582

Academic Editors: Antonin Ponsich, Mariona Vila Bonilla and Bruno Domenech

Received: 8 February 2023
Revised: 17 March 2023
Accepted: 21 March 2023
Published: 24 March 2023

1. Introduction

Unquestionably, the water distribution service directly affects the quality of life of people around the world [1,2]. However, this process constitutes one of the most significant expenses in the budget of any city [3,4]. The main reason is the high energy consumption of the water distribution network (WDN). Specifically, pumping stations (PSs) require a significant amount of energy to transport water to consumers [4]. Therefore, improving the efficiency of these systems allows for significant energy and economic savings [2,5]. Furthermore, the price of electricity has been increasing globally. Consequently, optimizing PS design and operations is crucial for achieving a cost-effective WDN.

The design of PSs has short- and long-term consequences [6]. In the immediate term, determining the investment cost for constructing physical structures and acquiring equipment can be done. In the long term, establishing most of the operating conditions throughout the life of the project is performed. Thus, the design of PSs must be optimized after considering the operational variables in the network [7,8]. The design of a PS includes selecting the number, model, type of pump, accessories, and control system [6]. This set of decisions involves many possible combinations and, therefore, can be expressed as a mathematical optimization problem to minimize the investment and operating costs [7].

Different optimization problems were proposed to achieve a minimal amount of energy consumed by PSs from different perspectives [1,2,7]. These models differ in the decision variables used to build their objective functions and constraints. For this purpose, the authors used different approaches, such as the trade-off method between reliability and energy efficiency [9], maximizing energy production [10], minimizing maintenance and energy costs [11,12], location and minimization of the leakage [13], multi-criteria-based approach to minimize the operational costs and operational lack of service and maximize the pressure uniformity and network resilience [14], individual analysis for each design and operation option available by using binary variables [15], and/or calculating the equivalent flow and equivalent volume to approximate the annual costs [8], among other techniques. However, despite all efforts, this problem has not been fully resolved due to its complexity, high level of non-linearity, and the vast space of solutions restricted by hydraulic conditions that ensure a minimum quality of service (minimum pressure) [1,2,16]. Consequently, the development of effective operational strategies in addition to appropriate mathematical models based on comprehensive knowledge about the system and efficient computationally guided search methods are essential for the application of these techniques to real-world WDN problems.

More recently, interest in the determination of the optimal flow distribution as an effective tool in the design of PSs has been growing [5,7,17], mainly when this calculation is driven through the setpoint curve (SC) [18]. The setpoint curve represents the required dynamic head (Hc) for every flow rate (Q) in the PS to satisfy the minimum pressure service in the nodes of the network. The main characteristic of this curve is that the resistance produced by consumption nodes is replaced by a constant value that is the minimum service pressure for consumption nodes at any time instant. More details about its mathematical construction can be found in [19].

In particular, Gutiérrez et al. [7] implemented a novel methodology based on the concurrent minimization of capital expenditures (CAPEX) and operating expenses (OPEX) using the optimization of flow distribution. They proposed a mathematical optimization model with three types of decision variables: (1) the fractions of flow provided by each PS, (2) the model of the pumps, and (3) the number of fixed and variable speed pumps. To solve the model, a pseudo-genetic algorithm proposed by [20] was used. The authors presented a case study to demonstrate the advantages of the method. The results indicate that despite the large number of combinations presented in the network, it appears to be possible to find feasible solutions, avoid oversizing the pumps, and adjust the flows contributed by the PSs to the changes in the consumption pattern of the network for 24 h. However, this methodology still has room for improvement. First, using the SC in each objective function evaluation ensures that the solution fulfills the minimum head by using the minimum energy expenditure. Nevertheless, each evaluation of a solution involves examining all nodes in an iterative way. Thus, the computational cost is high, and the optimization algorithm loses search capacity as the network grows. Second, each solution to the problem includes a decision variable that determines the flow contribution of each pumping station. This variable can take values between 0% and 100% of the total flow demanded by the network in each period, but depending on the topology of the network, many of these solutions are a priori infeasible.

In general terms, real water distribution networks contain many nodes, pipes, and accessories. One of the significant challenges faced by state-of-the-art methods used to optimize the design of pumping stations is the application of the methods described above in networks of real size [1,2,16]. Traditionally, the pumps are selected based on an operation point and, later, their operation is optimized once the equipment is selected. In contrast, this work proposed an approximation to the operation mode of the pump in the planning phase based on optimizing the energy and cost in the WDN. This new approach can help to better select the pumping equipment. In this context, the method proposed by Gutiérrez-Bahamondes et al. [7] is limited by the size of the network. To avoid this problem, this study proposed an automatic pre-processing strategy to accelerate the heuristic search processes of evolutionary algorithms applied to the problem posed in [7]. Several main advantages of this strategy can be described: (1) it reduces the computational burden, (2) it rules out infeasible solutions during the evaluation process in any period, and (3) it finds a set of solutions close to the optimal design of the pumping station. The newly proposed method was validated using a real-world case study, and its performance was evaluated and compared with the original method. EPANET [21] was used to evaluate the hydraulic behavior of the hydraulic network, while the optimization model was solved using a pseudo-genetic algorithm (PGA).

The remainder of the paper is organized into several sections: (1) Section 2 describes the mathematical model of the original method and the infeasibility problem, followed by an explanation of the pre-processing strategy. The developed methodology was then applied to a case study, and an optimization method was implemented. (2) Section 3 provides the results in which 100 experiments were executed and analyzed. The use of the preprocessing strategy improved the quality of solutions and speed of convergence. (3) The conclusions of the research can be found in Section 4.

2. Materials and Methods

This work proposed a new method to accelerate the process of searching for solutions to the problem posed by [7] and includes several improvements:

- A new constraint was added to the mathematical optimization model. For each PS, this equation allows us to discard all pump models that, due to their specifications, did not manage to supply at least the maximum flow rate during the analysis period.
- The method used network preprocessing to determine in advance the maximum and minimum flow that each PS could provide. This procedure made it possible to limit the search space for solutions to the problem, thus eliminating areas of total infeasibility. An area of infeasibility is limited by ranges of the decision variable where it is impossible to meet the hydraulic constraints of the model. Our proposed method maps these ranges before starting the optimization process, accelerating the convergence of the algorithm using infeasibility maps (IMs).
- This study combined the use of the SC with the mapping of infeasibility zones to rule out unfeasible solutions during the evaluation process in any period, thus avoiding unnecessary hydraulic simulations when it was detected that part of the solution was not viable. Consequently, the IMs reduce the search space of the optimization algorithm. Reducing the search space to increase computational efficiency is a significant challenge faced when optimizing water networks

2.1. Mathematical Model

First, for a better understanding, the mathematical optimization model proposed by [7] is briefly presented in this section. This model was based on the optimization of flow rate injection, which was based on the use of the SC concept. The SC curve can be defined as a theoretical curve that indicates the minimum energy (in terms of pressure head) required for pumping stations to meet the minimum pressure required for each demand in the network, namely, it is a representation of the pressure head versus the flow at a given point in the system.

Next, the main assumptions, simplifications, and limitations of the model are detailed. First, it should be understood that the SC concept does not deal explicitly with pumps as hydraulic machines; therefore, we started with the assumption that all the related curves (such as flow rate versus pumping head, efficiency, and power) were not known. The use of the SC allowed us to determine the energy required at the source without considering specific pump head-flow curves. That is, we were only dealing with the energy supplied by the pumping station. One SC curve was available for each supply source. Second, this work assumed that a direct injection network was optimized. That is, the network was fed directly from groundwater or did not have a high enough elevation for tanks to be installed. Third, the location of each possible pumping station was defined previously and was not part of the problem. Fourth, the method required a pump model database containing all the characteristic coefficients of each pump and the purchase costs of all accessories and control systems necessary to build the physical structure. Finally, it is important to note that to simplify the calculation of the total costs, the design of the stations was parameterized according to the established modular design. The mathematical notation, decision variables, objective function, and constraints are presented below.

2.1.1. Mathematical Notation

- N_t: total number of time steps in the optimization process.
- N_{ps}: total number of PSs in the network.
- N_b: total number of pump models available in the data set.
- F_a: amortization factor.
- r: interest rate.
- N_p: total number of project life periods.
- $H_{0,i}$, A_i: characteristic coefficients of the pump head installed in PS_i.
- E_i, F_i: characteristic coefficients of the performance curve of the pump installed in PS_i.
- $Q_{i,j,k}$: represents the discharge of pump k during time step j in PS_i.
- $p_{i,j}$: energy cost in PS_i during the time step j.
- Υ: specific gravity of water.
- Δt_j: discretization interval of the optimization period.
- $m_{i,j}$: the number of FSPs running in PS_i at time step j.
- $n_{i,j}$: the number of VSPs running in PS_i at time step j. These values depend on the selected pump model and the system selected to control the operation point.
- $N_{B,i}$: total number of total pumps.
- H_{Bmax}: maximum head of the largest pump available in the data set.
- $H_{max,i}$: maximum head supplied by PS_i during time analysis.
- $C_{pump,\,i}$: purchase cost of a pump installed in PS_i.
- n_i: number of frequency inverters in PS_i.
- $C_{facility,i}$: cost of accessories including pipes in PS_i.
- $C_{control,i}$: sum of a pressure transducer, flowmeter, and programmable logic controller cost for PS_i.

2.1.2. Decision Variables

- X_{ij}: percentage of the flow supplied from PS_i at each time step j.
- m_i: number of fixed speed pumps in PS_i.
- b_i: ID of the pump model to be installed in PS_i in the range $[1, N_b]$.

2.1.3. Objective Function

The optimization model minimizes the sum of the capital (CAPEX) and operational (OPEX) costs at the same time. Equation (1) presents the total annualized cost of the project in which *Fa* is the amortization factor, which applies an interest rate r during Np periods.

$$F = F_a \cdot CAPEX + OPEX \tag{1}$$

$$F_a = \frac{r \cdot (1 + r)^{Np}}{(1 + r)^{Np - 1}} \tag{2}$$

The CAPEX and OPEX are calculated according to Equations (3) and (4), respectively.

$$\text{CAPEX} = \sum_{i=1}^{N_{ps}} \left(N_{B,i} \cdot C_{\text{pump},i} + n_i \cdot C_{\text{inv},i} + C_{\text{facility},i} + C_{\text{control},i} \right) \tag{3}$$

$$\text{OPEX} = \sum_{j=1}^{N_t} \left\{ \sum_{i=1}^{N_{ps}} \left[\left(\sum_{k=1}^{m_{i,j}} \frac{\gamma \cdot \left(H_{o,i} - A_i \cdot Q_{i,jk}^2 \right)}{\left(E_i - F_i \cdot Q_{i,j,k} \right)} + \sum_{k=1}^{n_{j,i}} \frac{\gamma \cdot \left(H_{o,i} \cdot \alpha_{i,j,k} - A_i \cdot Q_{i,jk}^2 \right)}{\left(\frac{E_i}{\alpha_{i,j,k}} - \frac{F_i}{\alpha_{i,j,k}^2} \cdot Q_{i,j,k} \right)} \right) \cdot P_{i,j} \right] \Delta t_j \right\} \tag{4}$$

2.1.4. Constraints

The optimization model is restricted by continuity and momentum equations and by minimum head requirements in the demand nodes. Equations (5) and (6) guarantee that the total flow supplied by the PS will be equal to the flow demand.

$$x_{i,j} \geq 0 \forall i, j \tag{5}$$

$$\sum_{i=1}^{N_{ps}} x_{i,j} = 1 \forall j \tag{6}$$

$$H_{0,i} \geq H_{max,i} \forall PS_i \tag{7}$$

Equation (7) was incorporated into this work. The new method determines a subset made up only of the pump models capable of delivering the minimum head and flow to reach the service levels required by the network. The search range of the decision variable associated with the pump model, i.e., bi, is then limited to the previously defined set.

All intermediate details about the hydraulic calculations of the objective function can be found in reference [7].

2.2. Infeasibility Maps

The decision variable x_{ij} determines the fraction of flow that PS_i contributes during period j. This variable can have a range from 0 to 100 (expressed as a percentage) for which 0 indicates that the PS did not supply water in that period; in contrast, a value of 100 indicated that all the flow was supplied by a single PS in the period. Therefore, a huge number of possible combinations exist, and many of them are infeasible solutions.

The main causes of infeasibility are listed below:

1. The distribution of flow generates sectors of the network where it is not possible to reach the minimum required pressures.
2. Some of the PSs must provide a pressure greater than the maximum head of the largest pump that exists in the available catalog.
3. The sum of the flows supplied is greater than the demand.

Each additional evaluation of the objective function supposes an increase in the computational effort made by the optimization algorithm. For this reason, this study proposed to analyze the network previously used to establish minimum and maximum limits for the variable $x_{i,j}$. Thus, it was possible to avoid the evaluation of infeasible solutions, which could be ruled out using hydraulic criteria before starting the optimization process. Unfortunately, the non-linearity of the relationships between the hydraulic variables did not allow these values to be fixed, but this value could be expressed as a function of the piezometric head of a reference PS (PS_{ref}). This reference pumping station could be any of the pumping stations in the network. Furthermore, PS_{ref} supplied all the water that was not provided by the rest of the PSs.

Before executing the optimization algorithm, for each PS_i different from PS_{ref}, it was possible to build a graph called the "Infeasibility Map" (IM), such as the one presented in Figure 1.

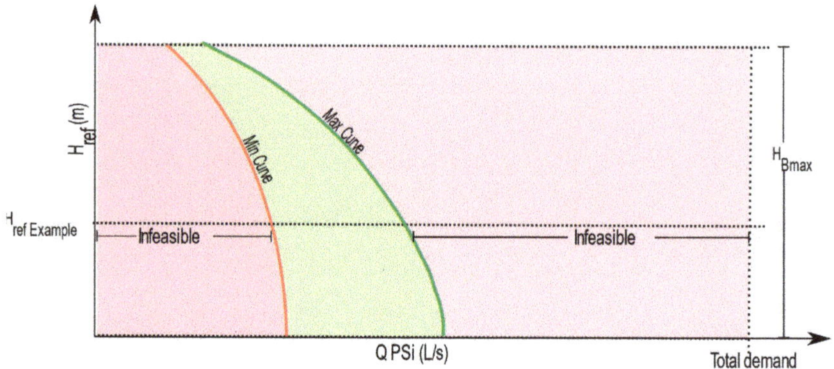

Figure 1. Infeasibility map for PS$_i$.

The horizontal axis represents the flow provided by PS$_i$ (QPSi). It could have values between zero and the total flow demanded by the network.

The vertical axis represents the head of PS$_{ref}$ (H$_{ref}$). For any H$_{ref}$, all points to the left of the minima curve (red color) are infeasible. This infeasibility was due to the fact that it was not possible to reach the minimum height required in all the nodes of the network. Similarly, at any point to the right of the maximum curve (green color), the head required by PS$_i$ always exceeded the maximum head of the largest pump available in the catalog (H$_{Bmax}$), and therefore, it would also be infeasible.

The green area of the graph is the bounded search region (BSR), which represents a set of combinations for which it was not possible to previously check infeasibility. That is, the BSR contained both feasible and infeasible points. Therefore, the optimization algorithm was in charge of traversing this space. Consequently, it was possible to use an IM to rule out a large number of combinations by previously limiting the range of the variable x_{ij}.

Finding these curves for all possible combinations can generate a high computational cost. Specifically, the number of hydraulic simulations increases significantly as the number of pumping stations grows. However, it is possible to estimate the maximum and minimum curves via randomly sampling combinations of pumping station flows using the Monte Carlo method [22]. Using this method greatly simplifies the proposed method.

The curves are used by the optimization algorithm in each evaluation of the objective function. In this new method, the value of x_{ij} represents a fraction of the difference between the highest value of the maximum flow curve and the lowest value of the minimum flow curve. It is important to note that this range is always less than the total demand. Therefore, it represents a search space reduction for any network, regardless of the topology.

Figure 2 shows the use of the IM with two PSs. The red line represents the SC calculated for a PS$_i$ different from PS$_{ref}$ [18]. From the solution, the intersection of the input flow x_{ij} and the respective SC could be obtained. If the resulting point was within the blue region, the solution could not be discarded. Otherwise, the solution was irrefutably infeasible. Outside this range, it would have been impossible to achieve a technically feasible solution.

The method presented in this work does not depend on the number of pumping stations. It could be applied to any problem with at least two PSs. For example, Figure 3 shows the decoding process in a network with three PSs: (1) PS$_1$, (2) PS$_2$, and (3) PS$_3$.

Analogous to the case of Figure 2, it was necessary to select a PS$_{ref}$. PS$_1$ was selected as the reference station after which the minimum and maximum curves for PS$_2$ and PS$_3$ were calculated. The limits defined by the curves allowed for generating the BSR for each PS$_i$. In Figure 3, blue and green areas represent the BSRs for PS$_3$ and PS$_2$, respectively. Regarding the flow supplied by each PS$_i$, x_{3j}, and x_{2j} represent the total percentage of flow supplied by PS$_3$ and PS$_2$, respectively. Consequently, PS$_1$ must supply the remaining flow with the head Hps_1. Note that each supply source had its own SC. Consequently, the SC was found for both pumping stations.

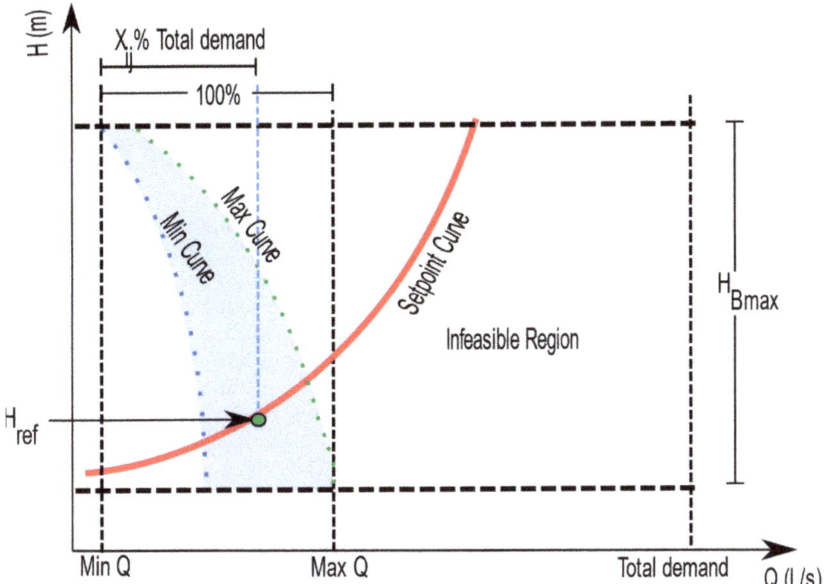

Figure 2. Decoding a solution for a 2-PS problem.

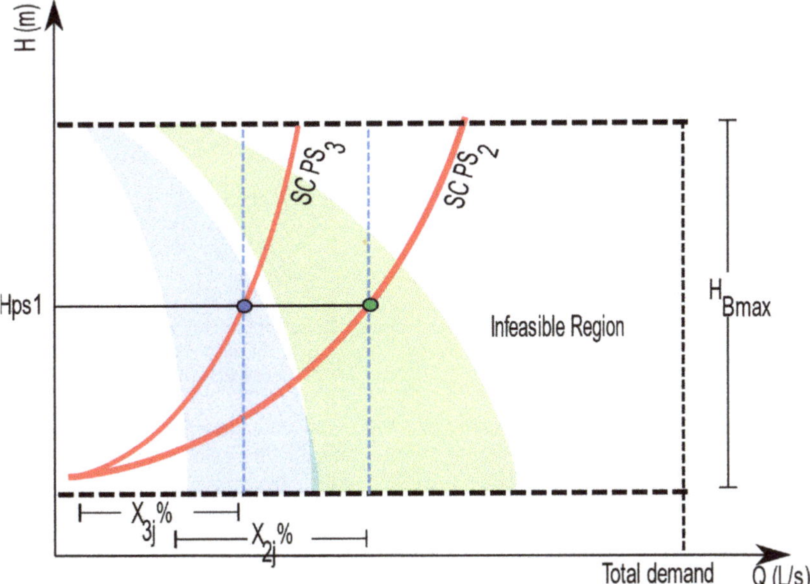

Figure 3. Decoding a solution for a 3-PS problem.

The use of IMs allows for reducing the number of hydraulic simulations carried out during the optimization process. The newly proposed method involves using IMs to rule out solutions during the evaluation of a solution when at least one period is infeasible. In this way, the hydraulic calculation stops, and the solution is penalized depending on the number of periods elapsed until finding the infeasibility.

The number of nodes and pipes in the network is the main cause of computational slowdown because the hydraulic motor calculates each node and each pipe in each iteration of the algorithm. This new procedure saves a large number of hydraulic simulations. This is important in the case of large networks because the computing time can be extremely high. Consequently, the application of the presented method is aimed at optimizing large networks. However, it could also be used in small networks without the need for changes.

2.3. Case Study

To apply the methodology described above, one case study was conducted. Figure 4a shows the topology of a WDN located in the city of Curicó (Chile). The network model was proposed by [23]. The network contained 7630 nodes and 8359 pipes. The network had 2 pumping stations working, PS_1 and PS_2. However, due to the growth of the city, the pumping equipment was old and susceptible to replacement. There is the possibility of putting a third water source into operation, located at PS_3. The node with the lowest elevation was 190 m, and the elevation of the largest node was approximately 295 m. The minimum operating pressure was 15 m for all network nodes. Information about the nodes and pipelines can be found in the Supplementary Materials.

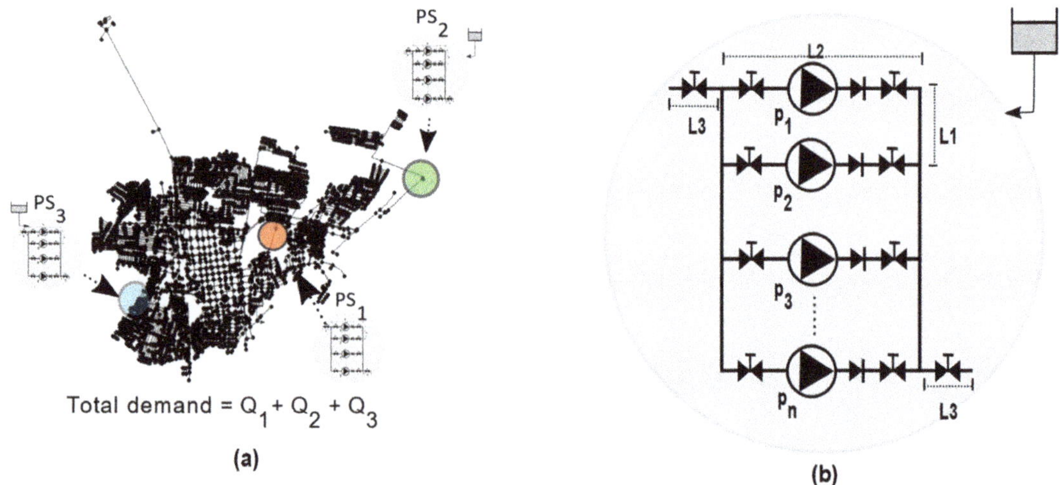

Figure 4. Case study. (**a**) Curicó water distribution network; (**b**) modular design for PS_i.

The total demand of the network was provided by the three pumping stations, where Q_1 was the flow provided by PS_1, Q_2 was the flow provided by PS_2, and Q_3 was the flow provided by PS_3. Figure 4b shows the modular design proposed by [7]. This scheme was used later to carry out the CAPEX calculations.

The hydraulic analysis was conducted for one day, and the time was divided into periods of one hour. A time pattern was used to characterize the time variation in demand, providing multipliers that were applied to the base demand to determine the actual demand in a given period. Figure 5 shows the 24 h use pattern.

To calculate the OPEX, Table 1 shows the hourly electricity rate used for each PS in the network. On the other hand, all the necessary coefficients to estimate the CAPEX were obtained from [7].

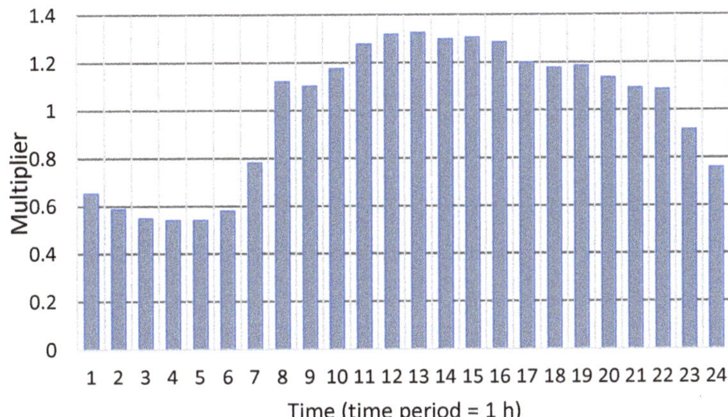

Figure 5. Demand pattern for the Curicó network.

Table 1. Electricity for the case study (EUR/kWh).

Time (h)	PS1	PS2	PS3
1–8	0.094	0.092	0.09
9–18	0.133	0.131	0.129
19–22	0.166	0.164	0.162
23–24	0.133	0.131	0.129

To perform the optimization process, a database with 67 pump models was used. The maximum flow rate of the pumps in the database varied between 9 L/s and 50.7 L/s. The annualized costs of these models were calculated using an interest rate of 5% per year and a projection time of 20 years. This led to an amortization factor of Fa = 7.92%.

2.4. Optimization Method

The solution space of the case study was 10^{104}. Consequently, the use of a computational method was required to solve the optimization model. Specifically, this work used a pseudo-genetic algorithm (PGA) developed by the authors of [20] to solve problems of an integer nature. Unlike a traditional genetic algorithm (GA), the PGA is based on an integer coding of the solution, and each decision variable can store different values represented by alphanumeric variables.

The objective of the work was to demonstrate that the use of IMs improves the performance of the optimization algorithm. For this comparison to be fair, the resolution of the proposed model was carried out using the same algorithm used by [7]. In this way, it was possible to directly compare the proposed methodologies and avoid unnecessary biases. In addition, the same parameters of population size (P), crossover frequency (Pc), and mutation frequency (Pm) recommended by the authors in previous works [24] were considered, specifically, population size (P = 100), crossover frequency (Pc = 90%), and mutation frequency (Pm = 5%).

The PGA was implemented using JMetalPy, which is an open-source Python library for solving single-objective and multi-objective optimization problems. It was inspired by the JMetal library, written in Java, and it implements evolutionary, local-search-based, and hybrid algorithms to solve various optimization problems [25]. Specifically, the Python 3.8 programming language was used. The objective function call was implemented according to the guidelines described in [12]. The hydraulic simulations were carried out using the programmer's toolkit of EPANET [21]. This system can conduct massive simulations and is integrated with the hydraulic network solver. To ensure a minimum level of statistical confidence in the results, 100 experiments were performed and analyzed.

Finally, to carry out the experiments, a computer with an Intel(R) Xeon(R) Gold 5218 CPU @ 2.30 GHz, 2300 Mhz, 16 main processors, and 32 logical processors equipped with the Windows 10 Pro operating system was used. The average time per execution was 21 h.

3. Results

The results compared the performance of the PS design method with and without IMs. First, PS_1 was selected as PS_{ref} and the IMs for the 24 periods were calculated for each PS_i. For example, Figure 6 shows the resulting IM for PS_2 in the period of greatest water demand (hour = 12). The orange zone represents the BSR and the remaining area represents hydraulically infeasible solutions that were not used by the PGA in the optimization process.

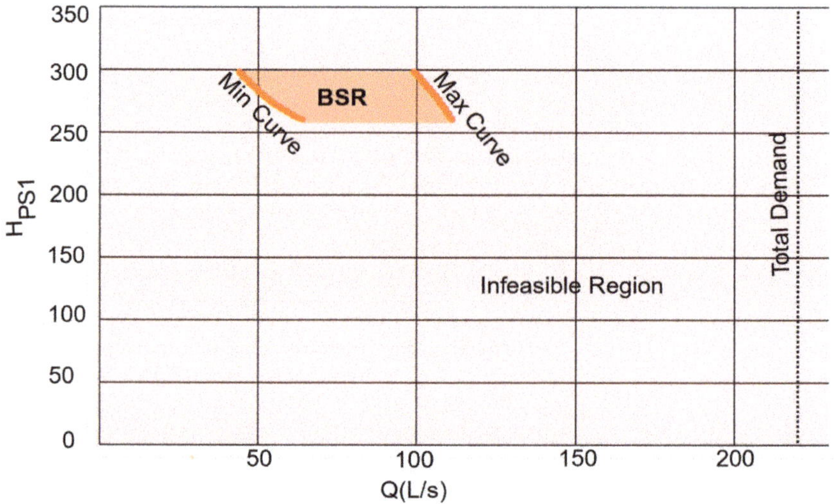

Figure 6. Calculated IM of PS_2.

Once the IMs were constructed, the algorithm searched only within the BSR. In this space, it was not possible to determine in advance whether the solutions were feasible/infeasible. This choice depended exclusively on the characteristics of the network and the hydraulic simulation to be executed. Both methods were compared. To increase the statistical reliability of the results, 100 experiments were performed. Each experiment ran 100,000 objective function evaluations, and the final values obtained for each method were compared. Note that the method presented in [7] performed one hydraulic simulation for each period analyzed. Consequently, each full day (24 h) led to 24 hydraulic simulations. However, the number of hydraulic simulations of the novel method presented in this work depended on the number of feasible periods of the analyzed solution. Figure 7 shows the results.

The blue dots represent the best solution obtained using the PGA with a search space limited by the IMs. The green dots show the best solution obtained using the PGA with the original method presented by [7], which utilized the complete search space. Additionally, the shaded area represents the complete distribution of all experiments simulated using IMs.

Note that the optimization algorithm converged to feasible solutions much faster when IMs were used, and the value of the objective function in all experiments when using IMs was less than the best solution using the complete search space. Consequently, the results demonstrate great improvement in the PGA performance when it was guided by IMs.

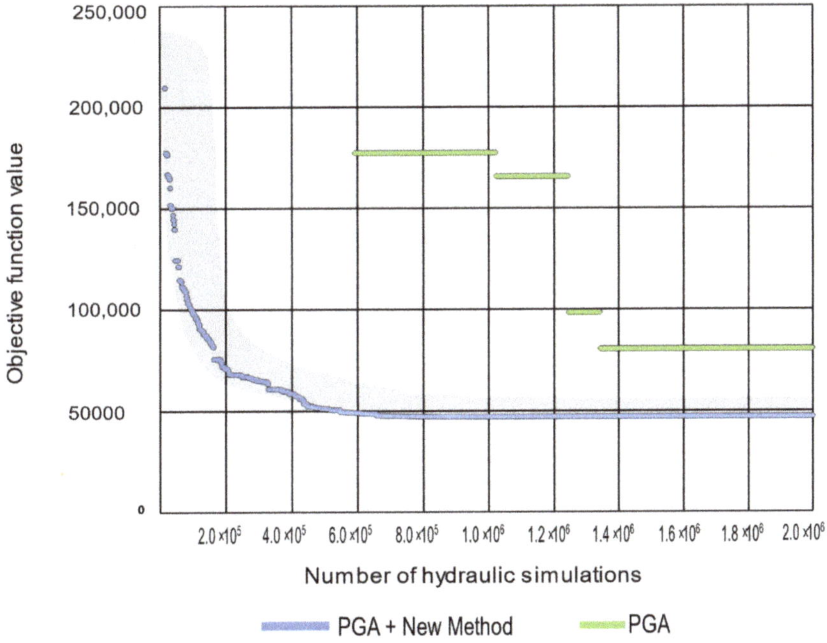

Figure 7. Comparison of the results with and without IMs.

It is important to highlight that the construction of the IMs also requires computational effort prior to the optimization process. The number of hydraulic simulations of this pre-processing will depend mainly on the number of pumping stations in the network. Specifically, in the case study, 4.0×10^5 hydraulic simulations were needed. However, this preprocessing was only executed once for the entire experiment and represented a small percentage of the total simulations. For example, for the case study in which 100 experiments were executed, the generation of IMs represented approximately 2% of the total number of simulations (2.0×10^8) and decreased in an inversely proportional manner with the number of experiments executed.

Previously, the monetary difference in the solutions obtained by each method was highlighted. Next, we focused on the hydraulic difference between both solutions. Figures 8 and 9 show the 24 h analysis of the pumping scheme of each PS_i for the best solution obtained by the PGA with and without IMs, respectively.

In both figures, the bars represent the number of active pumps during each period, and the dotted lines represent the total flow required by the network according to the consumption pattern in Figure 5.

The best solution obtained using the PGA without IMs (Figure 7) would have required the operation of the third pumping station to meet the operating conditions of the network. Notably, PS_1, PS_2, and PS_3 would have had to be at least ten, two, and two pumps, respectively. In contrast, when using the IMs, the PGA found many solutions in which only PS_1 and PS_2 were needed. Both solutions were hydraulically feasible, but the solution found using IMs was found to be more efficient, cheaper, and perfectly fit the network requirements. This feature is important for decision-making because if the search space is not correctly explored, unnecessary energy and building costs can be incurred.

Table 2 details the total yearly costs for the best solutions achieved by the PGA without and with IMs (Figures 8 and 9, respectively).

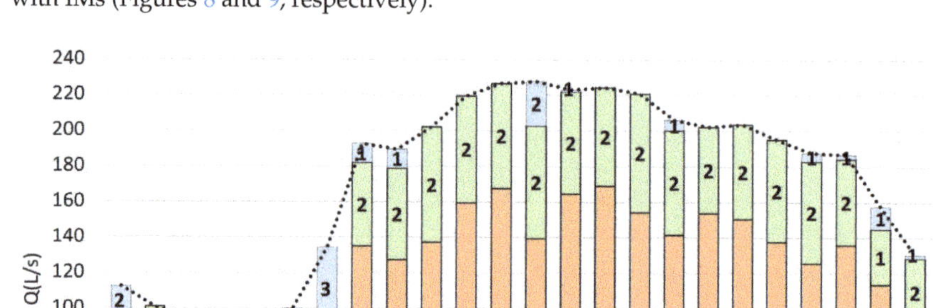

Figure 8. Best solution without IMs.

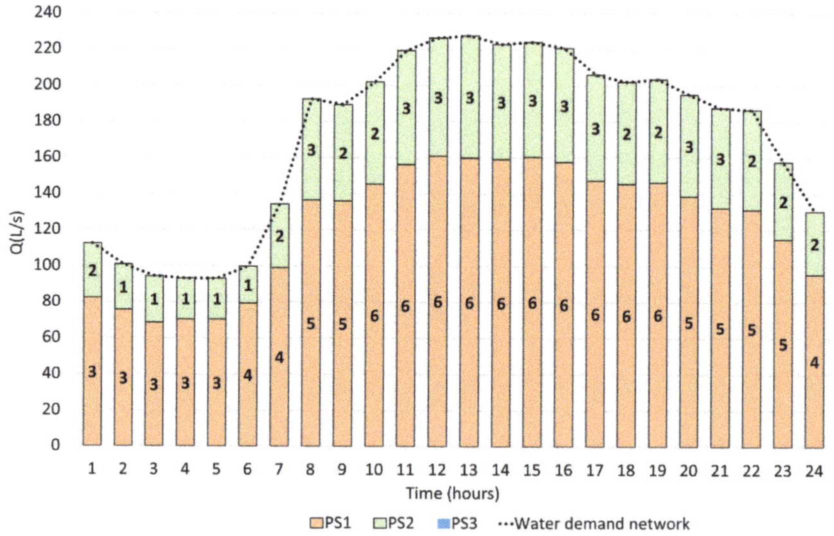

Figure 9. Best solution using IMs.

The use of IMs led to a reduction in the cost of the solution by 71% relative to previous solutions found without using a reduced search space. When analyzing the objective function in detail, the main difference occurred in the CAPEX term. First, pumping stations are expensive structures. Consequently, it is not profitable to activate the operation of PS_3 because it requires a high level of investment. Second, the optimized design without IMs required 10 pumps running on PS_1. This feature implies a high cost of purchasing this equipment.

Table 2. Cost comparison between the best solutions with and without IMs.

	OPEX		CAPEX		Fa • CAPEX + OPEX	
	PGA without IMs	PGA with IMs	PGA without IMs	PGA with IMs	PGA without IMs	PGA with IMs
PS1	EUR 121	EUR 80.7	EUR 190,756	EUR 94,470	EUR 54,710	EUR 34,002
PS2	EUR 34	EUR 31	EUR 34,544	EUR 37,077	EUR 13,754	EUR 13,054
PS3	EUR 16	-	EUR 85,618	-	EUR 12,107	-
				Total	EUR 80,571	EUR 47,056

Consequently, the best solution would supply all water demands from stations PS1 and PS2, and it would not be necessary to activate a third pumping station. According to the scheme in Figure 4, Table 3 shows the design specifications for pumping stations 1 and 2.

Table 3. Pump station designs for the case study.

		PS_1	PS_2
	ND_1	350	250
(mm)	ND_2	125	125
	ND_3	350	250
	L1	1.75	1.25
(m)	L2	3.75	3.75
	L3	3.50	2.50
	m_i	0	0
	n_i	6	3
	H_0	27.2632	27.2632
	A	−0.01416	−0.01416
	E	0.06929	0.06929
	F	0.00158	0.00158
	Model ID	GNI 50-13/7.5	GNI 50-13/7.5

ND_1, ND_2, and ND_3 are the nominal diameters of the corresponding pipe p, which is used for defining the diameters of elements such as isolation valves or check valves according to the modular design presented in Figure 4b. Similarly, L_1, L_2, and L_3 are the lengths of pipes. Furthermore, Table 3 shows the number of fixed-speed pumps (m_i) and the number of variable-speed pumps (n_i). H_0, A, E, and F represent the characteristic and efficiency curve coefficients. Finally, the last row displays the selected model pump from the database. It is important to note that the final solution only considered variable-speed pumps and ruled out fixed-speed pumps. The higher cost of this equipment could be offset by the reduction in energy consumption during the years of the project's life.

4. Conclusions

In the current context, improving the energy efficiency of pumping systems is a priority since these pumping systems represent a considerable percentage of the operating costs of any water supply company. Several approaches are described in the literature for optimizing the energy consumption of a PS. One possibility is approximating the operation mode of the pump(s) in the planning phase and optimizing the energy and cost in the WDN. This approach can help to make a better selection of pumping equipment. However, the computational cost is high, and the optimization algorithm loses search capacity as the network grows.

This work presents a new pumping station design method that considers the use of IMs for a better exploration of the search space. The method was applied to a real case study and was compared with the same design method without considering the use of IMs.

The use of IMs eliminates infeasible areas in the optimization process and improves the performance of the algorithm, both in terms of convergence speed and in the quality of the solutions. It is possible to highlight several findings after comparing the optimization methods with and without IMs:

- The exhaustive construction of the IMs required a significant number of hydraulic simulations. However, this procedure only needed to be done once, representing only 2% of the total number of simulations.
- When IMs were not used, the search space was too large, and the algorithm took a long time to find feasible regions, which were usually local minima. The use of IMs allowed for accelerating the convergence of the optimization algorithms, rapidly evolving toward better solutions. Specifically, the number of simulations required by the IM-guided algorithm managed to reduce the number of hydraulic simulations necessary to achieve convergence in the case study by 60%.
- The use of IMs in the case study achieved savings of 71% compared with the solutions obtained by the optimization algorithm when considering the complete search space. Additionally, the 100 experiments ran using IMs had better solutions than the best solution obtained using the PGA when no IMs were used. An inadequate exploration of the solution space implies unnecessary cost overruns and non-optimal solutions for a given problem.

Additionally, a new constraint was added to the model. In each evaluation of the objective function, the variable that determines the pump model in each PS only allows for selecting models from the catalog that have the maximum head required according to the flow distribution established by the solution. This mechanism allows the changes made by the PGA operators in these genes to always give rise to a new feasible solution.

The use of IMs guarantees that outside the bounded search region (BSR), there are no feasible solutions. However, it is not possible to determine the feasibility or infeasibility of solutions within this zone. In general terms, the search for global optima within the BSR continues to be a complex problem.

Finally, in small networks in which hydraulic simulation is not very computationally expensive, it could be possible to obtain solutions close to the global optimum by running a considerable number of evaluations of the objective function in a very short time. The construction of the IMs requires preprocessing of the network. Consequently, a limitation of the method could be a decrease in efficiency in small networks. Therefore, the use of IMs is only highly recommended when the analyzed network has a high number of nodes, pipes, and components.

Supplementary Materials: The following supporting information can be downloaded from https://www.mdpi.com/article/10.3390/math11071582/s1, Case Study S1.

Author Contributions: Conceptualization, J.H.G.-B., P.L.I.-R. and D.M.-M.; data curation, J.H.G.-B. and B.V.-M.; formal analysis, J.H.G.-B., D.M.-M., P.L.I.-R. and B.V.-M.; funding acquisition D.M.-M.; investigation, J.H.G.-B., D.M.-M., P.L.I.-R., B.V.-M. and F.S.-A.; methodology, J.H.G.-B., D.M.-M., P.L.I.-R. and B.V.-M.; project administration J.H.G.-B. and D.M.-M.; resources, J.H.G.-B., D.M.-M., P.L.I.-R., B.V.-M. and F.S.-A.; software, J.H.G.-B., D.M.-M. and B.V.-M.; supervision, D.M.-M. and P.L.I.-R.; validation, J.H.G.-B., D.M.-M., P.L.I.-R., B.V.-M. and F.S.-A.; visualization, J.H.G.-B., D.M.-M., P.L.I.-R., B.V.-M. and F.S.-A.; writing—original draft, J.H.G.-B. and D.M.-M.; writing—review and editing, J.H.G.-B., D.M.-M. and P.L.I.-R. All authors have read and agreed to the published version of the manuscript.

Funding: This research was funded by the Program Fondecyt Regular, grant number 1210410. It was also funded by the Ministry of Universities (Spain) and the Program European Union–Next Generation EU.

Data Availability Statement: Not applicable.

Acknowledgments: This work was supported by the Program Fondecyt Regular (project 1210410) of the Agencia Nacional de Investigación y Desarrollo (ANID), Chile. It was also supported by the Ministry of Universities (Spain) and the Program European Union–Next Generation EU and CONICYT PFCHA/DOCTORADO BECAS CHILE/2018—21182013.

Conflicts of Interest: The authors declare no conflict of interest.

References

1. Mala-Jetmarova, H.; Sultanova, N.; Savic, D. Lost in optimisation of water distribution systems? a literature review of system design. *Water* **2018**, *10*, 307. [CrossRef]
2. Mala-Jetmarova, H.; Sultanova, N.; Savic, D. Lost in optimisation of water distribution systems? a literature review of system operation. *Environ. Model. Softw.* **2017**, *93*, 209–254. [CrossRef]
3. Gupta, A.; Kulat, K.D. A selective literature review on leak management techniques for water distribution system. *Water Resour. Manag.* **2018**, *32*, 3247–3269. [CrossRef]
4. Bordea, D.; Pro, O.; Filip, I.; Drăgan, F.; Va, C. Modelling, Simulation and Controlling of a Multi-Pump System with Water Storage Powered by a Fluctuating and Intermittent Power Source. *Mathematics* **2022**, *10*, 4019. [CrossRef]
5. León-Celi, C.F.; Iglesias-Rey, P.L.; Martínez-Solano, F.J.; Savic, D. Operation of multiple pumped-water sources with no storage. *J. Water Resour. Plan Manag.* **2018**, *144*, 04018050. [CrossRef]
6. Blinco, L.J.; Simpson, A.R.; Lambert, M.F.; Marchi, A. Comparison of pumping regimes for water distribution systems to minimize cost and greenhouse gases. *J. Water Resour. Plan Manag.* **2016**, *142*, 04016010. [CrossRef]
7. Gutiérrez-Bahamondes, J.H.; Mora-Meliá, D.; Iglesias-Rey, P.L.; Martínez-Solano, F.J.; Salgueiro, Y. Pumping station design in water distribution networks considering the optimal flow distribution between sources and capital and operating costs. *Water* **2021**, *13*, 3098. [CrossRef]
8. Martin-Candilejo, A.; Santillan, D.; Iglesias, A.; Garrote, L. Optimization of the design of water distribution systems for variable pumping flow rates. *Water* **2020**, *12*, 359. [CrossRef]
9. Oshurbekov, S.; Kazakbaev, V.; Prakht, V.; Dmitrievskii, V. Improving reliability and energy efficiency of three parallel pumps by selecting trade-off operating points. *Mathematics* **2021**, *9*, 1297. [CrossRef]
10. Nagkoulis, N.; Katsifarakis, K.L. Minimization of Total Pumping Cost from an Aquifer to a Water Tank, Via a Pipe Network. *Water Resour. Manag.* **2020**, *34*, 4147–4162. [CrossRef]
11. Makaremi, Y.; Haghighi, A.; Ghafouri, H.R. Optimization of Pump Scheduling Program in Water Supply Systems Using a Self-Adaptive NSGA-II; a Review of Theory to Real Application. *Water Resour. Manag.* **2017**, *31*, 1283–1304. [CrossRef]
12. Gutiérrez-Bahamondes, J.H.; Salgueiro, Y.; Silva-Rubio, S.A.; Alsina, M.A.; Mora-Meliá, D.; Fuertes-Miquel, V.S. jHawanet: An open-source project for the implementation and assessment of multi-objective evolutionary algorithms on water distribution networks. *Water* **2019**, *11*, 2018. [CrossRef]
13. Fecarotta, O.; McNabola, A. Optimal location of pump as turbines (pats) in water distribution networks to recover energy and reduce leakage. *Water Resour. Manag.* **2017**, *31*, 5043–5059. [CrossRef]
14. Carpitella, S.; Brentan, B.; Montalvo, I.; Izquierdo, J.; Certa, A. Multi-criteria analysis applied to multi-objective optimal pump scheduling in water systems. *Water Sci. Technol. Water Supply* **2019**, *19*, 2338–2346. [CrossRef]
15. Weber, J.B.; Lorenz, U. Optimizing Booster Stations. In Proceedings of the Genetic and Evolutionary Computation Conference Companion, Berlin, Germany, 15–19 July 2017; pp. 1303–1310. [CrossRef]
16. Predescu, A.; Truică, C.O.; Apostol, E.S.; Mocanu, M.; Lupu, C. An advanced learning-based multiple model control supervisor for pumping stations in a smart water distribution system. *Mathematics* **2020**, *8*, 887. [CrossRef]
17. Gil, F.A.A.; Iglesias-Rey, P.L.; Martínez-Solano, F.J.; Cortes, J.V.L.; Mora-Meliá, D. Methodology for Projects Of Pumping Stations Directly Connected To The Network Considering The Operation Strategy. In Proceedings of the 22nd International Congress on Project management and Engineering, Madrid, Spain, 11–13 July 2018; 2018; pp. 551–563. Available online: http://dspace.aeipro.com/xmlui/handle/123456789/1728 (accessed on 20 March 2023).
18. León-Celi, C.F.; Iglesias-Rey, P.L.; Martínez-Solano, F.J.; Mora-Melia, D. The Setpoint Curve as a Tool for the Energy and Cost Optimization of Pumping Systems in Water Networks. *Water* **2022**, *14*, 2426. [CrossRef]
19. Briceño-León, C.X.; Iglesias-Rey, P.L.; Martínez-Solano, F.J.; Creaco, E. Integrating Demand Variability and Technical, Environmental, and Economic Criteria in Design of Pumping Stations Serving Closed Distribution Networks. *J. Water Resour. Plan Manag.* **2023**, *149*, 4023002. [CrossRef]
20. Mora-Melia, D.; Iglesias-Rey, P.L.; Martinez-Solano, F.J.; Fuertes-Miquel, V.S. Design of water distribution networks using a pseudo-genetic algorithm and sensitivity of genetic operators. *Water Resour. Manag.* **2013**, *27*, 4149–4162. [CrossRef]
21. Rossman, L.A. *EPANET 2.0 User's Manual (EPA/600/R-00/057)*; National Risk Management Research Laboratory: Cincinnatti, OH, USA, 2000.
22. Lučin, I.; Lučin, B.; Čarija, Z.; Sikirica, A. Data-driven leak localization in urban water distribution networks using big data for random forest classifier. *Mathematics* **2021**, *9*, 672. [CrossRef]
23. Negrete, M. *Modelación Computacional en EPANET de un Sector de la Red de Abastecimiento de Agua Potable de Curicó*; Universidad de Talca, Facultad de Ingeniería: Curicó, Chile, 2021.

24. Mora-Melia, D.; Iglesias-Rey, P.L.; Martínez-Solano, F.J.; Muñoz-Velasco, P. The efficiency of setting parameters in a modified shuffled frog leaping algorithm applied to optimizing water distribution networks. *Water* **2016**, *8*, 182. [CrossRef]
25. Benítez-Hidalgo, A.; Nebro, A.J.; García-Nieto, J.; Oregi, I.; Del Ser, J. jMetalPy: A Python framework for multi-objective optimization with metaheuristics. *Swarm Evol. Comput.* **2019**, *51*, 100598. [CrossRef]

 *mathematics*

Article

Optimizing PV Microgrid Isolated Electrification Projects—A Case Study in Ecuador

Bruno Domenech [1,2], Laia Ferrer-Martí [1,3], Facundo García [4], Georgina Hidalgo [4], Rafael Pastor [1,2] and Antonin Ponsich [1,2,*]

1   Supply Chain and Operations Management Research Group, Escola Tècnica Superior d'Enginyeria Industrial, Universitat Politècnica de Catalunya-BarcelonaTech, 08028 Barcelona, Spain; bruno.domenech@upc.edu (B.D.); laia.ferrer@upc.edu (L.F.-M.); rafael.pastor@upc.edu (R.P.)
2   Department of Management, Universitat Politècnica de Catalunya-BarcelonaTech, 08028 Barcelona, Spain
3   Department of Mechanical Engineering, Universitat Politècnica de Catalunya-BarcelonaTech, 08028 Barcelona, Spain
4   Catalan Association of Engineering Without Borders, 08027 Barcelona, Spain; dantega78@gmail.com (F.G.); georginahm92@gmail.com (G.H.)
*   Correspondence: antonin.sebastien.ponsich@upc.edu

Abstract: Access to electricity for the rural and indigenous population of Ecuador's Amazon Region (RAE) is considered a critical issue by the national authorities. The RAE is an isolated zone with communities scattered throughout the rainforest, where the expansion of the national grid is not a viable option. Therefore, autonomous electrification systems based on solar energy constitute an important solution, allowing the development of indigenous populations. This work proposes a tool for the design of stand-alone rural electrification systems based on photovoltaic technologies, including both microgrid or individual supply configurations. This tool is formulated as a Mixed Integer Linear Programming model including economic, technical and social aspects. This approach is used to design electrification systems (equipment location and sizing, microgrid configurations) in three real communities of the RAE. The results highlight the benefits of the developed tool and provide guidelines regarding RAE's electrification.

Keywords: rural electrification; mathematical programming; Ecuador's Amazon region; photovoltaic energy; microgrid

Citation: Domenech, B.; Ferrer-Martí, L.; García, F.; Hidalgo, G.; Pastor, R.; Ponsich, A. Optimizing PV Microgrid Isolated Electrification Projects—A Case Study in Ecuador. *Mathematics* **2022**, *10*, 1226. https://doi.org/10.3390/math10081226

Academic Editors: Ripon Kumar Chakrabortty and Armin Fügenschuh

Received: 16 December 2021
Accepted: 5 April 2022
Published: 8 April 2022

Publisher's Note: MDPI stays neutral with regard to jurisdictional claims in published maps and institutional affiliations.

1. Introduction

Currently, access to electricity remains a critical issue for around 1.1 billion people worldwide, generally affecting those living in rural contexts, where poverty levels are high. Indeed, both aspects (energy access and poverty) are closely connected, as proved by the Seventh Sustainable Development Goal stated by the United Nations ("ensure access to affordable, reliable sustainable and modern energy for all") [1]. This goal explicitly views access to energy as key to improving the living conditions of the most underprivileged populations, "a backbone of any modern economy" [2]. In this sense, several recent studies highlight the fact that electricity access is linked to increased incomes and productivity (different to agricultural activities [3]), as well as benefits in education (increased study time) and health (decrease in respiratory diseases due to lower kerosene usage) [4]. In order to improve electricity coverage, the extension of the national grid has been the main strategy for providing access to electricity. However, in areas with rough topography or remote population centers, the expansion of the national distribution grid may become infeasible [5]. In those cases, the development of off-grid electrification systems and microgrids is among the most economical solutions and has been successfully adopted in many practical frameworks [6]. Besides this, for such stand-alone generation systems, the current concern about global warming and the resulting interest in alternative energy sources have led to the intensive use of renewable resources. In particular, solar energy remains one of the most widely employed sources [7].

In this framework, different kinds of electrification systems have been tackled in the devoted literature, in many distinct worldwide contexts. For instance, regarding large-scale electrification (national or regional), Ehsan and Yang [8] highlight the tendency to formulate the integration and planning of generation systems based on renewable energy as an optimization problem. The model proposed in [9] for renewable allocation planning in large-scale power systems considers a dynamic environment leading to the expansion/reduction of the current network. Hassan [10] uses simulation to optimize solar photovoltaic-based generation systems in Iraq, considering two configurations, namely off-grid and on-grid versions (when possible). Taye et al. [11] use Geographical Information Systems (GIS) and a multi-criteria decision-making technique for rural electrification planning, allowing for an evaluation of the adequacy of renewable energy in Ethiopia. Through the application of the Analytic Hierarchy Process, they conclude that wind and particularly solar energies should be preferred rather than grid extension. In addition, several computational tools were recently developed, in particular for rural electrification planning. For instance, the REM (Reference Electrification Model [12]) focuses on the use of off-grid generation systems in order to plan electricity networks and has been used in several countries (India, Colombia, Kenya, Rwanda, etc.). Furthermore, OnSSET [13] is an open-source tool using GIS that is designed to complement energy planning models not supported by geographical analysis and that quantifies the investment, technology type and geo-referencing of national electrification projects based either on conventional or renewable energy technologies. Also obeying an open-source philosophy, recently developed Python-based tools represent free access alternatives. For instance, PyPSA (Python for Power System Analysis [14]) is a free toolbox for simulating and optimizing modern power systems, accounting for alternative working modes (conventional generators, variable wind and solar generation, storage units, mixed alternating and direct current networks, etc.) and with an improved scalability for dealing with large networks and long time series. Another recent computational tool for microgrid design, Sandia's Microgrid Design Toolkit [15], allows for the generation of design optimizations according to several criteria (investment and operation costs, reliability, performance levels) in order to produce a Pareto frontier of efficient configurations promoting the microgrid over individual supply systems. Based on advanced optimization and modeling approaches, it has been used to provide electricity in several military and public infrastructures worldwide.

At a local level, Bahramara et al. [16] review works using the well-established commercial tool HOMER [17], particularly focusing on the design of hybrid renewable energy systems. For instance, in [18], such a tool is used to design and analyze the robustness of environmental-friendly systems to be installed in Malaysian islands. In addition, interest is increasingly devoted to the design of microgrids, which are reduced-size grids connecting several users isolated from the national grid. Several applications have demonstrated that the design of self-sufficient microgrids produces higher benefits than those obtained by individual systems [19] and also reduces the life cycle environmental impact of the electrification systems [20]. Despite the additional complexity associated with the design of microgrid topologies, this strategy allows the energy supply to be independent from the resources available at demand points, cost savings thanks to economies of scale for shared equipment and supply flexibility in case of an increase in demand [21]. Accordingly, many studies have tackled the design of microgrid distribution structures around the world. The reader is referred to the recent surveys of Peters et al. [22] as well as Castilla et al. [23] for a perspective on the use of microgrids in Latin America, Mahomed et al. [24] in Uganda, Tenenbaum et al. [25] in sub-Saharan Africa or Lukuyu et al. [26] in East Africa. These studies insist on the need for adequate design and planning tools for microgrid-based electrification projects, supported by objective demand projection methodologies and developed in collaboration with local actors to account for the population's specific requirements.

Regarding the solving procedures, mathematical models have been developed with the aim of providing effective configurations (in terms of cost or any other performance criterion). Indeed, these models are adapted to the particular application addressed and may involve distinct configurations (individual/microgrid), energy sources or other features specific to each case study and to the communities to be electrified [27]. For instance, Leithon et al. [28] develop a model for energy allocation policy that minimizes energy usage. In [29], an optimization approach is developed for the electrification of highland communities in Peru through a model integrating social constraints associated with system management and community benefits. Ranaboldo et al. [30] develop optimization models including the particular considerations of electrification systems in Cabo Verde. Heuristic procedures have also been used. For instance, several matheuristics are introduced in [31], representing computationally efficient tools to optimize rural electrification systems involving both microgrids and individual supply. Moreover, different multi-criteria approaches (VIKOR and AHP) are compared in [32] for microgrid design in Venezuela, including economic (investment, maintenance and operation costs), social and environmental criteria. Finally, Python-based free access libraries have also been developed in this context, such as MicrogridsPy [33]. This tool tackles the problem of generation equipment sizing (Li-ion batteries, diesel generators and PV panels) and energy dispatching in remote and isolated contexts by minimizing the Levelized Cost of Energy (LCOE).

The above-mentioned studies emphasize the possibility of designing economically efficient and environmentally resilient off-grid generation systems based on renewable energies. This is particularly relevant for the so-called "last-mile" electrification (as opposed to mass electrification, for which different strategies such as grid expansion may provide better results). "Last-mile" electrification projects are not only useful in the context of, for example, small islands [34], where diesel generators might be replaced by photovoltaic and wind energies, as they have a great impact on greenhouse gas emissions while producing significant economic benefits. In several Latin American countries, both the Amazon (rainforest) and Andean (highlands) regions typically show rough terrains and represent massive challenges, such as reported for Brazil [35], Colombia [36] or Bolivia, Peru and Argentina [37]. In this framework, Ecuador is a country with a wide-spread national grid and high global access to electricity, but the indigenous populations of the Amazon basin are scattered over large areas covered by rainforest, leading to prohibitive costs for expanding the national grid. Furthermore, the difficulty in reaching these isolated communities and their fragile economic resources make electrification projects neither profitable for private distributors nor sustainable for governments. So, the aforementioned stand-alone, renewable-based systems appear as a viable electrification strategy. To the best of our knowledge, only one study has reported on the design of such systems in Ecuador [38]. Carried out in the Santa Elena province, this work reports the design of hybrid wind–photovoltaic systems through HOMER, concluding that most of the energy is supplied by PV cells. However, the paper does not account for microgrid formation, although such configurations are promoted among the guidelines stated by Ecuador's Ministry of Electricity and Renewable Energies (MEER) [39].

Thus, the present work addresses the development of autonomous electrification systems for isolated communities in the Amazon Region of Ecuador (RAE) by optimizing the design of PV-based systems involving microgrids. Thanks to a detailed analysis of the relevant local factors to be accounted for, a Mixed Integer Linear Program (MILP) is introduced as a computational tool for the automatic design of such electrification systems. This model extends several state-of-the-art approaches. In particular, Ferrer-Martí et al. [40] set out the basis of rural electrification system modeling involving hybrid generation and microgrids. Domenech et al. [29], in addition, include management and social constraints in the design phase. Despite considering some elements of these previous studies, our model accounts for characteristics that are specific to the RAE and therefore represents a novel design tool, which could yet be extrapolated to other contexts. In particular, the contribution in terms of new modeling features is threefold:

i.   A set of potential connections is established, indicating impossible wiring between different geographical points. This new constraint, motivated by local factors (explained in the following sections), is addressed through a decomposition strategy of the original problems that allows a more efficient solution process.

ii.   A new objective function now incorporates the parametrized ponderation of the costs of microgrids versus individual systems. This novel feature is motivated by the electrification policies dictated by Ecuador's national government, promoting microgrid configurations. However, the versatility of the formulation proposed here allows either one or the other configuration to be favored according to any policy makers' decisions.

iii.   For cultural reasons in the RAE, items shared by the members of a community cannot be stored on private ground. This requirement is reflected in the model by new constraints, which prevent microgrid generation units from being located at demand points.

Thus, this model constitutes a tool adapted to RAE conditions for assisting project promoters in the design of electrification systems. Three case studies addressing the electrification of indigenous communities in the RAE are subsequently solved in order to validate the proposed tool.

The remainder of this work is organized as follows. Section 2 presents a context analysis regarding electrification processes in Ecuador, a description of the electrification systems accounted for and their specific conditioning features in the RAE. The proposed mathematical model is described in Section 3, while Section 4 presents the case studies addressed and the numerical results obtained. Finally, some conclusions and prospects for future works are provided in Section 5.

2. Context Analysis for the Design of Stand-Alone Electrification Systems in the RAE

2.1. Overview of the Electrification Process in Ecuador

The Republic of Ecuador is a Latin American country with an area of 283,561 km^2 and about 17 million inhabitants. Despite the growth of its Gross Domestic Product, strong inequalities mean that more than 20% of Ecuadorians suffer from poverty, particularly in rural and indigenous communities. Besides, Ecuador has some of the greatest biodiversity in the world, meaning that environmental protection is a prominent feature of the country's landscape. In this context, government policies for the development of the national electricity system must find a trade-off between socio-economic development and environmental conservation [41]. Accordingly, efforts have been made by national authorities to guarantee a reliable and competitive electricity supply, supervised by the Regulatory Agency and Electricity Control. These measures have led to a significant increase in the total electricity generated [42] and the expansion of the national grid, reaching one of the best coverages in the sub-continent: 97.33% national access and 94% of the rural population in 2017 [43]. However, the development of Ecuador's electrification process still faces two important challenges.

First, some diversification is needed regarding the generation matrix. As stressed in [44], the only consistent trend in Ecuadorian energy policies has been the development of hydroelectricity, which represented more than 73% of the generation matrix in 2017 [42]. However, this strategy is criticized by indigenous communities and environmentalist associations due to its environmental impact [45]. Several studies have also emphasized the great potential of solar and wind energies and their advantages regarding socio-economic development and environmental conservation [41]. Adopting photovoltaic technologies may yield long-term benefits in terms of pollution abatement and climate change mitigation [39]. Accordingly, the Fund for Electrification of Rural and of Marginal Urban Areas (created in 2004, in order to improve electricity coverage in disadvantaged areas) initiated rural electrification programs relying on PV generation in the Amazonian and highland regions or hybrid systems in the Galapagos islands [46]. More recently, the current administration is promoting PV-powered microgrid designs for remote areas rather than individual systems [39]. Despite these efforts, solar and wind energies did not represent more than 0.5% of the primary sources used for electricity generation in 2017.

On the other hand, there are still some glaring inequalities in electricity access in the RAE. In spite of the efforts made to increase coverage in the corresponding provinces (Pastaza, Sucumbíos, Orellana, Napo, Morona and Zamora), their impact was undermined by the limited amount of economic resources. Indeed, the RAE (40% of Ecuador's total area) is characterized by the highest poverty levels in Ecuador and has the lowest electrification rates, particularly in rural areas, since the grid expansion strategy has left a great part of this territory uncovered [43]. Renewable energy technologies have been promoted, but several electrification programs were not considered profitable by the distribution operators and were thus interrupted. Indeed, the maintenance operations of stand-alone systems in rough terrain (rainforest) would either be too costly or require training programs for community members. In addition, the scarce economic resources of these mainly indigenous communities mean that users often cannot afford the electricity service costs, making these projects unsustainable for governments.

Thus, the six RAE provinces produce electricity almost entirely by thermal generation. In the Pastaza province, where the present study was carried out, the electricity coverage currently equals 89.3% (compared to 97.33% at national level), but the rural electrification rate is much lower (65.9% in 2010 [42]). This situation is particularly unsustainable in one of the most prolific zones for solar energy, where the transportation, firewood consumption and electricity needs (more than 1000 GWh/year) could be entirely covered by PV technologies [47]. Accounting for these general considerations, the systemic approach introduced in this work aims to provide an automatic tool for the design of electrification systems in rural and disadvantaged areas of the RAE, taking advantage of the high PV potential available and promoting the formation of microgrid-based systems.

2.2. Technical Description of Stand-Alone Systems

In [47], the authors demonstrate that while the RAE benefits from high solar resources, on the contrary, wind does not constitute a promising energy source. Furthermore, due to the prevailing dense rainforest vegetation, the construction and installation of wind turbines raises practical issues. Therefore, the electrification systems represented here are only based on photovoltaic technology, as illustrated in Figure 1 (with microgrid-based distribution). The electricity is produced by the PV panels, while the controllers protect batteries from overloads and deep discharges. The electricity is then stored in batteries to bridge the gap between generation and consumption. Next, the inverters transform the direct current from batteries into alternating current, which is more suitable for most electrical appliances. Finally, the electricity is distributed to demand points (households, schools, health centers, etc.) via microgrids or individual systems (individual systems are devoted to supplying single-user consumption).

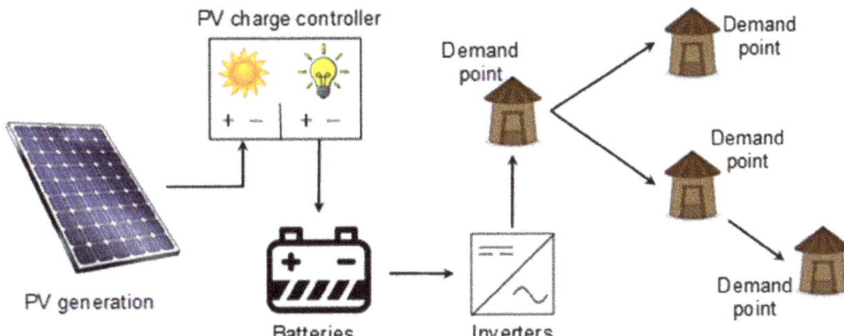

Figure 1. Scheme of a PV system with microgrid distribution.

Regarding the microgrid topology, a structure ensuring minimal costs is chosen, respecting the following conditions [48]: (i) power generation is centralized at a single point and (ii) the microgrid has a radial structure (each user can receive electricity from only one point). This structure is illustrated in Figure 1, where the generation units deliver power to four users connected according to a tree-like configuration.

2.3. Conditioning Factors for Stand-Alone Electrification Systems in the RAE

This section analyzes the main features conditioning the design of rural electrification projects with PV technology for indigenous communities in the RAE. This analysis involved studying and understanding the concrete reality of the targeted communities through three main action lines. First, the information obtained from the literature was enhanced with technical documents [49,50] produced by Ecuador's MEER in coordination with the NGO "Engineers Without Borders" (EWB). Second, information on the local context and needs was comprehensively collected through a 5-month field survey performed by some of the authors of this work. This stay allowed the indigenous population's lifestyle to be observed as well as the economic, social, environmental or cultural features relevant to the installation of autonomous systems for these communities to be identified. It also made possible an evaluation of the logistics needed for equipment transportation within rainforest conditions. Third, several key actors of rural electrification in Ecuador were interviewed to identify and validate the conditioning features adapted to the RAE's reality. This phase involved three technicians from the MEER, the leader of the renewable energies area in the power company Empresa Eléctrica Ambato S.A. (Pastaza province) and three coordinators, technicians and volunteers from the energy line of EWB.

The resulting list of conditioning factors encompasses all the features to be accounted for within the design process of electrification systems in the RAE. These considerations are captured in the mathematical model developed for the solution of the addressed problem (see next section).

(a) National and regional policies contemplate social aspects, such as opting for electrification designs including microgrids rather than individual systems. Indeed, the community-based management of joint installations provides social benefits, such as the coordination and cooperation of families sharing the same objectives. In order to encourage microgrid formation, priority is given to designs including such configurations, even though they entail a higher cost than individual supply systems (up to 20% higher, as proposed by MEER).

(b) The institutional framework of electrification projects may ensure economies of scale (for instance, when equipment is purchased for district or regional projects) but may also restrict the technical characteristics of power generation and distribution items. In the case study presented here, the limiter boxes allow only two output cables, which may have an impact on the microgrid structure.

(c) The communication paths available (rivers, airways) and the transportation means to get to the community in question have an impact on the technical equipment employed. Moreover, the current state of these paths may also involve space and weight limitations for the equipment units to be shipped. For instance, the varying water depth in a river (or landing strip dimensions) can limit the size of canoes (or aircraft) that can be used.

(d) The property concept in some indigenous communities means that the equipment shared by the community cannot be physically installed at a demand point, which is private ground. So, non-demand points should be identified for potential microgrid generation. Furthermore, this means the construction of sheds for electric equipment storage within the area where PV panels are to be located in order to protect batteries, inverters and regulators from weather or animals. This incurs additional costs associated with the purchase and installation of these buildings.

(e) In the Low Amazon region, the increased concern for environmental aspects leads to the development of underground connections rather than air connections. Indeed, despite the advantages of air microgrids in both practical (avoiding obstacles such as rivers, small buildings, etc.) and economic (cheaper installation and maintenance) terms, they have a negative environmental impact due to tree clearing around the microgrid installations. Underground wiring is also better protected from external agents (rain, animals) and presents technical advantages [51]. However, this policy involves constraints due to physical obstacles (river, ravine, floodable area, landing strip, etc.) that may prevent cable installation.

It should be noted that, in the above-mentioned factors, the estimation of end-users' demand is not mentioned. Evaluating the demand constitutes a critical phase in the design of electrification systems, since this parameter strongly influences the final system configuration [52]. An overestimated demand would lead to oversized systems, both wasting generation capacity and increasing electricity prices [53]. Several studies have highlighted the complexity of this estimation task, involving qualitative and quantitative assessments of the target community and its surroundings, an analysis of the energy sources used prior to electrification, the planned electricity usage and the projection of the demand growth [22,54,55]. In the RAE, the MEER performed such a study, as verified by the field survey carried out in Ecuador by some of the authors of this work. In order to guarantee equal opportunities to every family, the MEER decided to set standardized consumption levels for energy and peak power for all users. In this context, the demand values accounted for in the present study are those determined and imposed by the MEER (see Section 4.2). Note that these demands were conceived as a constant value, so projects may initially be slightly oversized but will be adequate in the medium term, when demand growth occurs during the first years of implementation.

Hence, this analysis identified the key elements ensuring the sustainability of electrification projects in the RAE. This phase sets the conceptual framework necessary for the development of a solution procedure for the design of stand-alone electrification systems in this region.

3. A Mathematical Model for the Design of Autonomous Rural Electrification Systems in the RAE

Since optimization approaches have proven their effectiveness in the framework of electrification system design [56], a mathematical model for stand-alone electrification systems, including the considerations analyzed above, is introduced in this section. The mathematical formulation proposed in this work includes parts of the procedure presented in [57] and further accounts for all the conditioning factors described in the previous section. The proposed method for the development of the mathematical tool introduced here is shown in Figure 2.

This new MILP aims to determine the details of the electrification design (individual and/or microgrid configurations, equipment location and selection, etc.) in order to minimize the project cost. To further highlight the original features of the tackled problem and the novelty of the model developed accordingly, their main characteristics are listed in Table 1, and the comparison with similar works emphasizes that no other approach considers all of them. In particular, the versatility in promoting either microgrid or individual supply configurations and the restriction on microgrid generation equipment being located at demand points are introduced for the first time in this work. Therefore, the electrification solutions obtained with the model proposed here, which responds to RAE's social, cultural and environmental requirements, could not be obtained with other tools, thus confirming the contribution of the present work. At the same time, since these requirements may be found in plenty of contexts different from RAE's, such as many isolated regions in Latin America [58] and Africa [24,26], and since the model proposed here also includes the classical features needed for the design of electrification systems based on renewables, it can be re-used for other applications in distinct contexts.

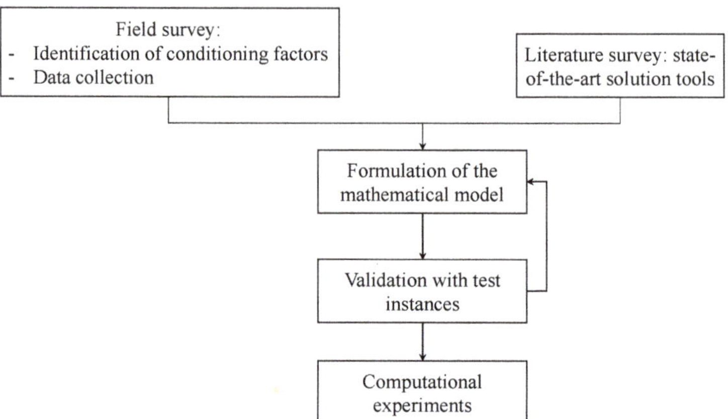

Figure 2. Flowchart of the proposed method.

Table 1. Comparison of main problem features addressed in this study and in previous related works.

Feature	a.*	b.*	c.*	d.*	e.*	f.*	g.*	h.*
PV generation	X		X	X	X	X	X	X
Battery storage	X		X	X	X	X	X	X
Microgrid distribution		X	X	X	X	X	X	X
Several microgrids			X	X	X	X	X	X
Individual supply		X	X	X	X		X	X
Versatile (microgrids vs. individual)								X
Forbidden connections				X				X
Restriction on microgrid generation points								X
Demand/non-demand points		X		X	X		X	X
Economic assessment	X	X	X	X	X	X	X	X
Additional installation costs	X				X	X		X

* where a. HOMER [17]; b. ViPOR [59]; c. Ferrer-Martí et al. [40]; d. REM [12]; e. Domenech et al. [57]; f. MicrogridsPy [33]; g. Sandia's Microgrid Design Toolkit [15]; h. The approach proposed in this work.

The numerical parameters of the model include information regarding demand point locations and requirements, the characteristics of the power generation and distribution equipment, as well as the data associated with the specific RAE features. These parameters are comprehensively defined in Table 2, but further explanations are provided here on the Matrix of Potential Connections.

As stated previously, only underground-wired microgrids are considered. However, the rainforest environment in which the indigenous communities live is normally characterized by floodable terrain. Besides this, obstacles such as rivers or landing strips can prevent underground wiring, thus impeding some direct connections between two points. Thus, all the allowed connections are concatenated in a binary matrix called the Matrix of Potential Connections (MPC), defined for each case study. This matrix of size $|P| \times |D|$ has elements equal to 1 if a connection is possible between the two points and 0 otherwise. It is worth mentioning that this matrix may highlight clusters of points completely isolated from the others (i.e., neither direct nor indirect connections can be established between any points of two different clusters).

This information offers two options: treating each community as a whole or breaking it down into several independently solved sub-problems. Indeed, there is no reason to simultaneously solve several physically separated zones, since they cannot be included in the same microgrid anyway. Additionally, due to the combinatorial nature of the MILP model presented hereafter, treating each sub-problem independently might allow significant CPU time savings when compared with solving the global problem. Thus, before using the optimization tool proposed here, a pre-processing step must be performed for each particular case in order to decide if the tackled community should be treated entirely or divided into isolated clusters to be solved independently one from another.

In addition, the decision variables have to account for the system configuration with its associated energy/power flows, as well as the number and type of equipment units to be used for power generation and distribution (see Table 3). The objective function consists in minimizing the total investment cost for wind turbines, PV panels, PV controllers, batteries, inverters, meters (to be installed at all the demand points in a microgrid, to grant equality of the provided services for all users) and wires. In addition, as stated in point d) of the conditioning factors (Section 2.3), a cultural requirement is that the generation equipment of (community-shared) microgrid systems cannot be set at demand points (private ground), meaning that additional costs corresponding to the installation of a shed for equipment storage have to be included.

Moreover, as explained in point (a) of Section 2.3, MEER's policy prioritizes microgrid-based designs over individual systems. Therefore, configurations including microgrids will be preferred even if their cost is up to α% higher than individual supply systems. Practically, the cost associated with all the items belonging to a microgrid will be multiplied by $1/(1 + \alpha/100)$, which is equivalent to reducing the corresponding cost by a factor $\alpha/(\alpha + 100)$%. The generation costs of individual systems (at demand points) can easily be distinguished from those of microgrid configurations by defining microgrid generation points as no-demand points. With regard to α, the MEER proposes 20%, but the influence of this value is also studied through a sensitivity analysis (see next section).

Table 2. Parameters of the mathematical model.

Parameter	Description	Unit
	Demand points	
P	Set of potential generation points, including the demand points.	-
D	Set of demand points, $D \in P$.	-
L_{pd}	Distance between two points p and d ($p \in P, d \in D$).	[m]
L_{max}	Maximum length of a wire segment of the microgrid.	[m]
MPC_{pd}	(p,d)-element of the matrix of potential connections ($p \in P, d \in D$). $\forall p \in P$, $\forall d \in D, MPC_{pd} \in \{0,1\}$.	-
Q_p	Subset of points to which point p can be directly connected with a wire segment ($p \in P$, $d \in D: p \neq d, MPC_{pd} = 1, L_{pd} \leq L_{max}$).	-
ED_p	Energy demand at p ($p \in D$).	[Wh/day]
PD_p	Power demand at p, considering the simultaneity factor ($p \in D$).	[W]
	PV generation	
S, NS	Set of PV panel types and maximum number of PV panels that can be placed at a point, respectively.	-
ES_s	Energy generated by a PV panel of type s ($s \in S$).	[Wh/day]
PS_s	Maximum power of a PV panel of type s ($s \in S$).	[W]
CS_s	Cost of a PV panel of type s ($s \in S$).	[US$]
Z	Set of PV controller types.	-
PZ_z	Maximum power of a PV controller of type z ($z \in Z$).	[W]
CZ_z	Cost of a PV controller of type z ($z \in Z$).	[US$]
	Electric equipment	
B	Set of battery types.	-
EB_b	Capacity of a battery of type b ($b \in B$).	[Wh]
CB_b	Cost of a battery of type b ($b \in B$).	[US$]
ηb	Battery efficiency.	[%]
DB	Maximum discharge proportion admitted for the batteries.	[%]
DA	Required autonomy of the batteries.	[days]
I	Set of inverter types.	-
PI_i	Maximum power of an inverter of type i ($i \in I$).	[W]
CI_i	Cost of an inverter of type i ($i \in I$).	[US$]
ηi	Inverter efficiency.	[%]
CL	Cost of an electric meter device.	[US$]

Table 2. *Cont.*

Parameter	Description	Unit
	Electricity distribution	
C	Set of wire types.	-
RC_c	Electric resistance (feed and return) of a wire of type c ($c \in C$).	[Ω/m]
IC_c	Maximum intensity of a wire of type c ($c \in C$).	[A]
CC_c	Cost of a wire of type c (feed and return), including the infrastructure ($c \in C$).	[US\$/m]
V_n	Nominal voltage.	[V]
V_{min}	Minimum voltage.	[V]
V_{max}	Maximum voltage.	[V]
ηc	Wire efficiency.	[%]
	Specific features for RAE electrification	
CA	Cost of a shed for equipment storage.	[\$US]
α	Accepted percentage of cost overhead of microgrids w.r.t. individual systems.	[%]
C_{max}	Maximum number of output connections from a microgrid point.	-

Table 3. Decision variables of the MILP formulation.

Variable	Description	Unit
	Integer non-negative variables	
xs_{ps}	Number of PV panels of type s placed at point p ($p \in P, s \in S$).	-
xz_{pz}	Number of PV controllers of type z placed at point p ($p \in P, z \in Z$).	-
xb_{pb}	Number of batteries of type b placed at point p ($p \in P, b \in B$).	-
xi_{pi}	Number of inverters of type i placed at point p ($p \in P, i \in I$).	-
	Float non-negative variables	
fe_{pd}	Energy flow between points p and d ($p \in P, d \in Q_p$).	[Wh/day]
fp_{pd}	Power flow between points p and d ($p \in P, d \in Q_p$).	[W]
v_p	Voltage at point p ($v_p \in [V_{min}, V_{max}], p \in P$).	[V]
	Binary variables	
xg_p	=1, if at least a wind turbine or PV panel is placed at point p ($p \in P$).	-
xc_{pdc}	=1, if there is a wire of type c between the points p and d ($p \in P, d \in Q_p, c \in C$).	-
x_{lp}	=1, if point p ($p \in D$) belongs to a microgrid (involving a meter device).	-

Taking into account these considerations, the objective function is formulated as indicated in Equation (1):

$$[MIN]Z = \sum_{p\in D}\sum_{s=1}^{S} CS_s \cdot xs_{ps} + \sum_{p\in D}\sum_{z=1}^{Z} CZ_z \cdot xz_{pz} + \sum_{p\in D}\sum_{b=1}^{B} CB_b \cdot xb_{pb} + \sum_{p\in D}\sum_{i=1}^{I} CI_i \cdot xi_{pi}$$

$$+\frac{1}{1+\frac{\alpha}{100}}\left[\begin{array}{l} \sum_{p\in P|p\notin D}\sum_{s=1}^{S} CS_s \cdot xs_{ps} + \sum_{p\in P|p\notin D}\sum_{z=1}^{Z} CZ_z \cdot xz_{pz} + \sum_{p\in P|p\notin D}\sum_{b=1}^{B} CB_b \cdot xb_{pb} \\[2mm] + \sum_{p\in P|p\notin D}\sum_{i=1}^{I} CI_i \cdot xi_{pi} + \sum_{p\in P|p\notin D} CA \cdot xg_p + \sum_{p\in D} CL \cdot xl_p + \sum_{p\in P}\sum_{d\in Q_p}\sum_{c=1}^{C} L_{pd} \cdot CC_c \cdot xc_{pdc} \end{array} \right] \quad (1)$$

Constraints (2) and (3) define the generation points (x_{gp} = 1) and bound the number of PV panels (Equation (2)) that can be installed at the same point.

$$\sum_{s=1}^{S} xs_{ps} \leq NS \cdot xg_p \quad p \in P \quad (2)$$

$$\sum_{s=1}^{S} xs_{ps} \geq xg_p \quad p \in P \quad (3)$$

Constraints (4) to (7) aim to cover the daily electricity needs of consumption points and capture the relationship between energy and power. Since the energy demanded by users is not constantly consumed at the same power level, both aspects are modeled with different constraints, as shown in the literature [29,40,57,60]. Hence, constraints (4) and (5) define the daily energy available at each consumption point, bounded by the energy produced by the installed PV panels (either individually or through a microgrid). In particular, constraint (4) enforces the conditions of energy conservation and demand satisfaction (demand points only): the energy arriving at point p plus the energy generated at p itself must be greater than (or equal to) the energy consumed by p plus the energy dispatched. The constraint includes the battery, inverter and wire efficiencies. Constraint (5) is equivalent to constraint (4) for no-demand points.

$$\sum_{q \in P | p \in Q_q} fe_{qp} + \sum_{s=1}^{S} ES_s \cdot xs_{ps} \geq \frac{ED_p}{\eta b \cdot \eta i} \left(\frac{1}{\eta c} + \left(1 - \frac{1}{\eta c} \right) xg_p \right) + \sum_{d \in Q_p} fe_{pd} \; p \in D \tag{4}$$

$$\sum_{s=1}^{S} ES_s \cdot xs_{ps} \geq \sum_{d \in Q_p} fe_{pd} \; p \in P; p \notin D \tag{5}$$

Constraints (6) and (7) are analogous to constraints (4) and (5), but for power demand. The location, type and quantity of inverters are determined according to user's demand and only consider the wires' efficiency.

$$\sum_{q \in P | p \in Q_q} fp_{qp} + \sum_{i=1}^{I} PI_i \cdot xi_{pi} \geq PD_p \left(\frac{1}{\eta c} + \left(1 - \frac{1}{\eta c} \right) xg_p \right) + \sum_{d \in Q_p} fp_{pd} \; p \in D \tag{6}$$

$$\sum_{i=1}^{I} PI_i \cdot xi_{pi} \geq \sum_{d \in Q_p} fp_{pd} \; p \in P; p \notin D \tag{7}$$

Constraint (8) and (9), associated with demand and no-demand points, respectively, force batteries to store enough energy to cover user demands, considering the required days of autonomy and the discharge factor.

$$\sum_{b=1}^{B} EB_b \cdot xb_{pb} + \left(\frac{DA}{DB} \sum_{j=1}^{D} \frac{ED_j}{\eta b \cdot \eta i \cdot \eta c} \right) (1 - xg_p) \geq \frac{DA}{DB} \left(\sum_{d \in Q_p} fe_{pd} + \frac{ED_p}{\eta b \cdot \eta i} \right) \; p \in D \tag{8}$$

$$\sum_{b=1}^{B} EB_b \cdot xb_{pb} + \left(\frac{DA}{DB} \sum_{j=1}^{D} \frac{ED_j}{\eta b \cdot \eta i \cdot \eta c} \right) (1 - xg_p) \geq \frac{DA}{DB} \sum_{d \in Q_p} fe_{pd} \; p \in P; p \notin D \tag{9}$$

Constraints (10) and (11), respectively, relate the energy and power flows with the existence of wires. If there is no wire between two points, the energy and power flows are zero; otherwise, they can take some value defined through constraints (4) to (7). The microgrid radial scheme is imposed in constraint (12): each non-generation point can have at most one input wire. Constraint (13) establishes the voltage drop between two points, according to the type of wires. In constraint (14), the flow intensity between two points connected by a wire is bounded by a maximum admissible intensity depending on the wire type.

$$fe_{pd} \leq \left(\sum_{j \in D} \frac{ED_j}{\eta b \cdot \eta i \cdot \eta c} \right) \sum_{c=1}^{C} xc_{pdc} \; p \in P; d \in Q_p \tag{10}$$

$$fp_{pd} \leq \left(\sum_{j \in D} \frac{PD_j}{\eta c} \right) \sum_{c=1}^{C} xc_{pdc} \; p \in P; d \in Q_p \tag{11}$$

$$\sum_{q \in P | p \in Q_q} \sum_{c=1}^{C} xc_{qpc} + xg_p \leq 1 \; p \in P \tag{12}$$

$$v_p - v_d \geq \frac{L_{pd} \cdot RC_c \cdot fp_{pd}}{V_n} - (V_{max} - V_{min})(1 - xc_{pdc}) \; p \in P; d \in Q_p; c \in C \qquad (13)$$

$$\frac{fp_{pd}}{V_n} - \left(\sum_{j \in D} \frac{PD_j}{V_{min} \cdot \eta c} \right)(1 - xc_{pdc}) \leq IC_c \; p \in P; d \in Q_p; c \in C \qquad (14)$$

The PV controllers must have an appropriate power determined by the maximum power of the PV panels installed at a certain point (15). Constraint (16) enforces the installation of inverters at generation points, while constraint (17) sets $x_{lp} = 1$ for the demand points belonging to a microgrid with an input connection (due to the radial configuration and microgrid power generation performed at no-demand points).

$$\sum_{z=1}^{Z} PZ_z \cdot xz_{pz} \geq \sum_{s=1}^{S} PS_s \cdot xs_{ps} \; p \in P \qquad (15)$$

$$\sum_{i=1}^{I} xi_{pi} \leq \sum_{j \in D} \frac{PD_j}{\eta_c} \cdot xg_p \; p \in P \qquad (16)$$

$$\sum_{q \in P | p \in Q_q} \sum_{c=1}^{C} xc_{qpc} \leq xl_p \; p \in D \qquad (17)$$

Finally, as mentioned in point (b) of Section 2.3, the characteristics of the equipment to be considered may depend either on the institutional framework of the electrification project or on local availability. Here, the limiter boxes only admit two output cables, meaning that the number of wires from any point p to any demand point of the microgrid is bounded by $C_{max} = 2$ in constraint (18), therefore restricting the topology of potential microgrids.

$$\sum_{d \in Q_p} \sum_{c=1}^{C} xc_{pdc} \leq C_{max} \; p \in P \qquad (18)$$

4. Case Study: Three Communities in the RAE

The mathematical tool developed in this work is used for the design of rural electrification projects in three communities of the RAE: Conambo, Suraka and Santa Rosa. First, the communities are described in socioeconomic terms, and their main features are determined according to the analysis presented in Section 2. Once all the relevant data and numerical parameters are comprehensively collected, the designed tool (based on the solution of a MILP) is employed to propose a specific electrification design for each case.

4.1. General Description of the Communities

The three communities studied have populations from the Sápara ethnicity and are located in the basin of the Conambo river. Figure 3 shows a map of Pastaza province, as well as the air/river routes available (in yellow) to reach the three communities from the district capital, Puyo. The straight red line indicates the linear distance (130 km) to the closest point reached by the national electric grid. Due to this geographical location and the rainforest vegetation predominant in this region, grid expansion has never been considered a viable option by the MEER, since it would predictably involve huge installation costs, severe technical issues and a negative environmental impact.

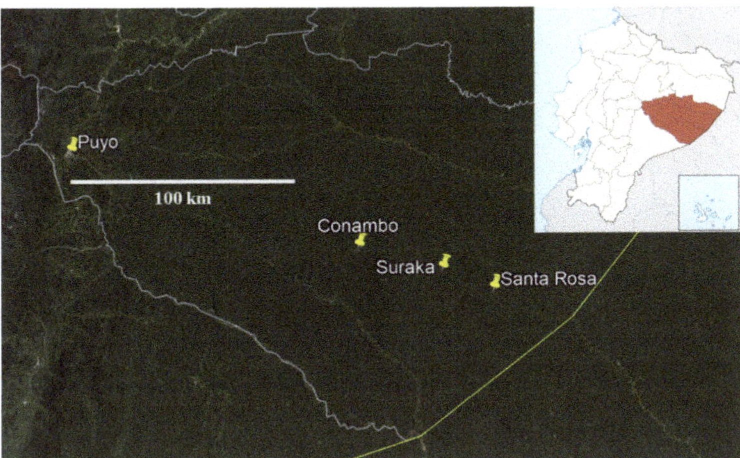

Figure 3. Map of the Pastaza district.

This map illustrates how Conambo, Suraka and Santa Rosa share comparable living standards and access to basic services, as well as very similar cultural features. With regards to basic services, none of the communities is provided with access to drinking water, drainage systems or electric energy. Santa Rosa is the only community with a small panel that feeds an internet-based communication system—the result of a recent national development program. On the other hand, Conambo and Suraka have radio stations to communicate with urban centers, in cases of medical emergency.

4.2. Problem Data, Point Distribution and Pre-Processing

The electrification of Conambo, Suraka and Santa Rosa is part of a global framework developed as a collaboration of different actors (EWB and MEER). As a consequence, all three communities share several conditioning features and numerical data. First, some guidelines set by Ecuador's national government mean that several parameters are determined. For instance, as mentioned in Section 2.3, standardized consumption levels are imposed by the MEER for all users in the RAE in order to guarantee equal opportunities to every family. Thus, energy (1000 Wh/day) and power (600 W) demands are considered equal for all demand points (either housing or community centers). These values were established from surveys of the local populations, taking into account their needs for lighting, telecommunications and small household appliances. Besides, as explained in point c) of Section 2.3, the type of generation and distribution items is restricted by the size of the transportation means available for shipping equipment to the targeted communities. In particular, the aircraft only allows one wire type (thicker cables do not fit in the aircraft), which may have an impact on microgrid structures (as observed next), because the voltage drop depends on wiring characteristics and could make some points unreachable. The values for the remaining parameters common to all three communities are synthesized in Table 4. Note that, in this table, all the techno-economic parameters (generation and distribution equipment) were gathered from commercial catalogs locally available in Ecuador.

Then, the geographic coordinates of demand points and potential generation locations at each community are defined, implicitly determining the distances L_{pd} between points $p, d \in P$. Additionally, the specific obstacles at each community allow MPC to be deduced, and thus the set of neighboring points ($p, d \in P$). Suraka consists of 12 demand points (9 houses, 2 communal centers and 1 school) and 3 potential generation locations for microgrids (points 13, 14 and 15, see Figure 4). Santa Rosa has 15 demand points (12 houses, 1 communal center, 1 school and 1 waiting room for air passengers) and 4 potential generation points (points 16–9, see Figure 5). Finally, Conambo is a wider community, having 60 demand points (48 houses, 1 communal center, 8 school classrooms, 2 community canteens and 1 waiting room close to the landing strip) and 6 potential generation points (points 61–66, see Figure 6). Note that in Figures 4–6, red points represent potential generation points, non-red points are for demand points, and the brown lines stand for landing strips.

Table 4. Numerical values of the parameters shared by the three communities.

Description	Parameter	Value	Unit
Electric equipment			
Batteries: types	$\vert B \vert$	2	-
Batteries: capacity	EB_b $(b \in B)$	1800; 3600	[Wh]
Batteries: cost	CB_b $(b \in B)$	300; 850	[US$]
Batteries: efficiency	ηb	85	[%]
Batteries: maximum discharge	DB	60	[%]
Batteries: required autonomy	DA	3	[days]
Inverters: types	$\vert I \vert$	2	-
Inverters: maximum power	PI_i $(i \in I)$	600; 3600	[W]
Inverters: cost	CI_i $(i \in I)$	400; 2000	[US$]
Inverters: efficiency	ηi	85	[%]
Meter devices: cost	CL	50	[US$]
Demand points			
Maximum length of wire segments	L_{max}	300	[m]
Energy demand	ED_p $(p \in D)$	1000	[Wh/day]
Power demand	PD_p $(p \in D)$	600	[W]
PV generation			
PV panel: types	$\vert S \vert$	1	-
PV panel: maximum number	NS	40	-
PV panel: energy generated	ES_s $(s \in S)$	1178.8	[Wh/day]
PV panel: maximum power	PS_s $(s \in S)$	330	[W]
PV panel: cost	CS_s $(s \in S)$	350	[US$]
PV controllers: types	$\vert Z \vert$	2	-
PV controllers: maximum power	PZ_z $(z \in Z)$	80; 2880	[W]
PV controllers: cost	CZ_z $(z \in Z)$	300; 700	[US$]
Distribution equipment			
Wires: types	$\vert C \vert$	1	-
Wires: electric resistance	RC_c $(c \in C)$	0.0016	[Ω/m]
Wires: maximum intensity	IC_c $(c \in C)$	60	[A]
Wires: cost	CC_c $(c \in C)$	3.94	[US$/m]
Nominal voltage	V_n	110	[V]
Minimum voltage	V_{min}	105	[V]
Maximum voltage	V_{max}	116	[V]
Wires: efficiency	ηc	90	[%]
RAE's specific features			
Shed cost	CA	1500	[$US]
Cost overhead (microgrids vs. individual systems)	α	$-20, 0, 20$	[%]
Maximum output connections in microgrids	C_{max}	2	-

It is now possible to set the elements of the matrix of potential connections (*MPC*). Due to the previously explained reasons, obstacles may hinder some connections, which necessitates a pre-processing stage to determine if each community should be treated as a whole or divided into several sub-problems to be solved independently. For Suraka, the landing strip blocks some direct connections, but any point may be indirectly connected to any other, so it can be considered as a single problem. On the other hand, Santa Rosa is settled on both sides of the river. Hence, two sub-problems can be naturally defined (see Figure 5):

- Right side (SR-R), with 4 demand points and 1 potential generation point.
- Left side (SR-L), with 11 demand points and 3 potential generation points.

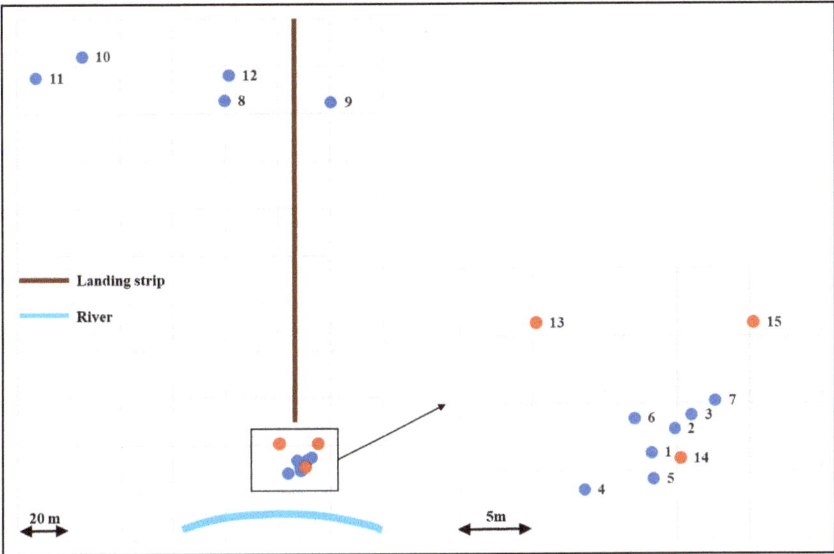

Figure 4. Demand and potential generation points in Suraka. Numbers refer to the households (blue) and potential generation points (red).

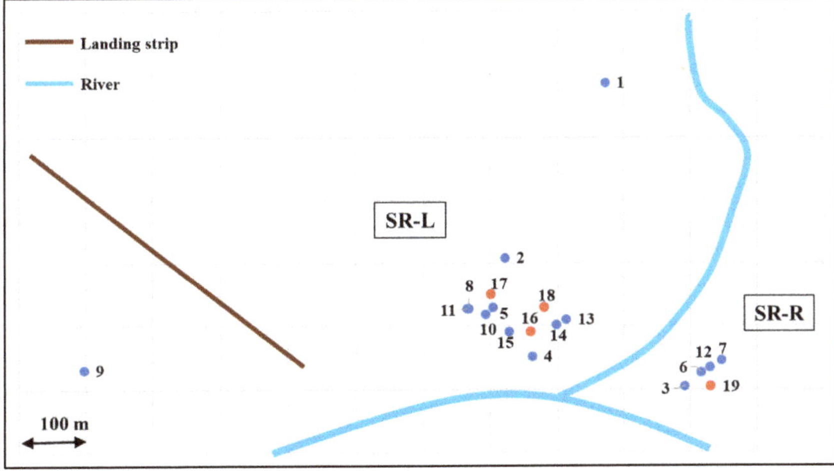

Figure 5. Demand and potential generation points in Santa Rosa. Numbers refer to the households (blue) and potential generation points (red).

Finally, Conambo is divided by the meanders formed by the Conambo river (Figure 6), and the landing strip also represents an obstacle, making some connections impossible through underground wiring. Consequently, the community is divided into five sub-problems (identified by different colors in Figure 6):

- Left side (C-L: orange points), 9 demand points and 1 potential generation point.
- Right side A (C-RA: green points), 10 demand points and 2 potential generation points.
- Right side B (C-RB: grey points), 15 demand points and 1 potential generation point.
- Right side C (C-RC: blue points), 20 demand points and 1 potential generation point.
- Right side D (C-RD: lilac points), 6 demand points and 1 potential generation point.

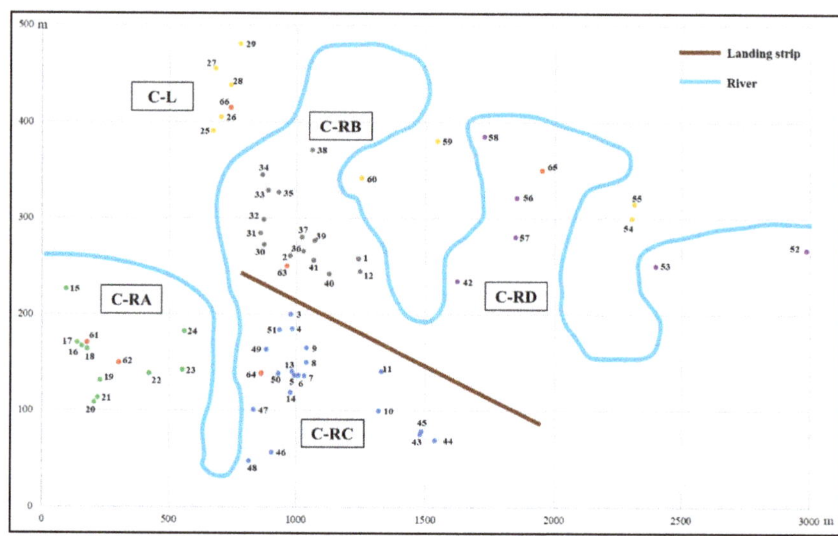

Figure 6. Demand and potential generation points in Conambo. Numbers refer to the households (blue) and potential generation points (red).

4.3. Experimental Results

As indicated previously, one contribution of the mathematical tool developed in Section 3 is the parameter α which, in the objective function, denotes the percentage of cost overhead accepted per microgrid with regard to individual systems. The introduction of this parameter was due to MEER's desire to promote grid formation, concretely proposing a value α of 20% (a microgrid configuration is preferred if its cost is up to 20% higher than the cost of individual systems). However, the aim is to design a versatile tool that may be used in both ways—i.e., either to promote or penalize microgrid formation over individual systems, depending on the policy makers' decisions. In order to perform a sensitivity analysis using this parameter and compare the configurations obtained in different cases, three executions were performed for each community, with different α values: 20% (microgrid formation is promoted as proposed by the MEER), 0% (microgrids are neither penalized nor promoted) and -20% (microgrid formation is penalized, with an inverse amount compared to that proposed by the MEER). Note that other values could be chosen for α in order to further promote/penalize grid formation, but the three selected values (-20%, 0%, 20%) seem to be enough to demonstrate the ability of the proposed tool to produce different kinds of configurations.

The computational experiments presented here were carried out solving the previously described MILP problem, using IBM ILOG CPLEX 12.2, run on an IntelCore i7-6700 3.40 GHz processor (16 Gb RAM). The MILP solver employed in CPLEX is based on a Branch and Cut algorithm. The corresponding OPL code of our model, as well as complete input data (already available in Section 4.2) and the detailed solutions obtained in every optimization process (decision variables and objective function), are available from the following website (free public access): https://gitioc.upc.edu/ioc/2022_mathematics_equador (accessed on 6 April 2022). In this way, the validity of the solutions obtained can be checked by the reader, and this study is completely reproducible. All the executions are performed using a relative optimality gap of 10–6 with a 1 h time limit, which proved to be enough to solve all the tackled instances optimally except the C-RC sub-problem (independently from the value of α). Note that, for the Santa Rosa and Conambo communities, optimality could be obtained for the corresponding sub-problems thanks to the "divide-and-conquer" strategy proposed here, which decomposes communities into sub-problems solved independently and allows for a reduction of instance sizes. This is why a classical solution tool such as CPLEX could be applied here, despite the complexity of the MILP model, without the need to develop a new ad-hoc solution technique. In the case of sub-problem C-RC, the remoteness of certain demand points with regard to the potential generation location allowed the problem to be further divided. The partial solutions obtained were reused within the complete model to obtain good quality, feasible solutions whose optimality could be subsequently demonstrated. Specific comments regarding the microgrid structures obtained for this instance are provided next.

The results are presented in Table 5, which shows, in addition to the size of each sub-problem, the objective function value and the real cost of the solution obtained (which may be different from the objective function according to α value) as well a description of the proposed configuration. In addition, the microgrid topologies produced (for a value $\alpha = 20\%$, such as that proposed by the MEER) are presented in Figures 7–9 for Suraka, Santa Rosa and Conambo, respectively. Please note that, in these three figures, yellow/white dots stand for demand points with/without individual generation systems. In addition, yellow squares represent the locations used as generation points for microgrids while gray-outlined white squares are for non-used potential generation points.

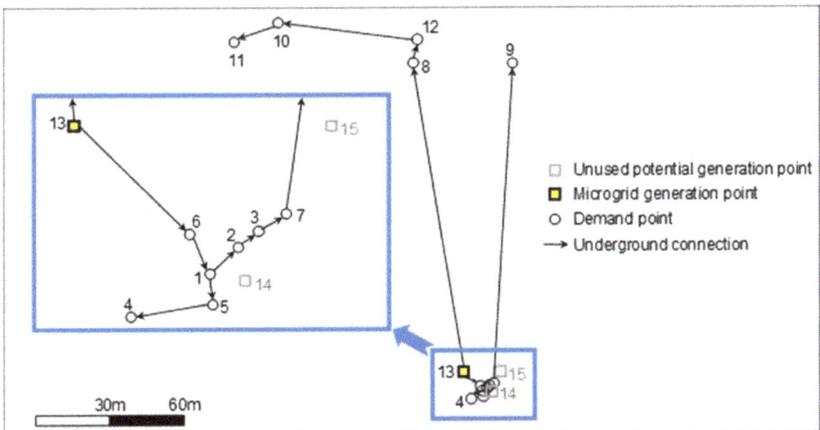

Figure 7. Microgrid obtained for Suraka.

First, for every sub-problem, different configurations are determined depending on the value of α (see Table 5), proving the versatility of the formulation developed, which is capable of adapting to user's preferences regarding microgrid or individual systems. However, apart from this global conclusion, different behaviors emerge from a closer analysis. For the Suraka and C-RA sub-problems, a microgrid involving all demand points is determined when $\alpha = 0$ or 20%, while all users are supplied by individual systems when $\alpha = -20\%$. A comparable trend is observed in SR-L and C L (one-microgrid configuration when $\alpha = 0$ or 20%), but, in these cases, some demand points located far away or isolated by obstacles are supplied by individual systems. In contrast, a different trend is observed in the remaining instances. For SR-R and C-RD, the minimum cost configurations only involve individual supplies when $\alpha = 0$ or -20%, while microgrids are proposed for $\alpha = 20\%$. These microgrids may include all users (SR-R) or exclude users isolated by river meanders (C-RD). Finally, C-RB and C-RC further demonstrate the versatility of the computational tool developed, which proposes three different configurations according to the α value. When $\alpha = -20\%$, all users are supplied through individual systems; when $\alpha = 0$, a combination of individual supplies and one microgrid is designed. The previous microgrid is expanded when α increases to 20%, including all users for C-RB and part of them for C-RC.

Table 5. Solutions obtained for the three communities.

Community	Sub-Problem	Demand Points	α (%)	Obj. Func. (USD)	Real Cost (USD)	Configuration
Suraka		12	−20	34,800	34,800	12 individual systems
			0	31,110	31,110	One microgrid (all 12 users)
			20	25,925		
Santa Rosa	SR-R	4	−20	11,600	11,600	4 individual systems
			0			
			20	10,231	12,277	One microgrid (all 4 users)
	SR-L	11	−20	31,900	31,900	11 individual systems
			0	29,848	29,848	One microgrid (9 users) and 2 individual systems
			20	25,840		
Conambo	C-L	9	−20	26,100	26,100	9 individual systems
			0	25,925	25,925	One microgrid (5 users) and 4 individual systems
			20	23,538		
	C-RA	10	−20	29,000	29,000	10 individual systems
			0	27,463	27,463	One microgrid (all 10 users)
			20	22,886		
	C-RB	15	−20	43,500	43,500	15 individual systems
			0	39,353	39,353	One microgrid (13 users) and 2 individual systems
			20	32,897	39,477	One microgrid (all 15 users)
	C-RC	20	−20	58,000	58,000	20 individual systems
			0	53,649	53,649	One microgrid (15 users) and 5 individual systems
			20	46,789	53,827	One microgrid (16 users) and 4 individual systems
	C-RD	6	−20	17,400	17,400	6 individual systems
			0			
			20	17,252	19,543	One microgrid (4 users) and 2 individual systems

With respect to the microgrid topologies shown in Figures 7–9, they illustrate behaviors clearly influenced by the conditions defined in Section 2.3. For instance, only one microgrid is formed in the solution obtained for Suraka, SR-L and C-RA, while another option would have been to create more independent microgrids (several potential generation points are available). Such two-grid configurations are discarded due to the shed required for equipment storage, which is more expensive than the additional wiring necessary to expand the microgrid. Besides, regarding C-RC, the configuration of the microgrid obtained with $\alpha = 20\%$ (Figure 9) is restricted by the maximum number of output connections from limiter boxes, $C_{max} = 2$ (using $C_{max} = 3$ allows cheaper configurations). In addition, some connections are limited by the type of wires available (due to the space limitations in the aircraft transporting the equipment), which prevents further expansion of the microgrid while respecting the allowed voltage drop.

In short, the results highlight that, in four cases, using $\alpha > 0$ allows microgrids to be created or expanded to replace the individual systems obtained when $\alpha = 0$, which confirms the soundness of the incentive proposed by the MEER (in the four other cases, microgrids are already found for $\alpha = 0$). Besides, the difference between the "real" cost of microgrid configurations obtained with $\alpha = 20\%$ and the cost of individual systems (designed when $\alpha = 0$) is on average lower than 5%. Therefore, the value proposed by the MEER (20%) seems high enough to promote the formation of microgrid-based systems. It might even be decreased in order to appear more attractive from an economic viewpoint without affecting the final results.

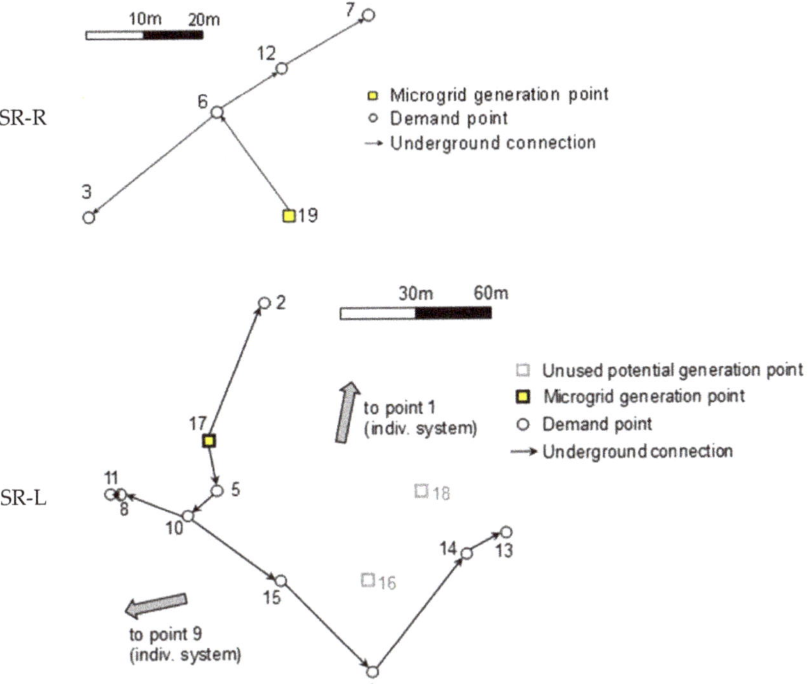

Figure 8. Microgrid obtained for Santa Rosa.

The comparison with classical processes for electrification systems such as those described in Section 1 further highlights the scientific contribution of the approach introduced here and the benefits obtained accordingly. Indeed, the particular features included in the mathematical model and derived from the conditioning factors proper to the RAE allowed the design of specific configurations. For example, in some instances for which several microgrids may have been installed, the requirement of building a shed for generation equipment led to a single microgrid. In addition, as mentioned before, thanks to the versatility of the design tool allowed by the α parameter, the systems proposed show more or larger microgrids than those that would have been obtained by generic tools. Therefore, the solution strategy developed here is justified by its contribution and, in turn, the new features introduced here may be included in standard tools in order to expand their application scope.

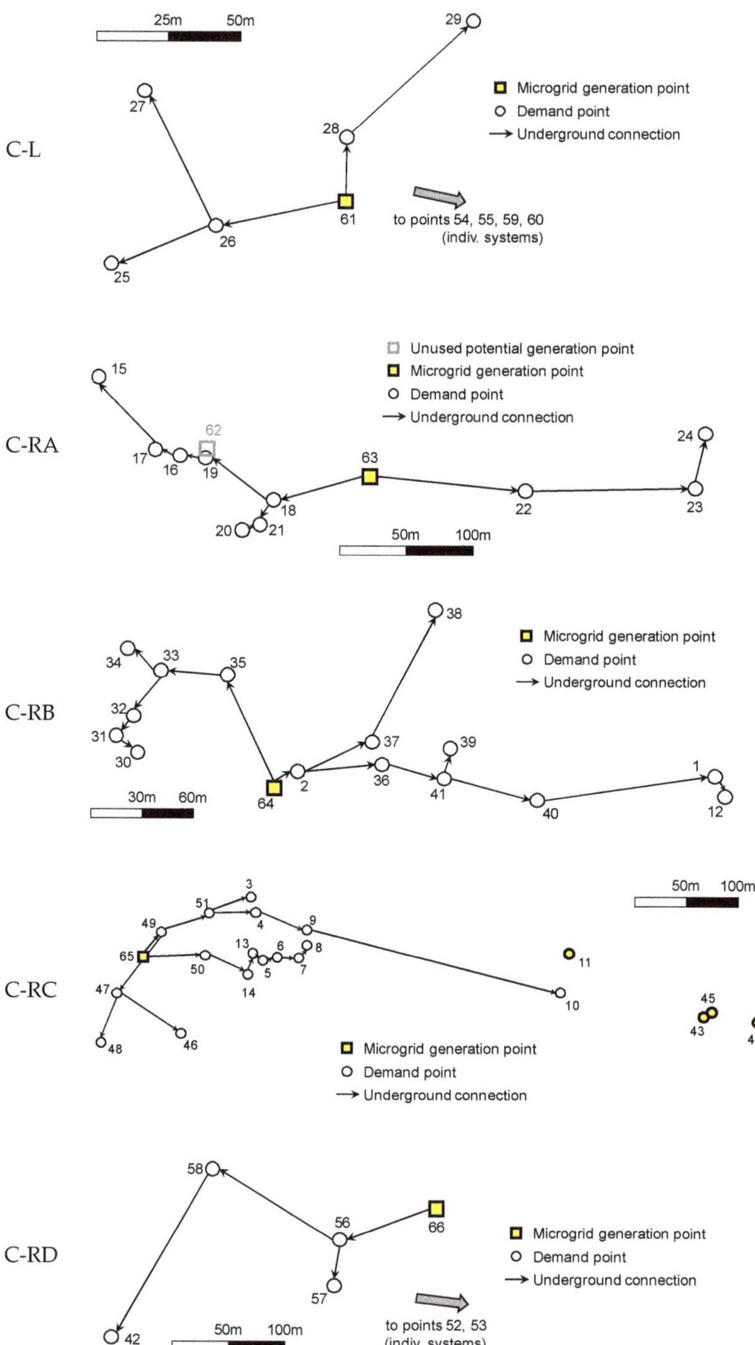

Figure 9. Microgrid/individual supply obtained for Conambo.

5. Conclusions

Despite global electrification rates roughly comparable to those of developed countries, Ecuador shows enormous disparities regarding access to this service, particularly in the rural areas of its Amazon basin. As a proposal to overcome these inequalities, this study presents a tool for the design of stand-alone electrification systems based on solar energy specially adapted to the case of the RAE. Developing this tool required a deep understanding of the considered community's way-of-life, as well as close collaboration with national institutions and local actors involved in the rural electrification process. This first step led to the formulation of an MILP model to design systems adapted to the specific conditioning factors in the RAE, accounting for either individual systems, microgrids or a combination of both, as well as the location, type and size of the equipment employed.

This mathematical tool was used to determine minimal cost designs for three communities in the RAE, leading to several conclusions. First, a matrix of potential connections was introduced to allow the possible breaking-down of large problems into sub-problems, whose size is tractable through mathematical programming. Besides, the introduction of RAE-specific features had important consequences on the configurations proposed. In the first place, microgrid generation equipment cannot be located at demand points, and the number of output connections from limiter boxes is bounded, influencing the final microgrid structures. More significantly, the inclusion of preferences regarding microgrid or individual systems highlighted the versatility of the solution tool, which proposes different configurations when varying the corresponding parameter α. Numerical experiments provided general guidelines to support electrification policy making, in particular regarding the adequacy of microgrid promotion and the setting of α, which may be decreased with regard to MEER's proposal (20%) to appear more attractive. In any case, the tool introduced here can be used to provide efficient solutions independent of the α value established by policy makers, with obvious benefits for the RAE's isolated communities.

More generally, this work shows that electrification of the RAE is technically feasible, provides higher supplies than the benchmark references (1000 Wh/day instead of the International Energy Agency threshold of 685 kWh/day [5]) and is economically competitive, with costs ranging from 2500 to 3000 USD per consumer. Therefore, the tool introduced here for electrification systems design has a significant impact on the lifestyle of the target populations, improving their life quality through access to electricity services, while respecting their customs and social traditions since the proposed model accounts for these latter issues. In addition, the electrification systems designed here are based on environmentally friendly technologies based on renewable energies, thus representing a movement towards the Sustainable Development Goals stated by the United Nations [1]. Finally, even though the mathematical tool proposed for electrification systems design is developed according to RAE's features, it is not limited to the case studies treated here. Rather, accounting for IEA's projections that microgrid-based systems using renewables should account for 30% of the connections to be installed in the next years [61], the model is formulated and implemented in a generic manner to improve its versatility and allow its adaptation to distinct geographic contexts (in particular, rural areas of Latin America or Africa) and policy makers' priorities.

Regarding perspectives for future work, a first guideline may be a study of the robustness of the configurations designed—sxin particular, accounting for possible variations of the energy and power demands. On the other hand, the environmental impacts should also be included in the evaluation of such small-scale electrification projects. Besides, it is worth recalling that, typically, different factors that play a role in the design of electrification systems are not always known with certainty. In particular, the amount of PV energy collected (which depends on meteorological factors) and user's energy/power demands (which are often difficult to estimate) are the parameters most frequently considered as uncertain in the specialized literature. Therefore, accounting for this uncertainty (either through stochastic programming or fuzzy set theory) might allow the design of more robust systems and constitutes a promising perspective.

Author Contributions: Conceptualization, B.D., L.F.-M.; methodology, B.D., G.H.; investigation, B.D., G.H., A.P.; resources, F.G.; validation, L.F.-M., F.G., R.P.; project administration: L.F.-M.; writing—original draft preparation: G.H., A.P.; writing—reviewing and editing: A.P.; supervision, R.P. All authors have read and agreed to the published version of the manuscript.

Funding: This research was funded by the Spanish Ministry of Science and Innovation (RTI-2018-097962-B-I00) and the Centre for Cooperation Development (CCD) of the Universitat Politècnica de Catalunya-BarcelonaTech (UPC).

Institutional Review Board Statement: Not applicable.

Informed Consent Statement: Informed consent was obtained from all subjects involved in the study.

Data Availability Statement: The data used to support the findings of this study are available at https://gitioc.upc.edu/ioc/2022_mathematics_equador (accessed on 6 April 2022).

Acknowledgments: The authors are very grateful to the NGO Engineering Without Borders (Spain) and the Energy21 program of University College Dublin (UCD) for all the support and assistance given during the development of this research.

Conflicts of Interest: The authors declare no conflict of interest.

References

1. United Nations. Transforming Our World: The 2030 Agenda for Sustainable Development. 2015. Available online: https://sdgs.un.org/goals (accessed on 18 January 2022).
2. Blimpo, M.P.; Cosgrove-Davies, M. *Electricity Access in Sub-Saharan Africa: Uptake, Reliability, and Complementary Factors for Economic Impact*; World Bank, Africa Development Forum: Washington, DC, USA, 2019; Available online: https://openknowledge.worldbank.org/handle/10986/31333 (accessed on 18 January 2022).
3. Jeuland, M.; Fetter, T.R.; Li, Y.; Pattanayak, S.K.; Usmani, F.; Bluffstone, R.A.; Chávez, C.; Girardeau, H.; Hassen, S.; Jagger, P.; et al. Is energy the golden thread? A systematic review of the impacts of modern and traditional energy use in low- and middle-income countries. *Renew. Sustain. Energy Rev.* **2021**, *135*, 110406. [CrossRef]
4. Peters, J.; Sievert, M. Impacts of rural electrification revisited-the African context. *J. Dev. Eff.* **2016**, *8*, 327–345. [CrossRef]
5. International Energy Agency World Energy Outlook: Electricity Access Database. 2017. Available online: https://www.iea.org/reports/world-energy-outlook-2017 (accessed on 19 January 2022).
6. Sovacool, B.K. Deploying off-grid technology to eradicate energy poverty. *Science* **2012**, *338*, 47–48. [CrossRef] [PubMed]
7. Hernández-Callejo, L.; Gallardo-Saavedra, S.; Alonso-Gómez, V. A review of photovoltaic systems: Design, operation and maintenance. *Sol. Energy* **2019**, *188*, 426–440. [CrossRef]
8. Ehsan, A.; Yang, Q. Optimal integration and planning of renewable distributed generation in the power distribution networks: A review of analytical techniques. *Appl. Energy* **2018**, *210*, 44–59. [CrossRef]
9. Slednev, V.; Bertsch, V.; Ruppert, M.; Fichtner, W. Highly resolved optimal renewable allocation planning in power systems under consideration of dynamic grid topology. *Comput. Oper. Res.* **2018**, *96*, 280–292. [CrossRef]
10. Hassan, Q. Evaluation and optimization of off-grid and on-grid photovoltaic power system for typical household electrification. *Renew. Energy* **2021**, *164*, 375–390. [CrossRef]
11. Taye, B.Z.; Workineh, T.G.; Nebey, A.H.; Kefale, H.A. Rural electrification planning using Geographic Information System (GIS). *Cogent Eng.* **2020**, *7*, 1836730. [CrossRef]
12. Ellman, D. The Reference Electrification Model: A Computer Model for Planning Rural Electricity Access. Master's Thesis, Massachusetts Institute of Technology, Boston, MA, USA, 2015.
13. Mentis, D.; Howells, M.; Rogner, H.; Korkovelos, A.; Arderne, C.; Zepeda, E.; Siyal, S.H.; Taliotis, C.; Bazilian, M.; de Roo, A.; et al. Lighting the World, The first global application of an open source, spatial electrification tool (OnSSET), with a focus on Sub-Saharan Africa. *Environ. Res. Lett. Focus Energy Access Sustain. Dev.* **2017**, *12*, 085003.
14. Brown, T.; Hörsch, J.; Schlachtberger, D. PyPSA: Python for Power System Analysis. *J. Open Re. Soft.* **2018**, *6*, 4. [CrossRef]
15. Eddy, J.; Miner, N.E.; Stamp, J. Sandia's Microgrid Design Toolkit. *Electr. J.* **2017**, *30*, 62–67. [CrossRef]
16. Bahramara, S.; Parsa Moghaddam, M.; Haghifam, M.R. Optimal planning of hybrid renewable energy systems using HOMER: A review. *Renew. Sustain. Energy Rev.* **2016**, *62*, 609–620. [CrossRef]
17. Lal, D.K.; Dash, B.B.; Akella, A.K. Optimization of PV/wind/micro-hydro/diesel hybrid power system in HOMER for the study area. *Int. J. Electr. Eng. Inform.* **2011**, *3*, 307–325.
18. Rozlan, M.B.M.; Zobaa, A.F.; Abdel Aleem, S.H.E. The optimisation of stand-alone hybrid renewable energy systems using HOMER. *Int. Rev. Electr. Eng.* **2011**, *6*, 1802–1810.
19. Hossain, E.; Kabalci, E.; Bayindir, R.; Perez, R. Microgrid testbeds around the world: State of art. *Energy Convers. Manag.* **2014**, *86*, 132–153. [CrossRef]
20. Wang, R.; Lam, C.M.; Hsu, S.C.; Chen, J.H. Life cycle assessment and energy payback time of a standalone hybrid renewable energy commercial microgrid: A case study of Town Island in Hong Kong. *Appl. Energy* **2019**, *250*, 760–775. [CrossRef]

21. Zhou, W.; Lou, C.; Li, Z.; Lu, L.; Yang, H. Current status of research on optimum sizing of stand-alone hybrid solar-wind power generation systems. *Appl. Energy* **2010**, *87*, 380–389. [CrossRef]
22. Peters, J.; Sievert, M.; Toman, M.A. Rural electrification through mini-grids: Challenges ahead. *Energy Policy* **2019**, *132*, 27–31. [CrossRef]
23. Castilla, M.; Gómez, M.; Mercado, P.; Moreira, C.; Negroni, J.J.; Sosa, J.; Zambroni de Sousa, A.C. The growing state of distributed generation and microgrids in the Ibero-American region: A view from the RIGMEI network. In Proceedings of the 2014 IEEE Power & Energy Society Transmission and Distribution Latin American Conference & Exposition, Medellín, Colombia, 10–13 September 2014.
24. Mahomed, S.; Shirley, R.; Tice, D.; Phillips, J. Business Model Innovations for Utility and Mini-Grid Integration: Insights from the Utilities 2.0 Initiative in Uganda. Energy and Economic Growth, Energy Insight Report. 2020. Available online: https://www.energyeconomicgrowth.org/publication/business-model-innovations-utility-and-mini-grid-integration-insights-utilities-20 (accessed on 19 January 2022).
25. Tenenbaum, B.; Greacen, C.; Siyambalapitiya, T.; Knuckles, J. *From the Bottom Up: How Small Power Producers and Mini-Grids Can Deliver Electrification and Renewable Energy in Africa*; World Bank Group Report; Directions in Development—Energy and Mining: Washington, DC, USA, 2014; Available online: https://openknowledge.worldbank.org/handle/10986/16571 (accessed on 19 January 2022).
26. Lukuyu, J.; Fetter, R.; Krishnapriya, P.P.; Williams, N.; Taneja, J. Building the supply of demand: Experiments in mini-grid demand stimulation. *Dev. Eng.* **2021**, *6*, 100058. [CrossRef]
27. Rezaei, M.; Dowlatabadi, H. Off-grid: Community energy and the pursuit self-sufficiency in British Columbia's remote and First Nations communities. *Local Environ.* **2015**, *21*, 789–807. [CrossRef]
28. Leithon, J.; Werner, S.; Koivunen, V. Energy optimization through cooperative storage management: A calculus of variations approach. *Renew. Energy* **2021**, *171*, 1357–1370. [CrossRef]
29. Domenech, B.; Ferrer-Martí, L.; Pastor, R. Including management and security of supply constraints for designing stand-alone electrification systems in developing countries. *Renew. Energy* **2015**, *80*, 359–369. [CrossRef]
30. Ranaboldo, M.; Domenech, B.; Vilar, D.; Ferrer-Martí, L.; Pastor, R.; García-Villoria, A. Renewable energy projects to electrify rural communities in Cape Verde. *Appl. Energy* **2014**, *118*, 280–291. [CrossRef]
31. Triadó-Aymerich, J.; Ferrer-Martí, L.; García-Villoria, A.; Pastor, R. MILP-based heuristics for the design of rural community electrification projects. *Comput. Oper. Res.* **2016**, *71*, 90–99. [CrossRef]
32. Rojas-Zerpa, J.C.; Yusta, J.M. Application of multicriteria decision methods for electric supply planning in rural and remote areas. *Renew. Sustain. Energy Rev.* **2015**, *52*, 557–571. [CrossRef]
33. Balderrama, S.; Lombardi, F.; Riva, F.; Canedo, W.; Colombo, E.; Quoilin, S. A two-stage linear programming optimization framework for isolated hybrid microgrids in a rural context: The case study of the "El Espino" community. *Energy* **2019**, *188*, 116073. [CrossRef]
34. Blechinger, P.; Cader, C.; Bertheau, P.; Huyskens, H.; Seguin, R.; Breyer, C. Global analysis of the techno-economic potential of renewable energy hybrid systems on small islands. *Energy Policy* **2016**, *98*, 674–687. [CrossRef]
35. Gómez, M.F.; Silveira, S. The last mile in the Brazilian Amazon—A potential pathway for universal electricity access. *Energy Policy* **2015**, *82*, 23–37. [CrossRef]
36. Garces, E.; Tomei, J.; Franco, C.J.; Dyner, I. Lessons from last mile electrification in Colombia: Examining the policy framework and outcomes for sustainability. *Energy Res. Soc. Sci.* **2021**, *79*, 102156. [CrossRef]
37. Fernández-Fuentes, M.H.; Eras-Almeida, A.A.; Egido-Aguilera, M.A. Characterization of Technological Innovations in Photovoltaic Rural Electrification, Based on the Experiences of Bolivia, Peru, and Argentina: Third Generation Solar Home Systems. *Sustainability* **2021**, *13*, 3032. [CrossRef]
38. Barzola, J.; Espinoza, M.; Pavón, C.; Cabrera, F. Solar-wind renewable energy system for off-grid rural electrification in Ecuador. In Proceedings of the 14th LACCEI International Multi-Conference for Engineering, Education, and Technology, "Engineering Innovations for Global Sustainability", San José, Costa Rica, 20–22 July 2016.
39. Feron, S.; Heinrichs, H.; Cordero, R.R. Are the electrification efforts in the Ecuadorian Amazon sustainable? *Sustainability* **2016**, *8*, 443. [CrossRef]
40. Ferrer-Martí, L.; Domenech, B.; García-Villoria, A.; Pastor, R. A MILP model to design hybrid wind–photovoltaic isolated rural electrification projects in developing countries. *Eur. J. Oper. Res.* **2013**, *226*, 293–300. [CrossRef]
41. Escribano, G. Ecuador's energy policy mix: Development versus conservation and nationalism with Chinese loans. *Energy Policy* **2013**, *57*, 152–159. [CrossRef]
42. Agencia de Regulación y Control de Electricidad de Ecuador (ARCONEL). *Revista Digital: Estadística del Sector Eléctrico Ecuatoriano 2017*; Ministerio de Electricidad y Energía Renovable: Quito, Ecuador, 2017. Available online: https://www.controlrecursosyenergia.gob.ec/estadisticas-del-sector-electrico-ecuatoriano-buscar (accessed on 19 January 2022). (In Spanish)
43. Cortés Valencia, I.V. Regulatory and Planning Approach to Rural Electrification in Isolated Areas of the Amazon: Bolivia, Brazil, Colombia, Ecuador and Peru. Master's Thesis, Escuela Técnica Superior de Ingeniería, Universidad Pontífica Comillas, Madrid, Spain, 2016.
44. Peláez-Samaniego, M.R.; García-Pérez, M.; Cortez, L.A.B.; Oscullod, J.; Olmedo, G. Energy sector in Ecuador: Current status. *Energy Policy* **2007**, *35*, 4177–4189. [CrossRef]

45. Finer, M.; Jenkins, C.N. Proliferation of hydroelectric dams in the Andean Amazon and implications for Andes-Amazon connectivity. *PLoS ONE* **2012**, *7*, e0035126. [CrossRef]

46. Robalino-López, A.; Mena-Nieto, A.; García-Ramos, J.E. System dynamics modeling for renewable energy and CO_2 emissions: A case study of Ecuador. *Energy Sustain. Dev.* **2014**, *20*, 11–20. [CrossRef]

47. Posso, F.; Sánchez, J.; Espinoza, J.L.; Siguencia, J. Preliminary estimation of electrolytic hydrogen production potential from renewable energies in Ecuador. *Int. J. Hydrogen Energy* **2016**, *41*, 2326–2344. [CrossRef]

48. Ferrer-Martí, L.; Garwood, A.; Chiroque, J.; Ramírez, B.; Marcelo, O.; Garfí, M.; Velo, E. Evaluating and comparing three community small-scale wind electrification projects. *Renew. Sustain. Energy Rev.* **2012**, *16*, 5379–5390. [CrossRef]

49. Tratural (Cia. Ltda). *Estudio Homologación de Unidades de Propiedad (UP) y Unidades de Construcción (UC) de Sistemas Fotovoltaicos No Conectados a la Red*; Ministerio de Electricidad y Energía Renovable: Quito, Ecuador, 2016. (In Spanish)

50. Enginyeria Sense Fronteres (ESF); Empresa Eléctrica Ambato, Sociedad Anónima. *Conceptos para la Gestión de Sistemas Fotovoltaicos Aislados*; Ministerio de Electricidad y Energías Renovables: Quito, Ecuador, 2016. (In Spanish)

51. Khan, A.A.; Malik, N.; Al-Arainy, A.; Alghuwainem, S. A review of condition monitoring of underground power cables. In Proceedings of the 2012 IEEE International Conference on Condition Monitoring and Diagnosis, Bali, Indonesia, 23–27 September 2012; pp. 909–912.

52. Trotter, P.A. A low-carbon way to increase energy access: How to scale mini-grids in developing countries. In Proceedings of the Applied Energy Symposium, MIT A+B, Boston, MA, USA, 22–24 May 2019.

53. Taneja, J. *If You Build It, Will They Consume? Key Challenges for Universal, Reliable, and Low-Cost Electricity Delivery in Kenya*; Working Paper No. 491; Center for Global Development: Washington, DC, USA, 2018.

54. Cader, C.; Blechinger, P.; Bertheau, P. Electrification planning with focus on hybrid mini-grids—A comprehensive modelling approach for the Global South. *Enrgy. Proced.* **2016**, *99*, 269–276. [CrossRef]

55. Galleguillos-Pozo, R.; Domenech, B.; Ferrer-Martí, L.; Pastor, R. Design of stand-alone electrification systems using fuzzy mathematical programming approaches. *Energy* **2021**, *228*, 120639. [CrossRef]

56. Baños, R.; Manzano-Agugliaro, F.; Montoya, F.G.; Gil, C.; Alcayde, A.; Gómez, J. Optimization methods applied to renewable and sustainable energy: A review. *Renew. Sustain. Energy Rev.* **2011**, *15*, 1753–1766. [CrossRef]

57. Domenech, B.; Ranaboldo, M.; Ferrer-Martí, L.; Pastor, R.; Flynn, D. Local and regional microgrid models to optimise the design of isolated electrification projects. *Renew. Energy* **2018**, *119*, 795–808. [CrossRef]

58. Tomei, J.; Cronin, J.; Agudelo Arias, H.D.; Córdoba Machado, S.; Mena Palacios, F.M.; Toro Ortiz, Y.M.; Borja Cuesta, Y.E.; Palomino Lemus, R.; Murillo López, W.; Anandarajah, G. Forgotten spaces: How reliability, affordability and engagement shape the outcomes of last-mile electrification in Chocó, Colombia. *Energy Res. Soc. Sci.* **2020**, *59*, 101302. [CrossRef]

59. Lambert, T.W.; Hittle, D.C. Optimization of autonomous village electrification systems by simulated annealing. *Sol. Energy* **2000**, *68*, 121–132. [CrossRef]

60. Ferrer-Martí, L.; Pastor, R.; Capó, G.M.; Velo, E. Optimizing microwind rural electrification projects. A case study in Peru. *J. Glob. Opt.* **2011**, *50*, 127–143. [CrossRef]

61. International Energy Agency (IEA). World Energy Outlook: Electricity Access Database. 2021. Available online: https://www.iea.org/reports/world-energy-outlook-2021 (accessed on 19 January 2022).

 *mathematics*

Article

Balancing Cost and Demand in Electricity Access Projects: Case Studies in Ecuador, Mexico and Peru

Rosa Galleguillos-Pozo [1], Bruno Domenech [1,2,*], Laia Ferrer-Martí [1,3] and Rafael Pastor [1,2]

[1] Institute of Industrial and Control Engineering, Universitat Politècnica de Catalunya—BarcelonaTech, 08028 Barcelona, Spain; rosa.galleguillos@upc.edu (R.G.-P.); laia.ferrer@upc.edu (L.F.-M.); rafael.pastor@upc.edu (R.P.)

[2] Department of Management, Universitat Politècnica de Catalunya—BarcelonaTech, 08028 Barcelona, Spain

[3] Department of Mechanical Engineering, Universitat Politècnica de Catalunya—BarcelonaTech, 08028 Barcelona, Spain

* Correspondence: bruno.domenech@upc.edu; Tel.: +34-934-017076

Abstract: Rural areas in developing countries have the highest concentrations of unelectrified communities. There is a clear link between electricity consumption and the Human Development Index, as highlighted by the 7th Development Goal of the United Nations. Estimating the energy needs of the previously nonelectrified population is imprecise when designing rural electrification projects. Indeed, daily energy demand and peak power assessments are complex, since these values must be valid over the project's lifetime, while tight budgets do not allow for the systems to be oversized. In order to assist project promoters, this study proposes a fuzzy mixed integer linear programming model (FMILP) for the design of wind–PV rural electrification systems including uncertainty in the demand requirements. Two different FMILP approaches were developed that maximized the minimum or the average satisfaction of the users. Next, the FMILP approaches were applied to six Latin American communities from three countries. Compared with the deterministic MILP (where the energy and peak power needs are considered as specific values), the FMILP results achieved a better balance between the project cost and the users' satisfaction regarding the energy and peak power supplied. Regarding the two approaches, maximizing the users' minimum satisfaction obtained globally better solutions.

Keywords: microgrids; rural electrification; fuzzy optimization; developing countries; case studies

MSC: 90C90

Citation: Galleguillos-Pozo, R.; Domenech, B.; Ferrer-Martí, L.; Pastor, R. Balancing Cost and Demand in Electricity Access Projects: Case Studies in Ecuador, Mexico and Peru. *Mathematics* 2022, *10*, 1995. https://doi.org/10.3390/math10121995

Academic Editor: Anatoliy Swishchuk

Received: 25 April 2022
Accepted: 7 June 2022
Published: 9 June 2022

Publisher's Note: MDPI stays neutral with regard to jurisdictional claims in published maps and institutional affiliations.

1. Introduction

"Ensuring access to affordable, reliable, sustainable and modern energy for all" has been recognized as the 7th Sustainable Development Goal of the United Nations [1]. Indeed, a connection exists between the Human Development Index (HDI) and energy access [2]: for less developed regions, slight increases in electricity consumption lead to huge socioeconomic growth, significantly improving the population's living standards. However, a significant proportion of the population in rural areas of developing countries still lack such a service [3].

Extending electricity access through the national grid can have important technoeconomic limitations in rural and remote areas because of the dispersion of demand and low end-user consumption. In contrast, standalone systems based on renewable energy are appropriate for isolated communities [4]. In particular, hybrid wind–photovoltaic (PV) systems can reduce costs and improve supply quality in comparison with single-technology projects [5]. Hybrid systems have proven to be suitable to address the electricity needs of residential clusters [6]. Additionally, the combination of individual supplies and microgrids can help medium-dispersed communities achieve a proper balance between extension lines and cost increases [7]. However, the whole design is complex, having to study many locations and sizes for generators, along with all possible connections among demand points to form the corresponding microgrids [8]. Hence, decision support tools are recommended for designing electrification systems correctly [8,9].

There is ample literature dealing with design tools for rural electrification systems including mathematical models and heuristic algorithms [10–12]. Many works focus on dimensioning a combination of generation technologies to cover the demand at minimum cost. The most used software is HOMER [13], which includes a detailed analysis of the demand, the energy resources and the equipment. For instance, Raji and Luta [14] used HOMER to design a community microgrid in South Africa, obtaining a technically and economically viable solution. Other optimization methods, such as integer linear programming, have also been used to evaluate wind–PV systems [15]. On the other hand, the distribution of electricity from generators to end users has been less studied [16], with the particular context of medium-dispersed communities receiving even less attention. ViPOR [17] considers, through simulated annealing, the location and electricity needs of each demand point to evaluate whether microgrid extension or individual supply is less expensive. García-Villoria et al. [18] developed a heuristic process to find the minimum cost combination of wind and PV technologies as well as microgrids and individual systems to distribute electricity in remote and medium-dispersed communities.

In the above works, demand was considered as a deterministic value, and the results are, therefore, subject to the quality of its estimation [19]. Consequently, the estimation of demand becomes critical, since an underestimation can leave the inhabitants dissatisfied, while an overestimation can unnecessarily increase the project costs. Inexact predictions will negatively impact the socioeconomic development of the area and/or produce economically unsustainable solutions. However, the real demand can be influenced by several factors such as [11] the local climate and geographic characteristics, the economy and culture, or the typology of consumers and their lifestyle. Therefore, the estimation of demand is inevitably subject to uncertainty [20,21].

In order to obtain robust designs regarding demand uncertainty, different approaches have been developed [22]. A relevant research area has focused on developing predictive algorithms for future demand estimation. For instance, genetic algorithms have been used to forecast the electricity requirements of populations in Turkey [23], Iran [24] and Mauritius [25]. For this purpose, social, economic and environmental indicators are gathered, and optimization algorithms aim to minimize deviation indicators. These algorithms have also been combined with artificial neural networks to improve the prediction results [26]. Under a different approach, Domenech et al. [27] developed an optimization–multicriteria methodology to design wind–PV electrification projects, which, first, generates a set of solutions for different demand scenarios and then selects the best one in terms of several criteria. Nevertheless, the project promoter still has to quantify the demand scenarios as unique values. From a different perspective, fuzzy logic can help solve complex problems with data uncertainty in the energy sector [28]. For instance, Onar et al. [29] developed a decision model with multiple fuzzy criteria for different experts to aid investors in selecting the most appropriate energy technology. Li et al. [30] proposed a fuzzy programming approach for planning an electrical energy generating system. Mohammadi et al. [31] introduced fuzzy elements in an MILP model to help in planning energy systems managing demand uncertainty. The results can help to achieve a balance between the guaranteed energy, the system cost and environmental problems. Vahedipour-Dahraie et al. [32] proposed a risk-averse probabilistic framework to schedule virtual power plants, taking into account demand response and uncertainty. The model helps to mitigate the negative impacts of uncertainty on the plant's performance. Wang et al. [33] developed a stochastic multiobjective model to design hybrid energy systems, considering demand and solar radiation uncertainty through probability distributions.

The reviewed works focused mainly on large- or medium-sized energy systems, while the analysis of demand uncertainty in the context of small-scale systems for newly electrified populations is scarce (Hossain et al.) [34]. As indicated by Domenech et al. [8], ad hoc tools considering the specific details of end users are required in order to improve the medium- and long-term sustainability of energy systems for these populations. Galleguillos-Pozo et al. [35] developed and compared five fuzzy MILP (FMILP) models, considering different assumptions, to design PV systems that balance the project cost and the demand satisfaction. This paper combined wind energy, controllers and batteries as well as detailed novel electrical features to make the most efficient FMILP model for exploring a wider range of solutions and obtaining better and more detailed electrification options. Hence, two FMILP models are proposed for designing wind–PV rural electrification projects, defining the best location and size of equipment for distribution through microgrids and individual supplies.

Consequently, the project promoters obtained a very powerful tool to assist in decision making when implementing projects in developing countries as well as robust solutions that are not dependent on the exact estimation of demand. Two modeling assumptions were considered and compared for the FMILP model: (a) to ensure that the least satisfied user was as satisfied as possible; (b) to ensure that the global satisfaction of all users was as high as possible. To validate the proposed solution procedure, six case studies were solved: six real communities from three Latin American countries (i.e., Ecuador, Mexico and Peru). The characteristics of the regions studied vary significantly (i.e., forest, semi-arid and highland), which tested the model's performance in different contexts. The solutions obtained (with FMILP) were compared with those that would have been obtained without considering demand uncertainty (with MILP). Compared to MILP, the FMILP results achieved a better balance between the project cost and the users' satisfaction in terms of the energy and peak power supplied. Regarding the modeling approaches, maximizing the minimum satisfaction obtained globally better solutions.

The remainder of the paper is organized as follows: Section 2 describes the specific problem including the design of wind–PV systems and uncertainty in users' demand estimation; Section 3 details the FMILP models for balancing the cost and the demand supplied; Section 4 presents the six case studies and the input data for the validation; Section 5 discusses the results of the case studies; finally, Section 6 highlights the main conclusions.

2. Problem Description

This section first describes the technical considerations of PV–wind electrification systems (Section 2.1); then, the complexity of estimating the electricity demand of end users is highlighted (Section 2.2).

2.1. Systems Design

Figure 1 shows the elements of the electrification systems dealt with in this paper (adapted from [36]). The population was dispersed among the demand points (houses, schools, health centers, etc.), each at a different location and having its own energy and peak power requirements. PV panels and wind turbines were used in order to supply the demand. Controllers protected the charge and discharge of the batteries, where the energy was stored for supply during non-generation periods. Next, inverters transformed the DC from the batteries into AC, which is better suited for most appliances. All this equipment was placed at a generation point, which was the only demand point in the case of individual systems or one of the demand points in the case of microgrids. The electricity was distributed at low voltage (LV) among microgrid demand points, using a radial structure suitable for rural areas in developing countries [17]. In addition, meters were installed at microgrid points to track users' consumption.

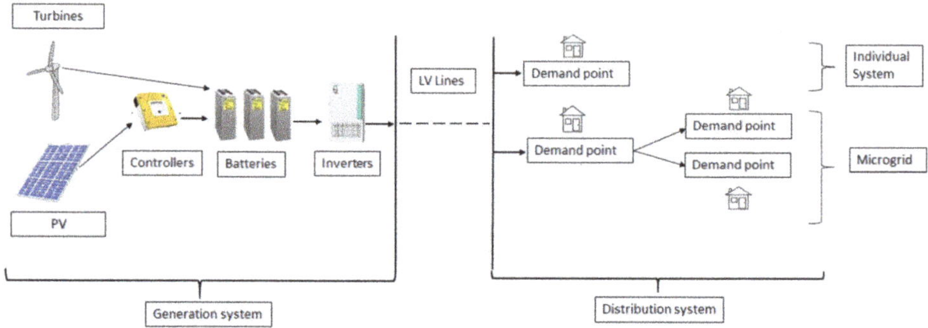

Figure 1. Scheme of the rural electrification systems (adapted from [36]).

2.2. Demand Estimation

Determining the energy and peak power demand of end users is complex and involves quantitative and qualitative information regarding the population as well as the energy sources prior to electrification [20]. In order to gather such information, local and regional databases can be consulted, end-users surveyed and interviewed and meetings held with specific categories of the population (women, children, elders, etc.). In addition, the surroundings of the community must be examined to identify any other characteristics, such as climatology or nearby villages, that can influence consumption [37]. Finally, the future expectations and productive activities to be developed during the project's lifetime must also be evaluated [38].

With the above information, the energy and peak power consumption of each demand point must be assessed; this is a complex task that is, logically, subject to uncertainty. Moreover, economies of scale and the staggered nature of equipment can lead to small variations in the demand having a significant impact on the project cost (and vice versa). Consequently, rather than defining unique values, it is easier for project promoters to determine both an essential demand, below which the project would not satisfy users' essential needs and an improved demand, above which the project would be too expensive [35]. A balance has to be sought between these two scenarios, maximizing the energy and peak power supplied on the one hand, while minimizing the project cost on the other.

2.3. Problem Formulation

Considering the above, the model developed to address the problem described must consider the following elements:

- As input data: The location and electricity requirements of demand points as well as the cost and technical characteristics of the equipment;
- As variables: The detailed solution including the equipment to be installed at each point and the microgrid connections between points;
- As an objective function: The maximization of end-users' satisfaction, considering the project cost as well as the energy and peak power supplied;
- As constraints: The satisfaction of users' electricity requirements taking into account uncertainty and the technical relationships between the equipment installed and the structure of the distribution microgrids.

3. Mathematical Modeling

In this work, two FMILP models are proposed for designing rural electrification projects, defining the best location and size of equipment as well as the distribution through microgrids and individual supplies. The models balance the project cost and the energy and peak power within the limits defined by the essential and improved demands. In order to introduce this balance into the models, the end-users' satisfaction regarding the energy, peak power and cost are included by means of several variables, normalized on a

0–1 scale. Hence, the solutions defined the satisfaction values for each of the three issues examined. For the essential demand (or lower values), the minimum energy and peak power were supplied to end users; thus, satisfaction was 0. In contrast, the project entailed the minimum cost; thus, satisfaction was 1. For the improved demand (or higher values), the maximum energy and peak power were supplied to end users; therefore, satisfaction was 1. In contrast, the project incurred the maximum cost; thus, satisfaction was 1. Finally, a linear progression from 0 to 1 was assumed for intermediate scenarios. This behavior was modeled as in the literature [39,40] and was validated by electrification experts [35].

Next, two FMILP models were developed to optimally design standalone wind–PV electrification systems for rural communities in developing countries, balancing the project cost and the demand supplied. The deterministic (nonfuzzy) model can be found in Ferrer-Martí et al. [36], although slight changes were made to better represent solutions: a wind controller was added to each wind turbine for the proper tracking of these devices, and the efficiency of the batteries and inverters was adjusted.

As explained before, the balance between the project cost and the demand supplied was introduced through several satisfaction variables: λ_C for the cost; λ_E for the energy; λ_P for the peak power. However, balancing these three issues can be conceived under different approaches, depending on the relative importance given to each one. Galleguillos-Pozo et al. [35] compared diverse approaches for a simpler problem (neither considering wind energy, controllers and batteries nor technical aspects such as voltage drops or equipment efficiencies, as done here), concluding that the best option is to directly compare the cost satisfaction (which tends toward cheap and low-demand solutions) with the average energy and peak power satisfaction (which tends toward expensive and high-demand solutions), without calibration parameters (which simplifies decision making for project promoters).

It must be noted that two modeling approaches were proposed regarding energy and peak power satisfaction. First (Section 3.1) was the maximization of the minimum satisfaction: the least satisfied demand point was focused on, assuming that if this point was satisfied, the remaining ones would also be more or equally satisfied. Second (Section 3.2) was the maximization of the average satisfaction: the focus was on satisfying all of the demand points as much as possible. The results were then be compared to identify those better representing the end-users' preferences.

3.1. Minimum Satisfaction Fuzzy Model

The approach modeled in this section assumed the maximization of the minimum satisfaction, i.e., the satisfaction of the least satisfied demand point of the community. The input data, variables, objective function and constraints are described below. In each subsection, the data and the constraints that introduce fuzziness are highlighted.

3.1.1. Input Data

- Indices:

a	Used to go through wind turbine options;
b	Used to go through battery options;
c	Used to go through LV line options;
d	Used to go through demand points (when referring to downstream points);
i	Used to go through inverter options;
p	Used to go through demand points;
q	Used to go through demand points (when referring to upstream points);
s	Used to go through PV panel options;
z	Used to go through PV controller options.

- General parameters:

A	Number of wind turbine options ($a = 1, \ldots, A$);
B	Number of battery options ($b = 1, \ldots, B$);
C	Number of LV line options ($c = 1, \ldots, C$);

CA_a — Cost (USD) of wind turbine a, including the support structure and a controller ($a = 1, \ldots, A$);

CB_b — Cost (USD) of battery b ($b = 1, \ldots, B$);

CC_c — Cost (USD/m) of line c, including the support structure ($c = 1, \ldots, C$);

CI_i — Cost (USD) of inverter i ($i = 1, \ldots, I$);

CM — Cost (USD) of a meter;

CS_s — Cost (USD) of panel s, including the support structure ($s = 1, \ldots, S$);

CZ_z — Cost (USD) of controller z ($z = 1, \ldots, Z$);

DB — Maximum depth of discharge (unit fraction) allowed for the batteries;

$EA_{p,a}$ — Energy (Wh/day) provided by wind turbine a located at point p ($p = 1, \ldots, N$; $a = 1, \ldots, A$);

EB_b — Capacity (Wh) of battery b ($b = 1, \ldots, B$);

ES_s — Energy (Wh/day) provided by panel s ($s = 1, \ldots, S$);

I — Number of inverter options ($i = 1, \ldots, I$);

IC_c — Maximum admissible intensity (A) of line c ($c = 1, \ldots, C$);

L^{MAX} — Maximum distance [m] at which 2 microgrid points can be directly connected;

$L_{p,d}$ — Distance (m) between points p and d ($p = 1, \ldots, N$; $d = 1, \ldots, N$);

N — Number of demand points (houses, schools, health centers, etc.);

NA — Maximum number that can be installed at the same point;

NS — Maximum number that can be installed at the same point;

PI_i — Peak power (W) of inverter i ($i = 1, \ldots, I$);

PS_s — Nominal power (W) of panel s ($s = 1, \ldots, S$);

PZ_z — Peak power (W) of controller z ($z = 1, \ldots, Z$);

Q_p — Set of points d that can be the destination of a microgrid line from point p ($p = 1, \ldots, N$; $d = 1, \ldots, N$: $p \neq d$ and $L_{p,d} \leq L^{MAX}$);

RC_c — Electrical resistance (Ω/m) of line c ($c = 1, \ldots, C$);

S — Number of PV panel options ($s = 1, \ldots, S$);

VB — Requested self-sufficiency (days) of the batteries;

V^{MAX} — Maximum voltage (V) above which demand points cannot be supplied;

V^{MIN} — Minimum voltage (V) below which demand points cannot be supplied;

V^N — Nominal voltage (V);

Z — Number of PV controller options ($z = 1, \ldots, Z$);

α — Calibration parameter for the objective function;

ηB — Efficiency (unit fraction) of the batteries;

ηC — Efficiency (unit fraction) of the lines;

ηI — Efficiency (unit fraction) of the inverters.

- Parameters that model fuzziness:

C^{MAX} — Maximum project cost. This value can be determined solving the deterministic model for the improved demand (E_p^{MAX} and P_p^{MAX}) [36];

C^{MIN} — Minimum project cost. This value can be determined solving the deterministic model for the essential demand (E_p^{MIN} and P_p^{MIN}) [36];

E_p^{MAX} — Improved energy demand (Wh/day) requested by demand point p ($p = 1, \ldots, N$);

E_p^{MIN} — Essential energy demand (Wh/day) requested by demand point p ($p = 1, \ldots, N$);

P_p^{MAX} — Improved peak power demand (W) requested by demand point p ($p = 1, \ldots, N$);

P_p^{MIN} — Essential peak power demand (W) requested by demand point p ($p = 1, \ldots, N$);

ΔC — Project cost range. $\Delta C = C^{MAX} - C^{MIN}$;

ΔE_p — Energy demand (Wh/day) range of point p ($p = 1, \ldots, N$). $\Delta E_p = E_p^{MAX} - E_p^{MIN}$;

ΔP_p — Peak power demand (W) range of point p ($p = 1, \ldots, N$). $\Delta P_p = P_p^{MAX} - P_p^{MIN}$.

3.1.2. Variables

- Integer non-negative:

$xa_{p,a}$ — Number of wind turbines type a installed at point p ($p = 1, \ldots, N$; $a = 1, \ldots, A$);

$xb_{p,b}$ — Number of batteries type b installed at demand point p ($p = 1, \ldots, N$; $b = 1, \ldots, B$);

$xi_{p,i}$ Number of inverters type i installed at demand point p ($p = 1, \ldots, N$; $i = 1, \ldots, I$);

$xs_{p,s}$ Number of PV panels type s installed at demand point p ($p = 1, \ldots, N$; $s = 1, \ldots, S$);

$xz_{p,z}$ Number of controllers type z installed at demand point p ($p = 1, \ldots, N$; $z = 1, \ldots, Z$).

- Real non-negative:

    ed_p Energy (Wh/day) supplied to demand point p ($p = 1, \ldots, N$);

    $fe_{p,d}$ Energy flow (Wh/day) between demand points p and d ($p = 1, \ldots, N$; $d \in Q_p$);

    $fp_{p,d}$ Power flow (W) between demand points p and d ($p = 1, \ldots, N$; $d \in Q_p$);

    pd_p Peak power (W) supplied to demand point p ($p = 1, \ldots, N$);

    v_p Voltage at demand point p ($p = 1, \ldots, N \mid v_p \in (V^{MIN}; V^{MAX})$).

- Binary:

    $xc_{p,d,c} \in \{0; 1\}$ One if a line type c directly connects demand points p and d; 0 otherwise ($p = 1, \ldots, N$; $d \in Q_p$; $c = 1, \ldots, C$);

    $xg_p \in \{0; 1\}$ One if at least one generator (wind turbine and/or PV panel) is installed at demand point p; 0 otherwise ($p = 1, \ldots, N$);

    $xm_p \in \{0, 1\}$ One if demand point p belongs to a microgrid ($p = 1, \ldots, N$).

- Dimensionless real non-negative that model fuzziness:

    λ_C Satisfaction with regards to the project cost;

    λ_E Satisfaction of the least satisfied point regarding the energy supplied;

    λ_P Satisfaction of the least satisfied point regarding the peak power supplied.

3.1.3. Objective Function

The objective function (1) aims to maximize the global satisfaction of end users with the solution obtained. This function includes, on the one hand, the project cost satisfaction (which tends toward cheap and low-demand solutions) and, on the other, the average between the energy and peak power satisfactions (which tend toward expensive and high-demand solutions). In addition, the objective function is calibrated through the α parameter, which allows for assigning more or less importance to one or another element, depending on the case study examined. This parameter also enables carrying out sensitivity analyses to examine the importance of the cost satisfaction vs. the energy and peak power satisfactions. In this paper, a value of $\alpha = 0.5$ was considered, according to previous works [35]. Finally, note that λ_C, λ_E and λ_P are dimensionless variables that represent the satisfaction of end users in regard to the solution on a 0–1 scale, as in the literature [33]. Their values are determined after solving the model (Section 5).

$$[MAX]\alpha \cdot \lambda\_C + \frac{1}{2}(1 - \alpha)(\lambda\_E + \lambda\_P) \tag{1}$$

3.1.4. Constraints

- General constraints

This is example two of an equation: Constraints (2), (3) and (4) define the generation points ($xg_p = 1$), as those are where the wind turbines and/or PV panels are located. Constraints (2) and (3) also limit the number of generators that can be installed at the same point. Constraint (5) sizes the batteries installed at each generation point so that they cover the demand of the point (ed_p, defined later in the fuzzy constraints) plus the dependent points through the output LV lines, taking the self-sufficiency requested, the depth of discharge and the efficiencies into account. Constraints (6) and (7) link the energy and power flows with the existence of an LV line between any two demand points, p and d. Constraint (8) establishes the radial structure of the microgrids: demand points can only have an input LV line, except for generation points, which cannot have any. Constraints (9) and (10), respectively, define the voltage drop between any two connected demand points and the maximum intensity that can flow. Constraint (11) sizes solar controllers according to the nominal power of the PV panels installed at each generation point. Constraint (12) means that inverters can only be installed at generation points. Finally, constraints (13) and (14) force meters to be installed at microgrid-connected points.

$$\sum_{a=1}^{A} xa_{p,a} \leq NA \cdot xg_p \qquad p = 1,\ldots,N \qquad (2)$$

$$\sum_{s=1}^{S} xs_{p,s} \leq NS \cdot xg_p \qquad p = 1,\ldots,N \qquad (3)$$

$$\sum_{a=1}^{A} xa_{p,a} + \sum_{s=1}^{S} xs_{p,s} \geq xg_p \qquad p = 1,\ldots,N \qquad (4)$$

$$\sum_{b=1}^{B} EB_b \cdot xb_{p,b} \frac{DB \cdot \eta B \cdot \eta I}{VB} + \sum_{j=1}^{N} \frac{E_j^{MAX}}{\eta C}(1 - xg_p) \geq ed_p + \sum_{d \in Q_p} fe_{p,d} \qquad p = 1,\ldots,N \qquad (5)$$

$$fe_{p,d} \leq \left(\sum_{j=1}^{N} \frac{E_j^{MAX}}{\eta C}\right) \sum_{c=1}^{C} xc_{p,d,c} \qquad p = 1,\ldots,N; d \in Q_p \qquad (6)$$

$$fp_{p,d} \leq \left(\sum_{j=1}^{N} \frac{P_j^{MAX}}{\eta C}\right) \sum_{c=1}^{C} xc_{p,d,c} \qquad p = 1,\ldots,N; d \in Q_p \qquad (7)$$

$$\sum_{q=1|p \in Q_q}^{N} \sum_{c=1}^{C} xc_{q,p,c} + xg_p \leq 1 \qquad p = 1,\ldots,N \qquad (8)$$

$$v_p - v_d \geq \frac{L_{p,d} \cdot RC_c \cdot fp_{p,d}}{VN} - \left(V^{MAX} - V^{MIN}\right)\left(1 - xc_{p,d,c}\right) \qquad p = 1,\ldots,N; d \in Q_p; c = 1,\ldots,C \qquad (9)$$

$$\frac{fp_{p,d}}{VN} - \left(\sum_{j=1}^{N} \frac{P_j^{MAX}}{VMIN \cdot \eta C}\right)\left(1 - xc_{p,d,c}\right) \leq IC_c \qquad p = 1,\ldots,N; d \in Q_p; c = 1,\ldots,C \qquad (10)$$

$$\sum_{z=1}^{Z} PZ_z \cdot xz_{p,z} \geq \sum_{s=1}^{S} PS_s \cdot xs_{p,s} \qquad p = 1,\ldots,N \qquad (11)$$

$$xi_{p,i} \leq \left(\sum_{j=1}^{N} \frac{P_p^{MAX}}{PI_i}\right) xg_p \qquad p = 1,\ldots,N; i = 1,\ldots,I \qquad (12)$$

$$\sum_{d \in Q_q} \sum_{c=1}^{C} xc_{p,d,c} \leq \left(P_p^{MAX} - 1\right) xm_p \qquad p = 1,\ldots,N \qquad (13)$$

$$\sum_{q=1|p \in Q_q}^{N} \sum_{c=1}^{C} xc_{q,p,c} \leq xm_p \qquad p = 1,\ldots,N \qquad (14)$$

- Constraints that model fuzziness

Constraint (15) defines the cost satisfaction variable (λ_C). The cost of the equipment installed (left side of the inequality: wind turbines, PV panels, controllers, batteries, inverters, meters and LV lines) ranges between the minimum cost (C^{MIN}, for full satisfaction $\lambda\_C = 1$) and the maximum cost ($C^{MAX} = C^{MIN} + \Delta C$, for null satisfaction $\lambda\_C = 0$). Constraint (16) carries out an energy balance at each demand point. The energy supplied to a point through the input lines or the generators installed at that point (left side of the inequality) must be higher than or equal to the energy consumed by the point (ed_p) plus the energy supplied to the dependent points through the output lines (last element). Constraints (17) and (18) define the energy consumption of each demand point. The consumption of a point ranges between the essential demand (E_p^{MIN}, for null satisfaction $\lambda\_E = 0$) to the improved demand ($E_p^{MAX} = E_p^{MIN} + \Delta E_p$, for full satisfaction $\lambda\_E = 1$). In addition, the efficiency of the LV lines must be considered (or not) depending on whether it is a point supplied by a microgrid (or a generation point). Considering this, the sum in brackets is included in both constraints as an upper bound to activate/disable one or another. Hence, in the case of generation points ($xg_p = 1$), constraint (17) is activated and (18) disabled. Therefore, the consumption of the point (ed_p) will be directly a value between E_p^{MIN} and $E_p^{MIN} + \Delta E_p$, depending on the value taken by λ_E. In contrast, for points supplied through a microgrid, constraint (17) is disabled and (18) activated; therefore, the consumption (ed_p) still ranges between E_p^{MIN} and $E_p^{MIN} + \Delta E_p$ but also considers the LV lines' efficiency (ηC). Additionally, note that the inequalities are defined in such a way that

λ_E takes the satisfaction value of the least satisfied demand point among the N points. Constraints (19), (20) and (21) are analogous to (16), (17) and (18), respectively, except for the peak power demand.

$$\sum_{p=1}^{N}\sum_{a=1}^{A} CA_a \cdot xa_{p,a} + \sum_{p=1}^{N}\sum_{s=1}^{S} CS_s \cdot xs_{p,s} + \sum_{p=1}^{N}\sum_{z=1}^{Z} CZ_z \cdot xz_{p,z} + \sum_{p=1}^{N}\sum_{b=1}^{B} CB_b \cdot xb_{p,b} +$$
$$+ \sum_{p=1}^{N}\sum_{i=1}^{I} CI_i \cdot xi_{p,i} + \sum_{p=1}^{P} CM \cdot xm_p + \sum_{p=1}^{N}\sum_{d\in Q_p}\sum_{c=1}^{C} L_{p,d} \cdot CC_c \cdot xc_{p,d,c} \leq C^{MIN} + \Delta C(1 - \lambda\_C) \tag{15}$$

$$\sum_{q=1\,|\,p\in Q_q}^{N} fe_{q,p} + \eta B \cdot \eta I \left(\sum_{a=1}^{A} EA_{p,a} \cdot xa_{p,a} + \sum_{s=1}^{S} ES_s \cdot xs_{p,s} \right) \geq ed_p + \sum_{d\in Q_p} fe_{p,d} \qquad p = 1, \ldots, N \tag{16}$$

$$ed_p \geq E_p^{MIN} + \Delta E_p \cdot \lambda\_E - \left(\sum_{j=1}^{N} \frac{E_j^{MAX}}{\eta C} \right)(1 - xg_p) \qquad p = 1, \ldots, N \tag{17}$$

$$ed_p \geq \frac{E_p^{MIN} + \Delta E_p \cdot \lambda\_E}{\eta C} - \left(\sum_{j=1}^{N} \frac{E_j^{MAX}}{\eta C} \right) xg_p \qquad p = 1, \ldots, N \tag{18}$$

$$\sum_{q=1\,p\in Q_q}^{N} fp_{q,p} + \sum_{i=1}^{I} PI_i \cdot xi_{p,i} \geq pd_p + \sum_{d\in Q_p} fp_{p,d} \qquad p = 1, \ldots, N \tag{19}$$

$$pd_p \geq P_p^{MIN} + \Delta P_p \cdot \lambda\_P - \left(\sum_{j=1}^{N} \frac{P_j^{MAX}}{\eta C} \right)(1 - xg_p) \qquad p = 1, \ldots, N \tag{20}$$

$$pd_p \geq \frac{P_p^{MIN} + \Delta P_p \cdot \lambda\_P}{\eta C} - \left(\sum_{j=1}^{N} \frac{P_j^{MAX}}{\eta C} \right) xg_p \qquad p = 1, \ldots, N \tag{21}$$

3.2. Average Satisfaction Fuzzy Model

Unlike the above model, which considered the maximization of the least satisfied demand point, now the maximization of satisfaction of all points is taken into account. Consequently, a specific satisfaction variable is considered for each point, and the objective function and some constraints are modified as described below.

- Dimensionless real non-negative variables that model fuzziness:

  λ_E_p Satisfaction of demand point p regarding the energy supplied ($p = 1, \ldots, N$);
  λ_P_p Satisfaction of demand point p regarding the peak power supplied ($p = 1, \ldots, N$).

- Objective function

The objective function (1′) substitutes (1) in order to maximize the global satisfaction of end users. This function includes, on the one hand, the project cost satisfaction and, on the other, the average between the energy and peak power satisfactions for all the demand points. In addition, a calibration parameter α is included and, in this paper, a 0.5 value was considered [35].

$$[MAX]\alpha \cdot \lambda\_C + \frac{1}{2N}(1 - \alpha)\left(\sum_{p=1}^{N} \lambda\_E_p + \sum_{p=1}^{N} \lambda\_P_p \right) \tag{1′}$$

- Constraints

Constraints (17′), (18′), (20′) and (21′), respectively, substitute (17), (18), (20) and (21). Note that instead of λ_E and λ_P, now λ_E_p and λ_P_p are used.

$$ed_p \geq E_p^{MIN} + \Delta E_p \cdot \lambda\_E_p - \left(\sum_{j=1}^{N} \frac{E_j^{MAX}}{\eta C} \right)(1 - xg_p) \qquad p = 1, \ldots, N \tag{17′}$$

$$ed_p \geq \frac{E_p^{MIN} + \Delta E_p \cdot \lambda\_E_p}{\eta C} - \left(\sum_{j=1}^{N} \frac{E_j^{MAX}}{\eta C} \right) xg_p \qquad p = 1, \ldots, N \tag{18′}$$

$$pd_p \geq P_p^{MIN} + \Delta P_p \cdot \lambda\_P_p - \left(\sum_{j=1}^{N} \frac{P_j^{MAX}}{\eta C} \right)(1 - xg_p) \qquad p = 1, \ldots, N \qquad (20')$$

$$pd_p \geq \frac{P_p^{MIN} + \Delta P_p \cdot \lambda\_P_p}{\eta C} - \left(\sum_{j=1}^{N} \frac{P_j^{MAX}}{\eta C} \right) xg_p \qquad p = 1, \ldots, N \qquad (21')$$

4. Case Studies

Six case studies from three different Latin American countries were examined in order to evaluate the above FMILP models. The main characteristics of the communities and their population are described as follows: two from the Ecuadorian Amazon (Section 4.1), two from a semi-arid Mexican area (Section 4.2) and two from the Peruvian highlands (Section 4.3). Note that the characteristics of the communities varied significantly in order to test the performance of the proposed solving procedure in different contexts. Finally, the techno-economic parameters of the equipment considered for the analysis are detailed (Section 4.4).

4.1. Ecuadorian Communities

The two communities studied were Suraka (2°02′21″ S–76°21′29″ W) and Conambo (2°00′22″ S–76°27′08″ W) (Figure 2). Both have similar standards of living, access to basic services and cultural characteristics. Regarding basic services, neither of them has access to drinking water, sewage systems or electricity. Suraka had 12 demand points: nine houses, two community centers and one school. In contrast, Conambo is a particularly large community, with 61 demand points: 49 houses, 8 school classrooms and 4 community centers (i.e., one meeting room, two dining rooms and one waiting room). Finally, according to the project promoters, wind turbines were not considered for the Amazon communities because of this technology's negative environmental impact (mainly, tree felling).

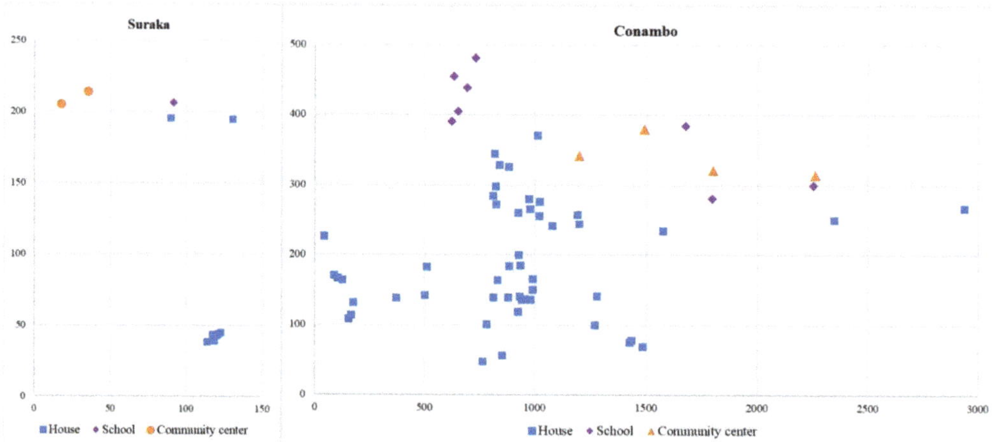

Figure 2. Layout of the Ecuadorian communities.

4.2. Mexican Communities

The communities studied were Tuzal (16°42′11″ N–93°55′02″ W) and Villa del Rio (16°44′42″ N–93°55′13″ W) (Figure 3), located in the state of Chiapas. This state is in the south of the country and has the lowest HDI: 0.667; there are approximately 6000 communities without access to electricity [41]. Tuzal is 90 km from the regional capital and had 14 houses, 1 school, 1 community center, 1 store and 1 church. None of the houses have drinking water; therefore, it must be carried from a nearby well. Access to this community is difficult because of the mountainous relief. Villa del Rio is 100 km from the regional capital and had 20 houses, 1 school, 1 community center, 2 stores and 2 churches. Access to the community is also difficult because of the mountainous relief and dirt roads.

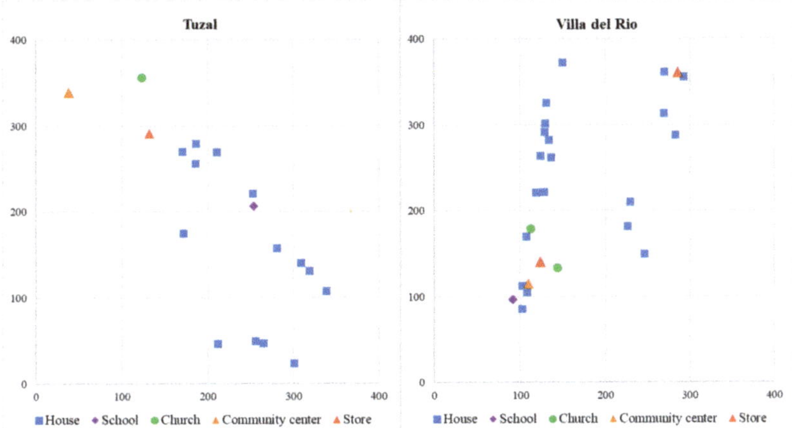

Figure 3. Layout of the Mexican communities.

4.3. Peruvian Communities

The two studied communities were El Alumbre (6°52′57″ S–78°26′23″ W) and Alto Peru (6°54′25″ S–78°37′24″ W) (Figure 4). The former had 33 houses, 1 school and 1 health center, widely dispersed. Alto Peru had 26 houses, 50% of them concentrated in 30% of the territory. The wind resource in both communities was variable; in some parts of the community, the wind resource was high, while other parts had low to moderate wind resource. The solar resource was highly significant, constant and the same for both communities.

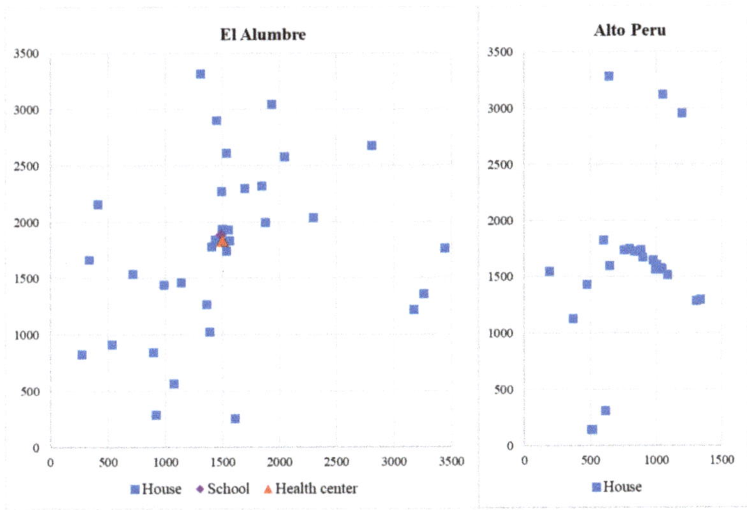

Figure 4. Layout of the Peruvian communities.

4.4. Input Data

Table 1 summarizes the data used for evaluating the proposed solving procedure. The data are different for the three countries studied. Regarding the essential and improved demands, values were defined together with experts from each country, according to the specific needs of each region's population. Based on these scenarios, the maximum (C^{MAX}) and minimum (C^{MIN}) costs for each community are calculated using the deterministic MILP model [36]. The other data were gathered from commercial catalogues, the literature review and consultations with project promoters. The models were solved with the ILOG CPLEX 12.6 on a 2.40 GHz, CPU Intel Core 15-1135G7 computer with 12 GB of RAM.

Table 1. Input data for each community and country.

Community		Ecuador		Mexico		Peru	
		Suraka	Conambo	Tuzal	Villa del Rio	El Alumbre	Alto Peru
Demand Points	Demand points N	12	61	18	26	35	26
	Maximum distance L^{MAX} (m)	500		500		500	
Energy Demand	Essential $E_p{}^{MIN}$ (Wh/day)	1000 (all)		100 (other) 750 (houses) 1500 (churches)		280 (houses) 975 (other)	280 (houses)
	Improved $E_p{}^{MAX}$ (Wh/day)	1500 (all)		150 (other) 1125 (houses) 2250 (churches)		420 (houses) 1463 (other)	420 (houses)
Peak Power Demand	Essential $P_p{}^{MIN}$ (W)	600 (all)		50 (other) 300 (houses) 750 (churches)		200 (houses) 600 (school) 1000 (health c.)	200 (houses)
	Improved $P_p{}^{MAX}$ (W)	900 (all)		75 (other) 450 (houses) 1125 (churches)		300 (houses) 900 (school) 1500 (health c.)	300 (houses)
Wind Turbines	Options A	n.a.		6		4	
	Maximum number NA	n.a.		28		28	
	Energy EA_{pa} (Wh/day)	n.a.		180 to 121,487		61 to 16,464	
	Cost CA_a (USD)	n.a.		1565 to 40,242		974 to 5132	
PV Panels	Options S	1		5		4	
	Maximum number NS	40		52		52	
	Energy ES_s (Wh/day)	1179		403 to 1048		217 to 652	
	Nominal power PS_s (W)	330		100 to 260		50 to 150	
	Cost CS_s (USD)	350		197 to 245		451 to 800	
PV Controller	Options Z	2		4		4	
	Peak power PZ_z (W)	480 to 2880		50 to 200		50 to 200	
	Cost CZ_z (USD)	300 to 700		67 to 125		67 to 125	
Batteries	Options B	2		4		4	
	Capacity EB_b (Wh)	1800 to 3600		24,422 to 63,360		1500 to 3000	
	Cost CB_b (USD)	300 to 850		132 to 387		225 to 325	
	Discharge DB (u.f.)	0.60		0.60		0.60	
	Self-sufficiency VB (days)	3		2		2	
	Efficiency ηB (u.f.)	0.85		0.85		0.85	
Inverters	Options I	2		5		4	
	Peak power PI_i (W)	600 to 3600		450 to 3000		300 to 3000	
	Cost CI_i (USD)	400 to 2000		60 to 582		377 to 2300	
	Efficiency ηI	0.85		0.85		0.85	
Meters	Cost CM (USD)	50		50		50	
LV Lines	Options C	2		3		2	
	Resistance RC_c (Ω/m)	0.0016 to 0.0030		0.0017 to 0.0027		0.0017 to 0.0027	
	Intensity IC_c (A)	60 to 96		89 to 101		89 to 101	
	Cost CC_c (USD/m)	3.94 to 6.03		4.90 to 5.25		4.90 to 5.00	
	Nominal voltage V^N (V)	220		220		220	
	Minimum voltage V^{MIN} (V)	210		210		210	
	Maximum voltage V^{MAX} (V)	230		230		230	
	Efficiency ηC (u.f.)	0.90		0.90		0.90	

5. Results and Discussion

This section, first of all, discusses the results obtained for the six studied communities in regard to how balanced solutions were obtained with the proposed FMILP models (Section 5.1). Then, the two modeling assumptions (i.e., minimum satisfaction and average satisfaction) are compared to identify the most suitable one (Section 5.2).

5.1. Results for the Six Case Studies

Figure 5 shows the results of satisfaction regarding the cost λ_C (blue), energy λ_E (green) and peak power λ_P (red) for the six studied communities. The results are organized in three images per country, using dashed (i.e., Suraka, Tuzal and El Alumbre) or dotted lines (i.e., Conambo, Villa del Rio and Alto Peru). The results are shown for the four solutions obtained in each community. The results of the deterministic MILP model are presented at the extremes of the figure: essential demand (top left) and improved demand (top right). The results of the FMILP models are presented in the middle: minimum satisfaction (mid-left) and average satisfaction (mid-right). Hence, for instance, in Suraka (Ecuador), Figure 5 shows the values of $\lambda\_C = 1.00$, $\lambda\_E = 0.22$ and $\lambda\_P = 0.16$ obtained for the essential demand with the deterministic MILP model and $\lambda\_C = 0.84$, $\lambda\_E = 0.13$ and $\lambda\_P = 1.00$ obtained with the minimum satisfaction FMILP model.

Regarding the MILP results, the essential demand solutions have full cost satisfaction and very low energy satisfaction with, occasionally, high power satisfaction. Indeed, the essential demand solutions were not limited to null energy and power satisfaction (equal to 0.0). The reason for this is the staggered nature of the equipment and economies of scale, which means that, in some cases, a higher energy and/or peak power demand than needed is supplied without increasing the cost. This varies depending on the community; for instance, in Tuzal the essential demand MILP model obtained an energy satisfaction of 0.14 and a power satisfaction of 0.90. In contrast, the energy satisfaction in Suraka was 0.22 and the peak power was 0.16. However, improved demand solutions always obtain full satisfaction for the energy and peak power and null satisfaction when it came to the cost.

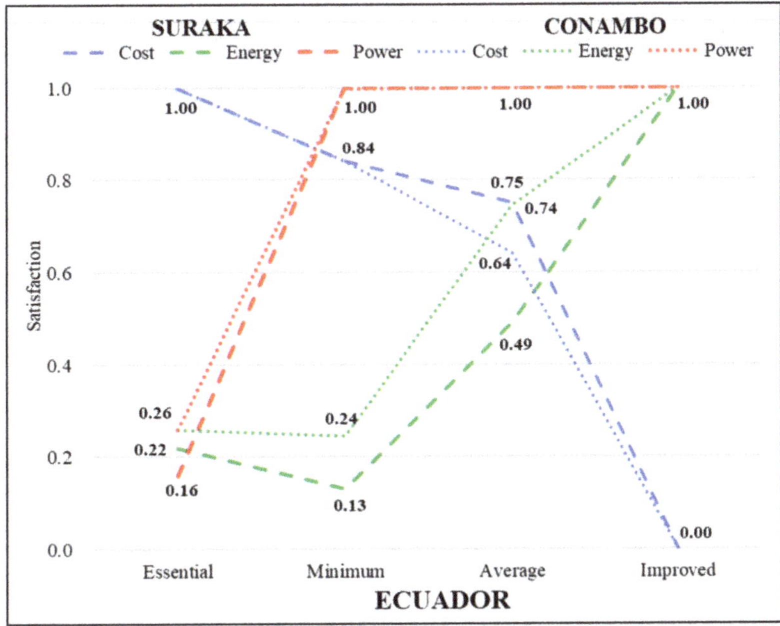

Figure 5. *Cont.*

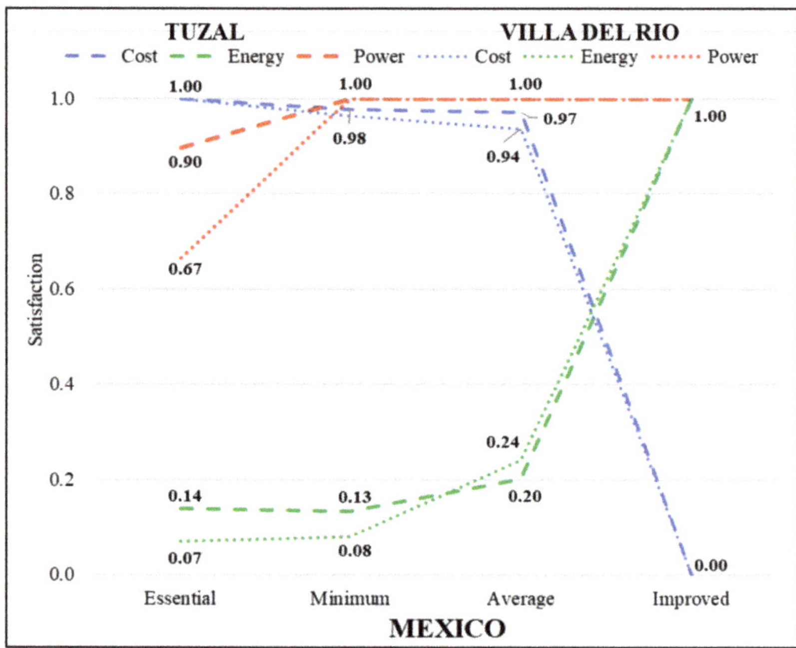

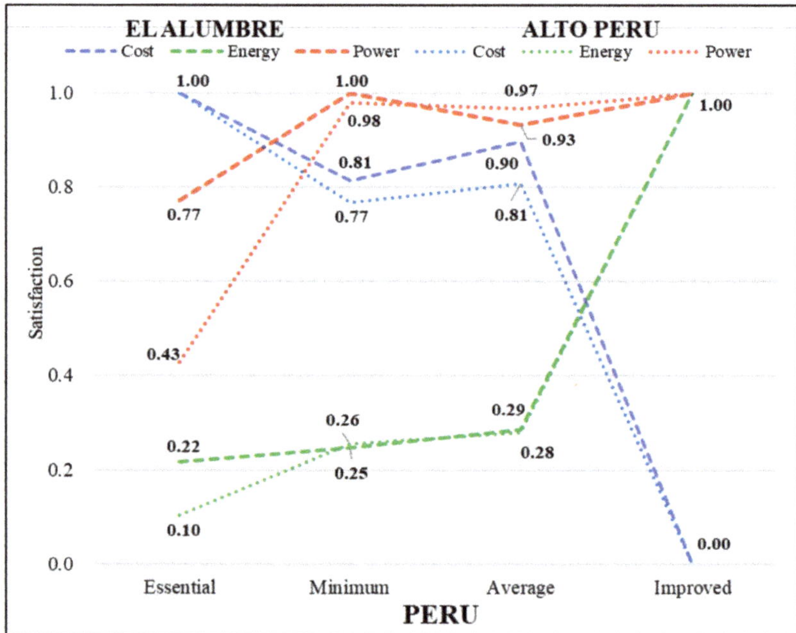

Figure 5. Results for the 6 case studies.

Regarding the FMILP solutions, when compared to the essential demand solutions, they provided similar energy satisfaction (slightly lower or higher, depending on the community) but with significantly higher peak power satisfaction in exchange for slightly more expensive solutions. For instance, in Alto Peru energy satisfaction increased from 0.10

(essential) to 0.26 (minimum) or 0.28 (average) and the peak power from 0.43 (essential) to 0.98 (minimum) or 0.97 (average). These improvements were accomplished with a small reduction in the cost satisfaction from 1.00 (essential) to 0.77 (minimum) or 0.81 (average). When compared to the improved demand solutions, the FMILP solutions provided similar peak power satisfaction but less energy satisfaction to obtain a much higher cost satisfaction.

Thus, the results confirm that the solutions obtained with the FMILP were more balanced than those from the deterministic MILP. In general, the FMILP solutions compensate for a reduction in one of the satisfaction variables with an increase in one of the other two satisfaction indicators. Therefore, the use of the FMILP models reduces the negative effects of uncertainty and obtains robust and globally better solutions in terms of satisfaction.

It is also worth noting that, in all of the Ecuadorian and Mexican communities, the average satisfaction assumption showed results with an energy satisfaction and equal peak power satisfaction similar to the minimum satisfaction assumption, but the average assumption solutions had a higher cost. In contrast, the Peruvian communities showed the opposite situation. Therefore, the comparison between these two modeling assumptions is not straightforward and is further examined in Section 5.2.

5.2. Comparison of Assumptions

The above section showed the most balanced solutions obtained with the FMILP models rather than with the deterministic model. However, the discussion regarding the minimum satisfaction and the average satisfaction assumptions needs to be examined in more detail. In this regard, note that the objective functions of the FMILP models (see Equations (1) and (1′)) balance the cost satisfaction with the average of the energy and peak power satisfactions. Therefore, the comparison of assumptions in Figure 5 is not straightforward, since variations in cost satisfaction are not directly proportional to variations in energy or peak power satisfaction.

In order to deal with this, both assumptions were compared. Figure 6 shows the 12 solutions examined (i.e., six communities with two assumptions per community). For each solution, two values were calculated: the minimum satisfaction objective function (1), top image; the average satisfaction objective function (1′), bottom image. For instance, in Suraka, the minimum satisfaction FMILP was solved, and the obtained value of the objective function (1) was 1.40 (top). For this solution, the value of the other objective function (1′) was calculated manually, obtaining 1.41 (bottom). Additionally, also for Suraka, the average satisfaction FMILP was solved, and the obtained value of the objective function (1′) was 1.49 (bottom). For this solution, the value of the other objective function (1) was calculated manually, obtaining 1.25 (top).

As shown in Figure 6, logically, the minimum satisfaction solutions (red bars) in the top image are higher than the average satisfaction solutions (green bars) for all of the communities; the opposite occurs in the bottom image. However, the differences between bar sizes were significantly higher for the minimum satisfaction objective function (top) than for the average satisfaction objective function (bottom). For instance, in Conambo the difference was 0.19 for the minimum satisfaction objective function (1.33 vs. 1.14), while it was only 0.05 for the average satisfaction objective function (1.46 vs. 1.51). In El Alumbre, the differences were even higher: 0.54 (1.44 vs. 0.90) and 0.07 (1.44 vs. 1.51), respectively. Consequently, the average satisfaction solutions logically obtained top values for their objective function (1′), but their performance on the minimum satisfaction objective function (1) was limited. In contrast, the minimum satisfaction solutions are more recommendable, since they logically obtained the top values in their objective function (1) and, in addition, they achieved close-to-top values in terms of the average satisfaction objective function (1′).

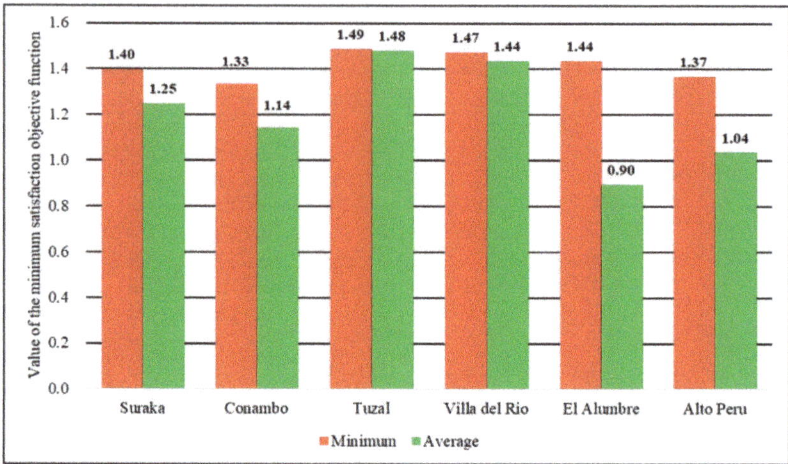

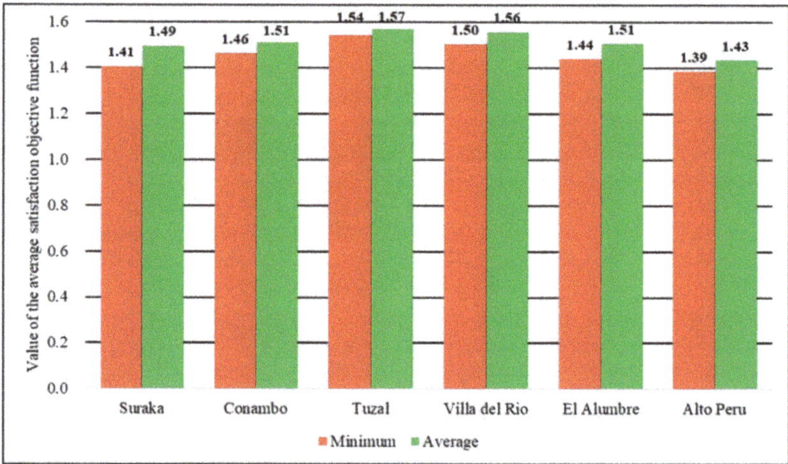

Figure 6. Comparison of the objective function values between assumptions.

In short, as a recommendation for project promoters willing to design an electrification project for a community without electricity, where demand is difficult to estimate, the authors suggest solving the two proposed FMILP models; the obtained solutions would balance, in a different way, the satisfaction regarding the cost, energy and peak power. Then, a choice can be made between these two solutions based on the very specific details of each one, taking into account the opinion of experts and the community. However, the general recommendation is that the minimum satisfaction FMILP model obtains globally better solutions.

6. Conclusions

Estimating demand in settlements accessing electricity for the first time is complex and subject to uncertainty. With the current tools, project developers must obtain a unique electrification solution (the quality of which logically depends on the estimated demand) or examine different demand values (each one leading to a different solution with a different cost) and then manually analyze the best one. In both cases, the decision-making process has limitations that might impact on the performance of the finally implemented solution.

To overcome this situation, this work developed a tool that enables satisfaction with regard to the project cost and the energy and peak power supplied to end users to be balanced.

In order to do so, a novel FMILP model was proposed, based on a modeling approach tested in the literature as efficient, to balance the cost satisfaction and the energy and peak power satisfaction. Hence, project promoters have a powerful tool for designing rural electrification projects in developing countries, combining wind and PV technologies as well as microgrids and individual systems, while taking into account the uncertainty in demand estimation. Rather than being subject to a specific demand value, the novel FMILP model enables a range of values to be specified and the most balanced solution is returned. In addition, two assumptions were modeled: maximizing the minimum satisfaction (focus on the least satisfied demand point) and maximizing the average satisfaction (global satisfaction of all points).

The validation of the proposed solving procedure was performed using six case studies from three Latin American countries (i.e., Ecuador, Mexico and Peru). In particular, two demand scenarios were defined: an essential demand, to cover basic end-user needs, and an improved demand, above which solutions would be considered too expensive. The FMILP solutions (one for each assumption) were compared with those obtained with a deterministic MILP model. The results show that the MILP models led to low-supply or expensive solutions, while the FMILP models allowed for a balance between the cost, energy and peak power to be achieved. Finally, the results for the FMILP models under the two assumptions were compared. Although the two models can be easily solved and the best option can then be selected based on specific details, the minimum satisfaction FMILP model is recommended in the case of promoters wanting a unique solution, since it obtains the top values for minimum satisfaction as well as close-to-top values for average satisfaction.

Prior to this work, project promoters could obtain a unique electrification solution, subject to the quality of the estimation of end-users' demand. In the case of wanting to test different demand scenarios, they had to solve each one manually through the deterministic MILP model and then select the best one after a discussion that might not be straightforward. In contrast, with the proposed FMILP models, this process is simplified. Now, project promoters only have to delimit the range of demand values, and the most balanced solution is directly obtained with each of the two FMILP models developed.

Author Contributions: Conceptualization, B.D. and L.F.-M.; methodology, R.G.-P. and L.F.-M.; software, R.G.-P. and B.D.; validation, R.P.; resources, B.D. and L.F.-M.; data curation, R.G.-P.; writing—original draft preparation, R.G.-P. and B.D.; writing—review and editing, L.F.-M. and R.P.; supervision, R.P. All authors have read and agreed to the published version of the manuscript.

Funding: This research was funded by the Spanish Ministry of Science and Innovation (RTI-2018-097962-B-I00) and the Centre for Cooperation Development (CCD) of the Universitat Politècnica de Catalunya-BarcelonaTech (UPC).

Institutional Review Board Statement: Not applicable.

Informed Consent Statement: Not applicable.

Data Availability Statement: Not applicable.

Acknowledgments: The authors are very grateful to the NGO Engineering Without Borders (Spain), Practical Action (Peru) and the National University of Sciences and Arts of Chiapas (Mexico). The author B.D. would like to thank the Serra Hunter Fellowship of the Generalitat de Catalunya.

Conflicts of Interest: The authors declare no conflict of interest. The funders had no role in the design of the study; in the collection, analyses, or interpretation of data; in the writing of the manuscript, or in the decision to publish the results.

References

1. United Nations (UN). *2030 Agenda for Sustainable Development*; United Nations: New York, NY, USA, 2015.
2. Eras-Almeida, A.A.; Egido-Aguilera, M.A. What is still necessary for supporting the SDG7 in the most vulnerable contexts? *Sustainability* **2020**, *12*, 7184. [CrossRef]
3. International Energy Agency (IEA). *Global Status Report. Towards a Zero-Emission, Efficient and Resilient Buildings and Construction Sector*; International Energy Agency: Paris, France, 2018; ISBN 978-92-807-3729-5.
4. Madriz-Vargas, R.; Bruce, A.; Watt, M. The future of community renewable energy for electricity access in rural Central America. *Energy Res. Soc. Sci.* **2018**, *35*, 118–131. [CrossRef]
5. Mehrjerdi, H. Modeling, integration, and optimal selection of the turbine technology in the hybrid wind-photovoltaic renewable energy system design. *Energy Convers. Manag.* **2020**, *205*, 112350. [CrossRef]
6. Sima, C.A.; Lazaroiu, G.C.; Dumbrava, V.; Tirsu, M. A hybrid system implementation for residential cluster. In Proceedings of the 2017 International Conference on Electromechanical and Power Systems, Iasi, Romania, 11–13 October 2017; pp. 275–280. [CrossRef]
7. Hirsch, A.; Parag, Y.; Guerrero, J. Microgrids: A review of technologies, key drivers, and outstanding issues. *Renew. Sustain. Energy Rev.* **2018**, *90*, 402–411. [CrossRef]
8. Domenech, B.; Ferrer-Martí, L.; Pastor, R. Comparison of various approaches to design wind-PV rural electrification projects in remote areas of developing countries. *WIREs Energy Environ.* **2019**, *8*, e332. [CrossRef]
9. Mavromatidis, G.; Orehounig, K.; Carmeliet, J. A review of uncertainty characterization approaches for the optimal design of distributed energy systems. *Renew. Sustain. Energy Rev.* **2018**, *88*, 258–277. [CrossRef]
10. Anoune, K.; Bouya, M.; Astito, A.; Abdellah, A.B. Sizing methods and optimization techniques for PV-wind based hybrid renewable energy system: A review. *Renew. Sustain. Energy Rev.* **2018**, *93*, 652–673. [CrossRef]
11. Liu, Y.; Yu, S.; Zhu, Y.; Wang, D.; Liu, J. Modeling, planning, application and management of energy systems for isolated areas: A review. *Renew. Sustain. Energy Rev.* **2018**, *82*, 460–470. [CrossRef]
12. Lyden, A.; Pepper, R.; Tuohy, P.G. A modelling tool selection process for planning of community scale energy systems including storage and demand side management. *Sustain. Cities Soc.* **2018**, *39*, 674–688. [CrossRef]
13. Soukeyna, M.; Ramdhane, I.B.; Ndiaye, D.; Elmamy, M.; Menou, M.; Yahya, A.M.; Mahmoud, A.K.; Youm, I. Feasibility analysis of hybrid electricity generation system by HOMER for Mauritanian northern coast. *Int. J. Phys. Sci.* **2018**, *13*, 120–131. [CrossRef]
14. Raji, A.K.; Luta, D.N. Modeling and optimization of a community microgrid components. *Energy Procedia* **2019**, *156*, 406–411. [CrossRef]
15. Sima, C.A.; Popescu, M.O.; Popescu, C.L.; Alexandru, M.; Popa, L.B.; Dumbrava, V.; Panait, C. Energy management of a cluster of buildings in a university campus. In Proceedings of the 2021 12th International Symposium on Advanced Topics in Electrical Engineering, Bucharest, Romania, 25–27 March 2021. [CrossRef]
16. Rojas-Zerpa, J.C.; Yusta, J.M. Application of multicriteria decision methods for electric supply planning in rural and remote areas. *Renew. Sustain. Energy Rev.* **2015**, *52*, 557–571. [CrossRef]
17. Bhagavathy, S.M.; Pillai, G. PV microgrid design for rural electrification. *Designs* **2018**, *2*, 33. [CrossRef]
18. García-Villoria, A.; Domenech, B.; Ferrer-Martí, L.; Juanpera, M.; Pastor, R. Ad-hoc heuristic for design of wind-photovoltaic electrification systems, including management constraints. *Energy* **2021**, *212*, 118755. [CrossRef]
19. Ciupageanu, D.A.; Barelli, L.; Lazaroiu, G. Real-time stochastic power management strategies in hybrid renewable energy systems: A review of key applications and perspectives. *Electr. Power Syst. Res.* **2020**, *187*, 106497. [CrossRef]
20. Riva, F.; Ahlborg, H.; Hartvigsson, E.; Pachauri, S.; Colombo, E. Electricity access and rural development: Review of complex socio-economic dynamics and causal diagrams for more appropriate energy modelling. *Energy Sustain. Dev.* **2019**, *43*, 203–223. [CrossRef]
21. Peters, J.; Sievert, M.; Toman, M.A. Rural electrification through mini-grids: Challenges ahead. *Energy Policy* **2019**, *132*, 27–31. [CrossRef]
22. Good, N.; Ellis, K.A.; Mancarella, P. Review and classification of barriers and enablers of demand response in the smart grid. *Renew. Sustain. Energy Rev.* **2017**, *72*, 57–72. [CrossRef]
23. Ozturk, H.K.; Canyurt, O.E.; Hepbasli, A.; Utlu, Z. An application of genetic algorithm search techniques to the future total exergy input/output estimation. *Energy Sources Part A* **2006**, *28*, 715–725. [CrossRef]
24. Azadeh, A.; Ghaderi, S.F.; Tarverdian, S.; Saberi, M. Integration of artificial neural networks and genetic algorithm to predict electrical energy consumption. *Appl. Math. Comput.* **2007**, *186*, 1731–1741. [CrossRef]
25. Badurally Adam, N.R.; Dauhoo, M.Z.; Elahee, M.K. A simulated-based genetic algorithm for the forecasting of monthly peak electricity demand. *Int. J. Oper. Res.* **2017**, *28*, 164–182. [CrossRef]
26. Badurally Adam, N.R.; Elahee, M.K.; Dauhoo, M.Z. Forecasting of peak electricity demand in Mauritius using the non-homogeneous Gompertz diffusion process. *Energy* **2011**, *36*, 6763–6769. [CrossRef]
27. Domenech, B.; Ferrer-Martí, L.; Pastor, R. Hierarchical methodology to optimize the design of stand-alone electrification systems for rural communities considering technical and social criteria. *Renew. Sustain. Energy Rev.* **2015**, *51*, 182–196. [CrossRef]
28. Suganthi, L.; Iniyan, S.; Samuel, A. Applications of fuzzy logic in renewable energy systems—A review. *Renew. Sustain. Energy Rev.* **2015**, *48*, 585–607. [CrossRef]

29. Onar, S.C.; Oztaysi, B.; Otay, İ.; Kahraman, C. Multi-expert wind energy technology selection using interval-valued intuitionistic fuzzy sets. *Energy* **2015**, *90*, 274–285. [CrossRef]
30. Li, G.; Sun, W.; Huang, G.H.; Lv, Y.; Liu, Z.; An, C. Planning of integrated energy-environment systems under dual interval uncertainties. *Int. J. Electr. Power* **2018**, *100*, 287–298. [CrossRef]
31. Mohammadi, M.; Noorollahi, Y.; Mohammadi-ivatloo, B. Fuzzy-based scheduling of wind integrated multi-energy systems under multiple uncertainties. *Sustain. Energy Technol. Assess.* **2020**, *37*, 100602. [CrossRef]
32. Vahedipour-Dahraie, M.; Rashidizadeh-Kermani, H.; Anvari-Moghaddam, A.; Siano, P. Risk-averse probabilistic framework for scheduling of virtual power plants considering demand response and uncertainties. *Int. J. Electr. Power* **2020**, *121*, 106126. [CrossRef]
33. Wang, J.; Qi, X.; Ren, F.; Zhang, G.; Wang, J. Optimal design of hybrid combined cooling, heating and power systems considering the uncertainties of load demands and renewable energy sources. *J. Clean. Prod.* **2021**, *281*, 125357. [CrossRef]
34. Hossain, M.A.; Chakrabortty, R.K.; Ryan, M.J.; Pota, H.R. Energy management of community energy storage in grid-connected microgrid under uncertain real-time prices. *Sustain. Cities Soc.* **2021**, *66*, 102658. [CrossRef]
35. Galleguillos-Pozo, R.; Domenech, B.; Ferrer-Martí, L.; Pastor, R. Design of stand-alone electrification systems using fuzzy mathematical programming approaches. *Energy* **2021**, *228*, 120639. [CrossRef]
36. Ferrer-Martí, L.; Domenech, B.; García-Villoria, A.; Pastor, R. A MILP model to design hybrid wind-photovoltaic isolated rural electrification projects in developing countries. *Eur. J. Oper. Res.* **2013**, *226*, 293–300. [CrossRef]
37. Pillot, B.; Muselli, M.; Poggi, P.; Dias, J.B. Historical trends in global energy policy and renewable power system issues in Sub-Saharan Africa: The case of solar PV. *Energy Policy* **2019**, *127*, 113–124. [CrossRef]
38. López-González, A.; Domenech, B.; Gómez-Hernández, D.; Ferrer-Martí, L. Renewable microgrid projects for autonomous small-scale electrification in Andean countries. *Renew. Sustain. Energy Rev.* **2017**, *79*, 1255–1265. [CrossRef]
39. Bilgen, B. Application of fuzzy mathematical programming approach to the production allocation and distribution supply chain network problem. *Expert Syst. Appl.* **2010**, *37*, 4488–4495. [CrossRef]
40. Mendel, J.M.; Korjani, M.M. A new method for calibrating the fuzzy sets used in fsQCA. *Inf. Sci.* **2018**, *468*, 155–171. [CrossRef]
41. Gómez-Hernández, D.F.; Domenech, B.; Moreira, J.; Farrera, N.; López-González, A.; Ferrer-Martí, L. Comparative evaluation of rural electrification project plans: A case study in Mexico. *Energy Policy* **2019**, *129*, 23–33. [CrossRef]

 *mathematics*

Article

An Adaptive Protection Scheme Based on a Modified Heap-Based Optimizer for Distance and Directional Overcurrent Relays Coordination in Distribution Systems

Mohamed Abdelhamid [1], Salah Kamel [1], Emad M. Ahmed [2,*] and Ephraim Bonah Agyekum [3]

[1] Department of Electrical Engineering, Faculty of Energy Engineering, Aswan University, Aswan 81528, Egypt; mohammed.abdulhameed@aswu.edu.eg (M.A.); skamel@aswu.edu.eg (S.K.)
[2] Department of Electrical Engineering, College of Engineering, Jouf University, Sakaka 72388, Saudi Arabia
[3] Department of Nuclear and Renewable Energy, Ural Federal University Named after the First President of Russia Boris Yeltsin, 19 Mira Street, 620002 Ekaterinburg, Russia; agyekumephraim@yahoo.com
* Correspondence: emamahmoud@ju.edu.sa

Abstract: This paper proposes an adaptive protection scheme (APS) based on the original heap-based optimization (HBO) and a modified HBO (MHBO). APS is used to solve protection relays coordination problems that include directional overcurrent relays (DOCRs) as well as the distance relay's second zone times. The complexity of the coordination problem increases with the impact of distributed generators (DGs) switching (ON/OFF). Topological changes in grid configuration frequently occur in distributing networks, equipped with DGs, causing changes in the values and direction of short circuit currents. This issue becomes a challenge for protection systems to avoid relays miscoordination and save a network's reliability. In the proposed MHBO, the Original HBO is modified by three points, population are divided into subgroups, then they are unified into one group gradually, those subgroups are exchanging some search agents between themselves, these search agents are called travelling agents, and the last one is about, upgrading an internal equation in the original algorithm. For validating the proposed relays coordination, the IEEE 8-bus test system, and the IEEE 14-bus distribution network are selected as case studies. The obtained simulated results of the proposed algorithm show better performance compared with those obtained by the previous algorithms.

Keywords: adaptive protection scheme; direction overcurrent relays; distance relays; distribution generators; heap-based optimizer; united sub-groups

Citation: Abdelhamid, M.; Kamel, S.; Ahmed, E.M.; Agyekum, E.B. An Adaptive Protection Scheme Based on a Modified Heap-Based Optimizer for Distance and Directional Overcurrent Relays Coordination in Distribution Systems. *Mathematics* **2022**, *10*, 419. https://doi.org/10.3390/math10030419

Academic Editors: Antonin Ponsich, Mariona Vila Bonilla and Bruno Domenech

Received: 20 December 2021
Accepted: 25 January 2022
Published: 28 January 2022

1. Introduction

The area of protection is currently one of the most important fields in power systems. To protect transmission lines, both directional overcurrent relays (DOCRs) and distance relays are generally used. Transmission lines are monitored by these protection relays from both ends. The occurrence of faults causes relays to activate trip scenarios [1].

Overcurrent relays (OCRs) generally operate based on the magnitude of the fault current, which is selected within parameters of the relay, whereas DOCRs incorporate the direction of the current flowing through the transmission line. A potential transformer is used to determine the direction of the voltage phasor. DOCRs are thus more costly than traditional OCRs. However, they are more advantageous than OCRs. Those kinds of relays must be set to operate as the backup, with a time delay greater than that of the primary relay [2].

Distance relay has two main zones. After detecting a fault, the first one begins working immediately. To avoid calculation errors, 80 percent of the transmission line is covered by this zone. The second zone then covers up to 120 percent of the transmission line by delay time. This large area also includes a portion of another transmission line [3].

The main issue in this paper is about the reduction of the operating time of the protection relay in order to provide the protection devices with the ability to isolate the fault area. This extends the lifespan of the components of the power system, making the system more reliable and healthier. However, because of constraints between DOCRs pairs and DOCRs and distance relay pairs, DOCRs and distance relays have more constraints and complex coordination problems. The miscoordination of these protection relays overlaps protection operations and fails to take advantage of the benefits of both distance and DOCR relays [4,5].

The contribution of RES-based distributed generators (DGs) to a distribution system is important. RES such as solar and wind energy are integrated into power systems. Many challenges are presented by DGs and problems with coordination, some of which include the change in the flow of the direction of fault currents and their magnitude [6].

Due to the impact of DGs on distribution networks, the protection system necessitates a flexible structure. In order to solve the problem of protection relays coordination, this research presents the adaptive protection scheme (APS) as a solution for this challenge. APS enables the changing of relay settings for both DOCRs and distance relays in response to any changes in the state of a network, based on the DG's ON/OFF states, using predetermined settings. APS, as a component of information and communication technologies (ICT), is primarily dependent on the communication network between smart grid components, or on SCADA. These communication networks enable APS to remotely set relay settings. APS is tested with a variety of scenarios that are most likely to trip in-network, and the best protection relay setting in each scenario is determined. This enables the protection system to reduce miscoordination and malfunction. The primary benefit of APS is to improve the selectivity and reliability of the protection system over traditional or fixed systems. The APS configures a group of protection relays that are determined by calculating optimal settings for each scenario using an optimization algorithm based on the DG's states [7,8].

Adaptive systems are designed to work on real-time systems. They need fast methods to rearrange their system's items. Hence, the APS uses the optimization algorithm due to its fast performance. APS is addressed in many research papers, which were developed based on optimization algorithms such as particle swarm optimization (PSO) in [9], genetic algorithm (GA) in [10], differential evolution algorithm (DEA) in [11], ant colony optimization (ACO) in [12], firefly algorithm (FA) in [13], gravitational search algorithm (GSA) in [14], manta ray foraging optimization (MRFO) in [7], and hybrid Harris hawks optimization (HHO) in [15] in order to coordinate the process of DOCRs.

In [1], APS was used to coordinate DOCRs and distance relays using school-based optimization algorithm (SBO) and its modified algorithm (MSBO). In this paper a new APS is suggested to solve the same coordination problem between DOCRs and distance relays but with a better optimal solution.

Usually, Metaheuristic optimization algorithms start with initial values, which are generated randomly to form their population, but this population between search space is limited. The optimization algorithm is used to improve the fitness of that population. Always metaheuristic optimization algorithms are formed by intrapopulation collaboration as the standard form.

Collaborative multi-population is a term that aims to introduce the SBO. This term is based on dividing the population into subgroups. This step is useful to increase its exploration performance.

As presented in SBO, which is a collaborative multi-population framework utilized by TLBO, the proposed modified algorithm is based on its original idea, which gives it the capability to increase its exploration performance [16].

This research work suggests a novel idea, which is about the collection of sub-groups into one main group after exploration is exploited. This idea improved the exploitation part by the search for an optimal solution with all populations of sub-groups. This idea balances both optimization algorithms performances exploration and exploitation. This balance is conducted with a new factor called M_{factor}.

There are many challenges for this idea; one of them is determining of the M_{factor}, which depends on the user's experience to face the problem of balance between exploitation and exploration, and the other one is about how to determine the subgroups' number. These challenges are faced with the experience of users or trial and error to have a good performance of the optimization algorithm.

HBO is presented in [17] and applied in many other engineering optimization problems, such as solar cell estimation [18], reactive power dispatch [19], Micro-grid design and sizing [20], and Proton exchange membrane fuel cell [21]. HBO solves these optimization problems with effectiveness.

There are other methods that are used to build APS, such as an environment APS based on Q-learning as in [22] and multi-agents as in [23,24].

Contributions of this paper are as follows:

- The proposed algorithm's response and convergence characteristics are improved by modifying the original HBO algorithm. There are three main points that were modified: subgroups were divided and then united, traveler agents were placed between subgroups, and an equation in the original HBO was modified. This algorithm would be useful in addressing other critical issues in other branches of the power system, such as microgrid, DG sizing, load frequency control, and solar cell parameter estimation.
- As a solution to the DG impact, an adaptive protection scheme was designed based on HBO and MHBO. That APS was used to coordinate both DOCRs and distance relays. In addition to the impact of DGs, the effect of distance relays complicates this co-ordination problem in the DOCR's coordination process.
- To verify the effectiveness of the proposed protection system, it was tested on IEEE 8-bus and IEEE 14-bus distribution networks, taking into account the effect of DG on/off states.

The following is the rest of the paper: the coordination problem and its mathematical modeling are presented in Section 2. The proposed protection scheme is presented in Section 3. The performance of HBO and MHBO with IEEE 8-bus and IEEE 14-bus distribution networks to solve the coordination problem is then presented in Section 4. Finally, Section 5 has the conclusions.

2. The Mathematical Modelling of Coordination Problem

The primary goal of this paper is to achieve the best possible coordination of DOCRs and distance relays. The objective function (OF) is the total operation times of the DOCRs at both near (T_{Near}) and far (T_{Far}) ends, as well as the second time zone of the distance relays (T_{Z2}). That OF is the shortest total operation times as described in [1]:

$$OF = min\left(\sum_{i=1}^{n} T_{Near}i + \sum_{i=1}^{n} T_{Far}i + \sum_{i=1}^{n} T_{Z2}i + F^{Pen} \right) \tag{1}$$

The international electro-technical commission (IEC) standard presents the standard time inverse of DOCRs characteristics by the following equation [2]:

$$T_i = \frac{\propto *TDS_i}{\left(\frac{I_f}{I_{pi}}\right)^{\beta} - \gamma} \tag{2}$$

where T_i is the relay's operation time of DOCRs for ith relay, TDS is the relay's time dial setting, and Ip is the relay's pickup current. The other constants α, β, and γ have values of 0.14, 0.02, and 1, respectively [25].

2.1. Problem's Limiters

The maximum operation time (T_{max}) is the primary limitation of any protection relay. In order to save the components of the power system from damage, this time should not exceed 2 s [26].

Relay's settings are limited with minimum and maximum values for each setting, as shown in the following equations [27]:

$$TDS_{min} \leq TDS \leq TDS_{max}, \tag{3}$$

$$Ip_{min} \leq Ip \leq Ip_{max} \tag{4}$$

$$T_{Z2min} \leq T_{Z2} \leq T_{Z2max} \tag{5}$$

2.2. The Problem's Constraints

Through the constraints between the primary and backup pair of DOCRs, as well as between the pair of DOCRs and distance relays at both ends, the proposed optimization problem becomes a higher constraint problem. These constraints are used to prevent miscoordination, which can occur when protection relays fail.

As shown in Figure 1, the relationship between DOCRs pair relays must deal with the backup relay (t_b), which operates with a delay time on the primary relay (t_p). This period of delay time is referred to as the coordination time interval (*CTI*). The *CTI* value is determined by the type of protection relays. The *CTI* value for electromagnetic relays must be greater than 0.3 s, while digital relays must be greater than 0.2 s [1]. In this research, digital relays are used. These constraints are depicted in the following equations [27,28]:

$$t_b{}^{F1} - t_p{}^{F1} > CTI \tag{6}$$

$$t_b{}^{F2} - t_p{}^{F2} > CTI \tag{7}$$

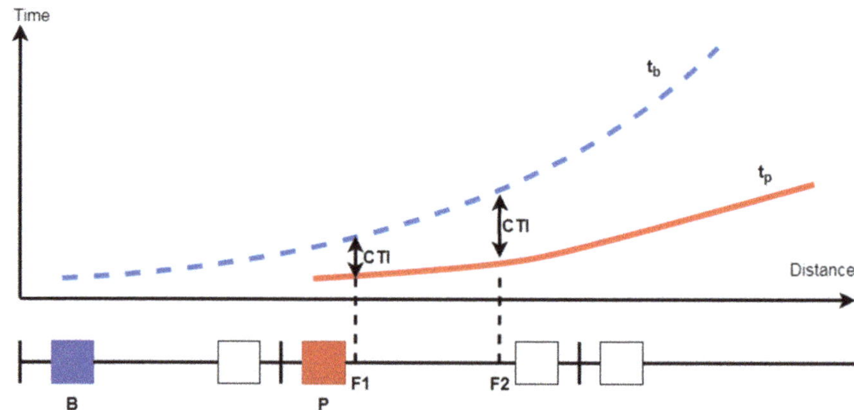

Figure 1. The relationship between DOCRs pair relays.

Figure 2 depicts the relationship between DOCRs and distance pair relays. At the near end, the backup distance relay liaises with the primary DOCRs relay, and T_{Z2b} must delay $t_p{}^{F1}$ with the *CTI* as described in Equation (8). While Equation (9) describes the distance and DOCRs relationship at the far end. The second zone of the primary distance relay (T_{Z2p}) must delay the primary DOCRs operation time ($t_p{}^{F1}$) with *CTI* at the far end [27].

$$T_{Z2b} - t_p{}^{F1} > CTI \tag{8}$$

$$T_{Z2p} - t_p{}^{F2} > CTI \tag{9}$$

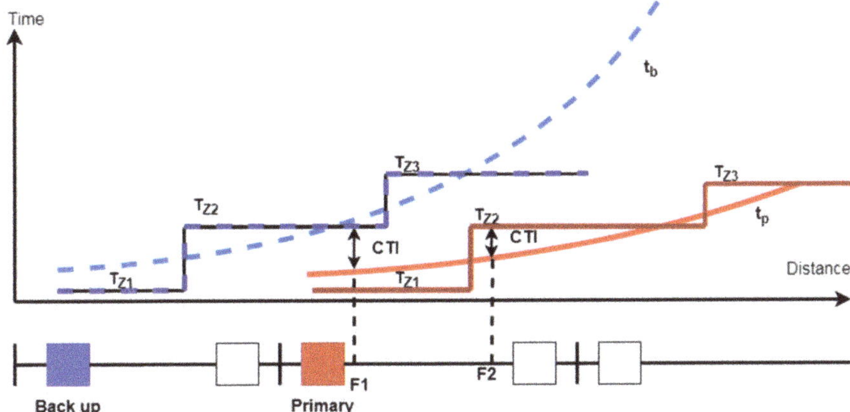

Figure 2. The relationship between Distance and DOCRs pair relays.

Based on the operation time of the primary relay at near and far ends. The relationship is developed to specify the minimum value of the distance relay's second zone. As shown in Equation (10), the maximum value of these equations is used as the time for the specific second zone of distance relay. This point contributes to the reduction of the penalty and constraints [1].

$$T_{Z2} = max\left(T_{Z2b}, T_{Z2p}\right) \tag{10}$$

For eliminating miscoordination between pairs relays, as recommended, the penalty function is developed as in the following equation [29]:

$$F^{pen} = \mu * \begin{cases} 1 \; if \; T^{backup} - T^{primary} < CTI \\ 0 \; if \; T^{backup} - T^{primary} \geq CTI \end{cases} \tag{11}$$

where μ is the penalty function's weighting factor.

When there is a miscoordination between relays pair, F^{pen} extends the total time of *OF*. As a result, the optimization algorithm tunes the relays setting parameters to reduce the size of *OF* to eliminate the miscoordination.

3. The Proposed Protection Scheme

3.1. Adaptive Protection Scheme (APS)

The proposed scheme in this paper is developed based on the optimal solutions obtained through the use of optimization algorithms. In addition, the HBO algorithm was used to evaluate the optimal solutions. Moreover, it is modified to improve its convergence characteristics and its ability to find better optimization solutions.

Figure 3 depicts the flow diagram of APS while taking into account the impact of DG. The obtained data from supervisory control and data acquisition (SCADA) was optimized using the centralized processing server. These data will be generated by the proposed algorithms in the APS to reset the DOCRs and distance relays. The main points of the proposed APS flow chart can be described as follows:

- The first point defines the actual topology of the distribution network, specifically the location, state, and size of DGs. Examine the distribution network topology for any changes. If nothing changes, the APS keeps with current protection relays settings. If the topology changes, the APS proceeds to the next point.
- In the second point, the APS calculates short circuit currents through CBs. For this mission, ETAP is used. Then, APS test the current relay settings' ability to save the protection system without losing the coordination of protection relays. If the current

settings of the relays are not able to protect the distribution network, the APS moves to the next point. Otherwise, APS returns to the previous step.

- In the third point, APS calls up the proposed optimization algorithm. The algorithm will seek the optimal solution that is suitable for covering changes in the distribution network while avoiding miscoordination. Finally, the APS reports the best solution for protection relay settings and sends it via ICT to the network operator or the intelligent electronic devices (IEDs) [30].

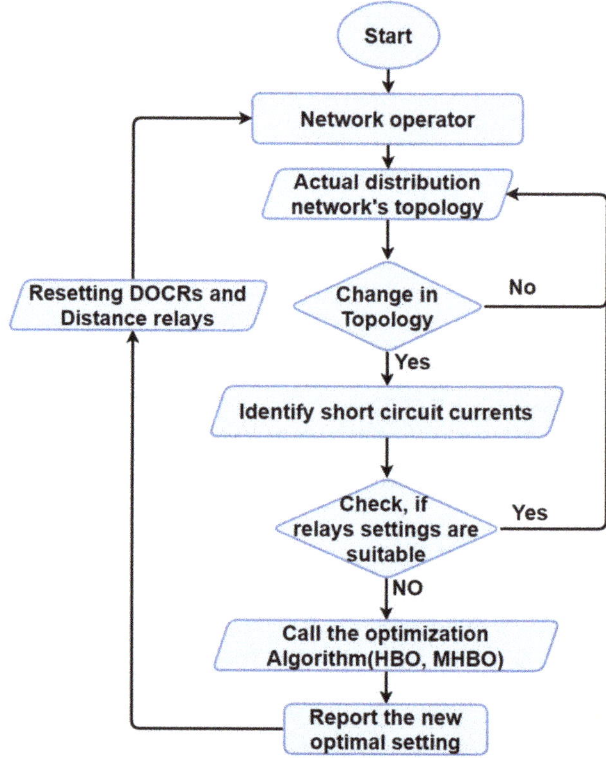

Figure 3. The Flow diagram of the proposed APS.

3.2. Original Heap-Based Optimization Algorithm

HBO is a novel metaheuristic algorithm, which is categorized as a human-based algorithm. HBO simulates the corporate rank hierarchy (CRH) in a very distinctive style. HBO is presented mathematically based on modeling three states of employee's relationships:

- Between the subordinates and their immediate supervisor.
- Between colleagues
- Self-contribution of employees

The use of the heap data structure in the CRH mapping allows for organizing the solutions in a hierarchical manner based on their fitness and the use of the arrangement in the algorithm's position-updating process in a very specific way. The mapping of the entire concept is modeled into the following steps:

- Modeling CRH
- Modeling the relationship between subordinates and the immediate supervisor
- Modeling the interaction between colleagues
- Modeling an employee's self-contribution to task execution.

- Overall update of the position of search agent uses the following equation:

$$
x_i^k(t+1) = \begin{cases}
x_i^k(t), & p \leq p_1 \\
B^k + \gamma \lambda^k \left| B^k - x_i^k(t) \right|, & p_1 < p \leq p_2 \\
S_r^k + \gamma \lambda^k \left| S_r^k - x_i^k(t) \right|, & p_2 < p \leq p_3 \ \& \ f\left(\vec{S}_r\right) < f\left(\vec{x}_i(t)\right) \\
x_i^k + \gamma \lambda^k \left| S_r^k - x_i^k(t) \right|, & p_2 < p \leq p_3 \ \& \ f\left(\vec{S}_r\right) \geq f\left(\vec{x}_i(t)\right)
\end{cases}
\tag{12}
$$

where $x_i^k(t+1)$ is the updated position, $x_i^k(t)$ is the current position, B^k is the parent position, S_r^k is the colleague position, p is a random value with in [0,1], $f\left(\vec{x}_i(t)\right)$ is the fitness value of the current position, $f\left(\vec{S}_r\right)$ is the fitness value of the colleague position, γ, λ, p_1, p_2, and p_3 as the following equations:

$$
\gamma = \left| 2 - \frac{t \bmod \frac{T}{c}}{\frac{T}{4c}} \right|
\tag{13}
$$

$$
\lambda^k = 2r - 1
\tag{14}
$$

$$
p_1 = 1 - \frac{t}{T}
\tag{15}
$$

$$
p_2 = p_1 - \frac{1 - p_1}{2}
\tag{16}
$$

$$
p_3 = p_2 - \frac{1 - p_1}{2} = 1
\tag{17}
$$

where t is the current iteration value, T is the total iteration number, r is a random value with in [0,1], and C as the following equation:

$$
C = T/25
\tag{18}
$$

3.3. Modified HBO Algorithm

The modified part suggests three main points to improve the exploration and exploitation performance of the original algorithm. These points are: update Equation (14), which describes λ, establishing many small companies then united in one big company, and traveling agents between companies.

3.3.1. The First Point: Developing λ Factor

This point is about developing the exploitation performance of the original algorithm. This point deals with the λ factor to modified by adding a term to connect λ with the iteration number as indicated in the following equation:

$$
\lambda^k = 0.5 * \left(1 - \frac{t}{T}\right) * (2r - 1)
\tag{19}
$$

This term gives the original HBO a chance to have more exploitation without effect on the exploration performance.

3.3.2. The Second Point: Sub-Group and M_{factor}

This point is about establishing small groups from search agents, these groups use HBO as an individual unit, then they are united in bigger groups until united in one group. As shown in Figure 4. The number of subgroups ($N_{subgroup}$) are determined by the user.

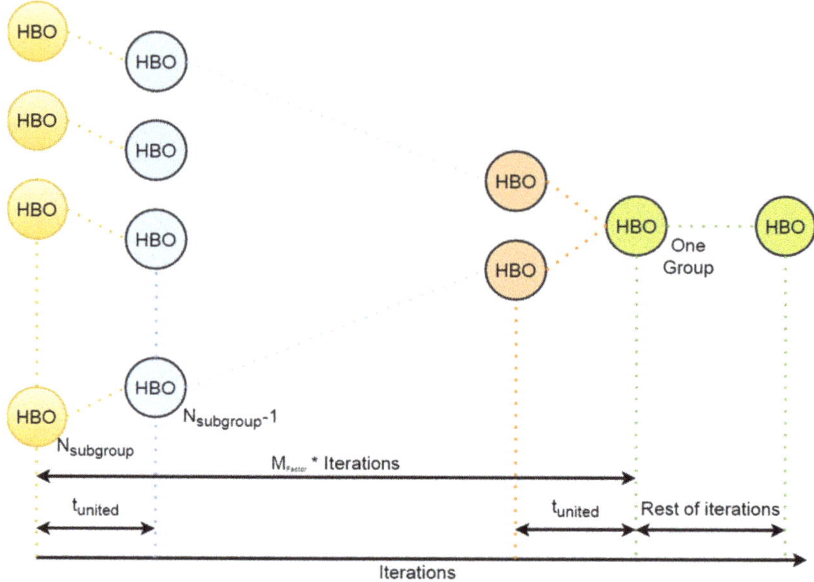

Figure 4. The diagram of united subgroups idea.

M_{factor} is a factor used to determine the number of iterations that are necessary to begin the unity (t_{united}). As described in the following equation:

$$t_{united} = round\left(M_{factor} * \frac{T}{N_{subgroup}} \right) \tag{20}$$

In addition, M_{factor} is determined by the user with a percentage value depending on the optimization problem in this coordination problem determined by 10% to increase the exploration performance of the HBO algorithm.

This point is very important to discover the search area adding to lose constraints, which is an important goal in an optimal problem. That point gives the algorithm all the search agents for exploitation along with other algorithm's iterations after uniting in one company.

3.3.3. The Third Point: Travelers

In the final point, in any company there are Travelers, they move between companies, those travelers give a chance to exchange skills between sub-groups. Travelers are unpredictable, thus, they are chosen randomly. The number of travelers can be limited by the user as a percentage value from the sub-group members.

4. Results and Discussion

In this research work, APS used both HBO and MHBO to tune optimal relay coordination problem in all cases of test systems. These relay's settings were TDS, Ip, and T_{Z2}. DOCRs have normal characteristic values such as 0.14, 0.02, and 1.0 for α, β, and γ constants, respectively. In addition to the maximum and minimum values of TDS and PS as 1.1 s and 0.1 s for TDS and 4 and 0.5 for PS. Moreover, the maximum operating time for the primary DOCRs or distance relays was 1.5 s [27].

The test systems were IEEE 8-bus test system and IEEE 14-bus distribution network. The test system's cases were the normal grid topological, and the other was a switch on the DGs on the grid. Optimal settings were used to reduce the operation time of relays and also for passing the system's constraints in both the near end and far end. These constraints were between DOCRs and Distance relays. These protection devices were assumed as digital relays with *CTI* equal to 0.2 s [1].

The proposed algorithms that were used in this paper have a population, max iteration, maximum travel percentage, society, and M_{factor} with values 840, 1000, 10%, 8, and 10%, respectively. MATLAB R2016a was used to run these algorithms. While ETAP 12.6.0 was used for the validation of the relay's operation times and the calculated 3 phase fault currents.

4.1. Test System I: IEEE 8-Bus Test System

The IEEE 8-bus test system, shown in Figure 5, consists of 7 transmission lines connected between 8 buses and feeds 4 loads from two synchronous generators. These generators feed the network by power transforms *T1*, and *T2*. This configuration will be considered as the normal topology. In order to investigate the performance of the proposed APS for relays coordination, an external 400 MW microgrid (EG) will be integrated into the system at the fourth bus (B4). The test system has 14 CB, each transmission line has two circuit breakers (CBs), that are activated by the APS. Furthermore, the protection settings are allowed to be changed according to the change of the grid topology [31].

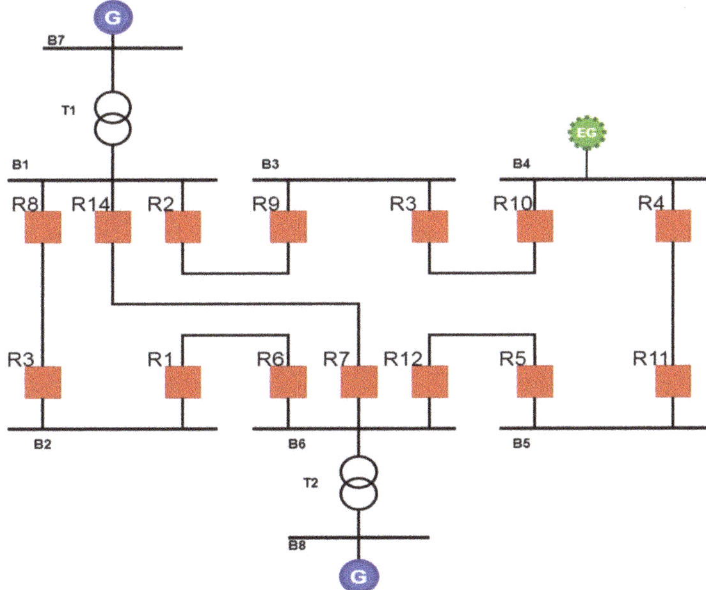

Figure 5. The single line diagram of IEEE8-bus test system.

The optimization problem aims at tuning 42 design variables. In addition to 40 constraints between DOCRs and distance relays, and 32 constraints between DOCRs in normal grid topology, while in the other case is 34 constraints. This makes that optimization problem a highly constrained problem adding to it is a non-linear problem. Each variable design is limited with maximum and minimum limiters.

Three-phase fault currents and CT values are extracted from [1].

The optimal values of variables designed for protective relays on the normal grid topology using MHBO and HBO are listed in Table 1. Additional to the other case about the external grid is the switching on of the optimal solution. Then Table 2 shows that the proposed algorithm's optimal solution passed the constraints between DOCRs, and between DOCRs and Distance. Whiles constraints pass in both between DOCRs, and between DOCRs and Distance, as shown in Table 3.

Table 1. IEEE 8-bus test system's relays setting.

| Relay | Traditional Topological Grid | | | | | | DG Switching ON | | | | | |
| | Original HBO | | | Modified HBO | | | Original HBO | | | Modified HBO | | |
	TDS	Ip	T_{Z2}	TDS	Ip	T_{Z2}	TDS	Ip	T_{Z2}	TDS	Ip	T_{Z2}
1	0.207	120.000	0.942	0.132	227.671	0.915	0.178	229.478	1.037	0.128	308.392	0.957
2	0.161	697.206	0.906	0.159	638.955	0.861	0.216	585.949	1.023	0.187	592.934	0.918
3	0.168	318.967	0.793	0.137	436.578	0.775	0.329	80.000	0.868	0.161	411.669	0.854
4	0.108	338.315	0.734	0.114	279.030	0.696	0.124	611.776	0.826	0.112	692.277	0.822
5	0.100	169.630	0.973	0.100	134.289	0.812	0.100	473.748	0.946	0.113	380.271	0.880
6	0.192	423.086	0.911	0.133	533.546	0.764	0.149	773.098	0.917	0.143	582.907	0.774
7	0.196	249.312	0.941	0.132	426.497	0.911	0.240	263.656	1.036	0.198	311.138	0.956
8	0.174	456.754	0.903	0.154	462.292	0.828	0.188	366.357	0.814	0.163	427.398	0.775
9	0.100	163.739	0.944	0.100	142.484	0.846	0.134	300.808	0.887	0.106	378.627	0.858
10	0.157	182.776	0.718	0.114	295.219	0.695	0.132	591.540	0.836	0.109	699.363	0.795
11	0.154	425.250	0.821	0.100	712.703	0.781	0.128	676.078	0.914	0.106	750.863	0.842
12	0.182	592.216	0.905	0.175	524.187	0.834	0.205	522.155	0.919	0.200	465.759	0.863
13	0.221	120.000	0.993	0.176	149.631	0.916	0.141	290.036	0.997	0.118	328.308	0.942
14	0.175	335.151	0.992	0.124	468.610	0.920	0.195	338.603	0.985	0.155	443.496	0.942
OF	28.609			25.983			30.192			27.799		

Table 2. IEEE 8-bus test system's operation times of Relay's pairs in Traditional grid by MHBO.

| Pair | Near-End | | | | | | Far-End | | | | | |
| | DOCRs | | | D&DOCR | | | DOCRs | | | D&DOCR | | |
	T_p	T_b	CTI	T_p	T_{Z2B}	CTI	T_p	T_b	CTI	T_p	T_{Z2P}	CTI
1	0.363	0.564	0.201	0.363	0.764	0.401	0.715	2.207	1.491	0.715	0.915	0.2
2	0.510	0.715	0.205	0.510	0.915	0.405	0.661	3.007	2.347	0.661	0.861	0.2
3	0.510	0.711	0.200	0.510	0.911	0.400	0.661	2.899	2.239	0.661	0.861	0.2
4	0.460	0.661	0.200	0.460	0.861	0.400	0.575	0.875	0.299	0.575	0.775	0.2
5	0.375	0.575	0.200	0.375	0.775	0.400	0.496	0.832	0.336	0.496	0.696	0.2
6	0.295	0.496	0.201	0.295	0.696	0.401	0.612	1.992	1.380	0.612	0.812	0.2
7	0.407	0.612	0.205	0.407	0.812	0.405	0.564	——	——	0.564	0.764	0.2
8	0.407	0.720	0.313	0.407	0.920	0.513	0.564	——	——	0.564	0.764	0.2
9	0.392	0.612	0.220	0.392	0.812	0.420	0.711	——	——	0.711	0.911	0.2
10	0.392	0.716	0.324	0.392	0.916	0.524	0.711	——	——	0.711	0.911	0.2
11	0.443	0.711	0.268	0.443	0.911	0.468	0.628	——	——	0.628	0.828	0.2
12	0.443	0.646	0.204	0.443	0.846	0.404	0.628	——	——	0.628	0.828	0.2
13	0.294	0.495	0.200	0.294	0.695	0.400	0.646	2.327	1.681	0.646	0.846	0.2
14	0.379	0.581	0.202	0.379	0.781	0.402	0.495	0.976	0.481	0.495	0.695	0.2
15	0.433	0.634	0.201	0.433	0.834	0.401	0.581	0.807	0.226	0.581	0.781	0.2
16	0.514	0.716	0.202	0.514	0.916	0.402	0.634	1.447	0.813	0.634	0.834	0.2
17	0.514	0.720	0.206	0.514	0.920	0.406	0.634	2.510	1.876	0.634	0.834	0.2
18	0.425	0.628	0.203	0.425	0.828	0.403	0.716	1.905	1.189	0.716	0.916	0.2
19	0.384	0.715	0.332	0.384	0.915	0.532	0.720	——	——	0.720	0.920	0.2
20	0.384	0.646	0.263	0.384	0.846	0.463	0.720	——	——	0.720	0.920	0.2

Figure 6 shows the convergence characteristics curves of HBO and MHBO in the case of the original case of the grid, whiles Figure 7 deals with the other case. The penalty is shown in Figure 8. This is for HBO and MHBO in the original case of the grid, while in the other case, the penalty is shown in Figure 9.

Table 3. IEEE 8-bus test system's operation times of relay's pairs in case external grid switching on by MHBO.

Pair	Near-End						Far-End					
	DOCRs			D&DOCR			DOCRs			D&DOCR		
	T_p	T_b	CTI	T_p	T_{Z2B}	CTI	T_p	T_b	CTI	T_p	T_{Z2P}	CTI
1	0.373	0.574	0.201	0.373	0.774	0.401	0.757	1.860	1.103	0.757	0.957	0.2
2	0.556	0.759	0.203	0.556	0.957	0.402	0.718	3.693	2.975	0.718	0.918	0.2
3	0.556	0.758	0.203	0.556	0.956	0.400	0.718	1.574	0.857	0.718	0.918	0.2
4	0.512	0.718	0.206	0.512	0.918	0.406	0.654	0.971	0.317	0.654	0.854	0.2
5	0.453	0.654	0.201	0.453	0.854	0.401	0.622	1.192	0.570	0.622	0.822	0.2
6	0.421	0.622	0.201	0.421	0.822	0.401	0.680	1.415	0.735	0.680	0.880	0.2
7	0.416	0.680	0.264	0.416	0.880	0.464	0.574	1.577	1.003	0.574	0.774	0.2
8	0.416	0.742	0.325	0.416	0.942	0.525	0.574	——	——	0.574	0.774	0.2
9	0.479	0.680	0.201	0.479	0.880	0.401	0.756	——	——	0.756	0.956	0.2
10	0.479	0.743	0.264	0.479	0.942	0.463	0.756	——	——	0.756	0.956	0.2
11	0.418	0.758	0.340	0.418	0.956	0.538	0.575	——	——	0.575	0.775	0.2
12	0.418	0.658	0.240	0.418	0.858	0.440	0.575	1.960	1.385	0.575	0.775	0.2
13	0.389	0.595	0.206	0.389	0.795	0.406	0.658	1.500	0.841	0.658	0.858	0.2
14	0.437	0.642	0.205	0.437	0.842	0.405	0.595	1.761	1.167	0.595	0.795	0.2
15	0.456	0.663	0.207	0.456	0.863	0.407	0.642	0.855	0.213	0.642	0.842	0.2
16	0.539	0.743	0.204	0.539	0.942	0.403	0.663	2.973	2.310	0.663	0.863	0.2
17	0.539	0.742	0.202	0.539	0.942	0.402	0.663	1.747	1.084	0.663	0.863	0.2
18	0.366	0.576	0.210	0.366	0.775	0.409	0.742	1.353	0.610	0.742	0.942	0.2
19	0.429	0.759	0.329	0.429	0.957	0.528	0.742	——	——	0.742	0.942	0.2
20	0.429	0.658	0.229	0.429	0.858	0.429	0.742	——	——	0.742	0.942	0.2

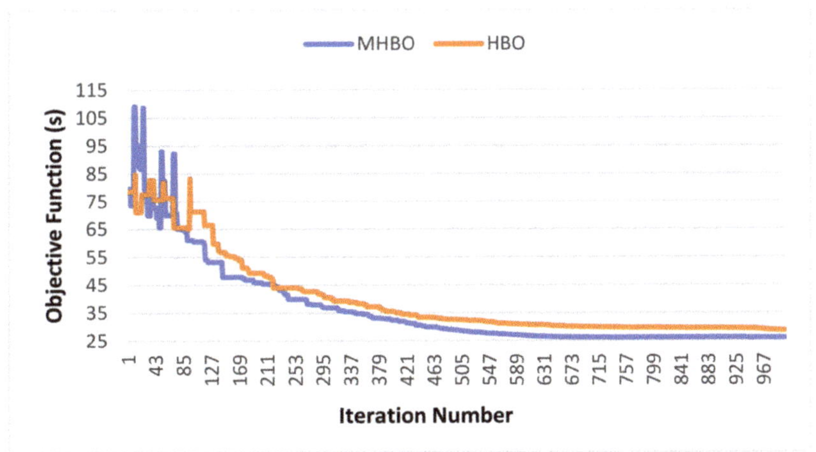

Figure 6. HBO and MHBO convergence characteristics in the traditional case of IEEE 8-bus.

Previous results proved the ability of the proposed APS to coordinate protection relays at IEEE 8-bus with reliability and suitable settings. In addition to avoiding miscoordination within limiters. APS has a more effective performance based on the MHBO than based on HBO. That is shown by the convergence characteristics. The convergence of MHBO is faster and better than the original HBO. In addition, the modified algorithm avoids constraints faster, as presented by the penalty meter. That proved the ability of the modified algorithm to increase its exploitation and exploration performances.

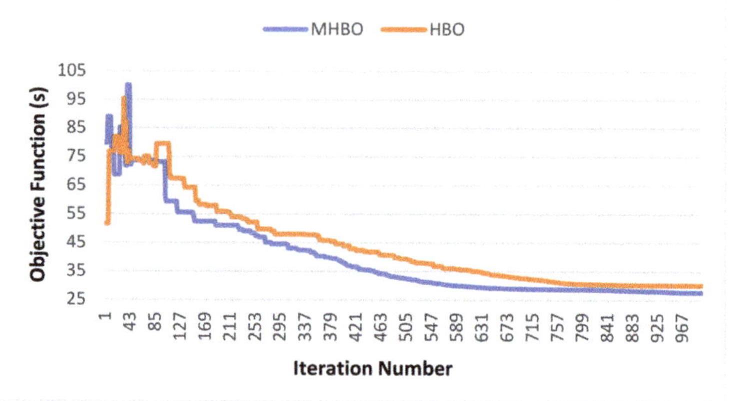

Figure 7. HBO and MHBO convergence characteristics in case the external grid switching on of IEEE 8-bus.

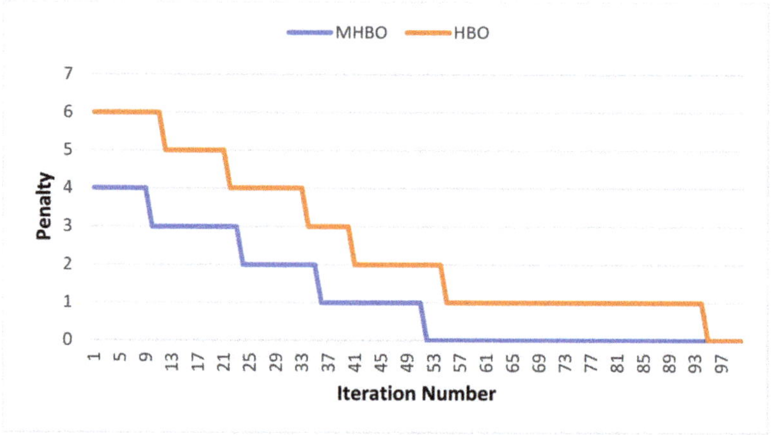

Figure 8. HBO and MHBO's penalty of the traditional IEEE 8-bus test system's grid.

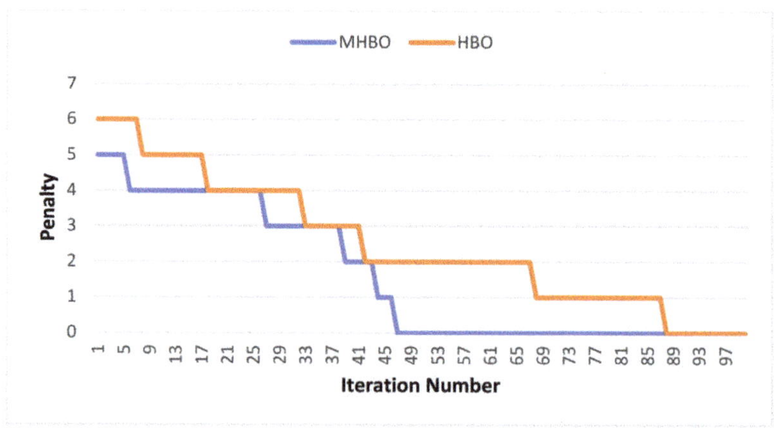

Figure 9. HBO and MHBO's Penalty of the IEEE 8-bus test system with external grid switching on.

4.2. Test System II: IEEE 14-Bus Distribution Network

The IEEE 14-bus distribution network is a downstream section of the IEEE 14-bus test system, as shown in Figure 10 [32]. This distribution network has 16 CBs to save its transmission lines, adding it is developed by adding 2 DGs, which are connected at the fifth bus and seventh bus. These DGs are synchronous generators with 5 MVA power rated and power factor of 0.9 lagging. That developed network is discussed in [33].

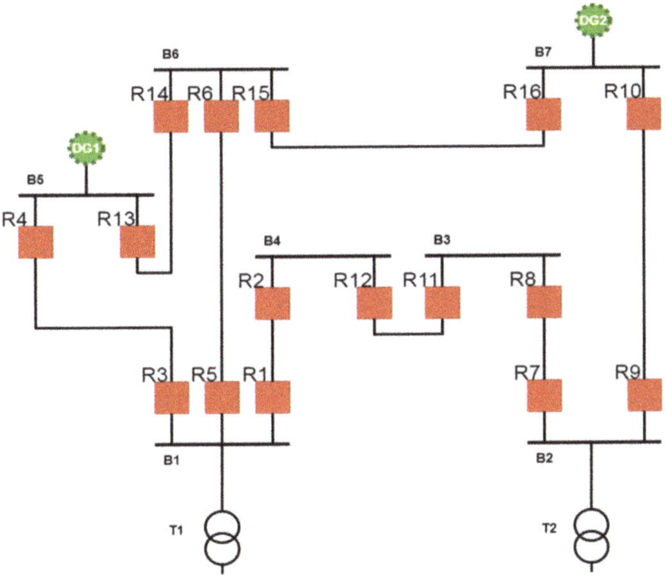

Figure 10. The single line diagram of 14-bus distribution network.

Three phases short circuit values and CT values are shown as in [28,32].

In this distribution network, protective relays have 48 variables design, which is required to be tuned by APS in both cases. The normal grid topology and the 2 DGs are switched on. Those variables design limited the minimum and maximum limiters. In addition to that, the coordination problem is constrained by 41 between DOCRs and 44 between DOCRs, and Distance relays. These constraints formed in the near end and the far end.

Optimal values tuned using HBO and MHBO are tabulated in Table 4. They are in cases of the original topology of the grid and after the DGs are switched on. Table 5 shows that the APS passed the constraints between DOCRs and between DOCRs and Distance relays, respectively. The APS passed constraints between DOCRs, and between DOCRs, and distance relays, as shown in Table 6.

The convergence characteristics curves of HBO and MHBO are shown in Figures 11 and 12, which occurred in the traditional grid and after switching DGs on, respectively. The penalty of HBO and MHBO are shown in Figures 13 and 14 to present the penalty in the traditional grid and the other case, respectively.

As demonstrated through the results and performance of the proposed APS in the coordination process of protection relays at the IEEE 14-bus distribution network, the following can be stated: APS tuned settings of distance and DOCRs with suitable settings, the protection system has reliability, effectiveness, and fast performance. APS based on MHBO has better convergence characteristics and better solutions than APS based on HBO. MHBO has better convergence and needs less iteration to avoid miscoordination based on penalty than HBO. This proves that MHBO improved its exploitation and exploration performance.

Table 4. The IEEE 14-bus distribution network's relays setting.

| Relay | Traditional Topological | | | | | | DGs Switching ON | | | | | |
| | Original HBO | | | Modified HBO | | | Original HBO | | | Modified HBO | | |
	TDS	Ip	T_{Z2}	TDS	Ip	T_{Z2}	TDS	Ip	T_{Z2}	TDS	Ip	T_{Z2}
1	0.162	220.260	0.924	0.103	331.824	0.831	0.396	425.225	1.642	0.334	400.742	1.377
2	0.100	231.140	0.981	0.109	157.897	0.794	0.229	388.320	1.419	0.207	345.712	1.207
3	0.169	136.227	0.840	0.140	135.702	0.729	0.272	447.128	1.352	0.253	322.388	1.086
4	0.126	48.240	0.887	0.100	53.731	0.795	0.341	73.162	1.567	0.242	99.886	1.389
5	0.185	183.177	0.896	0.148	187.170	0.765	0.554	131.937	1.403	0.353	260.312	1.187
6	0.190	68.554	0.865	0.151	76.625	0.759	0.446	99.856	1.572	0.376	97.299	1.344
7	0.309	86.289	0.909	0.154	300.674	0.817	0.520	249.499	1.516	0.401	283.478	1.267
8	0.142	91.804	0.706	0.173	60.000	0.704	0.397	179.927	1.501	0.218	354.279	1.265
9	0.100	270.557	0.765	0.117	204.846	0.738	0.371	441.540	1.683	0.470	99.719	1.194
10	0.100	192.886	1.021	0.100	164.198	0.889	0.278	296.324	1.614	0.170	428.783	1.395
11	0.189	147.967	0.906	0.137	195.115	0.808	0.393	215.055	1.330	0.390	159.422	1.193
12	0.206	74.439	0.858	0.100	220.970	0.852	0.372	287.695	1.619	0.354	229.531	1.399
13	0.140	78.984	0.870	0.115	86.699	0.787	0.461	90.888	1.560	0.318	107.955	1.216
14	0.304	20.000	0.783	0.184	42.912	0.655	0.595	20.000	1.084	0.588	20.378	1.078
15	0.184	92.008	0.860	0.138	131.920	0.809	0.367	261.465	1.512	0.315	200.627	1.188
16	0.100	180.535	0.868	0.100	156.605	0.787	0.454	173.132	1.709	0.287	230.759	1.309
OF	31.709			27.809			60.429			49.483		

Table 5. IEEE 14-bus distribution network's operation times of Relay's pairs in traditional grid by MHBO.

| Pair | Near-End | | | | | | Far-End | | | | | |
| | DOCRs | | | D&DOCR | | | DOCRs | | | D&DOCR | | |
	T_p	T_b	CTI	T_p	T_{Z2B}	CTI	T_p	T_b	CTI	T_p	T_{Z2P}	CTI
1	0.359	0.595	0.236	0.359	0.795	0.436	0.631	1.481	0.850	0.631	0.831	0.2
2	0.359	0.559	0.200	0.359	0.759	0.400	0.631	1.037	0.406	0.631	0.831	0.2
3	0.407	0.608	0.200	0.407	0.808	0.400	0.594	0.906	0.312	0.594	0.794	0.2
4	0.315	0.594	0.279	0.315	0.794	0.479	0.529	1.524	0.995	0.529	0.729	0.2
5	0.315	0.559	0.245	0.315	0.759	0.445	0.529	______	______	0.529	0.729	0.2
6	0.255	0.455	0.200	0.255	0.655	0.400	0.595	1.874	1.280	0.595	0.795	0.2
7	0.394	0.594	0.200	0.394	0.794	0.400	0.565	0.996	0.431	0.565	0.765	0.2
8	0.394	0.595	0.201	0.394	0.795	0.401	0.565	______	______	0.565	0.765	0.2
9	0.387	0.587	0.200	0.387	0.787	0.400	0.559	0.886	0.886	0.559	0.759	0.2
10	0.387	0.587	0.200	0.387	0.787	0.400	0.559	0.773	0.214	0.559	0.759	0.2
11	0.484	0.689	0.205	0.484	0.889	0.405	0.617	1.287	0.670	0.617	0.817	0.2
12	0.451	0.652	0.201	0.451	0.852	0.401	0.504	0.832	0.328	0.504	0.704	0.2
13	0.304	0.504	0.200	0.304	0.704	0.400	0.538	1.484	0.946	0.538	0.738	0.2
14	0.406	0.607	0.201	0.406	0.809	0.403	0.689	1.342	0.653	0.689	0.889	0.2
15	0.417	0.617	0.200	0.417	0.817	0.400	0.608	1.010	0.402	0.608	0.808	0.2
16	0.430	0.631	0.201	0.430	0.831	0.401	0.652	1.119	0.466	0.652	0.852	0.2
17	0.329	0.529	0.200	0.329	0.729	0.400	0.587	1.580	0.993	0.587	0.787	0.2
18	0.328	0.565	0.238	0.328	0.765	0.438	0.455	3.532	3.077	0.455	0.655	0.2
19	0.328	0.587	0.259	0.328	0.787	0.459	0.455	1.169	0.714	0.455	0.655	0.2
20	0.365	0.565	0.200	0.365	0.765	0.400	0.609	1.400	0.791	0.609	0.809	0.2
21	0.365	0.587	0.221	0.365	0.787	0.421	0.609	2.098	1.489	0.609	0.809	0.2
22	0.338	0.538	0.200	0.338	0.738	0.400	0.587	1.463	0.876	0.587	0.787	0.2

Table 6. The IEEE 14-bus distribution network's operation times of relays pairs with DGs switching on by MHBO.

Pair	Near-End						Far-End					
	DOCRs			D&DOCR			DOCRs			D&DOCR		
	T_p	T_b	CTI	T_p	T_{Z2B}	CTI	T_p	T_b	CTI	T_p	T_{Z2P}	CTI
1	0.906	1.195	0.289	0.906	1.389	0.483	1.177	3.066	1.889	1.177	1.377	0.2
2	0.906	1.142	0.236	0.906	1.344	0.439	1.177	3.185	2.008	1.177	1.377	0.2
3	0.747	0.993	0.246	0.747	1.193	0.446	1.007	1.219	0.211	1.007	1.207	0.2
4	0.588	1.010	0.422	0.588	1.207	0.619	0.886	1.657	0.771	0.886	1.086	0.2
5	0.588	1.142	0.554	0.588	1.344	0.756	0.886	———	———	0.886	1.086	0.2
6	0.536	0.880	0.344	0.536	1.078	0.542	1.189	3.459	2.271	1.189	1.389	0.2
7	0.788	1.010	0.223	0.788	1.207	0.420	0.987	1.655	0.668	0.987	1.187	0.2
8	0.788	1.195	0.407	0.788	1.389	0.601	0.987	———	———	0.987	1.187	0.2
9	0.808	1.016	0.209	0.808	1.216	0.409	1.144	———	———	1.144	1.344	0.2
10	0.808	1.106	0.299	0.808	1.309	0.501	1.144	1.354	0.209	1.144	1.344	0.2
11	0.967	1.196	0.229	0.967	1.395	0.428	1.067	2.233	1.166	1.067	1.267	0.2
12	0.917	1.172	0.255	0.917	1.399	0.482	1.065	1.318	0.253	1.065	1.265	0.2
13	0.805	1.057	0.252	0.805	1.265	0.460	0.994	14.694	13.700	0.994	1.194	0.2
14	0.728	0.988	0.260	0.728	1.188	0.460	1.195	1.498	0.303	1.195	1.395	0.2
15	0.842	1.067	0.225	0.842	1.267	0.425	0.993	1.308	0.316	0.993	1.193	0.2
16	0.966	1.177	0.211	0.966	1.377	0.411	1.199	1.567	0.368	1.199	1.399	0.2
17	0.672	0.878	0.206	0.672	1.086	0.414	1.016	2.726	1.710	1.016	1.216	0.2
18	0.726	0.987	0.261	0.726	1.187	0.461	0.878	1.962	1.085	0.878	1.078	0.2
19	0.726	1.106	0.380	0.726	1.309	0.583	0.878	1.458	0.580	0.878	1.078	0.2
20	0.721	0.987	0.266	0.721	1.187	0.466	0.988	1.483	0.495	0.988	1.188	0.2
21	0.721	1.016	0.295	0.721	1.216	0.495	0.988	1.554	0.566	0.988	1.188	0.2
22	0.781	0.994	0.213	0.781	1.194	0.413	1.109	1.335	0.226	1.109	1.309	0.2

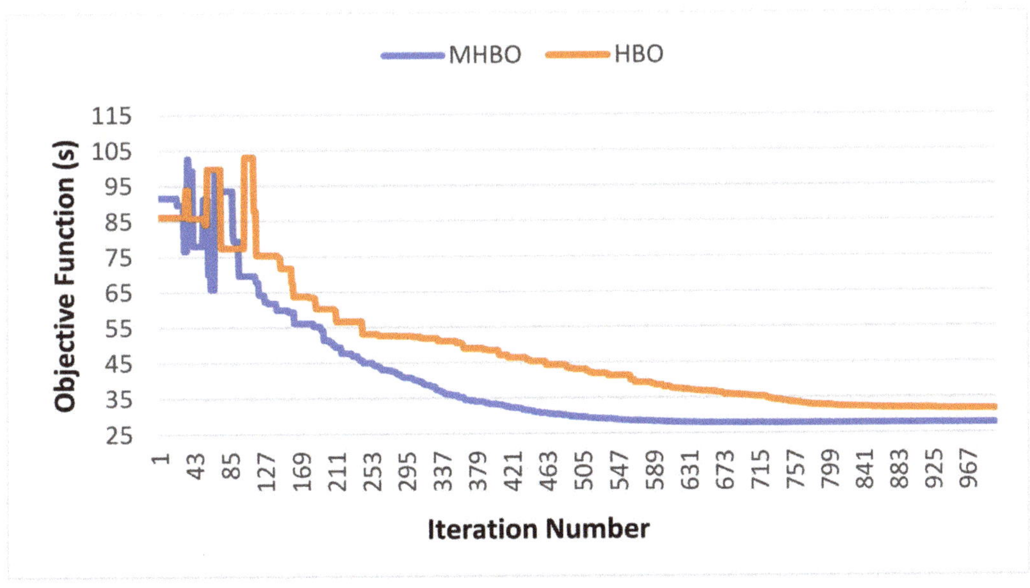

Figure 11. HBO and MHBO's convergence characteristics in traditional grid of IEEE 14-bus distribution network.

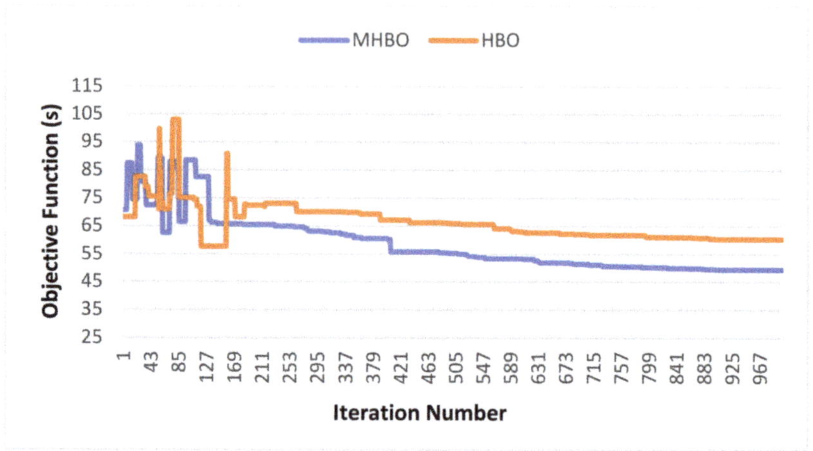

Figure 12. HBO and MHBO's convergence characteristics with DGs switching on case of IEEE 14-bus distribution network.

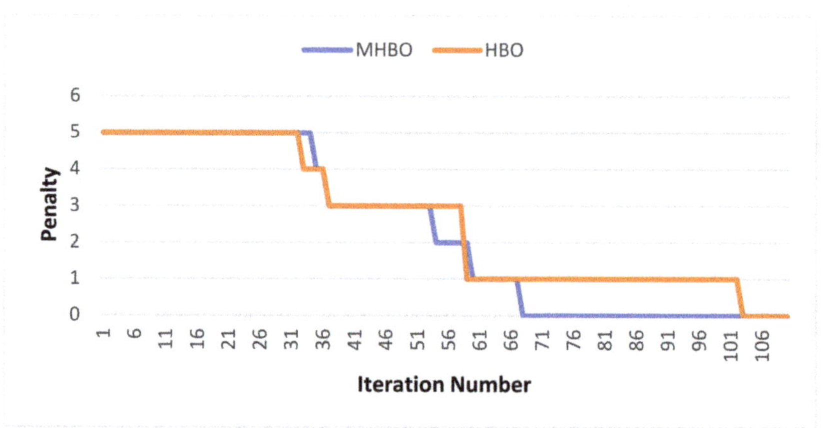

Figure 13. HBO and MHBO's Penalty of IEEE 14-bus distribution network's traditional grid.

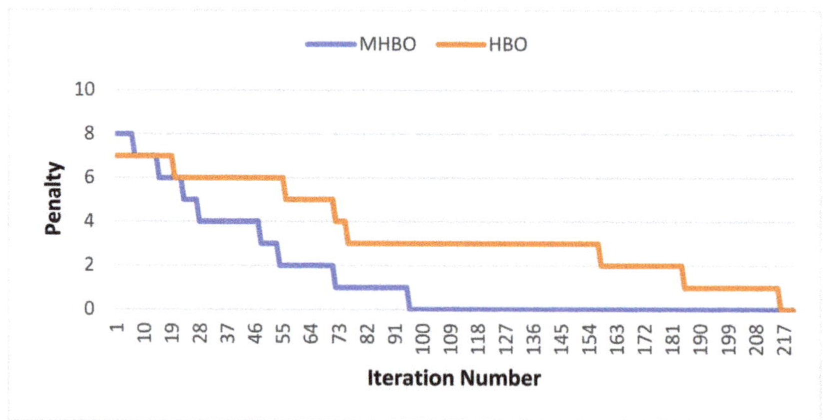

Figure 14. HBO and MHBO's penalty of IEEE 14-bus distribution network with DGs switching on.

4.3. Verification of MHBO Using ETAP 12.6.0

Results tuned by the MHBO algorithm are verified using the ETAP. All cases developed based on three-phase faults happened in both the near end and the far end of the following transmission lines:

The first case is at the transmission line between the 3rd and 4th bus-bars. As shown in Figure 15, ETAP's simulation is shown at the near end, operation times of pair relays 3 and 2 operate at 0.662 s, and 0.462 s, respectively, while at the far end, operating times are 0.875 s and 0.578, respectively.

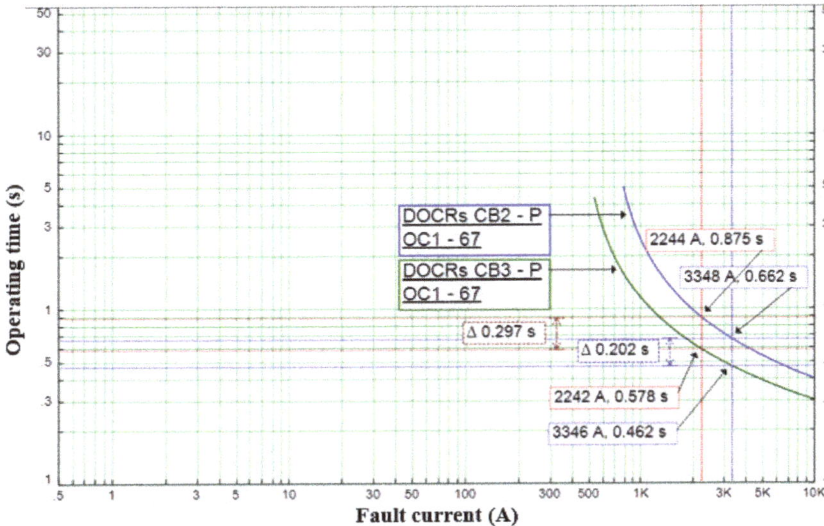

Figure 15. ETAP's simulation test of pair relays (3,2).

ETAP's simulation is also conducted, however, in this case, at the transmission line between 5th and 6th bus-bars. As shown in Figure 16, in this case, the operations time of pair relays 5 and 4 at the near end are 0.295 s and 0.496 s, respectively, while at the far end are 0.611 s and 1.98 s, respectively.

For the last case, the simulation is as shown in Figure 17, which is conducted at the transmission line between the 1st and 3rd bus-bars. That figure shows the operation times of pair relays 9 and 10 at the near end, which is 0.295 s and 0.494 s, respectively, whereas at the far end, they are 2.32 s and 0.611 s, respectively.

Simulations confirm that APS based on MHBO has the ability to coordinate protection relays without miscoordination between DOCRs since all *CTI* is equal or more than 0.2 s, and the operation times are within limits.

4.4. Statistical Results

Table 7 provides the statistical analysis of the proposed algorithms for HBO and MHBO. Each algorithm has a maximum value (Max), minimum value (Min), Mean of runs, and standard deviation (STD) of runs. The number of runs was 15 for each algorithm at each test case. These results proved that MHBO had better statistic parameters than HBO. Moreover, STD shows MHBO was more stable than HBO in all test cases. Therefore, MHBO has the ability to keep stable performance with more complex distribution networks. However, HBO has more variance with more complex distribution networks.

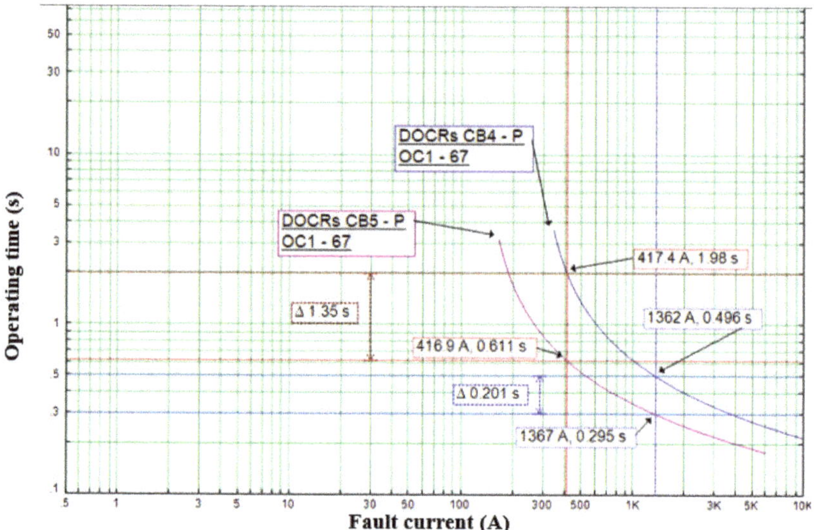

Figure 16. ETAP's simulation test of pair relays (5,4).

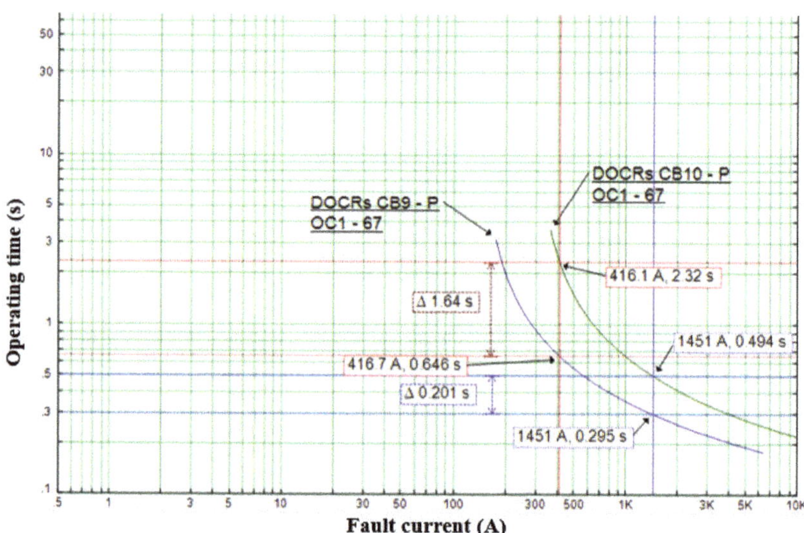

Figure 17. ETAP's simulation test of pair relays (9,10).

Table 7. Statistical parameters for proposed algorithms at all test cases.

| Statistic | IEEE 8-Bus Test System | | | | IEEE 14-Bus Distribution Network | | | |
| | Normal | | DG | | Normal | | DG | |
	MHBO	HBO	MHBO	HBO	MHBO	HBO	MHBO	HBO
Max	27.776	29.804	29.329	32.713	29.946	33.409	52.313	88.232
Min	25.983	28.608	27.798	30.191	27.809	31.708	49.483	60.429
Mean	26.894	29.142	28.483	31.183	28.996	32.367	50.715	69.651
STD	0.495	0.497	0.375	0.644	0.506	0.509	0.766	8.220

4.5. Comparison Study

In this paper, APS is designed to coordinate DOCRs and distance relays. That novel issue was discussed in [1] and presented APS based on the SBO algorithm and its modification. Those techniques are recent techniques and have effective performances to design APS. Table 8 presents the APS's results based on different optimization algorithms.

Table 8. Comparison between APS's based on optimization techniques at all test cases.

Optimization Technique	IEEE 8-Bus Test System		IEEE 14-Bus Distribution Network	
	Normal	DG	Normal	DG
MHBO	25.983	27.798	27.809	49.483
HBO	28.608	30.191	31.708	60.429
MSBO	28.072	32.601	34.806	51.068
SBO	33.705	35.388	36.86	57.268

From this table, MHBO has the best results in all test cases. At the same time, HBO has better solutions than MSBO and SBO in two cases. These cases are the IEEE 8-bus test system with EG switched ON case, and the normal IEEE 14-bus distribution network case and the other test cases HBO does not present an impact to the design of APS.

4.6. Applying APS in Real Power System

APS requires many hardware components to be applied in the real power system [34]. Those hardware components are the following:

- Protection relays include microprocessors.
- A central computer system to collect data from sensors, estimate DGs states, and apply algorithm to tune relay's setting.
- A communication infrastructure to connect protection relays with the central computer system.

5. Conclusions

In this research work, APS is proposed based on MHBO. The developed algorithm succeeded to overcome challenges in the area of coordination problem between protection relays. The simulated results show that APS has the ability to coordinate DOCRs and distance relays with suitable settings to solve the problem of distribution networks equipped with DGs. APS allows the power system to investigate both distance and DOCRs benefits with increased reliability. The modified algorithm (MHBO) makes APS more effective in resolving the coordination, as it has better convergence characteristics curves and optimal values than other previously suggested algorithms. The proposed algorithm reduces the relays time settings below the maximum operation times and within allowed limits. Finally, primary and backup relays are set without miscoordination at any end.

MHBO is an effective optimization algorithm but limited with the experience of users to identify its special parameters such as M factor, t_{united}, and $N_{subgroup}$. They are depended on the optimization problem.

For future works, those parameters will be used to modify other optimization algorithms. In Addition, MHBO will be used with other optimization problems. Moreover, we will try to design APS with better performance and characteristics to deal with the impact of DG, and tested in real large-scales networks.

Author Contributions: Conceptualization, S.K., E.M.A. and E.B.A.; methodology, M.A., S.K., E.M.A.; software, M.A. and E.B.A.; validation, M.A., S.K. and E.M.A.; formal analysis, M.A. and E.B.A.; investigation, S.K. and E.M.A.; resources, M.A. and S.K.; data curation, S.K. and E.M.A.; writing—original draft preparation, M.A., S.K., E.M.A. and E.B.A.; writing—review and editing, S.K. and E.M.A.; visualization, S.K. and E.B.A.; supervision, S.K. and E.M.A.; project administration, S.K.;

funding acquisition, S.K. and E.M.A. All authors have read and agreed to the published version of the manuscript.

Funding: This work was funded by the Deanship of Scientific Research at Jouf University under grant No (DSR-2021-02-0306).

Conflicts of Interest: The authors declare no conflict of interest.

Abbreviations

Acronym	Name
APS	adaptive protection scheme
HBO	heap- based optimization
MHBO	modified heap- based optimization
DOCRs	directional overcurrent relays
DGs	distributed generators
OCRs	Overcurent relays
ICT	information and communication technologies
PSO	particle swarm optimization
GA	genetic algorithm
DEA	differential evolution algorithm
ACO	ant colony optimization
FA	firefly algorithm
GSA	gravitational search algorithm
MRFO	manta ray foraging optimization
HHO	hybrid Harris hawks optimization
SBO	school-based optimization algorithm
MSBO	modified school-based optimization algorithm
IEC	The international electro-technical commission
SCADA	supervisory control and data acquisition
IEDs	intelligent electronic devices
CRH	the corporate rank hierarchy
EG	External grid
Notation	**Name**
OF	The objective function
T_{Near}	total operation times of the DOCRs at near end
T_{Far}	total operation times of the DOCRs at far end
T_{Z2}	the second time zone of the distance relays
T_i	the relay's operation time of DOCRs for ith relay
TDS	the relay's time dial setting
Ip	the relay's pickup current
$\alpha, \beta,$ and γ	Constant values
T_{max}	The maximum operation time
t_b	the backup relay's time
t_p	the primary relay's time
CTI	the coordination time interval
F^{pen}	the penalty function
μ	the penalty function's weighting factor.
$x_i^k(t+1)$	the updated position
$x_i^k(t)$	the current position
B^k	the parent position
S_r^k	the colleague position
p	a random value with in [0,1]
$f\left(\vec{x}_i(t)\right)$	the fitness value of current position
$f\left(\vec{S}_r\right)$	the fitness value of the colleague position

$\gamma, \lambda, C, T, p_1, p_2,$ and p_3 — Are special parameters of HBO algorithm

$N_{subgroup}$ — The number of subgroups

t_{united} — the number of iterations that are necessary to begin the unity

References

1. Abdelhamid, M.; Kamel, S.; Korashy, A.; Tostado-Véliz, M.; Banakhr, F.A.; Mosaad, M.I. An Adaptive Protection Scheme for Coordination of Distance and Directional Overcurrent Relays in Distribution Systems Based on a Modified School-Based Optimizer. *Electronics* **2021**, *10*, 2628. [CrossRef]
2. Abdelhamid, M.; Kamel, S.; Selim, A.; AMohamed, M.; Amed, M.; Elsayed, S. Development Of Bonobo Algorithm And Its Application For Optimal Coordination Of Directional Overcurrent Relays in Power Systems. *DYNA* **2021**, *96*, 492–497. [CrossRef]
3. Damchi, Y.; Sadeh, J.; Rajabi Mashhadi, H. Considering pilot protection in the optimal coordination of distance and directional overcurrent relays. *Iran. J. Electr. Electron. Eng.* **2015**, *11*, 154–164.
4. Khederzadeh, M. Back-up protection of distance relay second zone by directional overcurrent relays with combined curves. In Proceedings of the 2006 IEEE Power Engineering Society General Meeting, Montreal, QC, Canada, 18–22 June 2006. [CrossRef]
5. Perez, L.G.; Urdaneta, A.J. Optimal computation of distance relays second zone timing in a mixed protection scheme with directional overcurrent relays. *IEEE Trans. Power Deliv.* **2001**, *16*, 385–388. [CrossRef]
6. Sarwagya, K.; Nayak, P.K.; Ranjan, S. Optimal coordination of directional overcurrent relays in complex distribution networks using sine cosine algorithm. *Electr. Power Syst. Res.* **2020**, *187*, 106435. [CrossRef]
7. Akdag, O.; Yeroglu, C. Optimal directional overcurrent relay coordination using MRFO algorithm: A case study of adaptive protection of the distribution network of the Hatay province of Turkey. *Electr. Power Syst. Res.* **2021**, *192*, 106998. [CrossRef]
8. Ates, Y.; Uzunoglu, M.; Karakas, A.; Boynuegri, A.R.; Nadar, A.; Dag, B. Implementation of adaptive relay coordination in distribution systems including distributed generation. *J. Clean. Prod.* **2016**, *112*, 2697–2705. [CrossRef]
9. Vijayakumar, D.; Nema, R.K. A novel optimal setting for directional over current relay coordination using particle swarm optimization. *Int. J. Electr. Power Energy Syst. Eng.* **2008**, *1*, 1–6.
10. Singh, D.K.; Gupta, S. Optimal coordination of directional overcurrent relays: A genetic algorithm approach. In Proceedings of the 2012 IEEE Students' Conference on Electrical, Electronics and Computer Science, Bhopal, India, 1–2 March 2012; pp. 1–4.
11. Moirangthem, J.; Krishnanand, K.R.; Dash, S.S.; Ramaswami, R. Adaptive differential evolution algorithm for solving non-linear coordination problem of directional overcurrent relays. *IET Gener. Transm. Distrib.* **2013**, *7*, 329–336. [CrossRef]
12. Shih, M.Y.; Salazar, C.A.C.; Enríquez, A.C. Adaptive directional overcurrent relay coordination using ant colony optimisation. *IET Gener. Transm. Distrib.* **2015**, *9*, 2040–2049. [CrossRef]
13. Tjahjono, A.; Anggriawan, D.O.; Faizin, A.K.; Priyadi, A.; Pujiantara, M.; Taufik, T.; Purnomo, M.H. Adaptive modified firefly algorithm for optimal coordination of overcurrent relays. *IET Gener. Transm. Distrib.* **2017**, *11*, 2575–2585. [CrossRef]
14. Chawla, A.; Bhalja, B.R.; Panigrahi, B.K.; Singh, M. Gravitational search based algorithm for optimal coordination of directional overcurrent relays using user defined characteristic. *Electr. Power Compon. Syst.* **2018**, *46*, 43–55. [CrossRef]
15. ElSayed, S.K.; Elattar, E.E. Hybrid Harris hawks optimization with sequential quadratic programming for optimal coordination of directional overcurrent relays incorporating distributed generation. *Alex. Eng. J.* **2021**, *60*, 2421–2433. [CrossRef]
16. Degertekin, S.O.; Tutar, H.; Lamberti, L. School-based optimization for performance-based optimum seismic design of steel frames. *Eng. Comput.* **2021**, *37*, 3283–3297. [CrossRef]
17. Askari, Q.; Saeed, M.; Younas, I. Heap-based optimizer inspired by corporate rank hierarchy for global optimization. *Expert Syst. Appl.* **2020**, *161*, 113702. [CrossRef]
18. Rizk-Allah, R.M.; El-Fergany, A.A. Emended heap-based optimizer for characterizing performance of industrial solar generating units using triple-diode model. *Energy* **2021**, *237*, 121561. [CrossRef]
19. Elsayed, S.K.; Kamel, S.; Selim, A.; Ahmed, M. An Improved Heap-Based Optimizer for Optimal Reactive Power Dispatch. *IEEE Access* **2021**, *9*, 58319–58336. [CrossRef]
20. Kharrich, M.; Kamel, S.; Hassan, M.H.; ElSayed, S.K.; Taha, I.B.M. An Improved Heap-Based Optimizer for Optimal Design of a Hybrid Microgrid Considering Reliability and Availability Constraints. *Sustainability* **2021**, *13*, 10419. [CrossRef]
21. Abdel-Basset, M.; Mohamed, R.; Elhoseny, M.; Chakrabortty, R.K.; Ryan, M.J. An efficient heap-based optimization algorithm for parameters identification of proton exchange membrane fuel cells model: Analysis and case studies. *Int. J. Hydrog. Energy* **2021**, *46*, 11908–11925. [CrossRef]
22. Cui, Q.; Weng, Y. An environment-adaptive protection scheme with long-term reward for distribution networks. *Int. J. Electr. Power Energy Syst.* **2021**, *124*, 106350. [CrossRef]
23. Dos Reis, F.B.; Pinto, J.O.C.P.; dos Reis, F.S.; Issicaba, D.; Rolim, J.G. Multi-agent dual strategy based adaptive protection for microgrids. *Sustain. Energy Grids Netw.* **2021**, *27*, 100501. [CrossRef]
24. Sampaio, F.C.; Leão, R.P.S.; Sampaio, R.F.; Melo, L.S.; Barroso, G.C. A multi-agent-based integrated self-healing and adaptive protection system for power distribution systems with distributed generation. *Electr. Power Syst. Res.* **2020**, *188*, 106525. [CrossRef]
25. Abdelhamid, M.; Kamel, S.; Mohamed, M.A.; Aljohani, M.; Rahmann, C.; Mosaad, M.I. Political Optimization Algorithm for Optimal Coordination of Directional Overcurrent Relays. In Proceedings of the 2020 IEEE Electric Power and Energy Conference (EPEC), Edmonton, AB, Canada, 9–10 November 2020; pp. 1–7.

26. Abdelhamid, M.; Kamel, S.; Mohamed, M.A.; Rahmann, C. An Effective Approach for Optimal Coordination of Directional Overcurrent Relays Based on Artificial Ecosystem Optimizer. In Proceedings of the 2021 IEEE International Conference on Automation/XXIV Congress of the Chilean Association of Automatic Control (ICA-ACCA), Valparaíso, Chile, 22–26 March 2021; pp. 1–6.

27. Rivas, A.E.L.; Pareja, L.A.G.; Abrão, T. Coordination of distance and directional overcurrent relays using an extended continuous domain ACO algorithm and an hybrid ACO algorithm. *Electr. Power Syst. Res.* **2019**, *170*, 259–272. [CrossRef]

28. Rajput, V.N.; Adelnia, F.; Pandya, K.S. Optimal coordination of directional overcurrent relays using improved mathematical formulation. *IET Gener. Transm. Distrib.* **2018**, *12*, 2086–2094. [CrossRef]

29. Mohammadi, R.; Abyaneh, H.A.; Rudsari, H.M.; Fathi, S.H.; Rastegar, H. Overcurrent relays coordination considering the priority of constraints. *IEEE Trans. Power Deliv.* **2011**, *26*, 1927–1938. [CrossRef]

30. Shih, M.Y.; Conde, A.; Ángeles-Camacho, C.; Fernández, E.; Leonowicz, Z.; Lezama, F.; Chan, J. A two stage fault current limiter and directional overcurrent relay optimization for adaptive protection resetting using differential evolution multi-objective algorithm in presence of distributed generation. *Electr. Power Syst. Res.* **2021**, *190*, 106844. [CrossRef]

31. Yang, M.-T.; Liu, A. Applying Hybrid PSO to Optimize Directional Overcurrent Relay Coordination in Variable Network Topologies. *J. Appl. Math.* **2013**, *2013*, 879078. [CrossRef]

32. Christie, R. Power systems test case archive. *Electr. Eng. Dept. Univ. Washingt.* 2000. Available online: https://www2.ee.washington.edu/research/pstca (accessed on 20 November 2021).

33. Saleh, K.A.; Zeineldin, H.H.; Al-Hinai, A.; El-Saadany, E.F. Optimal coordination of directional overcurrent relays using a new time–current–voltage characteristic. *IEEE Trans. Power Deliv.* **2014**, *30*, 537–544. [CrossRef]

34. Ates, Y.; Boynuegri, A.R.; Uzunoglu, M.; Nadar, A.; Yumurtacı, R.; Erdinc, O.; Paterakis, N.G.; Catalão, J.P.S. Adaptive Protection Scheme for a Distribution System Considering Grid-Connected and Islanded Modes of Operation. *Energies* **2016**, *9*, 378. [CrossRef]

Article

Multi-Objective Optimal Design of a Hydrogen Supply Chain Powered with Agro-Industrial Wastes from the Sugarcane Industry: A Mexican Case Study

Luis Miguel Reyes-Barquet [1], José Octavio Rico-Contreras [2], Catherine Azzaro-Pantel [3], Constantino Gerardo Moras-Sánchez [1], Magno Angel González-Huerta [1], Daniel Villanueva-Vásquez [4,*] and Alberto Alfonso Aguilar-Lasserre [1,*]

[1] Graduate Studies and Research Division, Tecnológico Nacional de México/Instituto Tecnológico de Orizaba, Calle Oriente 9 Colonia Emiliano Zapata, Orizaba 94320, Mexico; luism.reyesbarquet@gmail.com (L.M.R.-B.); t_moras@yahoo.com.mx (C.G.M.-S.); magnogh@yahoo.com.mx (M.A.G.-H.)

[2] Grupo Porres Corporativo, Km 355 Carretera Federal Fortín de las Flores, Cordoba 94540, Mexico; jrico@gporres.com.mx

[3] Laboratoire de Génie Chimique, Université de Toulouse, U.M.R. 5503 CNRS/INP/UPS, 4 allée Emile Monso, CEDEX 4, 31432 Toulouse, France; catherine.azzaropantel@toulouse-inp.fr

[4] Departamento de Investigación y Posgrado, Tecnológico Nacional de México/Instituto Tecnológico Superior de Misántla, Km 1.8 Carretera a Lomas de Cojolite, Misantla 93821, Mexico

* Correspondence: dany.villavas@gmail.com (D.V.-V.); albertoaal@hotmail.com (A.A.A.-L.); Tel.: +52-(272)-725-7056 (ext. 114) (A.A.A.-L.)

Citation: Reyes-Barquet, L.M.; Rico-Contreras, J.O.; Azzaro-Pantel, C.; Moras-Sánchez, C.G.; González-Huerta, M.A.; Villanueva-Vásquez, D.; Aguilar-Lasserre, A.A. Multi-Objective Optimal Design of a Hydrogen Supply Chain Powered with Agro-Industrial Wastes from the Sugarcane Industry: A Mexican Case Study. *Mathematics* **2022**, *10*, 437. https://doi.org/10.3390/math 10030437

Academic Editors: Antonin Ponsich, Mariona Vila Bonilla and Bruno Domenech

Received: 15 December 2021
Accepted: 27 January 2022
Published: 29 January 2022

Publisher's Note: MDPI stays neutral with regard to jurisdictional claims in published maps and institutional affiliations.

Abstract: This paper presents an optimization modeling approach to support strategic planning for designing hydrogen supply chain (HSC) networks. The energy source for hydrogen production is proposed to be electricity generated at Mexican sugar factories. This study considers the utilization of existing infrastructure in strategic areas of the country, which brings several advantages in terms of possible solutions. This study aims to evaluate the economic and environmental implications of using biomass wastes for energy generation, and its integration to the national energy grid, where the problem is addressed as a mixed-integer linear program (MILP), adopting maximization of annual profit, and minimization of greenhouse gas emissions as optimization criteria. Input data is provided by sugar companies and the national transport and energy information platform, and were represented by probability distributions to consider variability in key parameters. Independent solutions show similarities in terms of resource utilization, while also significant differences regarding economic and environmental indicators. Multi-objective optimization was performed by a genetic algorithm (GA). The optimal HSC network configuration is selected using a multi-criteria decision technique, i.e., TOPSIS. An uncertainty analysis is performed, and main economic indicators are estimated by investment assessment. Main results show the trade-off interactions between the HSC elements and optimization criteria. The average internal rate of return (IRR) is estimated to be 21.5% and average payback period is 5.02 years.

Keywords: sugarcane bagasse; hydrogen energy; electrolysis; MILP; multi-criteria optimization; genetic algorithm; uncertainty; Monte Carlo simulation; TOPSIS

---

1. Introduction

In recent years, the popularity of hydrogen as a promising sustainable energy carrier has increased significantly to contribute to clean energy transition [1]. In particular, hydrogen has a noticeable role to play in the transport sector which requires large amounts of clean energy as an enabler of deep decarbonization of this difficult to abate sector. One of the advantages of using hydrogen is the availability of different production processes [2]. The biomass contained in some agro-industrial wastes can provide enough energy to be used for hydrogen production in a variety of processes [3]. Several paths can be followed in biomass resource exploitation, among which the selection of the most appropriate conversion technology is challenging. Agro-industrial wastes are commonly known as residues that offer little benefit to their producers, so their recovery can be an option to investigate. The use of agro-industrial waste for energy production can be an alternative end-of-life for these resources by creating sustainable and renewable systems that minimize pollutant emissions. The cogeneration of electricity and thermal power could provide energy autonomy for these companies and additional income from the sale of their energy overflows, while their waste gets a second use. Applying the necessary technologies for efficient use of the energy generated from renewable resources requires a comprehensive vision that includes the assessment of several factors for decision support at different levels [4]. The objective of this work is thus to include these options in the planning and design of a hydrogen supply chain network.

The electrical energy used for hydrogen production is generated with agro-industrial wastes in 50 sugar factories located in Mexico, where steam generators are powered by burning sugar cane bagasse. The electricity generated is used for self-consumption for the sugar companies and the excess is often sold to the national grid, but is commonly wasted because of low demand; thus, an HSC network where the excess energy can be exploited may turn out to be convenient. In the proposed model, the behavior of the electricity production systems is modeled using probability distributions, among other model parameters. The major contribution of this study is the integration of a multi-objective optimization model using a genetic algorithm (GA) with a hydrogen production system generated from agro-industrial waste for mobility purposes, integrating the proposed network with already existing infrastructure from the national energy industry. The model is evaluated with energy prices and geographic information from different regions across the country. (GA). The obtained solutions offer a variety of options for setting the HSC since a multi-criteria approach is adopted to optimize economic and environmental objectives simultaneously.

The presented mathematical model is inspired on three previously built formulations, whereby Parker [5] adopts the profit maximization approach for its flexibility in terms of resource utilization, and de León Almaráz [4] considers global warming criteria, and its mathematical formulation for transport and storage in an HSC is adopted by this study. Finally, Rico Contreras [6] presents the mathematical model for generation of electricity at sugar mills available for hydrogen production; integrating these approaches contributes to the formulation of the mixed integer linear program (MILP), and significant changes were made to adapt the mathematical formulation to the case examined in this study.

The optimal HSC network configuration is selected using a multi-criteria decision-making technique (MCDM). Due to the type of problem (input data), a multi-attribute decision-making (MADM) method is adopted. This type of technique calculates the distance between each alternative and a central point. VIKOR and TOPSIS methods were considered (differing by criteria normalization procedure). Both techniques use the CP method that seeks to obtain the closest alternative from the hypothetical optimal solution. The TOPSIS method was selected since it considers the distance to the ideal solution and the distance to the non-ideal solution, while VIKOR only considers the distance to the ideal solution.

2. Literature Review

The literature review identifies the tools, technologies, resources, and other important factors to consider when designing the hydrogen supply chain (HSC) for mobility purposes. The reviewed works were selected based on similar studies with MILP models and the main scientific objective regarding the design of HSC networks. A variety of case studies were analyzed to determine the most appropriate research path given the actual conditions of the field of study. The classification of the relevant studies is based on the objective functions, agro-industrial waste, raw materials, production technologies (alkaline/ Proton Exchange Membrane (PEM) electrolysis) and the region where the methodology is implemented.

A review of the different decision levels for HSC is presented in Azzaro C. et al. [7] on the different components related to hydrogen production, transportation, and distribution.

More than 40 authors contribute to a compilation of multiple case studies, where the most recent methodologies used for modeling the HSC supply chain are presented for design, planning and operation strategies, providing diverse tools that allow the design of complex systems using mathematical models involving economic, environmental and risk criteria.

Jiyong K. et al. [8] proposed a methodology for HSC infrastructure design including production, storage, and transportation with a generic optimization-based model. The network design is formulated as a MILP to identify the optimal configuration of the supply chain from various alternatives. The goal was to consider not only cost efficiency, but safety criteria as well. Since these two aspects are contradictory, multi-objective optimization techniques were required to find practical solutions. With this approach, the effects of uncertainty in demand can also be analyzed, and deterministic and stochastic analysis methods were compared.

The pioneering work presented in A. Almansoori and N. Shah [9] emphasizes the challenges of HSC design focused on three main factors: the presence of various links in the supply chain (including local hydrogen distribution and refueling stations), the high level of interaction between the components of the supply chain and their subsystems, and the uncertainty in hydrogen demand. In this work, the growing uncertainty in the variation of hydrogen demand in the long term was integrated into an existing generic optimization model, using a scenario-based approach. For both cases, the most feasible solution involves a centralized production with small or medium-sized storage facilities and distribution through tanker trucks. The performance of the model was evaluated using sensitivity and risk analysis.

In their latest work, Güler MG et al., 2020 [10] presented a design for an HSC in Turkey for 2021–2050 using a MILP modeling approach. A mathematical optimization model was adopted to evaluate the objective functions in Turkey. The results show decentralized production as one feasible alternative to fulfill the demand, and the local production rate exhibited a significant increase from 12% to 48% by the end of the planning horizon, revealing future considerations that must be considered. The analysis revealed that almost all regions either produce or import hydrogen, but do not do both.

The work by P. Gabrielli et al., 2020 [11] concerns the optimal design of a low-carbon Swiss HSC. The infrastructure design is performed by solving an optimization problem that determines the hydrogen, biomass, and CO_2 network configuration with a focus on production technologies. A national scale case study was analyzed to derive specific guidelines concerning the design of the HSC deploying carbon capture and storage. The impact of relevant design parameters was assessed, such as the location of CO_2 storage facilities, the techno-economic characteristics of CO_2 capture technologies and network losses. The study highlights the benefits of biomass and carbon capture and storage for decarbonizing HSC networks compared to the use of electrolysis for hydrogen production due to the high carbon intensity of the electricity mix.

C. Quarton and S. Samsatli, 2020 [12] present an optimization framework to determine how carbon dioxide and hydrogen technologies could fit into existing value chains in the energy and chemicals sector, analyzing how effectively these technologies can contribute to meet the climate change goals. The first study concerning the modeling and optimization of an integrated value chain for carbon dioxide and hydrogen is performed, providing assessment of the role of carbon capture, utilization and storage (CCUS), and hydrogen technologies. The results showed opportunities for CCUS to decarbonize existing power generation capacity and emphasize the need of renewable energy and hydrogen to achieve lower cost decarbonization and flexibility in the long term. The importance of negative emissions policies to encourage investors was also discussed.

An optimization-oriented review regarding HSC design is presented by Lei Li et al., 2019 [13]. Some drawbacks and missing aspects in the literature are identified, and key components of the HSC are presented. Models are classified based on several model features. It is highlighted that profit maximization has received less attention compared to other optimization criteria, and only two of the references reported profit as the HSC performance measure.

A social cost–benefit assessment is performed by Ochoa R. et al., 2020 as post-optimal analysis for HSC design and deployment [14]. The sequential application of an optimization strategy employing genetic algorithms and a multi-criteria decision-making tool at first determine the optimal solution for the HSC network design problem. The evaluation is then performed by a social cost–benefit analysis (SCBA) to estimate the impact of hydrogen mobility deployment on social welfare. A subsidy policy scenario was implemented where results showed that CO_2 abatement dominates the externalities, while platinum was the second largest externality.

Husna I. et al., 2016 [15] present a comparative study between biomass burning and gasification techniques. It is highlighted that direct burning of biomass and co-firing with coal is most used since it is the most economic convenient decision for the biomass power plant, while little plant modifications are required. On the gasification of biomass field, some points are made highlighting the benefits of chemical recovery to produce higher process steam and electricity efficiencies, reducing capital cost compared to conventional technologies.

Loong Lam H. et al., 2013 [16] proposed a methodological framework for designing waste-to-energy supply chains that considers efficient resources management and reduction of greenhouse gas emissions. A two-stage optimization model was developed, with MILP being used in both stages. Different technologies were considered for the whole exploitation of the resources in alternative forms. It was concluded that the green strategy adopted contributes significantly to the amount of power generated in existing power plants. Further studies concerning the integration of the available infrastructure and alternative energy technologies are required to determine opportunities for a more efficient resource exploitation.

The study by Gumte K. et al., 2021 [17], presents a nationwide analysis of a supply chain network fed with bioenergy; the study looks forward to integrating a fraction of the obtained biofuels with traditional fuels during the 2018–2026 horizon. A MILP is built to handle multiple types of raw materials, products and transport alternatives, while performing the techno, economic and environmental analysis, looking forward to making optimal operational and design decisions. The main findings remark that 43% and above biomass feed is needed for the supply chain network to survive.

Goodarzian F. et al., 2021 [18] propose the design of a three-echelon green medicine supply chain network through a fuzzy bi-objective MILP model, considering multiple periods, products, and transportation modes. The study measures the environmental impacts derived from establishing pharmacies and hospitals, aiming to reduce greenhouse gas emissions and to control environmental pollutants. Meta-heuristic algorithms are used to solve the model, including two novel hybrid algorithms known as Hybrid Firefly Algorithm and Simulated Annealing (HFFA-SA) and Hybrid Firefly Algorithm and Social Engineering Optimization (HFFA-SEO).

A bi-objective optimization model approach is proposed by Abdolazimi O. et al., 2020 [19], where a comparison of exact and meta-heuristic methods is performed. The main objective of this study is to improve the inventory grouping based on ABC analysis. The objective functions seek to maximize the total net profit of the items in the central stock, and in different locations. The aim is to simultaneously optimize the number of inventory groups, the number of items to be assigned and the service level. Statistical analysis besides the AHP and VIKOR techniques is implemented to compare the applied optimization techniques in terms of efficiency. To solve the model in different dimensions, two exact methods (LP-metric and ε-constraint) and two meta-heuristic methods (NSGA-II and MOPSO) are applied.

A systematic literature review on multi-criteria decision making methods applied in different areas of supply chain management is conducted by Paul A. et al., 2021 [20]. A total of 106 published journal articles were analyzed. It is highlighted that MCDM methods are commonly used for analyzing several factors of sustainable supply chain management. In this review, it is highlighted that most of the published articles combine only two MCDM methods, and integration with other techniques, such as simultaneous optimization and simulation, are missing in the literature.

A literature review presented by Tordecilla R. et al., 2021 [21], refers to existing literature on the use of simulation techniques in the formation of resilient supply chain networks (SCNs). Research opportunities have been identified for the inclusion of three criteria (such as financial, environmental, and social) during the process of marking and the application of a multidisciplinary approach to integrating metaheuristic algorithms, simulation, and machine learning methods to integrate uncertainty and dynamic conditions.

A multi-objective novel model was developed by Hosseini S. et al., 2020 [22]. The model deals with the design/reorganization of the wheat supply network, which includes different suppliers, existing warehouses, warehouse candidate locations, flour mills, and warehouses in an uncertain environment. The purpose of the proposed model is to reduce costs, non-resiliency, and the negative effects of social responsibility. The results show that considering the cost, durability, and social impact simultaneously can greatly help improve the performance of the wheat supply chain model.

The paper presented by Gital Y. et al., 2020 [23] discusses the appropriate design and planning of a biomass supply chain network that incorporates flows from poultry farms to biogas facilities. A multi-stage novel solution methodology is designed to solve the problem of designing a biomass supply chain network. Spatial information systems, as well as hierarchy processing techniques, are used to determine the candidate location of biogas infrastructure. The aim is to determine the total amount, location, and size of biogas facilities, alongside network flow, and the electricity generated. The sensitivity analysis shows both maximum distance parameters, and purchase prices have a significant impact on decisions, as well as financial benefit.

The aim of the research conducted by Rasi R. et al., 2021 [24] is to optimize economic and environmental dimensions in a sustainable supply chain (SSC) using a MILP model to incorporate both criteria simultaneously. According to the authors, the value of the work relies on the limited alternatives regarding the design and optimization of SSC networks. The research is among the first to integrate the selection of sustainable suppliers and the optimization of performance indicators. The differences between the genetic algorithms and the MILP methods can be explained by managing the issues and their various logic alternatives.

A review regarding the development of biomass-based cogeneration energy systems in Malasia is presented by Zailan R. et al., 2021 [25]. The aim of the analysis is to report recent improvements in co-firing technology using biomass in Malaysia with the optimization modeling role. The authors address technical issues concerning the key players of the technologies and the biomass supply chain, remarking the importance of biomass utilization for energy generation in regions where agro-industrial wastes are abundant.

The study presented by Nunes L. et al., 2020 [26] reviews the status of research on biomass supply chain modelling and highlights the growing importance of biomass as a renewable alternative energy source. The review identifies modeling as a critical step in improving comprehension leading to improved supply chain performance. It is said that research using supply chain models focuses on examining specific supply chain conditions, often with the aim of reducing costs.

Seung S. et al., 2020 [27] presented a study involving the development of a hydrogen supply chain optimization model using a centralized storage approach that integrates and combines the flow of different production facilities into integrated bulk storage. The results show that a hydrogen supply chain with a central storage approach improves the phase transition of the hydrogen-producing plants, while reducing the total annual cost of the network.

A techno-economic analysis review of biomass supply chain was conducted by Yuen S. et al., 2021 [28]. The study emphasizes the growing needs of biomass caused by the increased risk of climate change. The study aims to provide an overview of the different types of methods or techniques used to assess the feasibility of biomass-based industries from a technical point of view. The study also looks forward to describing the uncertainty of the supply chain that should be included in the model test using the Malaysian case study to show the impact of this uncertainty. In total, 78% of reviewed articles chose the method of testing the mathematical model with optimization. A minority have undergone stochastic tests that include systemic uncertainty.

Rafique R. et al., 2021 [29] introduces and develops a model to design a bioenergy supply chain with the aim of minimizing the energy gap under budget and the challenges of biomass availability. The dynamic features of the model capture interactions between people, size, energy demand, biomass availability, energy consumption and the overall domestic product. The analysis highlights that the cost of further development of the bioenergy system can vary greatly during the planning horizon. Complete configuration starts as a very central system and shifts to a decentralized system divided into areas where power plants emit biofuel and provide energy locally.

Li L. et al., 2019 [30] conducted a study focusing on developing a mathematical model that encompasses the entire hydrogen supply network. The model is integrated with a hydrogen fueling station planning approach to produce a new configuration. The proposed model looks at the supply of feedstock, installation and operation facilities, the operation of transportation modes, and a system for carbon capture and storage. The proposed model can study the interactions that exist between different parts of a hydrogen supply network. Therefore, many HSC building plans are guaranteed.

From the reviewed literature, it can be concluded that further research in terms of evaluating the economic and environmental benefits of utilizing alternative energy sources and technologies in the existing energy industry infrastructure might provide the sufficient arguments to determine whether it is convenient or not to look forward to the exploitation of agricultural wastes for these means in specific regions. A summary of the literature review is presented in Table 1. We classified the relevant studies based on the adopted objective function, feedstock types (energy sources), considered hydrogen production technologies, and analyzed case studies. This study assesses the economic and environmental behavior of a power-to-hydrogen supply chain through a stochastic modelling approach, where the existing energy and biomass infrastructure is integrated on a national scale. Electricity produced by biomass combustion is already available as an energy source across the country due to the large quantities of sugarcane bagasse generated annually by agro-industrial activities and the ready-to-use infrastructure located at biomass producer facilities for energy generation and self-consumption, although a considerable part of this energy may be wasted due to the lack of synchronization of supply and demand. The results can help provide alternatives for countries that rely heavily on primary and secondary activities where biomass is widely available and where national energy autonomy is a concern.

Table 1. Summary of the reviewed literature with a supply chain optimization approach.

Reference	Objective Function	Feedstock (Energy Source)	Hydrogen Production Technology	Case Study
[8]	Total cost minimization Total relative risk minimization	NG, renewable electricity	SMR, electrolysis	South Korea
[9]	Total cost minimization	NG, oil, coal, biomass, solar power	SMR, biomass and coal gasification, electrolysis	Great Britain
[10]	Total cost minimization	NG, coal, biomass, solar, wind, hydroelectric, geothermal	SMR, coal and biomass gasification, electrolysis	Turkey
[11]	Total cost minimization GWP minimization	NG, biomass, electricity	SMR, gasification, electrolysis	Swiss

Table 1. *Cont.*

Reference	Objective Function	Feedstock (Energy Source)	Hydrogen Production Technology	Case Study
[12]	NPV maximization Emissions minimization	NG, wind power	Electrolysis	Great Britain
[14]	Total Cost minimization GWP minimization	NG, renewable electricity, nuclear power	SMR, electrolysis	France (Midi-Pyrénées)
[15]	-	Coal, biomass	Electrolysis, gasification	Malaysia
[16]	NPV maximization Transport cost minimization	Biomass	-	Malaysia

NG = Natural Gas, SMR = Steam Methane Reforming, GWP = Global Warming Potential, NPV = Net Present Value.

The objective of this study is to evaluate the economic and environmental implications of using biomass wastes from sugar factories for energy generation, opening the scope to a non-conventional application according to the state of the art, which implies the utilization of already existing infrastructure, at the time that a resource commonly considered as waste is exploited. The innovation value of this contribution relies on the proposal of a wastes exploitation scheme that can be escalated in a variety of ranges, and can be applied to other energy sources, like biomass wastes originated from other agro-industrial sectors.

3. Materials and Methods

3.1. Methodological Framework

The methodological framework applied in this study is presented in three general frames; the first one concerns the input data used in the modelThe second aspect refers to the tools used to find the optimal solution for the proposed model, which implies the mathematical formulation, solving methods and solution selection technique. The last segment shows the outputs obtained from the applied methodology and its representation form, which implies a pareto front and graphic representations of the optimal supply chain configuration (Figure 1).

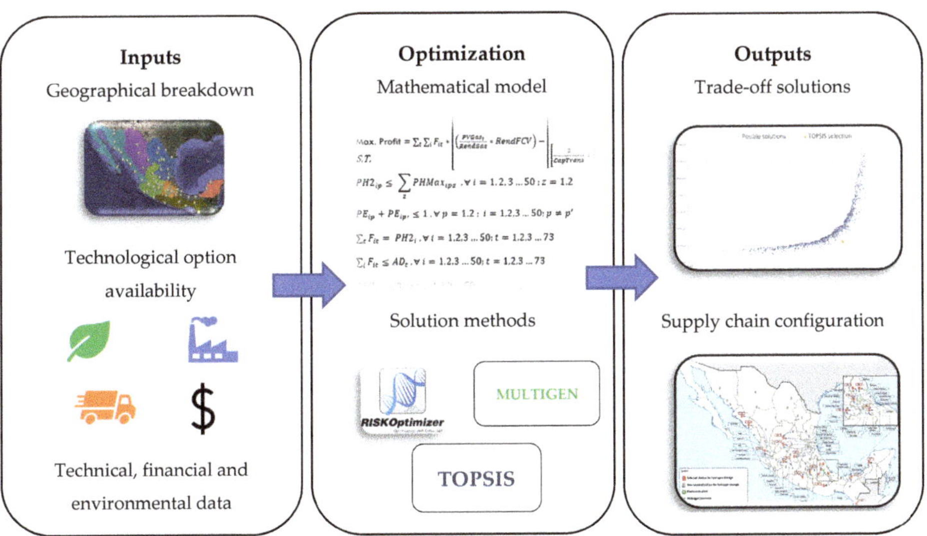

Figure 1. Methodological framework applied.

3.2. Modelling Assumptions

The actual model describes the optimal behavior of a hydrogen production system in the steady state, considering many aspects, such as the production, distribution, and storage operating and investment costs, the accessibility of the raw material, the selling price for hydrogen at distribution points, and the greenhouse gas emissions. The approach applied focuses on developing an optimization model that maximizes profit and minimizes greenhouse gas (GHG) emissions in a system where hydrogen is obtained using agro-industrial wastes from sugar factories in Mexico.

The model arrangement integrates several assumptions that serve as a starting point for the estimation of the economic and environmental indicators that support the decision-making process in the strategic planning of the HSC. These assumptions are as follows:

- The operating time of the system is divided into harvest and non-harvest periods, in which the behavior during the generation of electrical energy differs from one another.
- It is assumed that investments in land and construction have already been paid off. Therefore, these aspects are not considered in the required capital investment.
- Given amounts of available electric energy and storage capacities are considered as model constraints.

3.3. Optimization Model Structure

The proposed model structure is integrated through several calculation modules, which are mainly divided into the following areas: production, transport, and storage. Figure 2 shows the general structure of the model. A description of each module is presented later.

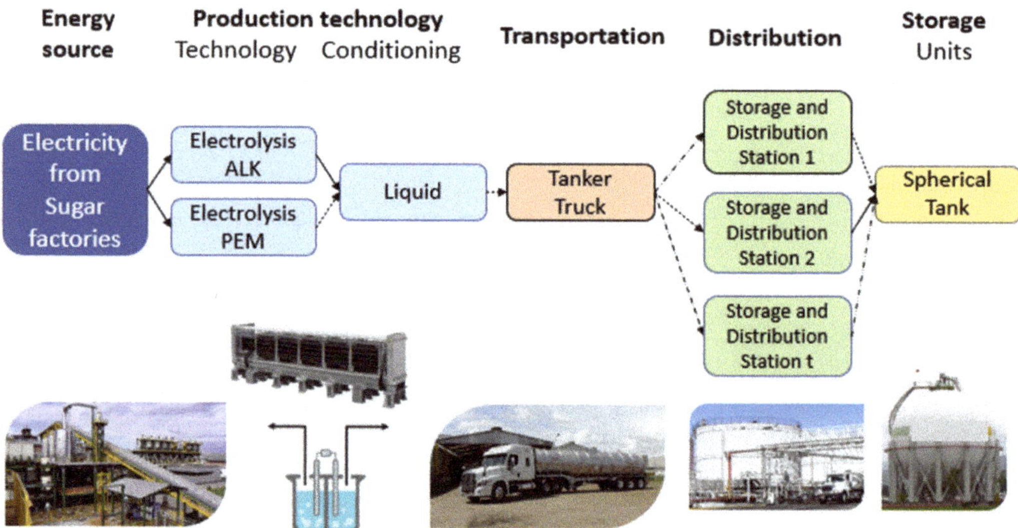

Figure 2. Hydrogen supply chain superstructure.

3.3.1. Hydrogen Production Module

The production module estimates the amount of hydrogen that is convenient to produce based on the availability of electrical energy generated in each of the sugar cane mills by burning bagasse, which is an uncertain parameter for every mill whose behavior responds through probability distributions. The major objective of these calculations is to estimate the operating and investment costs that will result from the production infrastructure. In this section of the model, the selection of the best production technology and the estimate of the amount of hydrogen to be produced by each sugar factory is evaluated (Figure 3).

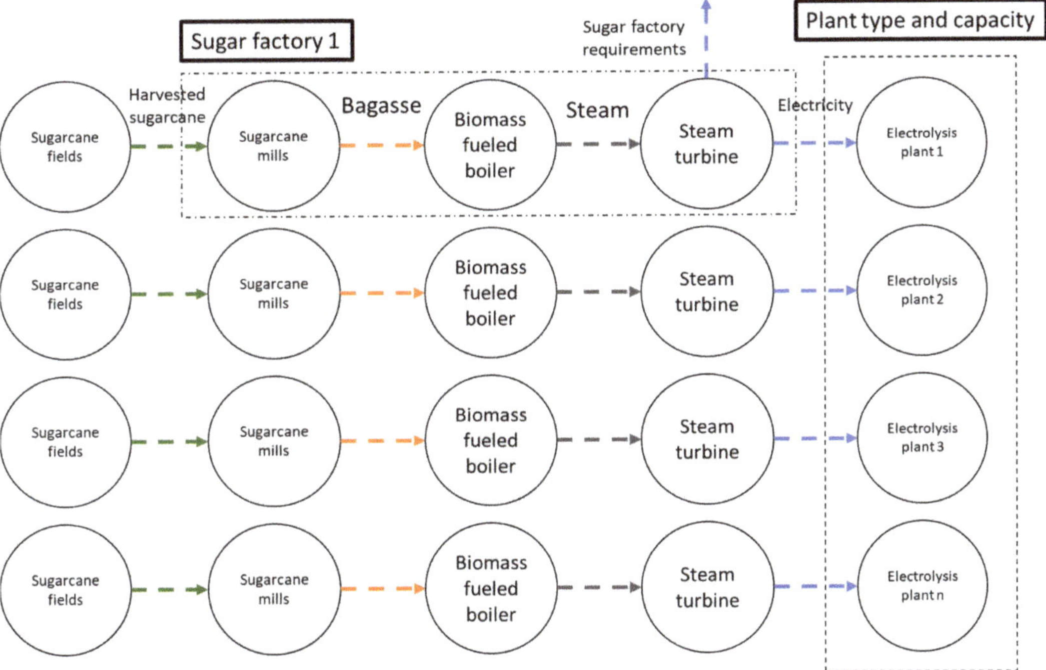

Figure 3. Hydrogen production scheme.

Hydrogen production is divided into two periods: the harvest season, when the greatest amount of H_2 is produced due to the enormous amount of electricity generated from the intensive operation of the sugar factories during this time of year; and the non-harvest season, during which mill operations are reduced due to the lack of raw sugarcane to be processed, thereby lowering the rate of electricity generation and the amount of energy available for hydrogen production. The length of each period is considered as an uncertain parameter according to probability distribution given in days per year [6].

Production cost estimations start by calculating the tons of raw sugar cane that will be processed by each mill during harvest times. The amount of bagasse obtained from sugar cane processing and the amount of moisture it contains are also measured. These values are unique for every sugar cane mill and are represented by probability distributions obtained from historical production records. Humidity measurement is used to determine bagasse energy potential [17]. The amount of bagasse that is used in each mill to generate steam in the boiler rooms during each period depends on the energy consumption behavior of the mill. The steam production dedicated to power generation in each period is estimated using the theoretical efficiencies of the boiler and the bagasse energy potential, also considering the fraction of the dead time operation. Using the amount of steam used to generate electricity, the amount of MWh generated in each period is calculated. Some of this electricity is used by the sugar factories for their daily activities, whereas the overflows are usually fed into the national electricity grid and sold to other organizations. In the proposed model, the energy overflows are used for hydrogen production, whereas their availability is different for the harvest and non-harvest periods.

Once the amount of electrical energy available for hydrogen production in each mill during each period is determined, the optimization model evaluates the most convenient means of production to convert the energy to hydrogen; the proposed technologies are alkaline water electrolysis and proton membrane exchange electrolysis, considering efficiency, investment capital and annual operating costs for each type of production facility. In addition, the variable production costs are calculated, considering the electricity and water prices for each region in which the hydrogen is produced.

3.3.2. Hydrogen Transportation Module

The hydrogen transport module focuses on estimating the capital and operating costs arising from hydrogen distribution activities throughout the supply chain, from the production facilities to the delivery of the hydrogen to the storage and dispatch stations (SDSs)—these are the endpoints where the hydrogen would be stored before they are delivered to the refueling stations (refueling stations are not considered in the actual model). In the analysis, the SDSs are considered as the final stage of the proposed supply chain design (as presented in Figure 4). The amount of greenhouse gas emissions caused by transport activities is also estimated. To achieve this, the optimization model determines the hydrogen flow in tons per year, considering the hydrogen that is generated in both harvest and non-harvest seasons. The model then evaluates the convenience of transporting the hydrogen generated in each electrolysis plant to each storage location; the most favorable network configuration relies on the active objective function. When optimizing with multiple destinations, two main factors influence this decision: the shipping distance (an aspect that has a direct impact on transport costs and equivalent CO_2 kg production), and the selling price of hydrogen at the storage locations, a value that relies on the SDSs' location selected to receive the determined amount of H_2, which has a direct impact on the income generated.

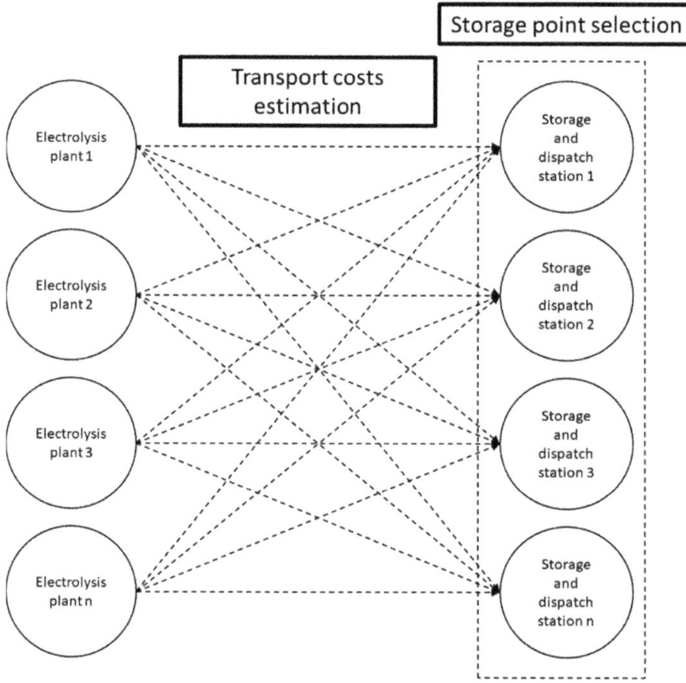

Figure 4. Hydrogen distribution scheme.

Once the annual hydrogen flow is estimated, the number of trips to be made by the transport trucks is calculated based on the vehicle's loading capacity. The time available for each transport vehicle is considered in the calculation, and the number of vehicles required for all distribution operations during the year is determined, thereby obtaining the transport investment cost. The transport operating costs are estimated considering the fuel consumption, the maintenance cost factors (whereby both costs depend directly on the travelling distance from the electrolysis plant to the SDS), the driver's wages, and the toll costs of the selected route. The distance and the toll costs of 50 sugar cane mills for each of the 73 SDSs are shown as two data fields that can be called up via the information system of the national communications and transport department. Finally, the amount of equivalent CO_2 emitted by the network is calculated.

3.3.3. Hydrogen Storage Module

Liquid hydrogen is stored at the SDS, these being the storage points selected by the model in the multiple solutions found. This module calculates the investment capital and operating costs required for the storage units. The number of storage units is determined by the model according to the maximum hydrogen inventory received at a given station during the year of operation. Within these costs, the conditioning energy required for hydrogen compression is calculated and its price depends on the region where the SDSs that have been selected for storing the hydrogen are located. Additionally, the storage costs per unit are considered, including the operating and maintenance costs of the storage unit. The above factors determine the total cost of storage, a value that is added to the cost of production and transportation to determine the final cost of hydrogen on each SDS. Moreover, the revenue generated at each station depends on the gasoline sales price at such SDS, as this price is used as a reference for establishing a competitive sales price for hydrogen, as both serve as mobility fuel for medium-sized vehicles.

3.4. Optimization Model Formulation

3.4.1. Model Notation and Decision Variables

Multiple acronyms definitions, as well as model variables and parameters are presented in Table 2.

Table 2. Glossary.

Nomenclature	Description
Alk	Alkaline electrolysis
CCUS	Carbon capture, utilization and storage
CONACYT	Consejo Nacional de Ciencia y Tecnología
CONADESUCA	Comité Nacional para el Desarrollo Sustentable de la Caña de Azúcar
FCEV	Fuel cell electric vehicle
GA	Genetic algorithm
GHG	Greenhouse gas
GWP	Global warming potential
HSC	Hydrogen supply chain
HSCN	Hydrogen supply chain network
MILP	Mixed integer linear programming
Min	Minimize
MW	Mega watt
MWh	Mega watt hour
NG	Natural gas
NPV	Net present value
O&M	Operation and maintenance
OF	Objective function
PEM	Proton exchange membrane electrolysis
SCBA	Social cost–benefit analysis
SDS	Storage and dispatch station

Table 2. *Cont.*

Nomenclature	Description
SMR	Steam methane reforming
TOPSIS	Technique for order of preference by similarity to ideal solution
Indices	
i	Sugar mills
p	Hydrogen production technology
r	Identification number for regions
t	Identification number for storage and dispatch stations
z	Production period
Decision Variables	
F_{it}	Hydrogen flow rate between sugar mill i and station t (ton/year)
PE_{ip}	Electrolysis plant type p at sugar mill i logic variable with values of 0 or 1
$PH2_{ipz}$	Hydrogen production rate during period z from plant type p at sugar mill i (ton/year)
Parameters	
AD_t	Available storage capacity at station t (m^3)
$AExp_t$	Total annual expenses of hydrogen stored at station t ($/year)
$AProf_t$	Annual profit generated at station t ($/year)
$AToll^C_{it}$	Annual toll costs between sugar mill i and storage station t ($/year)
C^{Alm}_t	Annual storage cost at station t ($/year)
$Capex_p$	Capital expenditures for electrolysis plant type p ($/MW)
Cap^{Inst}_{ip}	Installed capacity of plant type p at sugar mill i (MW)
Cap^{Trans}	Transportation mode capacity (ton)
C^{Comb}_{it}	Fuel transportation costs between sugar mill i and storage station t ($/year)
C^{Cond}_t	Conditioning cost per ton of hydrogen at station t ($/ton)
CFP_{ip}	Annual fixed production cost for plant type p at sugar mill i ($/year)
$CFUP_{ip}$	Fixed production costs per ton of hydrogen for plant type p at sugar mill i ($/ton)
CIP_{ip}	Production investment capital ($)
C^{Mant}_{it}	Maintenance expenses for transportation mode between sugar mill i and storage station t ($/year)
CMO_{it}	Annual transportation labor costs between sugar mill i and station t ($/year)
C^{Prod}_t	Annual hydrogen production costs stored at station t ($/year)
C^{Trans}_{it}	Transportation cost between sugar mill i and storage station t ($/year)
CU^{Alm}	Storage cost per ton of hydrogen at station t ($/ton)
CU^P_{ip}	Production cost per ton of hydrogen for plant type p at sugar mill i ($/ton)
CVU^P_{ip}	Variable production cost per ton of hydrogen for plant type p at sugar mill i ($/ton)
d_{it}	Distance between sugar mill i and storage station t (km)
DMT	Availability of transportation mode (days/year)
DOp_z	Operational days during period z (days)
EC	Fuel economy of transportation mode (km/L)
E^{Cons}_p	Electricity consumption per ton of hydrogen p (MW/ton)
$EnAc$	Conditioning energy required per ton of hydrogen (MW/ton)
$FCEV^{Perf}$	FCEV performance (km/ton of hydrogen)
FP_t	Fuel price per liter at station t ($/L)
Gas^{Perf}	Medium size combustion vehicle performance (km/L of gasoline)
GM	Maintenance expenses of transportation mode ($/km)
GWP^{Total}	System's annual total GWP (eq kg CO_2/year)
NUT_{it}	Number of transport units between sugar mill i and station t
$Opex_p$	Annual operating expense ratio to CAPEX of plant type p (%)
PCG^{Alm}	Storage GWP per ton of hydrogen (kg CO_2 eq/ton)
PCG^P	Production GWP per ton of hydrogen (kg CO_2 eq/ton)
PCG^{Trans}	Transportation GWP per ton of hydrogen (kg CO_2 eq/ton)
PEE_r	Electric power price at station t ($/MW)
p^{GWP}	Production GWP (eq. kg CO_2/year)
$PHMax_{ipz}$	Maximum hydrogen production during period z from plant type p at sugar mill i (ton)
PVA_r	Water cubic meter price at region r ($/$m^3$)
$PVGas_t$	Reference fuel price per liter at station t ($/L)
$PVH2_t$	Hydrogen selling price at station t ($/ton)
SC	Monthly driver wage ($/month)

Table 2. *Cont.*

Nomenclature	Description
S^{GWP}	Storage GWP (eq kg CO_2/year)
TCD	Charge and discharge time of transportation mode (h/trip)
T^{GWP}	Transportation GWP (eq. kg CO_2/year)
$TollC_{it}$	Toll cost for hydrogen transportation units per trip ($)
$TotalUt_t$	Annual total utilities at station t ($/year)
$Trips_{it}$	Annual trips amount required between sugar mill i and station t (trips/year)
TUW	Transport unit weight (ton)
Vm	Average speed for transportation Unit (km/h)
$W^{Cons}{}_p$	Water consumption per ton of hydrogen at plant type p (m^3/ton)

3.4.2. Production Constraints

Hydrogen production is limited by the amount of electrical energy available from sugar mills during the periods of harvesting and non-harvesting. The optimization model determines the most suitable amount of hydrogen to be produced annually. The annual amount of hydrogen that is generated in the type p electrolysis plant in the sugar mill i ($PH2_{ip}$) must be less than or equal to the sum of the maximum amount of produced hydrogen in both z periods, as described in Equation (1).

$$PH2_{ip} \leq \sum_z PHMax_{ipz} \, , \ \forall\, i = 1,\, 2,\, 3 \ldots 50\,; z = 1,2 \tag{1}$$

The electrolysis technology is selected by binary variable PE_{ip}, which takes on the zero value if no technology is selected at all, or takes the value 1 if it is selected to generate hydrogen in the sugar mill i. Since it is not possible to select both technologies for the same point of production, a constraint must be set to limit these events from being mutually exclusive. Equation (2) describes this limitation.

$$PE_{ip} + PE_{ip'} \leq 1 \, , \ \forall\, p = 1,2 \,; i = 1,2,3 \ldots 50; p \neq p' \tag{2}$$

The selection of one or the other electrolysis technology implies a difference in the conversion efficiency of electrical energy into hydrogen, both have different investment costs, annual operating, and maintenance costs.

3.4.3. Transportation Constraints

Produced hydrogen at each location should be distributed to the stations where it offers the highest economic and environmental benefits, considering the potential income, transportation costs, and CO_2 generation to make this decision. To achieve this, Equation (3) limits the flow rate of hydrogen per year distributed from sugar mill i to station t (F_{it}) to meet the amount of hydrogen transported to one or more stations with the amount produced at the supplier electrolysis plants ($PH2_i$).

$$\sum_t F_{it} = PH2_i \, , \ \forall\, i = 1,2,3 \ldots 50; t = 1,2,3 \ldots 73 \tag{3}$$

3.4.4. Storage Constraints

Each SDS has a limited storage capacity, so the sum of the hydrogen flows (F_{it}) resulting from the production points i and which are to be stored in each terminal t must be limited by the available storage volume (AD_t) at this station. To achieve this, Equation (4) limits the amount of hydrogen a station can receive from one or more electrolysis plants.

$$\sum_i F_{it} \leq AD_t \, , \ \forall\, i = 1,2,3 \ldots 50; t = 1,2,3 \ldots 73 \tag{4}$$

3.4.5. Non-Negativity Constraints

All continuous, integer and binary variables must be non-negative.

$$PH2_{ip} \geq 0 \, , \, \forall \, i = 1,2,3\ldots50 \tag{5}$$

$$PE_{ip} \geq 0, \, \forall \, i = 1,2,3\ldots50 \tag{6}$$

$$F_{it} \geq 0, \, \forall \, i = 1,2,3\ldots50; t = 1,2,3\ldots73 \tag{7}$$

3.5. Profit Maximization Objective Function

The total profit of the system is calculated as the difference between the revenue obtained in the storage station and the increase in production (C^{Prod}_t), transport (C^{Trans}_{it}), and storage costs (C^{Alm}_t) achieved in one year of operation. Equation (8) describes the calculation for this statement.

$$MAX: \ TotalProfit = \sum_t (Profit_t = incomes_t - outcomes_t), \, \forall \, t = 1,2,3, \ldots73 \tag{8}$$

The income parameter results from the multiplication of the tons of hydrogen that are intended for storage in station t by the hydrogen sales price ($PVH2_t$) determined for the respective station, as shown in Equation (9).

$$Incomes_t = \sum_i F_{it} * PVH2_t, \, \forall \, i = 1,2,3\ldots50; \, t = 1,2,3\ldots73 \tag{9}$$

Hydrogen sales prices ($PVH2_t$) are determined based on the sales price for gasoline at each station t considering the power offered by each type of vehicle. This is achieved by Equation (10), which estimates the cost per kilometer (US\$/km) it would cost to the end-user. The sales price of gasoline is divided by the average theoretical power that a gasoline engine (Gas^{Perf}_t) offers for the car used as a reference in this analysis, resulting in a cost in US\$/km. This value is then multiplied by the average power of a hydrogen fuel cell engine ($FCEV^{Perf}$), measured in km/kg H_2, which determines the hydrogen sales price in US\$/kg at each SDS.

$$PVH2_t = \frac{PVGas_t}{Gas^{Perf}} * FCEV^{Perf}, \, \forall \, t = 1,2,3\ldots73 \tag{10}$$

when calculating the total annual costs ($AExp_t$), the operating costs for the production, transport, and storage of hydrogen from generation in the electrolysis systems to storage at the SDSs are considered. This is represented by Equation (11).

$$AExp_t = C^{Prod}_t + C^{Trans}_{it} + C^{Alm}_t \tag{11}$$

3.5.1. Production Costs

The production cost (C^{Prod}_t) is calculated using Equation (12), where the hydrogen flows (F_{it}) from point i to endpoint t is multiplied by the production cost per unit (CU^P_{ip}) produced in sugar mill i.

$$C^{Prod}_t = \sum_i (F_{it} * CU^P_{ip}) \, ; \, \forall \, i; \, t \tag{12}$$

The estimate of the production costs in each electrolysis plant is determined by the sum of the variable production costs per unit (CVU^P_{ip}), which relates to the consumption of water and electricity in the process, and the fixed unit production costs ($CFUP_{ip}$), including the cost of operating and maintaining the production facilities as expressed in Equation (13).

$$CU^P_{ip} = CVU^P_{IP} + CFUP_{ip}, \, \forall \, i \tag{13}$$

CVU^P_{ip} (Equation (14)) results from costs of electricity and water volume required hydrogen production per ton. These costs vary depending on the prices of these resources (PEE_r and PVA_r) in each region r. The power consumption depends on the electrolysis technology selected at each point i, since each type of plant has a different transformation performance (Equation (14)).

$$CVU^P_{pr} = (PEE_r * E^{Cons}_p) + \left(PVA_r * W^{Cons}_p\right), \ \forall\, r,\, p \tag{14}$$

The fixed production costs (CFP_{ip}) comprise the operating and maintenance costs ($Opex_p$) in the production facilities, which are expressed as a percentage (%) of the investment capital and refer to an annual cost. Both the production investment capital (CIP_{ip}) and the operating and maintenance costs depend on the hydrolysis technology selected. The cost of capital estimate is based on the installed capacity (Cap^{Inst}_{ip}) of energy processing converted into hydrogen at point i (where an additional gap of 20% is considered to compensate for possible fluctuations in the electricity supply) multiplied by the cost of the capital per installed MW ($Capex_p$). The maximum electricity conversion capacity is estimated using the maximum amount of electricity per hour that will be achieved during the harvest season. This is shown in Equations (15)–(17).

$$Cap^{Inst}_{ip} = \frac{PH2_{ipz}}{OpD_z * 24} * E^{Cons}_P * 1.2; \ z = 1, \ \forall\, i, p \tag{15}$$

$$CIP_{ip} = Capex_p * CapInst_{ip}, \ \forall\, i,\, p \tag{16}$$

$$CFP_{ip} = CIP_{ip} * Opex_p, \ \forall\, i,\, p \tag{17}$$

The fixed unit production cost ($CFUP_{ip}$) is estimated by dividing the annual cost by the annual production (Equation (18)) during harvest and non-harvest periods.

$$CFUP_{ip} = \frac{CFP_{ip}}{PH2_{ipz} + PH2_{ipz'}}, \ \forall\, i,\, p,\, z \tag{18}$$

3.5.2. Transportation Costs

The transportation costs (C^{Trans}_{it}) consider the fuel consumption (C^{Comb}_{it}), the labor costs (CMO_{it}), and the maintenance costs (C^{Mant}_{it}) of the transport units, as well as the toll costs ($TollC_{it}$), the values of which are specific for the transport of the hydrogen produced in each plant location i and delivered to the stations t during the entire operating days. Equation (19) is used to illustrate these calculations.

$$C^{Trans}_{it} = \sum_i (C^{comb}_{it} + CMO_{it} + C^{Mant}_{it} + TollC_{it}); \ \forall\, i,\, t \tag{19}$$

First, the estimate of the number of trips required to distribute the hydrogen flow allocated from facilities i to stations t is obtained, dividing the annual hydrogen flow by the capacity of the transport units (Cap^{Trans}), as shown in Equation (20).

$$Trips_{it} = \frac{F_{it}}{Cap^{Trans}} \ ; \ \forall\, i,\, t \tag{20}$$

The fuel cost (C^{Comb}_{it}) used by the transport units to distribute the hydrogen is obtained by multiplying the estimated number of trips by twice the distance from point i to point t (d_{it}). This value is then multiplied by the fuel price ($PComb_t$) and divided by the fuel consumption (EC) in km/L. This concept is illustrated in Equation (21).

$$C^{comb}_{it} = \frac{PComb_t}{EC} * (2 * d_{it}) * Trips_{it}; \ \forall\, i,\, t \tag{21}$$

The labor cost is calculated using the number of transport units required for hydrogen distribution for all the days of operation. The number of transport units is estimated using Equation (22), where Vm relates to the average speed of the unit, TCD to the loading and unloading time, and DMT to the available time that the transport units consider disposed of. Both values are expressed in hours/year.

$$NUT_{it} = Trips_{it} * \left(\frac{2d_{it}}{Vm} + TCD \right) * \frac{1}{DMT} ; \forall\, i,\, t \tag{22}$$

The NUT_{it} parameter is multiplied by the driver's monthly salary (SC) and multiplied by 12 (months per year) to calculate the annual labor cost (CMO_{it}), as shown in Equation (23).

$$CMO_{it} = NUT_{it} * SC * 12; \forall\, i, t \tag{23}$$

The maintenance cost of the transport unit is calculated by multiplying the maintenance cost (GM) by the total distance in all working days. This is expressed in Equation (24).

$$CMant_{it} = GM * (2d_{it}) * Trips_{it}; \forall\, i,\, t \tag{24}$$

Finally, the annual toll costs ($AToll^C_{it}$) that must be covered to use the routes selected by the model for hydrogen distribution are calculated. This is achieved by considering the number of trips multiplied by the toll price ($TollP_{it}$), which is specific to each route, as shown in Equation (25).

$$AToll^C_{it} = Trips_{it} * TollP_{it}; \forall\, i,\, t \tag{25}$$

3.5.3. Storage Costs

The total storage costs comprise the storage costs per unit (CU^{Alm}), considering the O&M costs of the storage units and hydrogen conditioning cost per unit (C^{Cond}_t), a value that is a function of the electrical power required to liquefy the hydrogen ($EnAc$) to the desired conditions prevailing in the region in which the SDS is located. With this assumption, the conditioning cost per unit is calculated using Equation (26), while the total storage cost is calculated using Equation (27).

$$C^{Cond}_t = EnAc * PEE_r; \forall\, r,\, t \tag{26}$$

$$C^{Alm}_t = \sum_i F_{it} * \left(CU^{Alm} + C^{Cond}_t \right); \forall\, i,\, t \tag{27}$$

3.6. GWP Objective Function

The GWP parameter considered in this model includes the greenhouse gas emissions from hydrogen storage (S^{GWP}), and transport (T^{GWP}), which are generated during an entire year of system operation. Equation (28) is used to calculate the total amount of equivalent CO_2 kilograms for the entire operation.

$$Min\ GWP^{Total} = P^{GWP} + S^{GWP} + T^{GWP} \tag{28}$$

3.6.1. Production GWP

The greenhouse gas emissions from hydrogen production are determined by multiplying the total hydrogen produced in the year of operation by the amount of CO_2 produced per kilogram of hydrogen (PCG^P), as shown in Equation (29).

$$P^{GWP} = \sum_i PH2_i * PCG^P; \forall\, i \tag{29}$$

3.6.2. Transportation GWP

Hydrogen transport is a major contributor to emissions from the CO_2 and heavily depends on the distances between the production points and the storage stations selected by the model to store hydrogen. Equation (30) is used to estimate the calculation of the kilograms of equivalent CO_2 produced by transportation. These calculations start with the distances traveled in the year of operation of the system, with the number of trips made multiplied by twice the distance from the production site to the SDS. The resulting value is multiplied by the eq- CO_2 kg (PCG^{Trans}), and the weight of the transport unit (*WeightUT*) is also considered when estimating this parameter.

$$T^{GWP} = \sum_{it}(2 * d_{it} * Trips_{it}) * PCG^{Trans} * WeightUT \; ; \; \forall \, i, \, t \tag{30}$$

3.6.3. Storage GWP

The storage of hydrogen also generates a significant amount of equivalent CO_2, mainly related to energy conditioning and the operation of storage units. The estimate of the carbon dioxide emissions generated by the storage of hydrogen is determined using Equation (31), which uses the variable PCG^{Alm}, which refers to the equivalent CO_2 kg/ton of hydrogen, and which is multiplied by the total hydrogen tons accumulated in each terminal for the entire year of operation.

$$S^{GWP} = \sum_{it} F_{it} * PCG^{Alm} \; ; \; \forall \, i, \, t \tag{31}$$

3.7. Solution Methods

For solving MILP problems, the use of genetic algorithms appears to be one of the most effective methods to find a wide range of feasible solutions when solving similar mathematical problems according to the literature. For selecting the multi-objective optimization method, several alternatives were considered. The selected approach was a meta-heuristic technique, using MULTIGEN software which is a GA used by the research team in previous studies. In addition, multi-objective simulated annealing and multi-objective tabu search techniques were evaluated. At first, a mono-objective optimization method was applied to identify the behavior of the model concerning the optimal solutions for each objective function (to identify antagonism), then multi-criteria optimization was performed. MULTIGEN turned out to be convenient in terms of efficiency and convergence time. MULTIGEN has been applied by the research team in previous studies concerning multi-objective optimization of the HSC [14,31]. The optimization approach was performed in two stages. The first one focuses on the single optimization of each objective function. The second one is aimed to obtain a range of feasible solutions when both optimization criteria are considered simultaneously. For selecting the mid-point solution from the obtained pareto front, the multi-criteria decision-making technique TOPSIS was applied. The assignment of weights for each criterion was performed along the organization interested in the study, assigning equivalent weights for both criteria, since the company decided that both aspects were equally relevant in the decision making.

The GA applied for solving the mathematical model was built using the user interface, generated by the optimization software. The GA parameters were defined based on an iterative procedure, where different combinations were evaluated, selecting those with the smallest solving times. The TOPSIS method was applied using a spreadsheet that allows evaluation of the 1000 possible solutions.

3.8. Mathematical Model Optimization Framework

The mathematical model optimization was carried out with two GA's, the first regarding the independent optimization of each target using the Evolver optimization software in version 7.6 developed by PALISADE, obtaining the best value for each objective function. The second GA is a multi-objective optimization tool that implements a variant of NSGA II developed in the Chemical Engineering Laboratory at the Institut National Polytechnique de Toulouse (INPT). The MULTIGEN algorithm was set to optimize the optimization criteria at the same time. The optimization algorithms were calculated using an 8-core AMD Ryzen 7 2700X processor at 3.7 GHz.

3.8.1. Mono-Objective Optimization

The individual criteria optimization is carried out using the GA interface, which is integrated into the Evolver optimization software. With this software, the user can easily define an optimization model, prioritizing that the logic of the decision variables and the constraints correspond to the mathematical formulation.

The performance of a GA for finding optimal solutions can be influenced by its parameter configuration. Therefore, a sensitivity analysis was performed to define these elements and look for those that would give the best results in finding the optimal solution. These parameters are listed in Table 3 along with the stopping conditions considered for the mono-objective optimization, which were defined to obtain workable solutions until a significant improvement is found over a certain number of iterations.

Table 3. Genetic algorithm parameters and stopping conditions for mono-objective optimization.

Parameter	Value
Population	30,000
Crossing rate	0.5
Mutation rate	0.1
Solution method	Order
Stopping conditions	
Max. Change	0.005%
Max. Iterations without improvement	20,000

3.8.2. Multi-Objective Optimization

The multi-criteria optimization phase is carried out by MULTIGEN optimization software. The model formulation is introduced by generating the optimization interface in which the GA parameters, such as population size or the number of generations, can be defined. The selected configuration of the GA is shown in Table 4. These values are determined by a sensitivity analysis, from which the best configuration for the selected algorithm could be determined.

Table 4. Multi-objective genetic algorithm configuration.

Parameter	Value
Population	36,500
Number of generations	73,000
Crossing rate	0.9
Mutation rate	0.5

Different parameters were used in both algorithms since each of them responds differently to the parameter values. Several values were tried before finding the optimal configuration for each GA. When optimizing multiple objectives simultaneously, a Pareto front is generated with a set of different feasible solutions; then, the alternative that better meets both optimization criteria is selected using a decision-making technique (TOPSIS).

4. Case Study

4.1. Mexican Sugarcane Industry

Sugar cane is mainly used in Mexico to make refined sugar by extracting syrups from its stems. In the 2018/2019 harvest season, the National Committee for the Sustainable Development of Sugar Cane (CONADESUCA) reported a harvested area of 805.5 thousand hectares, around 57,036,700 tons of gross base cane and 6.4 million tons of sugar. The average yield per hectare at the national level is estimated at 70.81 tons in the industrialized acreage dedicated to grinding in the sugar mills [32,33].

The main activities of the sugar mills are divided into two periods: harvest or grinding period. This is when the harvested cane is processed for sugar production and the maintenance period, which coincides with the rainy season when farmers devote themselves to growing sugar cane. In the second phase, production in the mill is stopped to take over the dismantling, repair, and improvement of the factory to prepare for the next grinding period. The 2018/2019 harvest took place over 179 days with 50 sugar mills operating, mainly located in the west, the Gulf, and the south of the country.

Sugarcane Bagasse Generation and Characteristics

In this study, information of 50 sugar mills is taken from the sixth statistical report of the agro-industrial sugar cane sector in Mexico [34] by CONADESUCA, which provides data from the harvest period 2006/2007 to 2018/2019. The amount of bagasse available is modeled as a percentage of the tons of raw cane milled annually. Acting as model inputs, the amount of ground raw cane, the remaining bagasse fraction, and the moisture contained in the bagasse are considered as uncertain parameters and modeled using probability distribution. The mathematic formulation for calculating the fraction of bagasse that is available in the HSC for power generation is extracted from the work previously carried out by Rico Contreras, among the calculations for converting the bagasse into electricity [6]. This information is presented in Appendixes A and B.

4.2. Hydrogen in Mexico

4.2.1. Hydrogen Demand

The estimated hydrogen demand for mobility purposes has been determined based on the available capacity of each of the 76 SDSs, which are spread across Mexican territory and are currently used for fossil fuel storage and subsequent distribution at petrol stations for sale to the public [35].

4.2.2. Hydrogen Production

The proposed model considers two primary means of hydrogen production: alkaline electrolysis and the proton exchange membrane [36]. They are mainly considered due to their technological maturity and their availability in the international market. Each technology has different properties that can have a significant impact on the cost of hydrogen production [37]. These are shown in Table 5. Electricity and water prices were modeled using probability distributions, as listed in Appendix C.

Table 5. Production parameters.

Parameter	Alkaline	PEM	Reference
E^{Cons} (kWh/kgH$_2$)	49	52	
Performance (HHV) (%)	71	64	
CAPEX ($/kW)	507.8	740.5	[36]
Opex (%CAPEX/year)	3	2	
Lifetime (years)	20	20	
W^{Cons} (m^3/ton H$_2$)	9		

The variable cost of hydrogen produced by electrolysis is heavily influenced by the electricity and water prices of the region in which it is produced. Information on these prices has been compiled for each region considered in the study.

4.2.3. Hydrogen Storage

Capital costs of the storage units, the storage unit costs, and the parameters to produce greenhouse gases are presented in Table 6. Information concerning the storage capacity and availability for each SDS is presented in Appendix D.

Table 6. Hydrogen storage parameters.

Parameter	Storage Unit	
Minimum Capacity (kg)	500	
Maximum capacity (kg)	10,000	
Investment capital ($)	5,542,595	
C^{Alm} ($/kg H_2)	0.722	[7,9]
Lifetime (years)	20	
S^{GWP} (kg CO_2 per ton H_2)	704	
Maximum storage time (days)	10	Assumption

4.2.4. Hydrogen Transportation

This study uses real geographic information from the communications and transportation department to determine the shipping distances and toll costs of the selected routes and to find the optimal route configuration. The proposed transportation mode to be used in the hydrogen shipment are tanker trucks, as this is the transportation mode of fossil fuels currently used in Mexico [35]. The toll costs of the selected routes for the hydrogen distribution considers the type of truck used, which are 6-axis vehicles. The distances between each mill and the SDSs considered are collected as well [38]. To calculate the transport costs, these values must be multiplied by two to get the round-trip flight costs. The hydrogen transport parameters are listed in Table 7. Data sets used for distance and transportation costs calculations are listed in Appendix E.

Table 7. Hydrogen transportation parameters.

Parameter	Value	Scale	Reference
TUW	40	Ton	[9]
SC	736	$/month	[35]
EC	2.3	km/L	[7]
FP	-	-	Appendix D
TCD	2	Hours per trip	[7]
C^{Mant}	2.42	$/km	
Vm	67	km/h	[7]
DMT	18	Hours/day	Assumption
T^{GWP}	62	g CO_2 per ton-km	
Cap^{Trans}	3.5	Ton	[4]
$Trans^{Capex}$	293,756	$	[7]

4.2.5. Hydrogen Selling Price

The information for estimating the hydrogen sales price is given in Table 8. The annual distance traveled by a medium-sized private vehicle is also established to be used in the calculation of the hydrogen selling price.

Table 8. Hydrogen selling price parameters.

Parameter	Value
$FCEV^{Perf}$	0.98 kg H_2/100 km
Annual average distance traveled for medium size vehicles	15,000 km/year

5. Results and Discussion

5.1. Mono-Objective Optimization Results

Both objective functions were initially optimized independently of one another. With these results, it is possible to create a comparison table showing the resulting values from both selected criteria optimizations, as shown in Table 9.

Table 9. Mono-objective optimization results.

Parameter	Profit O.F.	GWP O.F.
Number of production units	50 ALK	50 ALK
Number of transport units	73	55
Number of storage units	275	286
Investment capital costs		
Production capital cost	$373,654,974	$373,654,974
Transport capital cost	$5,402,025	$4,070,019
Storage capital cost	$1,524,213,622	$1,585,182,167
Total capital cost	$1,903,270,621	$1,962,907,160
Operating costs		
Production	$188,692,213	$188,692,213
Transport	$5,682,987	$2,242,429
Storage	$27,354,603	$28,880,026
Total Outcome	$221,729,804	$219,815,777
Average cost per unit ($/kg H_2)	$3962	$3928
Profit estimation		
Total hydrogen production (ton/year)	55,965	55,965
Average selling price ($/ton)	$8938	$8782
Total income	$500,220,813	$491,490,525
Annual profit	$278,491,009	$271,675,857
Net profit margin	55.67%	44.72%
GWP (kg eq. CO_2)		
Production	-	-
Transport	39,399,360	39,399,360
Storage	19,783,361	7,015,414
Total GWP (kg eq.CO_2)	59,182,721	46,414,774
GWP per unit (kg eq. CO_2/ton H_2)	1057	829
Optimization time (s)	17,388	21,728

Based on the resulting values, it is determined that it is possible to produce hydrogen at the 50 locations of the sugar mill, which allows the system to produce 55,965 tons of hydrogen per year.

From the profit maximization O.F. obtained solution, 73 transportation units and 275 storage units are required to ensure the logistics demand of hydrogen. In contrast, in the GWP O.F. solution, only 55 transport units and 286 storage units are needed. Additionally, the capital expenditures for each element of the supply chain were estimated, resulting in US$1,903,270,621 for the first O.F., and US$1,962,907,160 for the second one. The obtained solutions put the annual operating cost of the entire system at US$221,729,804 and US$219,815,777 for each O.F., respectively. The production cost obtained in the first O.F. optimization contributes 85% to the final cost of hydrogen (Figure 5), while transportation and storage give 3% and 12%, respectively.

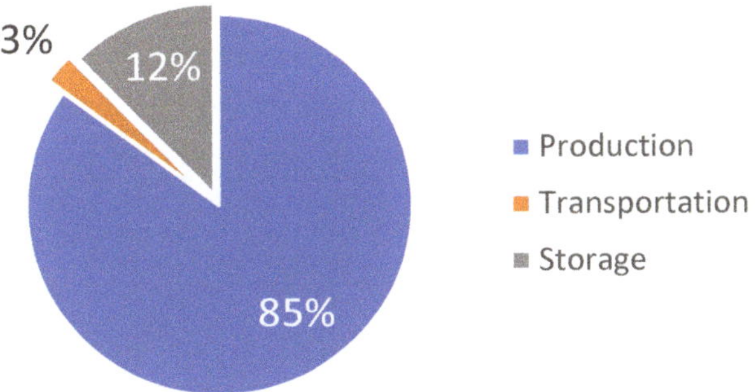

Figure 5. Pie chart of the hydrogen total cost composition obtained from Profit O.F.

In Figure 6, a pie chart shows the composition of the total cost of hydrogen obtained from the GWP O.F. optimization. It can be observed that the transportation costs reduced their participation on the total cost of hydrogen in the optimization of the second O.F. from 3% to 1%. This is expected since the GWP optimization looks mainly to deliver the hydrogen to the closet SDS to reduce the gases emitted by the network. The production and storage cost participation increased due to the previous statement.

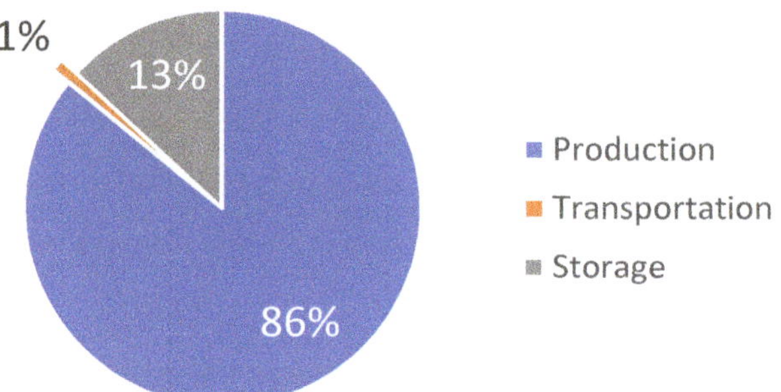

Figure 6. Pie chart of the hydrogen total cost composition obtained from GWP O.F.

Another clear difference is the average selling price of hydrogen which goes from US$8938/ton in the profit optimization to US$8782/ton in the GWP optimization, which was expected since the selling price of hydrogen is a critical factor for a SDS to be selected in the profit O.F. The annual profit of the system is estimated at US$278,491,009, which equates to a net profit margin of 55.67% for the first objective function and US$271,675,857 with a profit margin of 44.72% for the second O.F. Occupancy in the SDS's refers to the percentage of storage volume at the selected station in which hydrogen is stored, whereby a ratio of 4.49% is achieved.

A detailed economic report is shown in Table 10. In the first column are the names of the sugar factories where the electrolysis plants were located. The second column shows the names of the storage and dispatch stations where the hydrogen is stored, which are the locations for the storage units. In the rest of the columns, the information of the hydrogen flow from the PE to the storage location, the costs of production, transportation, and storage per unit are shown separately at first, then the total cost of hydrogen and the selling price per unit at each SDS. A profit estimation calculated from the difference of selling revenues and total cost is displayed.

Table 10. Profit O.F. detailed economic report.

E.P. Location	SDS	Hydrogen Flow (Ton/Year)	Production Cost ($/Ton)	Transportation Cost ($/Ton)	Storage Cost ($/Ton)	Total Cost per Unit ($/Ton)	Selling Price ($/Ton)	Profit ($/Year)
El Molino	Guamúchil	880	1984.82	265.82	290.91	2541.55	9198.38	5,858,008
Puga		1414	1984.82	266.11	290.91	2541.85	9198.38	9,412,337
El Dorado	Culiacán	479	1984.82	35.71	290.91	2311.44	9163.75	3,282,257
Quesería		1292	3269.16	367.73	290.91	3927.80	9163.75	6,764,828
Ameca	Tepic	1050	3269.16	81.73	290.91	3641.80	9085.17	5,715,544
Bellavista		641	3269.16	97.94	290.91	3658.01	9085.17	3,478,795
José Ma Morelos		648	3269.16	151.03	290.91	3711.10	9085.17	3,482,392
Melchor Ocampo		1162	3269.16	138.41	290.91	3698.48	9085.17	6,259,316
Tala		1714	3269.16	82.71	290.91	3642.83	9085.17	9,328,207
Aarón Sáenz	Zacatecas	1104	3456.53	171.07	500.74	4128.34	9030.35	5,411,801
El Mante		976	3456.53	172.05	500.74	4129.32	9030.35	4,783,390
San Miguel del Naranjo		1980	3456.53	163.51	500.74	4120.78	9030.35	9,720,987
Alianza Popular	Aguascalientes	1216	3456.53	161.64	500.79	4118.96	9032.66	5,975,092
Plan de Sal Luis		1400	3456.53	225.29	500.79	4182.61	9032.66	6,790,102

Table 10. *Cont.*

E.P. Location	SDS	Hydrogen Flow (Ton/Year)	Production Cost ($/Ton)	Transportation Cost ($/Ton)	Storage Cost ($/Ton)	Total Cost per Unit ($/Ton)	Selling Price ($/Ton)	Profit ($/Year)
Lázaro Cárdenas		273	3269.16	81.68	500.79	3851.62	9078.54	1,426,943
Pedernales	Zamora	436	3269.16	106.04	500.79	3875.98	9078.54	2,268,306
Santa Clara		655	3269.16	38.51	500.79	3808.45	9078.54	3,451,896
Tamazula		1566	3269.16	59.58	500.79	3829.52	9078.54	8,219,934
Plan de Ayala		1325	3456.53	201.18	500.79	4158.50	9013.80	6,433,280
El Higo	Celaya	1957	3436.98	182.47	500.79	4120.24	9013.80	9,576,710
Pánuco		1918	3436.98	254.86	500.79	4192.63	9013.80	9,247,004
Atencingo	Cuautla	1827	3617.09	25.44	557.81	4200.34	8944.01	8,666,645
Casasano		645	3617.09	19.30	557.81	4194.20	8944.01	3,063,613
Calipam		233	3617.14	53.49	557.81	4228.44	8872.10	1,081,978
El refugio	Tehuacán	475	3616.80	76.18	557.81	4250.79	8872.10	2,195,132
Constancia		886	3436.98	64.24	557.81	4059.04	8872.10	4,264,358
Motzorongo		1341	3436.98	58.99	557.81	4053.78	8872.10	6,461,356
Emiliano Zapata	Iguala	1187	3617.09	50.39	557.81	4225.29	8998.92	5,666,304
López Mateos	Oaxaca	1607	3616.80	76.47	557.81	4251.08	8933.10	7,523,971
Tres Valles		2396	3436.98	86.00	557.81	4080.80	8933.10	11,626,096
Huixtla	Tapachula	1202	3616.80	25.98	557.81	4200.59	8927.95	5,682,255
El Modelo	Perote	1079	3436.98	44.20	528.29	4009.48	8845.38	5,217,947
Mahuixtlán		436	3436.98	48.72	528.29	4014.00	8845.38	2,106,469
La Gloria	Xalapa	1581	3436.98	29.91	528.29	3995.19	8816.01	7,621,740
San Pedro		1273	3436.98	82.86	528.29	4048.13	8816.01	6,069,513
El Carmen		577	3436.98	19.79	528.29	3985.07	8797.40	2,776,722
El Potrero		1707	3436.98	21.91	528.29	3987.18	8797.40	8,211,057
La providencia		811	3436.98	30.11	528.29	3995.38	8797.40	3,894,444
Progreso	Escamela	913	3436.98	48.23	528.29	4013.51	8797.40	4,367,711
San Cristobal		560	3436.98	18.81	528.29	3984.09	8797.40	2,695,459
San Miguelito		525	3436.98	55.60	528.29	4020.87	8797.40	2,507,675
San Nicolas		1103	3436.98	23.48	528.29	3988.75	8797.40	5,303,941
La margarita		1226	3616.80	17.04	528.29	4162.13	8773.28	5,653,241
Cuatotolapan	Tierra Blanca	835	3436.98	60.31	528.29	4025.59	8773.28	3,964,315
San Cristobal		2584	3436.98	28.68	528.29	3993.96	8773.28	12,349,672
Benito Juárez	Villahermosa	1438	3436.98	26.18	528.29	3991.45	8733.89	6,819,600
Santa Rosalia		781	3436.98	27.31	528.29	3992.58	8733.89	3,702,945
Azsuremex		223	3436.98	166.31	547.35	4150.69	8760.07	1,027,891
La Joya	Campeche	826	3553.49	32.12	547.35	4132.96	8760.07	3,821,972
Pucte		1602	3553.49	103.05	547.35	4203.88	8760.07	7,298,984
-	Total	55,965	-	-	-	-	-	278,491,009
-	Average	1119	3352.11	94.50	486.00	3961.94	8938.11	5,569,820

It is possible to see significant differences in the contribution of the various elements of the supply chain to costs. For example, hydrogen from the El Molino and Puga generation points makes a higher contribution to the transport costs than the rest, as the reported production costs in these facilities are exceptionally low (US\$1984.82/ton of H_2) compared with other facilities. It is possible to distribute hydrogen over greater distances to stations with higher sales prices.

The hydrogen distribution for this solution is a decision that is heavily influenced by the selling price at the SDS for which it is intended. However, a SDS an extremely large distance from the electrolysis plant that supplies it would cause higher transport costs. Therefore, the model carries out an assessment and determines to which of the storage stations the hydrogen produced should be distributed.

The GWP for supply chain operations was then calculated. The electrical energy from the emissions balance of bagasse production is regarded as neutral due to its agricultural origin, so that the estimate of greenhouse gas emissions is limited to the transport and storage factors, the second one contributes majorly with a share of 67% of greenhouse gas emissions. On this basis, it is estimated that this configuration of the HCS generates 59,182,721 kg of equivalent CO_2, or 1057 kg of CO_2/ton of distributed and stored hydrogen.

The HSC configuration obtained from the profit objective function optimization is presented in Figure 7.

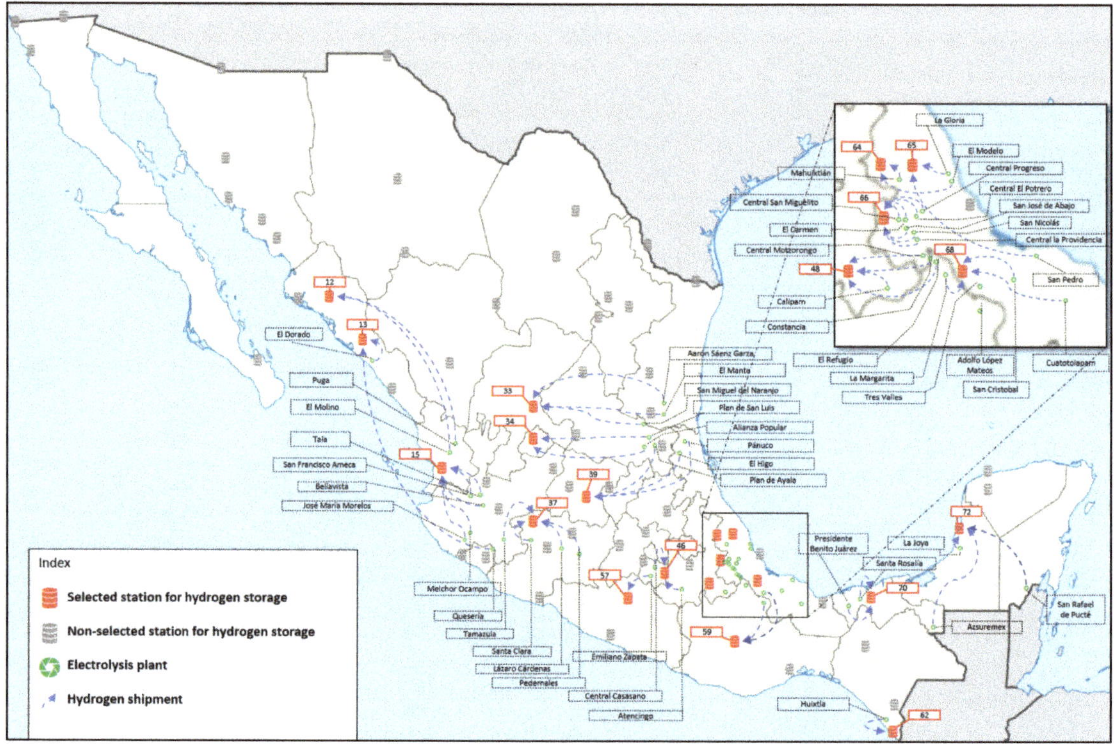

Figure 7. HSC configuration obtained by profit O.F. optimization.

Concerning the optimization of the GWP objective function, considerable differences can be observed compared to the profit optimization function. First, the number of transport units has been significantly reduced to 55, so the investment capital is also reduced. However, this configuration requires 286 storage units, a higher number than previous results, and while this is the factor that has the greatest impact on the capital cost. Thanks to this, the investment required to deploy the supply chain increases to US$1,962,907,160.

The production makes the largest contribution to operating costs but remained constant for both OFs. Besides, the operating costs for the transport are reduced by 60%, which is a consequence of the fact that the algorithm in this OF mainly focuses on the selection of the shortest distances from the hydrogen production points to the SDS and requires fewer transport units to carry out the distribution. As a result, the unit cost of hydrogen will be significantly reduced to an average of US$3928 per ton.

With respect to profit, the average selling price is US$8782/ton of hydrogen. Because of this, there are fewer economic benefits compared to the solution shown above, which in this case is US$271,675,857, resulting in a profit margin of 44.72%.

Table 11 shows the key results of the economic indicators for each station selected by the model for hydrogen storage and shows the unit cost of supply chain operations and the selling price at each SDS. In this case, the average final cost of hydrogen is reduced compared to the previous solution, assuming a value of US$3908/ton and an average sales price of US$8804/ton.

Table 11. GWP O.F. detailed economic report.

E.P. Location	SDS	Hydrogen Flow (Ton/Year)	Production Cost ($/Ton)	Transportation Cost ($/Ton)	Storage Cost ($/Ton)	Total Cost per Unit ($/Ton)	Selling Price ($/Ton)	Profit ($/Year)
El Dorado	Culiacán	479	1984.82	35.71	290.91	2311.44	9163.75	3,282,247
El Molino	Tepic	880	1984.82	12.03	290.91	2287.82	9085.17	5,981,676
Puga		1414	1984.82	12.18	290.91	2287.92	9085.17	9,611,305
Aarón Sáenz	Cd. Victoria	1104	3456.58	41.85	531.24	4029.67	8841.90	5,312,714
Alianza Popular		1216	3456.58	102.55	531.24	4090.37	8841.90	5,777,865
San Miguel del Naranjo		1562	3456.58	63.65	531.24	4051.47	8841.90	7,482,657
Pánuco		1918	3436.98	94.70	531.24	4062.92	8841.90	9,166,055
El Mante	Cd. Mante	976	3456.58	10.36	531.24	3998.18	8783.10	4,670,094
San Miguel del Naranjo		418	3456.58	39.24	531.24	4027.06	8783.10	1,988,030
Plan de Ayala	Cd. Valles	1325	3456.58	7.66	531.24	3995.48	8809.97	6,379,213
Plan de SL		1400	3456.58	19.55	531.24	4007.37	8809.97	6,723,660
El Higo		1957	3436.98	37.03	531.24	4005.26	8809.97	9,402,801
Ameca	Zapopan	1050	3269.16	30.40	500.79	3800.34	8990.47	5,449,620
Bellavista		641	3269.16	28.83	500.79	3798.77	8990.47	3,327,871
Tala		1714	3269.16	15.08	500.79	3785.02	8990.47	8,922,122
Santa Clara	Zamora	655	3269.16	38.56	500.79	3808.50	9078.54	3,451,876
Lázaro Cárdenas	Uruapan	273	3269.16	39.15	500.79	3809.09	9000.49	1,417,253
Pedernales		436	3269.16	69.30	500.79	3839.24	9000.49	2,250,304
Quesería	Colima	1292	3269.16	17.34	500.79	3787.28	8927.31	6,640,918
Tamazula		1566	3269.16	43.32	500.79	3813.26	8927.31	8,008,598
José María Morelos	Manzanillo	648	3269.16	90.77	500.79	3860.71	8667.29	3,114,665
Melchor Ocampo		1162	3269.16	98.62	500.79	3868.57	8667.29	5,576,116
Atencingo	Cuautla	1827	3617.14	25.44	557.81	4200.39	8944.01	8,666,588
Casasano		645	3617.14	19.30	557.81	4194.25	8944.01	3,063,593
Calipam	Tehuacán	233	3617.14	53.44	557.81	4228.39	8872.10	1,081,986
Emiliano Zapata	Cuernavaca	1187	3617.14	22.74	557.81	4197.69	8915.18	5,599,657
Huixtla	Tapachula	1202	3616.80	25.98	557.81	4200.59	8927.90	5,682,213
Mahuixtlán	Xalapa	436	3436.98	27.31	528.29	3992.58	8816.01	2,103,009
El Carmen	Escamela	577	3436.98	19.74	528.29	3985.02	8797.40	2,776,734
El Potrero		1707	3436.98	21.91	528.29	3987.18	8797.40	8,211,016
La Providencia		811	3436.98	31.58	528.29	3996.86	8797.40	3,893,227
Progreso		913	3436.98	48.23	528.29	4013.51	8797.40	4,367,679
San José de Abajo		560	3436.98	33.79	528.29	3999.07	8797.40	2,687,057
San Miguelito		525	3436.98	55.60	528.29	4020.87	8797.40	2,507,667
Adolfo López Mateos	Veracruz	1607	3616.80	62.97	528.29	4208.06	8522.45	6,933,222
El Modelo		1079	3436.98	28.44	528.29	3993.71	8522.45	4,886,503
La Gloria		1581	3436.98	29.32	528.29	3994.60	8522.45	7,158,529
Motzorongo		1341	3436.98	50.34	528.29	4015.62	8522.45	6,043,655
San Cristobal		2584	3436.98	68.22	528.29	4033.50	8522.45	11,599,444
San Nicolás		1103	3436.98	55.80	528.29	4021.07	8522.45	4,965,017
San Pedro		1273	3436.98	44.25	528.29	4009.53	8522.45	5,744,944
El Refugio	Tierra Blanca	475	3616.80	33.74	528.29	4178.83	8773.23	2,182,339
La Margarita		1226	3616.80	17.04	528.29	4162.13	8773.23	5,653,206
Constancia		886	3436.98	26.62	528.29	3991.90	8773.23	4,236,264
Tres Valles		2396	3436.98	12.13	528.29	3977.41	8773.23	11,490,797
Cuatotolapam	Minatitlán	835	3436.98	44.94	528.29	4010.22	8623.23	3,851,868
Azsuremex	Villahermosa	223	3436.98	109.48	528.29	4074.75	8733.89	1,038,987
Benito Juárez		1438	3436.98	26.18	528.29	3991.45	8733.89	6,819,623
Santa Rosalía		781	3436.98	27.31	528.29	3992.58	8733.89	3,702,960
La Joya	Campeche	826	3553.49	32.12	547.35	4132.96	8760.07	3,822,004
San Rafel Pucté	Yucatán	1602	3553.49	74.71	547.35	4175.54	8524.36	6,966,830
-	Total	55,965	-	-	-	-	-	271,675,857
-	Average	1097	3354	40.72	513.11	3907.96	8803.93	5,433,517

Finally, a significant decrease in the equivalent CO_2 tons emitted by the system can be observed, which corresponds to a reduced travel distance for the hydrogen distribution. As a result, the amount of CO_2 emitted per ton of hydrogen is significantly reduced, assuming values of 829 kg equivalent CO_2/ton of H_2, which corresponds to 78.42% of the value obtained in the previous solution. For this configuration, it was found that the contribution from transport to CO_2 emissions decreased from 33% to 15%.

The HSC configuration obtained from the optimization of the GWP objective function is shown in Figure 8. The model in this case is mainly committed to storing the hydrogen in the nearest SDSs from the production facilities, the major reason for the significant decrease in CO_2 emissions generated by the system.

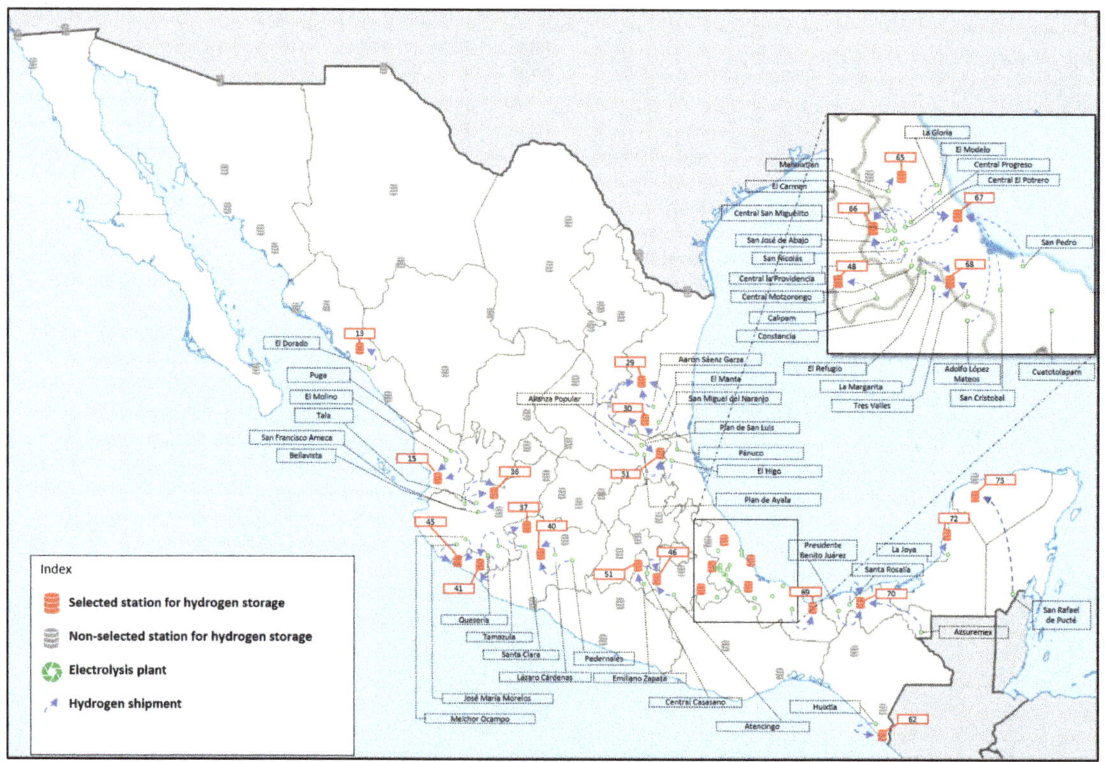

Figure 8. HSC configuration obtained by GWP O.F. optimization.

5.2. Multi-Objective Optimization Results

The simultaneous optimization of both objective functions carried out with the MULTI-GEN optimization software, through which it is possible to obtain a Pareto front with a set of 1000 possible solutions, the one that fulfills both criteria most satisfactorily. Figure 9 shows a Pareto front diagram and the solution chosen by the TOPSIS.

In most cases, the hydrogen storage terminals where higher profits would be made are not close to the points where hydrogen production takes place. However, at some point, the increase in profit is no longer proportional to the increase in emissions, which indicates that there are solutions whose emissions are considerably high ($<5.70 \times 10^7$) and whose contribution to profit is not as significant compared to other solutions found for the model.

The solution selected using the TOPSIS method that best meets both optimization criteria is highlighted in the diagram. With this configuration, a profit of US\$275,197,557/year is achieved, and 51,443,692 kg of equivalent CO_2 is emitted annually. Next, the HSC design based on this configuration is presented, in which important performance indicators were estimated.

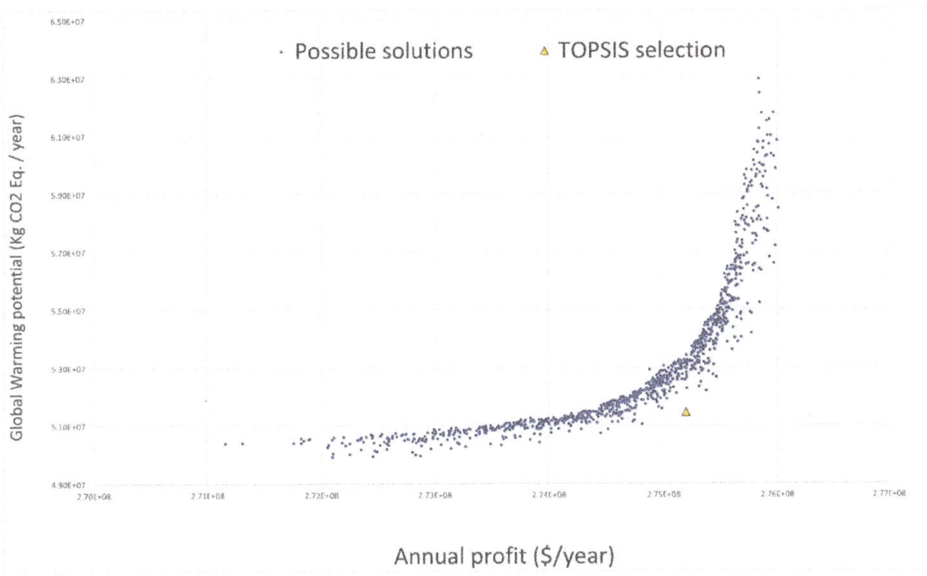

Figure 9. Pareto front chart and TOPSIS selected solution.

5.3. Optimal Hydrogen Supply Chain Configuration

Table 12 shows the results of the general economic and environmental system indicators for the optimal solution that TOPSIS selected from the Pareto front.

Table 12. Multi-objective optimization results.

Parameter	Values
Number of production units	50 ALK
Number of transport units	59
Number of storage units	279
Investment capital costs	
Production capital cost	$373,654,974
Transport capital cost	$4,366,020
Storage capital cost	$1,546,384,002
Total capital cost	$1,924,404,997
Operating costs	
Production	$188,692,213
Transport	$3,550,495
Storage	$29,250,926
Total outcome	$275,197,558
Average cost per unit ($/kg H_2)	$3958
Profit estimation	
Total hydrogen production (ton/year)	55,965
Average selling price ($/ton)	$8875
Total income	$496,691,192
Annual profit	$275,226,444
Net profit margin	55.40%
GWP (kg CO_2 eq.)	
Production	0
Transport	39,399,360
Storage	12,044,332
Total GWP (kg CO_2 eq.)	51,443,692
GWP per unit (kg CO_2/ton H_2)	919
Optimization time (s)	19,879

The average contribution of each element in the supply chain to the final cost of hydrogen in the storage station can be determined. The cost of hydrogen production adds an average of 85% to the total cost of the product in the supply chain. In this case, the transport costs add (on average) 2% to the total costs of hydrogen. Table 13 lists the economic details within the HSC, listing the SDSs selected for hydrogen storage and their supplier production points.

Table 13. Multi-objective optimal solution detailed economic report.

E.P. Location	SDS	Hydrogen Flow (Ton/Year)	Production Cost ($/Ton)	Transportation Cost ($/Ton)	Storage Cost ($/Ton)	Total Cost Per Unit ($/Ton)	Selling Price ($/Ton)	Profit ($/Year)
El Dorado	Culiacán	479	1984.82	35.71	290.91	2311.44	9163.75	3,282,247
El Molino	Tepic	880	1984.82	12.03	290.91	2287.82	9085.17	5,981,676
Puga		1414	1984.82	12.18	290.91	2287.92	9085.17	9,611,305
San Miguel del Naranjo	Matehuala	1980	3456.58	86.84	531.24	4074.66	8982.42	9,717,387
Aarón Sáenz	Cd. Victoria	1104	3456.58	41.85	531.24	4029.67	8841.90	5,312,714
Pánuco		1918	3436.98	94.70	531.24	4062.92	8841.90	9,166,055
El Mante	Cd. Mante	976	3456.58	10.36	531.24	3998.18	8783.10	4,670,094
Plan de Ayala	Cd. Valles	1325	3456.58	7.17	531.24	3994.99	8809.97	6,379,864
Alianza Popular		1216	3456.58	26.96	531.24	4014.78	8809.97	5,830,960
El Higo		1957	3436.98	37.03	531.24	4005.26	8809.97	9,402,801
Plan de SL	S.L.P.	1400	3456.58	126.67	531.24	4114.49	8835.66	6,609,652
Ameca	Zapopan	1050	3269.16	30.40	500.79	3800.34	8990.47	5,449,620
Bellavista		641	3269.16	28.83	500.79	3798.77	8990.47	3,327,871
José María Morelos		648	3269.16	84.53	500.79	3854.47	8990.47	3,328,129
Melchor Ocampo		1162	3269.16	64.73	500.79	3834.68	8990.47	5,991,035
Tala		1714	3269.16	15.08	500.79	3785.02	8990.47	8,922,122
Quesería	Zamora	1292	3269.16	124.41	500.79	3894.35	9078.54	6,697,967
Santa Clara		655	3269.16	38.56	500.79	3808.50	9078.54	3,451,876
Tamazula		1566	3269.16	59.58	500.79	3829.52	9078.54	8,219,962
Pedernales	Irapuato	436	3269.16	122.64	500.79	3892.58	9016.65	2,234,093
Lázaro Cárdenas	Uruapan	273	3269.16	39.15	500.79	3809.09	9000.49	1,417,253
Calipam	Tehuacán	233	3617.14	53.44	557.81	4228.39	8872.10	1,081,986
Constancia		886	3436.98	64.24	557.81	4059.04	8872.10	4,264,375
Motzorongo		1341	3436.98	58.99	557.81	4053.78	8872.10	6,461,367
Atencingo	Cuernavaca	1827	3617.14	44.01	557.81	4218.96	8915.23	8,580,084
Casasano		645	3617.14	24.66	557.81	4199.61	8915.23	3,041,575
Emiliano Zapata		1187	3617.14	22.74	557.81	4197.69	8915.23	5,599,716
Mahuixtlán	Toluca	436	3436.98	188.75	557.81	4183.55	8927.21	2,068,232
El Refugio	Azcapotzalco	475	3616.80	194.60	557.81	4369.20	8856.19	2,131,316
La Margarita		1226	3616.80	195.83	557.81	4370.43	8856.19	5,499,534
El Potrero	Añil	1707	3436.98	167.39	557.81	4162.18	8904.42	8,094,980
Progreso		913	3436.98	188.65	557.81	4183.45	8904.42	4,310,236
Adolfo López Mateos	Oaxaca	1607	3616.80	76.47	557.81	4251.08	8933.10	7,524,008
Tres Valles		2396	3436.98	86.00	557.81	4080.80	8933.10	11,626,131
Benito Juárez	Tuxtla Gutiérrez	1438	3436.98	79.47	557.81	4074.26	8781.48	6,768,982
Huixtla	Tapachula	1202	3616.80	25.98	557.81	4200.59	8927.95	5,682,272
El Modelo	Xalapa	1079	3436.98	29.96	528.29	3995.24	8816.01	5,201,617
La Gloria		1581	3436.98	29.91	528.29	3995.19	8816.01	7,621,725
El Carmen	Escamela	577	3436.98	19.79	528.29	3985.07	8797.40	2,776,706
La Providencia		811	3436.98	30.11	528.29	3995.38	8797.40	3,894,422
San José de Abajo		560	3436.98	33.79	528.29	3999.07	8797.40	2,687,057
San Miguelito		525	3436.98	55.60	528.29	4020.87	8797.40	2,507,667
San Nicolás		1103	3436.98	23.48	528.29	3988.75	8797.40	5,303,935
San Cristobal	Tierra Blanca	2584	3436.98	28.68	528.29	3993.96	8773.28	12,349,769
San Pedro		1273	3436.98	63.21	528.29	4028.49	8773.28	6,040,122
Cuatotolapam	Minatitlán	835	3436.98	44.94	528.29	4010.22	8623.23	3,851,868
Santa Rosalía	Villahermosa	781	3436.98	27.31	528.29	3992.58	8733.89	3,702,960
Azsuremex	Mérida	223	3436.98	208.10	547.35	4192.44	8524.41	966,030
La Joya		826	3553.49	79.57	547.35	4180.40	8524.41	3,588,160
San Rafel Pucté		1602	3553.49	74.71	547.35	4175.54	8524.41	6,966,908
-	Total	55,965	-	-	-	-	-	275,198,425
-	Average	1119	3352.11	66.40	519.01	3937.52	8874.71	5,503,968

The CO_2 emissions from transport and storage were estimated at 51,443,692 kg equivalent carbon dioxide per year, with transport processes contributing 23%. The optimal design of the HSC network is shown in Figure 10. The hydrogen produced is distributed across a larger number of storage terminals compared with the solution that minimized the GWP. On the other hand, it can also be observed that the distribution distances are usually shorter compared to the solution found, which maximizes the benefits of the system and reaches a central point from both limits.

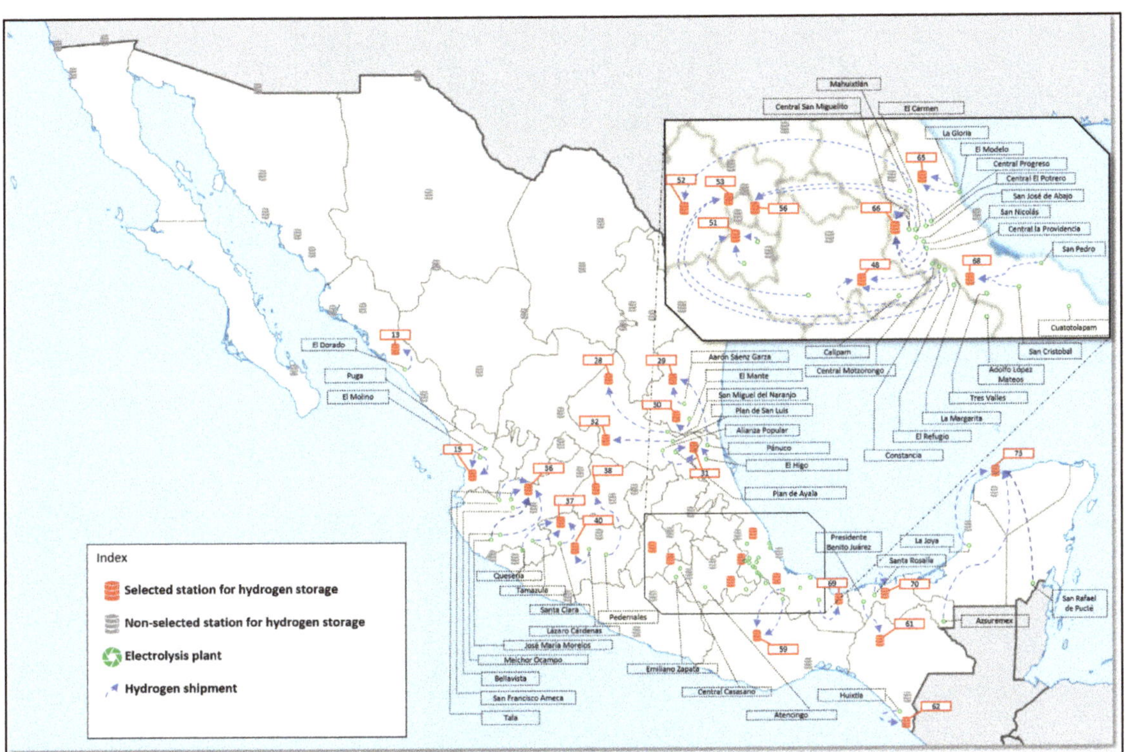

Figure 10. Optimal design for hydrogen supply chain network.

Investment Assessment and Uncertainty Analysis

An investment assessment within a horizon of 10 years was performed to estimate the internal rate of return (IRR) and payback period, using probability distributions for modeling the uncertain behavior within model inputs. The uncertainty analysis was performed using the Monte Carlo simulation methodology. In Figure 11, IRR ranges are estimated for each hydrogen receiving SDS, where it can be observed that Tepic's HSC is the most profitable case with an average of 28.90%, with minimum and maximum values oof about 15.10% and 34.20%, respectively, while Toluca's HSC is the least profitable one, with an average IRR of 15.80%, and minimum and maximum values of about 7.10% and 21%, respectively. The average IRR for all SDS is 21.50%, which is considered an acceptable value in terms of this study.

In terms of payback period, the average value for all SDS is 5.02 years. As expected, and according to the IRR, the case with the shortest payback period is Tepic, with an average value of 3.94 years, and minimum/maximum values about 3.45 and 6.11 years, respectively. In the case of the largest payback period, Toluca presented 6.12 years on average, and minimum/maximum values of 4.97 and 9.01 years, respectively. This information is presented in Figure 12.

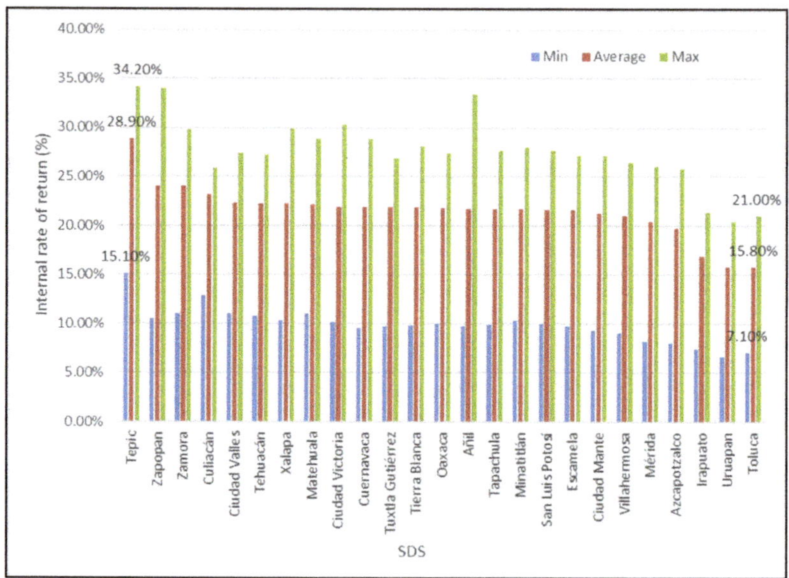

Figure 11. IRR for each SDS case.

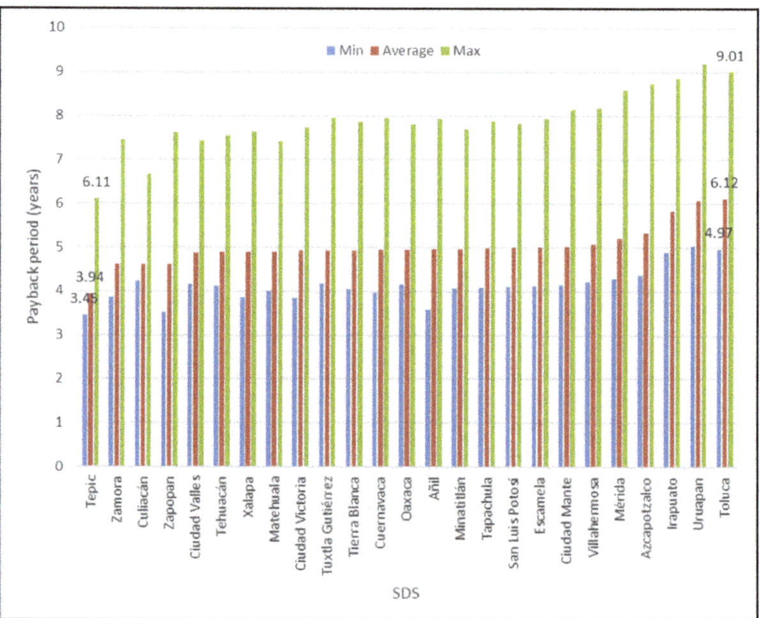

Figure 12. Payback period for each SDS Case.

From the uncertainty analysis it can be concluded that, in most cases, the HSC's deployment might turn out convenient in economic term, due to their acceptable IRR and short payback periods. There are some cases like Tepic's where the case is extremely convenient and others like Toluca's where economic indicators are not that favorable. The main reason why there are big differences between cases is the wide range of water, electricity, and fuel prices across the country, along with differences in raw sugar cane availability and quality.

6. Conclusions

The information gathered was used to develop a mathematical optimization model that estimates the main economic and environmental indicators of the HSC network operation; the optimization criteria are defined as the annual profits and GWP. The latter refers to the generation of equivalent carbon dioxide that comes from the HSC activities.

Once the optimization criteria were established, it was possible to find the optimal values of the mathematical model using the artificial intelligence tool known as GA, which was used as a first approximation under a single criteria approach to know the limits of the model and obtain the maximum and minimum values of the relevant parameters. Subsequently, both optimization criteria were optimized at the same time, so that many feasible solutions could be generated, from which the one that best met the specified criteria was selected. This optimal configuration selection was made using the TOPSIS multi-criteria decision technique. Based on these results, it was possible to observe the different configurations that the hydrogen supply chain can take, as well as the advantages and disadvantages associated with each solution. In addition, the proportion of the contribution of the many elements of the system to investment capital and operating costs, as well as their contribution to the equivalent CO_2 emissions, could be defined. The obtained results show that it turns out to be economically convenient to produce hydrogen in each of the 50 proposed production points for all the scenarios, the storage infrastructure layout distributed across strategic parts of the country exposes several advantages in terms of resource utilization, since the closeness of multiple storage points from each production plant location brings a wide scope of possible solution alternatives. Several differences can be observed between the solutions: in the profit maximization function the profit ratio of 55.67% and 1057 kg of CO_2 per ton of hydrogen is achieved, while the GWP minimization function offers an average profit ratio of 44.72% and 829 kg of CO_2 emitted due to direct hydrogen transportation and storage activities. An evaluation to quantify the economic benefits of using the available electric energy and the utilization of already existing infrastructure for hydrogen production, storage and transportation can be exposed as a starting point for considering the integration of hydrogen as an energy carrier in developing countries, with the infrastructure deployment being the most capital-intensive phase of the energy transition to a hydrogen economy.

The impact of the study relies on putting into perspective the economic and environmental benefits obtained from non-conventional energy sources, and its integration to the national energy grid, directing such energy to sectors with higher demand, like the transportation sector. The knowledge acquired supports the decision-making process during the exploration of new alternatives in the search for supplying the energy deficit in a specific region—Mexico, in this case. This paradigm opens the scope of research to new possibilities for considering economically and environmentally convenient solutions, under resource constraints and the uncertainty contained in the system. The proposed model was validated in a case study of the Mexican sugarcane industry.

Further research is recommended by adding refueling station location capabilities to the model to complete the final HSC echelon. It is also recommended to evaluate social risk by quantifying possible hazards and optimizing the risk criteria along the economic and environmental objective functions. It can be highlighted from the reviewed literature that there are few studies that integrate biomass waste utilization and hydrogen production, and even less studies using electrolysis in a biomass to power to hydrogen configuration using existing infrastructure in all the HSC echelons. As far as we know, this is the only study that considers this type of hydrogen production scheme applied to Mexican territory.

Some model limitations include that it was designed for evaluating an operation year that is divided in two periods. Moreover, the model was built for considering only the electrolysis process for hydrogen production, and existing storage infrastructure, which restricts the possibilities of the model in terms of specific location of such facilities.

Author Contributions: Conceptualization, L.M.R.-B. and A.A.A.-L.; Data curation, L.M.R.-B.; Formal analysis, L.M.R.-B.; Investigation, L.M.R.-B., J.O.R.-C. and D.V.-V.; Methodology, L.M.R.-B. and A.A.A.-L.; Project administration, J.O.R.-C. and A.A.A.-L.; Resources, J.O.R.-C. and D.V.-V.; Software, C.A.-P.; Supervision, C.G.M.-S., M.A.G.-H. and A.A.A.-L.; Validation, C.G.M.-S. and M.A.G.-H.; Writing—original draft, L.M.R.-B.; Writing—review and editing, L.M.R.-B., J.O.R.-C., C.A.-P., C.G.M.-S., D.V.-V. and A.A.A.-L. All authors have read and agreed to the published version of the manuscript.

Funding: This research was funded by Tecnológico Nacional de México grant number 7737.20P, and by Consejo Nacional de Ciencia y Tecnología through a scholarship given to Luis Miguel Reyes Barquet (Main author) with CVU: 920654. The APC was funded by Daniel Villanueva Vásquez.

Institutional Review Board Statement: Not Applicable.

Informed Consent Statement: Not Applicable.

Data Availability Statement: Public data sets available at SAGARPA, "Planeación Agrícola Nacional 2017–2030," 2016 (https://www.gob.mx/agricultura/acciones-y-programas/planeacion-agricola-nacional-2017--2030-126813 accessed on 22 March 2020); CONADESUCA, 6to. Informe Estadístico del Sector Agroindustrial de la Caña de Azúcar en México, zafras 2009–2010/2018–2019, Comité Nacional para el Desarrollo Sustentable de la Caña de Azúcar, 2019 (https://siiba.conadesuca.gob.mx/Archivos_Externos/6to_informe_estad%C3%ADstico.pdf accessed on 17 February 2020); Comisión Nacional de hidrocarburos, "Reservas de hidrocarburos en México conceptos fundamentales y análisis 2018" (https://www.gob.mx/cnh/documentos/analisis-de-informacion-de-las-reservas-de-hidrocarburos-de-mexico-al-1-de-enero-del-2018?idiom=es accessed on 25 February 2020); Mendoza A., Cadena A. and de Buen O., Estudio de pesos y transportes, Secretaría de comunicaciones y transporte, 2010.

Acknowledgments: We thank CONACYT, the Corporate Porres Group, Orizaba Institute of Technology, and all professor-researchers for their support, and the National Technology of Mexico for funding the project with reference number 7737.20-P, entitled "Multi-criteria Optimization of a Hydrogen Supply Chain Generated from Agro-industrial Waste".

Conflicts of Interest: The authors declare that they have no known competing financial interests or personal relationships that could have appeared to influence the work reported in this paper.

Appendix A. Calculations for Estimating Model Inputs

In Equation (A1), the bagasse availability is calculated in tons for each sugar mill i using the quantity of raw sugarcane and the mass fraction of bagasse, both represented by probability distributions.

$$AvBag_i = tCane_i * \%BagInCane_i, \quad \forall\, i = 1\ldots50 \tag{A1}$$

In Equation (A2) Operation hours parameter for each period z is calculated considering the number of operation days in each period z (modeled using probability distributions) and the downtime during operation.

$$OpHrs_z = (DOp_z * 24) * (100\% - \%Downtime), \quad \forall\, 1,\, 2 \tag{A2}$$

The quantity of bagasse per hour combusted in the boilers of each sugar mill i is calculated using Equation (A3).

$$BagBrn_{iz} = \frac{AvBag_i}{OpHrs_z}, \quad \forall\, z = 1,\, 2\, ; i = 1\ldots50 \tag{A3}$$

The lower bagasse energy content in cal/ton is estimated using Equation (A4) extracted from [32], where bagasse humidity ($BagHum_i$) is an uncertain parameter, modeled by probability distributions for each mill i.

$$BagECont_i = 17{,}799.3 - 20{,}305.98 * BagHum_i \; \forall \, i = 1 \ldots 50 \tag{A4}$$

Equation (A5) calculates the bagasse energy flow per hour.

$$BagEFlow_{iz} = BagBrn_{iz} * BagECont_i \, , \forall \, i = 1 \ldots 50 \tag{A5}$$

Steam production in tons at each sugar mill i is calculated using Equation (A6).

$$Steam_{iz} = \frac{BagEFlow_i \left(\frac{BoilerEf}{DEnthalpy} \right)}{1000}, \; \forall \, i = 1 \ldots 50 \tag{A6}$$

Electric power generation in MWh at each mill i is estimated using Equation (A7).

$$ElecPwr_{iz} = \frac{Steam_i - (\%SelfCons * Steam_i)}{GenPerf}, \; \forall \, i = 1 \ldots 50 \tag{A7}$$

Table A1. Complementary calculations glossary.

Variable	Description
%Downtime	Fraction of inactivity time (%)
%SteamSelfCons	Percentage of steam consumption (%)
$AvBag_i$	Available bagasse at each sugar mill i (tons)
$BagBrn_{iz}$	Bagasse burning flow at mill i during period z (tons/hour)
$BagECont_i$	Bagasse energy content at mill i (kcal/ton)
$BagEFlow_{iz}$	Bagasse energy content flow at mill i (kcal/hour)
$BagHum_i$	Mass fraction of humidity content at mill i (%)
$BagInCane_i$	Mass fraction of bagasse in sugar cane at each sugar mill i (bagasse tons/sugarcane tons)
BoilerEf	Boiler efficiency (%)
DEnthalpy	Steam delta enthalpy (kcal/cm^2)
$ElecPwr_{iz}$	Electric power generation at mill i during period z (MWh)
GenPerf	Electric generator turbine performance (steam tons/MWh)
DOp_z	Operation days during period z (days)
$OpHrs_z$	Operation hours during period z (hours)
$Steam_{iz}$	Steam production at mill i during period z (tons/hour)
$tCane_i$	Sugar cane available at each sugar mill i (tons)

Appendix B

Table A2. Probability distributions for bagasse availability modelling.

Sugar Mill	tCane (Tons)	BagInCane	BagHum (%)
Aaron Sáenz	RiskLaplace (1,062,951, 162,684.8)	RiskExtvalueMin (0.28208, 0.0052635)	RiskPareto (45.277, 50.01)
Alianza popular	RiskPareto (15.534, 1,091,755)	RiskPareto (17.647, 0.24674)	RiskUniform (42.853, 54.287)
Ameca	RiskUniform (1,032,772, 1,314,071)	RiskExtvalueMin (0.24318, 0.007397)	RiskPareto (47.183, 49.841)
Atencingo	RiskUniform (1,539,709, 1,931,089)	RiskExtvalueMin (0.28181, 0.0017849)	RiskPareto (227.42, 50.64)
Azsuremex	RiskUniform (111,320, 236,294)	RiskExtvalueMin (0.35416, 0.024192)	RiskExtvalueMin (51.1982, 0.88002)

Table A2. *Cont.*

Sugar Mill	tCane (Tons)	BagInCane	BagHum (%)
Bellavista	RiskUniform (544,556, 767,230)	RiskLaplace (0.26549, 0.0042446)	RiskExtvalueMin (51.7613, 0.39862)
Benito Juárez	RiskUniform (915,567, 1,669,420)	RiskExtvalueMin (0.29877, 0.0024705)	RiskExtvalueMin (51.2247, 0.46764)
Calipam	RiskLaplace (185,777.6667, 24,246.0872)	RiskPareto (17.107, 0.31175)	RiskExtvalueMin (50.8465, 0.70592)
Casasano La abeja	RiskPareto (17.203, 581,923)	RiskPareto (34.074, 0.25738)	RiskKumaraswamy (0.075606,0.18032, 46.1,51.18)
Constancia	RiskPareto (10.619, 751,826)	RiskLaplace (0.27543, 0.010389)	RiskPareto (98.361, 49.106)
Cuatolapam	RiskPareto (8.3168, 669,112)	RiskExtvalue (0.283955, 0.016257)	RiskUniform (49.9225, 51.9875)
El Carmen	RiskExtvalueMin (565,173.2923, 110,856.4894)	RiskExtvalueMin (0.323, 0.010938)	RiskKumaraswamy (0.078411, 0.19166, 50.629, 53.053)
El Higo	RiskNormal (1,758,914, 89,388)	RiskNormal (0.3233037, 0.0076643)	RiskUniform (51.7425, 56.0475)
El Mante	RiskUniform (606,942, 1,101,350)	RiskKumaraswamy (0.076156, 0.18217, 0.296446, 0.314114)	RiskLaplace (51.1, 0.44173)
El Modelo	RiskExtvalueMin (1,059,250.2819, 96,686.0013)	RiskPareto (25.15, 0.26806)	RiskTriang (48.7756, 50.41, 50.41)
El Molino	RiskPareto (5.0488, 681,227)	RiskPareto (77.099, 0.27102)	RiskPareto (135.6, 50.25)
El Potrero	RiskNormal (1,629,870, 78,703)	RiskPareto (66.285, 0.2666)	RiskTriang (47.8444, 50.61, 50.61)
El Refugio	RiskExtvalueMin (460,201.2784, 48,913.5247)	RiskPareto (145.56, 0.28926)	RiskPareto (63.715, 49.85)
El Dorado	RiskNormal (451,622, 124,580)	RiskPareto (20.357, 0.26842)	RiskTriang (48.5712, 51.865, 51.865)
Emiliano Zapata	RiskUniform (1,001,194, 1,241,654)	RiskPareto (12.091, 0.26608)	RiskKumaraswamy (0.079838, 0.18665, 48.426, 54.43)
Huixtla	RiskUniform (865,578, 1,386,963)	RiskLaplace (0.27892, 0.016637)	RiskLaplace (50.12, 0.52322)
José Ma Morelos	RiskLaplace (573,662, 97,203.5759)	RiskLaplace (0.30045, 0.0091253)	RiskTriang (48.274, 52.01, 52.01)
La Gloria	RiskExtvalue (1,387,788, 128,254)	RiskLaplace (0.27426, 0.0057259)	RiskKumaraswamy (0.073444, 0.19034, 47.59, 50.08)
La Joya	RiskPareto (6.2914, 662,566)	RiskUniform (0.260448, 0.28558)	RiskPareto (25.533, 48.01)
La Margarita	RiskExtvalueMin (1,114,659.5247, 65,442.6361)	RiskPareto (69.982, 0.29615)	RiskKumaraswamy (0.081137, 0.18753, 48.63, 51.85)
La providencia	RiskUniform (622,858, 921,585)	RiskPareto (20.115, 0.25945)	RiskKumaraswamy (0.074596, 0.18167, 47.5, 51.71)
Lázaro Cárdenas	RiskUniform (220,651, 420,987)	RiskPareto (25.779, 0.21863)	RiskKumaraswamy (0.074316, 0.18577, 49.732, 51.932)
López Mateos	RiskLaplace (1,552,596, 164,296.2606)	RiskExtvalue (0.2769587, 0.004824)	RiskPareto (51.682, 50.35)
Mahuixtlan	RiskUniform (345,480, 488,480)	RiskExtvalueMin (0.27271, 0.0014487)	RiskLaplace (49.9522, 0.10657)
Melchor Ocampo	RiskLaplace (1,110,585, 54,862.1928)	RiskLaplace (0.28742, 0.0042788)	RiskKumaraswamy (0.075628, 0.18143, 50.36, 53.11)
Motzorongo	RiskLaplace (1,301,433, 203,462.3613)	RiskPareto (24.532, 0.25684)	RiskLaplace (49.89, 0.33796)
Panuco	RiskUniform (1,299,749, 1,906,185)	RiskPareto (48.802, 0.31117)	RiskExtvalue (50.1014, 1.0208)

Table A3. Probability distributions for operation days and bagasse utilization.

Variable	Probability Distribution	Unit
OpDays during harvesting period (z = 1)	Pert (155,160,179)	Days
OpDays during non-harvesting season (z = 2)	Pert (30,32.82,35.65)	Days
AvBag for energy production (z = 1)	Pert (52%,52.42%,52.848%)	% de Bagazo
AvBag for energy production (z = 2)	Pert (7%,7.33%,7.68%)	% de Bagazo

Appendix C

Table A4. Probability distributions for electricity and water prices modelling.

Region (r)	Electricity Price ($/MW)	Water Price ($/m^3)
Northwest	Pert (26.23, 35.23, 44.19)	Pert (0.18, 0.40, 0.56)
North	Pert (26.23, 35.23, 44.19)	Pert (0.18, 0.40, 0.56)
Northeast	Pert (41.06, 64.33, 79.26)	Pert (0.07, 0.24, 0.73)
West	Pert (37.21, 60.66, 76.98)	Pert (0.13, 0.23, 0.44)
Center	Pert (42.99, 67.58, 86.21)	Pert (0.038, 0.11, 0.238)
South	Pert (42.99, 67.58, 86.21)	Pert (0.025, 0.093, 0.159)
Gulf	Pert (41.23, 64, 81.47)	Pert (0.105, 0.236, 0.236)
Southeast	Pert (42.42, 66.28, 81)	Pert (0.0951, 0.190, 0.190)

Appendix D

Table A5. Storage availability and probability distributions for fuel prices modelling.

Region	State	ID (t)	Name	Design Capacity (Barrels)	Utilization Rate	Fuel Price (MX$)
Northwest	B.C. Norte	1	ROSARITO	1,393,000	0.73	RiskLogistic (19.20514, 0.18998)
	B.C. Norte	2	ENSENADA	135,000	0.74	RiskLogistic (19.39158, 0.18992)
	B.C. Norte	3	MEXICALI	155,000	0.76	RiskLogistic (19.45041, 0.19028)
	Sonora	4	NOGALES	45,000	0.77	RiskLaplace (19.6776, 0.30941)
	Sonora	5	MAGDALENA	40,000	0.67	RiskLaplace (19.6675, 0.32126)
	Sonora	6	HERMOSILLO	125,000	0.69	RiskLaplace (19.3266, 0.32346)
	Sonora	7	GUAYMAS	750,000	0.71	RiskLaplace (19.1096, 0.32513)
	Sonora	8	CIUDAD OBREGÓN	170,000	0.66	RiskLaplace (19.3257, 0.32251)
	Sonora	9	NAVOJOA	35,000	0.72	RiskLoglogistic (15.3836, 4.3047, 24.893)
	B.C. Sur	10	LA PAZ	230,000	0.7	RiskExtvalueMin (19.6679, 0.37766)
	Sinaloa	11	TOPOLOBAMPO	760,000	0.71	RiskTriang (17.9917, 19.7924, 20.1903)
	Sinaloa	12	GUAMÚCHIL	105,000	0.71	RiskTriang (18.7036, 20.2588, 20.8076)
	Sinaloa	13	CULIACÁN	115,000	0.74	RiskTriang (18.8595, 20.0375, 20.6478)
	Sinaloa	14	MAZATLÁN	620,000	0.75	RiskWeibull (5.175, 1.5556)
	Nayarit	15	TEPIC	95,000	0.7	RiskLaplace (19.6781, 0.27458)
North	Chihuahua	16	CIUDAD JUÁREZ	245,000	0.75	RiskLaplace (18.6858, 0.32223)
	Chihuahua	17	CHIHUAHUA	420,000	0.8	RiskLaplace (19.1491, 0.30599)
	Durango	18	DURANGO	75,000	0.69	RiskLaplace (19.6863, 0.27829)
	Chihuahua	19	PARRAL	55,000	0.73	RiskLaplace (19.6639, 0.3026)
	Durango	20	GÓMEZ PALACIO	475,000	0.72	RiskLaplace (19.5364, 0.30492)
Northeast	Coahuila	21	SABINAS	100,000	0.73	RiskLaplace (19.5153, 0.319)
	Coahuila	22	MONCLOVA	235,000	0.77	RiskLaplace (19.4711, 0.33153)
	Tamaulipas	23	NUEVO LAREDO	75,000	0.78	RiskLaplace (19.34, 0.3101)
	Tamaulipas	24	REYNOSA	23,500	0.62	RiskLaplace (19.3046, 0.33903)
	Nuevo León	25	SANTA CATARINA	850,000	0.69	RiskLoglogistic (18.23, 1.0127, 6.1548)
	Nuevo León	26	SALTILLO	151,000	0.78	RiskLaplace (19.4162, 0.33261)
	Nuevo León	27	CADEREYTA	100,000	0.75	RiskLoglogistic (17.4049, 1.7244, 10.6)
	SLP	28	MATEHUALA	33,000	0.74	RiskLoglogistic (18.1427, 1.272, 7.2404)
	Tamaulipas	29	CIUDAD VICTORIA	195,000	0.75	RiskLoglogistic (17.8593, 1.2518, 7.2491)
	Tamaulipas	30	CIUDAD MANTE	21,000	0.71	RiskLaplace (19.0238, 0.35456)
	SLP	31	CIUDAD VALLES	75,000	0.74	RiskLoglogistic (17.792, 1.2502, 7.2677)
	SLP	32	SAN LUIS POTOSÍ	100,000	0.69	RiskLaplace (19.1377, 0.34971)

Table A5. *Cont.*

Region	State	ID (t)	Name	Design Capacity (Barrels)	Utilization Rate	Fuel Price (MX$)
West	Zacatecas	33	ZACATECAS	85,000	0.68	RiskLaplace (19.5594, 0.3408)
	Aguascalientes	34	AGUASCALIENTES	105,000	0.65	RiskLaplace (19.5644, 0.33496)
	Guanajuato	35	LEÓN	110,000	0.73	RiskLaplace (19.5183, 0.32495)
	Jalisco	36	ZAPOPAN	390,000	0.72	RiskLoglogistic (18.47193, 0.94869, 5.5621)
	Michoacán	37	ZAMORA	90,000	0.71	RiskLaplace (19.6637, 0.32359)
	Guanajuato	38	IRAPUATO	430,000	0.73	RiskLaplace (19.5297, 0.31447)
	Guanajuato	39	CELAYA	180,000	0.72	RiskLaplace (19.5235, 0.32444)
	Michoacán	40	URUAPAN	130,000	0.79	RiskLoglogistic (18.1592, 1.2971, 7.5307)
	Colima	41	COLIMA	55,000	0.79	RiskLoglogistic (18.1186, 1.1784, 7.112)
	Michoacán	43	MORELIA	135,000	0.73	RiskLaplace (19.5371, 0.30931)
	Jalisco	44	EL CASTILLO	345,000	0.64	RiskLoglogistic (18.52751, 0.91876, 5.1437)
	Michoacán	45	LÁZARO CÁRDENAS	830,000	0.73	RiskLaplace (18.7947, 0.33233)
	Colima	46	MANZANILLO	465,000	0.71	RiskLaplace (18.773, 0.31928)
Center	Morelos	47	CUAUTLA	60,000	0.75	RiskLaplace (19.3723, 0.31474)
	Puebla	48	PUEBLA	425,000	0.71	RiskLaplace (19.2147, 0.31217)
	Puebla	49	TEHUACÁN	45,000	0.72	RiskLaplace (19.2166, 0.32322)
	Querétaro	50	QUERÉTARO	230,000	0.72	RiskLaplace (19.4604, 0.31185)
	Edo. De México	51	SAN JUAN IXHUATEPEC	225,000	0.62	RiskLoglogistic (18.26004, 0.9894, 5.5995)
	Morelos	52	CUERNAVACA	135,000	0.76	RiskLoglogistic (18.0638, 1.2074, 7.239)
	Edo. De México	53	TOLUCA	195,000	0.69	RiskLoglogistic (17.5463, 1.7658, 11.077)
	CDMX	54	AZCAPOTZALCO	1,500,000	0.74	RiskLoglogistic (18.0401, 1.1, 6.6497)
	Hidalgo	55	PACHUCA	170,000	0.71	RiskLoglogistic (18.0877, 1.0409, 6.3148)
	CDMX	56	BARRANCA DEL MUERTO	125,000	0.73	RiskLoglogistic (18.26353, 0.99106, 5.6165)
	CDMX	57	AÑIL	235,000	0.67	RiskLoglogistic (18.24477, 0.99028, 5.7233)
South	Guerrero	58	IGUALA	60,000	0.7	RiskLaplace (19.4913, 0.30988)
	Guerrero	59	ACAPULCO	235,000	0.62	RiskLaplace (19.1366, 0.31701)
	Oaxaca	60	OAXACA	110,000	0.76	RiskLaplace (19.3487, 0.31066)
	Oaxaca	61	SALINA CRUZ*	1,479,000	0.76	RiskLogistic (18.86307, 0.18242)
	Oaxaca	62	SALINA CRUZ	205,000	0.75	RiskLogistic (18.86307, 0.18242)
	Chiapas	63	TUXTLA GUTIÉRREZ	105,000	0.71	RiskLogistic (19.02036, 0.17406)
	Chiapas	64	TAPACHULA*	24,500	0.62	RiskLaplace (19.3375, 0.30994)
	Chiapas	65	TAPACHULA II	65,000	0.78	RiskLaplace (19.3375, 0.30994)
Gulf	Veracruz	66	POZA RICA	55,000	0.7	RiskLaplace (18.8571, 0.31891)
	Veracruz	67	PEROTE	25,000	0.74	RiskLoglogistic (17.8551, 1.265, 7.42)
	Veracruz	68	XALAPA	45,000	0.6	RiskLoglogistic (17.8126, 1.2419, 7.1738)
	Veracruz	69	ESCAMELA	98,000	0.72	RiskLaplace (19.0548, 0.32629)
	Veracruz	70	VERACRUZ	536,000	0.66	RiskLaplace (18.4593, 0.32756)
	Veracruz	71	TIERRA BLANCA	71,000	0.69	RiskLaplace (19.0025, 0.31694)
	Veracruz	72	MINATITLÁN	10,000	0.59	RiskLogistic (18.67753, 0.18353)
	Tabasco	73	VILLAHERMOSA	328,500	0.72	RiskLaplace (18.9172, 0.31921)
Southeast	Yucatán	74	PROGRESO	280,500	0.71	RiskLaplace (18.4223, 0.32023)
	Campeche	75	CAMPECHE	265,000	0.79	RiskLaplace (18.9739, 0.31608)
	Yucatán	76	MÉRIDA	148,000	0.77	RiskLaplace (18.4635, 0.31978)

Appendix E

Distance Matrix (km) — Northwest (BC, BCS, Sonora, Sinaloa, Nayarit) and North (Chihuahua, Durango)

Region	City	ROSARITO	ENSENADA	MEXICALI	NOGALES	MAGDALENA	HERMOSILLO	GUAYMAS	CIUDAD OBREGÓN	NAVOJOA	LA PAZ	TOPOLOBAMPO	GUAMÚCHIL	CULIACÁN	MAZATLÁN	TEPIC	CD JUAREZ	CHIHUAHUA	DURANGO	PARRAL	GOMEZ PALACIO
Northwest	El Molino	2052	2070	1873	1430	1343	1150	1021	908	824	913	684	583	474	271	6.5	1449	1145	481	895	707
	El Dorado	1633	1652	1455	1012	924	732	603	484	405	495	275	168	55.5	175	448	1400	1058	432	847	658
	Puga	2059	2244	1880	1437	1350	1157	1028	909	830	920	700	590	481	278	19.2	1455	1152	488	902	713
Northeast	Aaron Saenz	2688	2753	2488	1989	1946	2155	2031	1905	1838	4119	1557	1444	1344	1181	862	1586	1184	774	1069	735
	Alianza Popular	2727	2792	2527	2219	2129	1940	1816	1690	1623	4189	1488	1374	1274	1061	799	1637	1268	709	1153	842
	El Monte	2714	2780	2515	2003	1960	2169	2045	1919	1852	4146	1545	1432	1332	1119	850	1581	1212	771	1097	764
	Plan de Ayala	2775	2839	2574	2265	2176	1987	1862	1737	1669	4235	1534	1421	1321	1108	838	1682	1313	755	1198	865
	Plan de SL	2682	2683	2710	2012	1973	1933	1803	1685	1610	1716	1475	1381	1257	1050	847	1628	1257	787	1129	807
	San Miguel	2876	2647	2674	1976	1937	1875	1746	1627	1552	1658	1418	1323	1199	992	789	1592	1221	729	1093	771
West	Ameca	2253	2272	2074	1629	1544	1351	1223	1104	1025	1115	895	783	676	473	192	2089	1237	702	1097	799
	Bellavista	2279	2279	2100	1655	1570	1377	1248	1129	1050	1140	920	810	701	498	226	1572	1203	667	1088	764
	Jose Ma Morelos	2442	2477	2278	1832	1747	1554	1425	1306	1231	1318	1097	1002	878	673	398	1758	1387	851	1259	949
	Lazaro Cardenas	2524	2560	2360	1914	1829	1636	1507	1388	1313	1400	1179	1084	960	756	480	1721	1550	814	1222	913
	Melchor Ocampo	2398	2434	2234	1788	1703	1511	1381	1262	1187	1274	1053	958	834	630	354	1714	1343	808	1215	906
	Pedernales	2632	2668	2468	2022	1937	1744	1615	1496	1421	1508	1287	1192	1068	864	588	1739	1368	832	1240	930
	Queseria	2383	2419	2219	1773	1688	1495	1366	1247	1172	1259	1038	943	819	615	339	1699	1328	793	1200	891
	Santa Clara	2442	2477	2278	1832	1747	1554	1425	1306	1231	1318	1097	1002	878	673	398	1677	1308	772	1179	870
	Tala	2232	2268	2068	1622	1538	1345	1215	1097	1022	1108	887	793	668	464	189	1567	1198	662	1083	759
	Tamazula	2374	2410	2211	1765	1680	1487	1358	1239	1164	1251	1029	935	811	606	331	1685	1316	781	1202	878
Center	Atencingo	2922	2988	2722	2307	2218	2029	1905	1779	1712	4353	1576	1463	1363	1150	881	1981	1612	1077	1498	1174
	Calipam	3021	3087	2822	2407	2317	2128	2004	1878	1811	4453	1676	1562	1463	1250	981	2067	1698	1163	1584	1260
	Casasano	2858	2923	2658	2243	2154	1965	1840	1715	1647	4289	1512	1399	1299	1086	817	1883	1514	979	1399	1076
	Emiliano Zapata	2878	2944	2679	2264	2174	1985	1861	1735	1668	4310	1533	1420	1320	1107	838	1909	1540	1005	1426	1102
South	El refugio	3114	3179	2914	2499	2410	2221	2096	1971	1903	4545	1768	1655	1555	1342	1073	2115	1764	1232	1622	1818
	Huixtla	3873	3939	3674	3259	3169	2980	2856	2730	2663	5305	2528	2414	2315	2102	1833	2875	2523	1991	2392	2077
	La margarita	3158	3223	2958	2543	2454	2265	2140	2015	1947	4589	1812	1699	1599	1386	1117	2158	1764	1232	1622	1318
	López Mateos	3210	3275	3010	2595	2506	2317	2192	2067	1999	4641	1864	1751	1651	1438	1169	2274	1922	1328	1781	1414
Gulf	Azsuremex	3733	3799	3534	3119	3029	2840	2716	2590	2523	5165	2388	2274	2174	1961	1693	2735	2383	1851	2242	1937
	Benito Juárez	3471	3537	3272	2857	2767	2578	2454	2328	2261	4903	2126	2012	1912	1699	1431	2473	2121	1589	1980	1675
	Constancia	3114	3179	2914	2499	2410	2221	2096	1971	1903	4545	1768	1655	1555	1342	1073	2115	1764	1232	1622	1318
	Cuatotolapan	3292	3357	3092	2677	2588	2399	2274	2149	2081	4723	1946	1833	1733	1520	1251	2293	1942	1410	1800	1496
	El Carmen	3038	3104	2839	2424	2334	2145	2021	1895	1828	4470	1693	1579	1479	1267	998	2040	1688	1156	1547	1242
	El Higo	2965	3030	2765	2350	2261	2072	1947	1822	1754	4396	1619	1506	1406	1193	924	1754	1402	870	1261	956
	El Modelo	3156	3222	2957	2542	2452	2263	2139	2013	1946	4588	1811	1697	1597	1384	1115	2136	1785	1253	1643	1339
	El Potrero	3064	3129	2864	2449	2360	2171	2046	1921	1853	4495	1718	1605	1505	1292	1023	2065	1714	1182	1572	1268
	La Gloria	3156	3222	2957	2542	2452	2263	2139	2013	1946	4588	1811	1697	1597	1384	1115	2136	1785	1253	1643	1339
	La providencia	3091	3156	2891	2476	2387	2198	2073	1948	1880	4522	1745	1632	1532	1319	1050	2092	1741	1209	1599	1295
	Mahuixtlán	3092	3157	2892	2477	2388	2198	2074	1949	1881	4523	1746	1633	1533	1320	1051	2072	1721	1188	1579	1275
	Motzorongo	3103	3168	2903	2488	2399	2210	2085	1960	1892	4534	1757	1644	1544	1331	1062	2104	1753	1221	1611	1307
	Panuco	2998	3063	2798	2383	2294	2105	1980	1855	1787	4429	1652	1539	1439	1226	957	1787	1435	903	1294	989
	Progreso	3278	3343	3078	2663	2574	2385	2260	2135	2067	4709	1932	1819	1719	1506	1237	2279	1928	1396	1786	1482
	San Cristobal	3212	3277	3012	2597	2508	2319	2194	2069	2001	4643	1866	1753	1653	1440	1171	2213	1862	1330	1720	1416
	San José de Abajo	3077	3142	2877	2462	2373	2184	2059	1934	1866	4508	1731	1618	1518	1305	1036	2078	1727	1195	1585	1281
	San Miguelito	3064	3129	3064	3129	2864	2449	2360	2171	2046	1921	1853	4495	1718	1605	1505	1292	1023	2065	1714	1182
	San Nicolas	3064	3129	3064	3129	2864	2449	2360	2171	2046	1921	1853	4495	1718	1605	1505	1292	1023	2065	1714	1182
	San Pedro	3278	3343	3078	2663	2574	2385	2260	2135	2067	4709	1932	1819	1719	1506	1237	2279	1928	1396	1786	1482
	Santa Rosalia	3471	3537	3272	2857	2767	2578	2454	2328	2261	4903	2126	2012	1912	1699	1431	2473	2121	1589	1980	1675
	Tres Valles	3153	3218	2953	2538	2449	2260	2135	2010	1942	4584	1807	1694	1594	1381	1112	2154	1803	1271	1661	1357
Southeast	La Joya	3838	3094	3639	3224	3134	2945	2821	2695	2628	5270	2493	2379	2280	2067	1798	2840	2488	1956	2347	2042
	Pucte	4105	4276	4011	3596	3506	3317	3193	3067	3000	5642	2865	2751	2651	2438	2169	3212	2860	2328	2719	2414

Northeast (NL, Tam, SLP, Coa) and West (Zacatecas, Aguascalientes, Jalisco, Colima, Michoacán y Gto)

SABINAS	MONCLOVA	NUEVO LAREDO	REYNOSA	SANTA CATARINA	SALTILLO	CADEREYTA	MATEHUALA	CIUDAD VICTORIA	CIUDAD MANTE	CIUDAD VALLES	SAN LUIS POTOSÍ	ZACATECAS	AGUASCALIENTES	LEÓN	ZAPOPAN	ZAMORA	IRAPUATO	CELAYA	URUAPAN	COLIMA	MORELIA	EL CASTILLO	LÁZARO CÁRDENAS	MANZANILLO
1103	991	1268	1274	1047	978	1096	737	872	850	839	561	527	412	417	191	356	488	538	467	380	471	229	702	460
1104	992	1270	1276	1049	979	1098	1219	1354	1332	1321	1042	733	850	856	630	795	892	976	906	819	910	668	1141	898
1103	991	1268	1274	1047	978	1096	737	872	850	839	561	539	423	429	203	368	500	549	479	392	483	241	714	471
748	602	627	437	412	471	381	304	112	31.1	125	311	483	475	498	653	651	503	548	721	848	679	645	942	927
844	738	837	610	606	561	554	306	276	146	50	246	418	410	428	588	573	438	483	656	783	542	580	877	862
717	610	655	465	440	499	409	301	132	4.4	99	308	480	472	495	650	636	500	545	718	845	604	642	939	924
890	784	752	562	536	607	505	352	229	97	1.6	291	460	550	420	635	619	487	498	702	823	587	621	918	908
810	724	746	567	540	546	512	291	231	99	44	318	488	584	501	663	647	514	526	730	851	615	649	946	935
794	688	707	528	501	510	473	255	192	60	84	260	430	426	444	605	589	456	498	672	793	557	591	888	878
1089	1089	1087	1072	1072	792	921	610	739	721	703	416	414	293	299	72	239	337	417	349	234	352	110	580	318
1054	948	1053	1037	821	757	887	576	705	688	669	382	381	259	265	49	202	299	383	312	176	315	66	500	260
1241	1134	1232	1221	1008	944	1069	757	886	896	850	563	562	440	446	231	381	481	565	493	222	496	247	476	131
1150	1043	1137	1075	911	866	972	611	740	723	704	428	525	403	297	296	140	228	234	18	32	132	282	214	565
1197	1091	1189	1177	965	901	1025	713	842	825	806	519	518	396	402	187	339	417	521	449	176	453	203	499	379
1168	1061	1155	1092	929	884	990	629	758	740	722	446	543	421	315	404	268	246	252	133	589	150	390	277	673
1182	1076	1174	1161	949	889	1010	698	827	810	791	504	503	381	387	172	324	422	506	270	34.5	438	188	407	119
1145	1018	1132	1069	906	861	967	606	735	717	699	423	482	360	236	314	81	225	275	82	219	226	200	313	301
1049	943	1046	1033	822	758	882	570	699	682	663	376	383	251	259	32.5	199	297	378	310	214	313	70.7	541	299
1168	1062	1165	1153	941	877	1001	690	819	801	783	496	503	373	379	206	175	287	336	213	88.4	309	180	398	173
1328	1222	1315	1075	1090	1045	1151	789	918	741	669	607	789	681	575	757	621	509	436	604	897	489	743	685	982
1417	1311	1404	1131	1179	1135	1240	870	862	789	718	687	870	762	656	838	702	590	537	684	1024	570	825	901	1115
1231	1124	1217	1012	992	947	1053	691	821	699	784	509	692	584	478	640	501	412	359	486	824	371	626	702	908
1257	1151	1244	1086	1116	974	1080	718	847	752	811	536	718	611	504	663	526	438	386	509	847	394	649	592	931
1460	1348	1444	1113	1223	1173	1272	914	878	769	684	737	928	820	713	896	759	647	595	742	1080	627	882	958	1164
2220	2108	2204	1792	1983	1933	2032	1674	1557	1448	1363	1497	1687	1584	1478	1661	1524	1412	1359	1507	1845	1392	1646	1723	1929
1504	1392	1488	1077	1267	1217	1316	958	841	732	648	781	945	837	731	913	777	665	612	759	1098	645	899	976	1182
1556	1444	1540	1129	1319	1269	1368	1010	893	784	700	833	1018	910	803	986	849	737	685	832	1170	717	971	1048	1254
2079	1973	2066	1641	1841	1795	1842	1540	1379	1307	1245	1357	1581	1455	1333	1500	1376	1266	1213	1363	1680	1254	1491	1605	1785
1812	1705	1799	1373	1573	1528	1575	1273	1112	1039	1039	1090	1319	1193	1071	1238	1114	1004	951	1101	1418	992	1229	1343	1523
1456	1349	1443	1065	1217	1172	1278	917	804	731	730	734	928	820	713	896	759	647	595	742	1080	627	882	958	1164
1625	1519	1612	1186	1350	1341	1388	1086	925	851	852	903	1139	1014	891	1059	934	825	771	922	1239	813	1050	1163	1344
1392	1285	1379	1106	1153	1108	1214	851	982	772	700	670	886	760	638	805	681	571	518	668	985	559	796	910	1090
973	867	820	601	613	589	585	434	313	182	108	371	568	568	574	732	707	559	436	820	912	711	723	1062	1017
1450	1148	1166	910	960	1167	932	911	669	596	595	729	1004	857	734	923	798	668	614	786	1082	677	914	1028	1187
1422	1315	1409	1057	1183	1152	1244	883	796	723	722	700	911	786	663	831	706	597	543	694	1011	585	822	935	1116
1450	1144	1162	926	956	1189	928	920	665	593	591	737	1004	857	734	923	798	668	614	786	1082	677	914	1028	1187
1430	1324	1417	1062	1192	1146	1252	891	801	728	591	708	938	813	690	858	733	624	570	721	1018	612	849	962	1143
1400	1294	1202	945	975	1116	947	861	684	612	611	678	939	793	670	858	734	601	550	722	1018	613	849	963	1123
1450	1343	1436	1071	1211	1166	1073	911	810	737	737	728	3103	3103	3103	3103	3103	3103	3103	3103	3103	3103	3103	3103	3103
871	764	783	543	576	636	548	476	264	184	149	413	601	601	607	765	740	592	647	846	945	737	756	1088	1050
1436	1329	1422	1014	1197	1152	1036	896	773	184	699	714	1125	1000	877	1045	920	811	757	908	1225	799	1036	1149	1330
1551	1444	1369	1112	1312	1267	1114	1012	851	779	778	829	1059	934	811	979	854	745	691	842	1159	733	970	1083	1264
1426	1320	1413	1059	1188	1142	1248	887	798	725	725	704	924	799	676	844	719	610	536	707	1024	598	835	948	1129
1572	1268	1197	1067	1172	1127	1233	871	806	734	733	689	911	786	663	831	706	597	543	694	1011	585	822	935	1116
1418	1311	1404	1077	1179	1134	1240	878	817	744	743	696	911	786	663	831	706	597	543	694	1011	585	822	935	1116
1594	1488	1324	1066	1096	1310	1068	1055	805	713	752	872	1125	1000	877	1045	920	811	757	908	1225	799	1036	1149	1330
1831	1727	1920	1394	1595	1549	1396	1294	1133	1061	999	1111	1319	1193	1071	1238	1114	1004	951	1101	1418	992	1229	1343	1523
1525	1419	1512	1087	1287	1242	1089	986	826	753	752	804	1000	875	752	920	795	686	632	783	1100	674	911	1024	1205
2185	2073	2169	1757	1948	1898	1997	1639	1522	1413	1328	1462	1686	1561	1438	1605	1481	1371	1318	1469	1785	1159	1596	1710	1890
2357	2445	2341	2129	2320	2370	2369	2010	1894	1785	1700	1833	2058	1932	1810	1977	1852	1743	1689	1840	2157	1731	1968	2082	2262

Figure A1. *Cont.*

Center (Morelos, Pue, Qro, Edo.MX, CDMX, Hid, Tlaxc)											South (Chiapas, Guerrero, Oaxaca)							
CUAUTLA	PUEBLA	TEHUACÁN*	QUERÉTARO	SAN JUAN IXHUATEPEC	CUERNAVACA	TOLUCA	AZCAPOTZALCO	PACHUCA	BARRANCA DEL MUERTO	AÑIL	IGUALA*	ACAPULCO	OAXACA	SALINA CRUZ*	SALINA CRUZ	TUXTLA GUTIÉRREZ	TAPACHULA	TAPACHULA II
832	859	981	561	721	786	667	742	755	742	742	861	1102	1198	1531	1531	1569	1875	1875
1314	1341	1463	1043	1203	1268	1149	1224	1237	1224	1224	1343	1583	1680	2013	2013	2051	2357	2357
832	859	981	561	721	786	667	742	755	742	742	861	1102	1198	1531	1531	1569	1875	1875
711	663	751	496	698	794	691	695	552	706	709	877	1091	1040	1169	1169	1207	1513	1513
749	757	877	430	633	729	626	629	650	641	644	807	1022	987	1117	1117	1155	1460	1460
711	663	751	493	614	791	688	692	553	703	630	865	1038	1017	1146	1146	1184	1490	1490
795	579	676	476	679	774	669	673	353	687	692	854	1068	932	1062	1062	1100	1405	1405
734	619	716	504	706	801	696	701	508	714	720	903	1091	931	1151	1151	1199	1550	1550
765	771	900	446	649	744	638	643	669	657	662	845	1033	1116	1174	1174	1223	1574	1574
723	746	875	471	624	696	557	618	641	611	622	748	665	1047	1296	1296	1401	1752	1752
686	709	838	401	587	659	520	581	606	574	585	711	948	1053	1302	1302	1408	1758	1758
867	890	1019	618	768	840	701	762	788	755	766	892	1130	1234	1484	1484	1589	1939	1939
502	525	654	298	404	476	336	398	423	390	402	528	481	870	1119	1119	1224	1575	1575
823	846	975	539	724	796	657	719	744	711	723	848	1086	1191	1440	1440	1545	1896	1896
537	543	673	317	421	493	354	416	441	408	420	546	544	889	1137	1137	1242	1593	1593
808	831	960	524	709	781	642	704	729	696	708	833	674	1176	1425	1425	1530	1881	1881
614	619	740	327	497	569	430	492	517	484	496	622	580	964	1213	1213	1318	1669	1669
701	706	835	431	584	656	517	579	604	571	583	709	946	1051	1300	1300	1405	1756	1756
697	702	832	388	581	653	513	575	600	567	579	705	665	1047	1296	1296	1401	1752	1752
67.7	112	229	422	196	129	224	194	224	180	180	191	354	444	694	694	799	1149	1149
281	171	51	511	317	342	357	315	313	309	288	405	559	220	469	469	710	904	904
18.4	134	251	325	121	35.9	151	120	181	107	107	157	303	467	716	716	821	1172	1172
97.8	182	299	352	145	40.3	174	143	226	130	130	110	262	515	764	764	869	1220	1220
330	220	173	561	367	391	406	364	362	358	345	454	616	412	476	476	525	876	876
1095	985	937	1325	1132	1156	1171	1129	1127	1123	1110	1219	1063	628	416	416	326	61.2	61.2
348	238	190	578	385	409	424	382	380	376	363	471	634	430	480	480	529	879	879
420	310	263	651	457	481	496	454	452	533	435	544	706	215	320	320	460	650	650
943	833	785	1173	979	1004	1019	977	975	971	958	1066	1229	1025	677	677	448	1066	1066
676	566	518	906	709	736	751	709	707	703	691	799	962	757	409	409	217	568	568
320	210	162	550	353	380	395	353	351	347	335	443	606	401	477	477	526	876	876
489	379	331	719	525	550	565	523	520	516	504	612	775	571	292	292	341	692	692
256	146	98	486	317	316	331	289	287	283	271	379	542	337	500	500	548	899	899
527	482	579	557	435	542	501	452	371	456	453	643	831	795	953	953	1002	1353	1353
345	235	255	544	375	406	420	378	346	372	360	468	631	494	525	525	574	925	925
286	176	128	516	319	346	361	378	317	313	301	409	572	367	472	472	521	872	872
353	243	267	553	384	414	429	387	354	381	369	477	639	506	537	537	586	937	937
294	184	136	524	328	355	370	328	325	321	309	417	580	376	470	470	519	870	870
294	184	218	494	325	355	370	343	296	322	310	418	580	458	604	604	653	1004	1004
314	204	156	544	350	374	389	347	345	341	329	437	599	395	483	483	532	882	882
592	547	645	598	500	607	566	514	436	521	518	766	897	860	1009	1009	1058	1409	1409
299	189	142	530	336	360	375	333	331	327	315	423	585	381	486	486	535	885	885
415	305	257	645	451	476	490	449	446	443	430	538	701	496	360	360	409	760	760
290	180	132	520	326	351	366	324	321	317	305	413	576	372	474	474	523	874	874
274	164	117	504	311	335	350	308	306	302	290	398	560	356	483	483	535	882	882
281	171	124	512	318	342	357	315	313	309	297	405	567	363	493	493	542	892	892
458	348	300	688	494	519	534	492	489	4585	473	581	744	540	401	401	450	800	800
697	587	539	927	733	758	773	731	728	724	712	820	983	779	430	430	205	556	556
389	279	231	619	450	450	465	423	421	417	404	513	675	246	351	351	454	805	805
1043	944	864	1267	1130	1086	1135	1068	1081	1068	1068	1182	1373	1128	803	803	569	875	875
1415	1315	1235	1639	1502	1458	1507	1439	1452	1439	1439	1554	1745	1500	1175	1175	941	1247	1247

Gulf (Ver, Tab)								Southeast (Yuc, Camp, Q.Roo)		
POZA RICA	PEROTE	XALAPA	ESCAMELA	VERACRUZ	TIERRA BLANCA	MINATITLÁN	VILLAHERMOSA	PROGRESO	CAMPECHE	MÉRIDA
935	1002	1051	998	1135	1112	1308	1480	2052	1862	2020
1417	1483	1533	1479	1617	1594	1790	1961	2534	2344	9580
935	1002	1051	998	1135	1112	1308	1480	2052	1862	2020
551	585	770	634	728	926	1092	1688	1690	1500	1658
347	514	555	896	597	691	889	1056	1638	1448	1606
394	551	592	770	634	728	927	1093	1667	1477	1635
309	537	578	655	558	721	919	1084	1583	1393	1551
349	577	557	695	598	760	959	1124	1716	1514	1677
449	607	642	795	689	790	989	1154	1739	1538	1701
833	835	875	896	1025	1006	1204	1369	1930	1740	1899
796	797	838	859	945	968	1167	1332	2052	1862	2020
958	1025	1074	1021	1159	1136	1332	1503	2075	1885	2044
604	671	720	667	804	781	977	1148	1721	1531	1689
911	978	1027	974	1112	1089	1285	1456	2028	1838	1997
600	666	716	662	800	777	973	1144	1717	1527	1685
902	950	997	964	1102	1079	1275	1446	2019	1829	1987
682	749	798	745	882	859	1055	1227	1799	1609	1767
783	850	900	846	984	941	1157	1328	1900	1710	1869
770	837	896	833	971	948	1144	1315	1887	1697	1856
589	220	260	250	367	360	558	728	1228	1038	1196
446	216	261	169	290	270	469	634	1118	928	1086
547	242	283	272	390	382	580	746	1298	1108	1266
400	290	331	320	438	430	628	794	1348	1158	1316
359	254	207	83.4	109	50.4	284	450	1057	867	1025
1087	982	935	848	937	765	543	526	1087	897	1055
363	257	210	101	112	32.7	288	453	1020	830	989
413	307	260	173	162	59.3	219	385	941	751	910
935	829	783	696	685	612	391	220	590	389	552
668	562	515	429	417	345	123	67	659	457	620
360	254	207	72.7	109	55.3	285	450	1042	840	1008
481	375	328	242	231	158	100	266	857	656	819
420	164	141	5.1	129	109	308	473	1064	863	1026
219	380	420	559	461	563	761	926	1518	1317	1479
225	119	72.1	166	33.7	135	334	499	1090	889	1052
352	194	199	38.6	101	82	280	445	1037	836	998
221	128	80.7	147	45.6	147	345	511	1102	901	1064
357	202	204	46.8	106	75.4	278	444	1035	834	997
240	68.8	23.3	128	112	214	412	577	1169	968	1131
366	222	213	66.5	115	61.5	291	456	1048	846	1009
278	436	477	624	518	619	818	983	1574	1373	1536
329	223	176	52.4	91.5	95.6	294	459	1051	849	1012
407	301	254	168	156	84.2	168	333	925	724	886
354	198	201	42.9	103	84.2	282	448	1039	838	1001
362	183	210	112	92.4	281	456	362	1048	846	1009
373	190	220	34.3	122	103	301	466	1058	857	1019
361	255	208	211	111	180	209	374	966	764	927
689	583	386	450	439	366	144	53.3	645	444	606
381	276	229	142	181	28.1	214	379	970	769	932
1063	957	910	823	812	740	518	346	253	52	215
1312	1206	1159	1073	1062	989	767	689	257	307	217

Figure A1. Distance matrix for hydrogen transportation.

Toll Cost / trip

Region	Location	Northwest (BC, BCS, Sonora, Sinaloa, Nayarit)															North (Chihuahua, Durango)				
		ROSARITO	ENSENADA	MEXICALI	NOGALES	MAGDALENA	HERMOSILLO	GUAYMAS	CIUDAD OBREGÓN	NAVOJOA	LA PAZ	TOPOLOBAMPO	GUAMÚCHIL	CULIACÁN	MAZATLÁN	TEPIC	CD JUAREZ	CHIHUAHUA	DURANGO	PARRAL	GOMEZ PALACIO
Northwest	El Molino	243	248	189	177	178	163	158	147	137	248	117	108	96	69	0	420	369	192	319	289
	El Dorado	143	149	90	77	74	64	58	48	38	149	18	0	0	27	96	378	327	150	277	241
	Puga	243	248	189	177	175	163	158	147	137	248	117	108	96	69	0	420	369	192	319	283
Northeast	Aaron Saenz	232	238	179	159	159	345	339	329	319	214	225	216	204	177	108	175	125	23	78	28
	Alianza Popular	244	249	190	293	280	280	274	264	254	334	234	225	213	186	117	207	156	32	105	49
	El Mante	229	329	170	159	159	345	339	329	319	229	225	216	204	177	108	175	123	23	78	28
	Plan de Ayala	254	260	201	304	300	290	285	274	264	344	244	235	223	196	127	217	166	42	116	59
	Plan de SL	254	260	201	304	300	290	285	274	264	344	244	235	223	196	127	217	166	42	116	59
	San Miguel	254	260	201	304	300	290	285	274	264	344	244	235	223	196	127	217	166	42	116	59
West	Ameca	267	272	213	201	197	187	182	171	161	272	141	132	120	93	24	444	393	216	343	307
	Bellavista	267	272	213	201	197	187	182	171	161	272	141	132	120	93	24	444	393	216	343	307
	Jose Ma Morelos	267	272	213	201	197	187	182	171	161	272	141	132	120	93	24	444	393	216	343	307
	Lazaro Cardenas	312	317	258	246	242	232	227	216	206	317	186	177	165	138	69	282	231	107	181	145
	Melchor Ocampo	267	272	213	201	197	187	182	171	161	272	141	132	120	93	24	444	393	216	343	307
	Pedernales	306	312	253	240	237	227	221	211	201	312	181	172	160	133	63	266	215	91	165	129
	Queseria	303	309	250	237	234	224	218	208	198	309	178	169	157	130	60	284	233	109	183	147
	Santa Clara	299	304	245	233	229	219	213	203	193	304	173	164	152	111	56	224	172	48	122	86
	Tala	267	272	213	201	197	187	182	171	161	272	141	132	120	93	24	444	393	216	343	307
	Tamazula	285	391	232	219	216	205	200	190	179	291	159	151	139	111	42	266	215	234	165	129
Center	Atencingo	431	437	378	365	362	351	346	336	325	437	305	297	285	257	188	279	228	104	178	142
	Calipam	458	463	404	392	388	378	373	362	352	463	332	323	311	284	215	306	254	131	204	168
	Casasano	400	406	347	335	331	321	315	305	295	406	275	266	254	227	158	271	267	95	169	133
	Emiliano Zapata	396	402	343	331	327	317	311	301	291	402	271	262	250	223	154	265	213	89	163	127
South	El refugio	496	501	442	430	426	416	410	400	390	501	370	361	349	322	253	344	292	168	242	206
	Huixtla	596	602	543	530	527	517	511	501	491	602	471	462	450	422	353	444	393	269	343	307
	La margarita	506	511	452	440	436	426	420	410	400	511	380	371	359	332	263	344	292	168	242	280
	López Mateos	512	518	459	446	443	432	427	417	406	518	386	378	366	338	269	306	254	185	204	223
Gulf	Azucaremex	566	572	513	501	497	487	481	471	461	572	441	432	420	393	324	415	363	239	313	277
	Benito Juárez	560	566	507	494	491	480	475	465	454	566	434	425	414	386	317	408	357	233	307	271
	Constancia	496	501	442	430	426	416	410	400	390	501	370	361	349	322	253	344	292	168	242	206
	Cuatotolapan	526	531	472	460	456	446	440	430	420	531	400	391	379	352	283	374	322	198	272	236
	El Carmen	487	493	434	421	418	407	402	392	381	493	361	353	341	313	244	335	284	160	234	198
	El Higo	370	375	316	304	300	290	285	274	264	375	244	235	223	196	127	217	166	42	116	80
	El Modelo	493	498	439	427	423	413	408	397	387	498	367	358	346	319	250	348	296	172	246	210
	El Potrero	496	501	442	430	426	416	410	400	390	501	370	361	349	322	253	344	292	168	242	206
	La Gloria	493	498	439	427	423	413	408	397	387	498	367	358	346	319	250	348	296	172	246	210
	La providencia	496	501	442	430	426	416	410	400	390	501	370	361	349	322	253	344	292	168	242	206
	Mahuixtlán	466	471	412	400	396	386	381	370	360	471	340	331	319	292	223	321	269	145	219	183
	Motzorongo	496	501	442	430	426	416	410	400	390	501	370	361	349	322	253	344	292	168	242	206
	Panuco	375	381	322	309	306	296	290	280	270	381	250	241	229	201	132	223	171	48	121	85
	Progreso	526	531	472	460	456	446	440	430	420	531	400	391	379	352	283	374	322	198	272	236
	San Cristobal	522	527	468	456	452	442	437	426	416	527	396	387	375	348	279	370	318	195	268	232
	San José de Abajo	496	501	442	430	426	416	410	400	390	501	370	361	349	322	253	344	292	168	242	206
	San Miguelito	493	498	439	427	423	413	408	397	387	498	367	358	346	319	250	348	296	172	246	210
	San Nicolas	493	498	439	427	423	413	408	397	387	498	367	358	346	319	250	348	296	172	246	210
	San Pedro	526	531	472	460	456	446	440	430	420	531	400	391	379	352	283	374	322	198	272	236
	Santa Rosalia	560	566	507	494	491	480	475	465	454	566	434	425	414	386	317	408	357	233	307	271
	Tres Valles	506	511	452	440	436	426	420	410	400	511	380	371	359	332	263	344	292	168	242	230
Southeast	La Joya	593	598	539	527	523	513	508	497	487	598	467	458	446	419	350	441	389	266	339	303
	Pucte	603	608	549	537	533	523	517	507	497	608	477	468	456	429	360	451	389	275	349	313

Northeast (NL, Tam, SLP, Coa)												West (Zacatecas, Aguascalientes, Jalisco, Colima, Michoacán y Gto)												
SABINAS	MONCLOVA	NUEVO LAREDO	REYNOSA	SANTA CATARINA	SALTILLO	CADEREYTA	MATEHUALA	CIUDAD VICTORIA	CIUDAD MANTE	CIUDAD VALLES	SAN LUIS POTOSÍ	ZACATECAS	AGUASCALIENTES	LEÓN	ZAPOPAN	ZAMORA	IRAPUATO	CELAYA	URUAPAN	COLIMA	MORELIA	EL CASTILLO	LÁZARO CÁRDENAS	MANZANILLO
289	289	135	136	111	320	113	96	108	108	127	89	85	80	79	24	56	56	78	56	60	80	24	114	88
248	248	302	303	278	278	280	192	204	204	223	185	161	176	175	120	152	152	175	152	157	177	120	210	185
289	289	135	136	111	320	113	96	108	108	127	89	85	80	79	24	56	56	78	56	60	80	24	114	88
25	2	25	0	0	0	0	0	0	0	0	11	11	27	43	84	28	19	40	67	120	53	84	116	148
36	36	59	8	36	36	37	20	8	8	8	20	20	36	52	93	37	28	49	76	129	62	36	125	157
25	2	25	0	0	0	0	0	0	0	0	11	11	27	43	84	28	19	40	67	120	53	84	116	148
46	46	25	0	0	0	0	0	0	0	0	31	31	46	66	103	52	38	60	86	140	73	103	136	167
46	46	25	0	0	0	0	0	0	0	0	31	31	46	66	103	52	38	60	86	140	73	103	136	167
46	46	25	0	0	0	0	0	0	0	0	31	31	46	66	103	52	38	60	86	140	73	103	136	167
87	87	111	112	87	87	89	72	84	84	103	65	61	56	55	0	32	32	54	32	36	56	0	36	64
87	87	111	112	87	87	89	72	84	84	103	65	61	56	55	0	32	32	54	32	36	56	0	36	64
87	87	111	112	87	87	89	72	84	84	103	65	61	56	55	0	32	32	54	32	36	56	0	36	64
84	84	107	109	84	84	85	68	80	80	99	61	95	90	61	45	11	61	69	2	81	16	45	47	109
87	87	111	112	87	87	89	72	84	84	103	65	61	56	55	0	32	32	54	32	36	56	0	36	64
68	68	91	93	68	68	70	53	64	64	84	45	80	75	45	39	5	45	53	14	76	0	39	63	104
124	124	147	149	124	124	125	109	120	120	140	101	98	93	92	36	18	68	91	18	0	93	36	0	28
32	32	55	57	32	32	34	17	28	28	48	9	37	32	9	32	0	0	22	0	18	25	32	35	46
87	87	111	112	87	87	89	72	84	84	103	65	61	56	55	0	32	32	54	32	36	56	0	36	64
106	106	129	131	106	106	107	90	102	102	121	83	80	74	73	18	0	9	31	0	18	34	18	18	46
115	115	139	102	115	115	117	100	111	102	88	93	167	161	164	164	132	132	110	121	201	108	164	171	229
142	142	165	120	142	142	144	127	138	120	107	119	193	188	159	191	159	159	136	148	170	135	191	198	198
107	107	130	132	107	107	108	92	103	89	122	84	158	153	123	134	102	123	101	91	170	77	134	62	198
101	101	124	126	101	101	102	86	97	97	116	78	152	147	117	130	98	117	95	87	166	73	130	55	194
180	180	203	74	180	180	181	165	74	74	60	157	231	226	197	229	197	197	174	186	265	172	229	235	265
280	280	304	158	280	280	282	265	158	158	144	258	332	327	297	329	298	297	275	286	366	273	329	336	366
190	190	213	67	190	190	191	175	67	67	53	167	231	226	197	229	197	197	174	186	265	172	229	233	265
196	196	220	73	196	196	198	181	73	73	60	173	247	242	159	245	213	159	136	202	282	189	245	252	282
251	251	131	128	251	251	252	235	128	128	114	228	302	297	267	300	268	267	245	257	316	243	300	306	364
244	244	125	122	244	244	246	229	122	122	108	222	296	290	261	291	261	261	239	250	330	237	291	300	358
180	180	203	74	180	180	181	165	74	74	60	157	231	226	197	229	197	197	174	186	265	172	229	235	265
210	210	233	65	210	210	65	195	65	65	51	187	261	256	227	259	227	227	204	216	295	202	259	265	323
176	176	200	77	176	176	178	161	77	77	63	153	227	222	193	225	193	193	170	182	262	169	225	232	289
46	46	25	0	0	0	0	0	0	0	0	31	11	46	66	103	52	38	60	86	140	73	103	136	167
67	67	60	34	34	67	34	51	34	34	21	65	228	230	201	226	194	201	178	183	269	170	226	233	297
180	180	203	77	180	180	181	165	77	77	63	157	231	226	197	229	197	197	197	186	265	172	229	235	265
67	67	60	34	34	67	34	51	34	34	21	65	228	230	201	226	194	201	178	183	269	170	226	233	297
180	180	203	74	180	180	181	165	74	74	60	157	231	226	197	229	197	197	197	186	265	172	229	235	265
157	157	173	34	150	157	152	135	34	34	21	127	201	203	174	199	167	174	151	156	242	143	199	206	270
180	180	203	74	180	180	181	165	74	74	60	157	231	226	197	229	197	197	174	186	265	172	229	235	265
15	13	35	10	10	51	10	36	10	10	5	36	36	52	68	108	53	44	65	151	145	138	108	201	173
210	210	233	65	210	210	65	195	65	65	51	187	261	256	227	259	227	227	200	216	295	202	259	265	323
206	206	229	83	206	206	208	191	83	83	69	183	257	252	223	255	223	223	200	212	291	199	255	262	319
180	180	203	74	180	180	181	165	74	74	60	157	231	226	197	229	197	197	186	265	172	229	233	235	265
180	180	203	77	180	180	181	165	77	77	63	157	228	230	201	226	194	201	178	183	269	170	226	233	297
180	180	203	77	180	180	181	165	77	77	63	157	228	230	201	226	194	201	178	183	269	170	226	233	297
210	210	233	65	210	210	65	195	65	65	51	187	261	256	227	259	227	227	200	216	295	202	259	265	323
244	244	125	122	244	244	246	229	122	122	108	222	296	290	261	291	261	261	239	250	330	237	291	300	358
190	190	213	67	190	190	239	175	67	67	53	231	231	226	197	229	197	174	186	265	172	229	235	265	265
277	277	300	154	277	277	279	262	154	154	140	254	320	123	294	326	294	294	271	283	362	370	326	333	390
287	287	310	164	287	287	288	272	164	164	150	264	338	333	304	336	304	304	281	293	372	279	336	342	400

Figure A2. *Cont.*

Center (Morelos, Pue, Qro, Edo.MX, CDMX, Hid, Tlaxc)											South (Chiapas, Guerrero, Oaxaca)							
CUAUTLA	PUEBLA	TEHUACÁN*	QUERÉTARO	SAN JUAN IXHUATEPEC	CUERNAVACA	TOLUCA	AZCAPOTZALCO	PACHUCA	BARRANCA DEL MUERTO	AÑIL	IGUALA*	ACAPULCO	OAXACA	SALINA CRUZ*	SALINA CRUZ	TUXTLA GUTIÉRREZ	TAPACHULA	TAPACHULA II
173	188	72	100	152	149	127	160	146	160	160	178	220	251	314	314	335	353	353
269	285	311	196	248	245	233	256	243	256	256	275	316	348	410	410	432	450	450
173	188	72	100	152	149	127	160	146	160	160	178	220	251	314	314	335	353	353
89	94	120	19	75	92	42	102	79	102	102	93	156	178	118	118	140	158	158
105	88	114	28	84	101	51	92	79	92	92	102	165	172	112	112	134	152	152
89	94	120	19	75	92	42	102	79	102	102	93	186	178	118	118	140	158	158
116	80	107	38	102	111	61	102	0	102	59	113	176	164	104	104	126	144	144
116	80	107	38	102	111	61	102	0	102	59	113	176	164	104	104	126	144	144
116	80	107	38	102	111	61	102	0	102	59	113	176	164	104	104	126	144	144
149	164	191	76	128	125	103	136	122	136	136	154	196	227	290	290	311	329	329
149	164	191	76	128	125	103	136	122	136	136	154	196	227	290	290	311	329	329
149	164	191	76	128	125	103	136	122	136	136	154	196	227	290	290	311	329	329
108	123	150	90	87	84	62	95	82	95	95	114	45	186	249	249	271	289	289
149	164	191	76	128	125	103	136	122	136	136	154	196	227	290	290	311	329	329
93	108	135	75	71	68	46	79	66	79	79	98	61	171	233	233	255	273	273
185	201	227	112	178	161	139	172	159	172	172	191	12	264	326	326	348	366	366
117	132	159	44	96	93	71	104	91	104	104	123	33	195	258	258	280	298	298
149	164	191	76	128	125	103	136	122	136	136	154	196	227	290	290	311	329	329
126	141	168	53	105	102	80	113	100	113	113	132	56	204	267	267	288	307	307
0	8	34	88	58	0	23	47	35	47	47	23	64	71	133	133	155	173	173
68	27	0	115	85	68	86	74	62	74	74	91	132	36	69	69	91	109	109
0	41	68	78	26	0	23	26	37	26	26	23	64	104	166	166	188	206	206
0	41	68	74	22	5	18	22	42	22	22	16	57	104	166	166	188	206	206
106	65	9	153	123	106	124	111	100	111	111	129	170	93	57	57	79	97	97
206	165	109	253	223	206	225	212	201	212	212	229	19	13	19	19	18	0	0
106	65	9	153	123	106	124	111	100	111	111	129	170	93	57	57	79	97	97
122	81	25	169	139	122	140	128	116	128	128	91	132	0	19	19	63	81	81
176	135	79	224	193	176	195	182	171	182	182	200	241	164	49	49	20	38	38
170	129	64	217	187	170	188	176	164	176	176	193	234	158	42	42	14	32	32
106	65	9	153	123	106	124	111	100	111	111	129	170	93	57	57	79	97	97
136	95	39	183	152	136	154	141	130	141	141	159	200	123	19	19	41	59	59
102	61	5	149	119	102	120	108	96	108	108	125	166	89	61	61	82	101	101
116	80	107	38	102	111	61	102	0	102	59	113	176	164	104	104	126	144	144
78	36	29	61	31	78	59	83	8	83	83	81	123	65	77	77	99	117	117
106	65	9	153	123	106	124	111	100	111	111	129	170	93	61	61	82	101	101
78	36	29	61	31	78	59	83	8	83	83	81	123	65	77	77	99	117	117
106	65	9	153	123	106	124	111	100	111	111	129	170	93	57	57	79	97	97
83	42	46	130	93	83	94	88	77	88	88	99	140	83	76	76	97	115	115
106	65	9	153	123	106	124	111	100	111	111	129	170	93	57	57	79	97	97
121	57	49	0	88	108	116	88	65	88	88	131	173	86	104	104	126	144	144
136	95	39	183	152	136	154	141	130	141	141	159	200	123	19	19	41	59	59
132	91	35	179	149	132	150	138	126	138	138	155	196	119	35	35	56	74	74
106	65	9	153	123	106	124	111	100	111	111	129	170	93	57	57	79	97	97
106	65	9	153	123	106	124	111	100	111	111	129	170	93	61	61	82	101	101
106	65	9	153	123	106	124	111	100	111	111	129	170	93	61	61	82	101	101
136	95	39	183	152	136	154	141	130	141	141	159	200	123	19	19	41	59	59
170	129	64	217	187	170	188	176	164	176	176	193	234	158	42	42	14	32	32
116	75	19	163	132	116	134	121	110	121	121	139	180	6	35	35	56	74	74
203	162	106	250	220	203	221	209	197	209	209	226	267	190	75	75	47	65	65
213	172	116	260	230	213	231	218	207	218	209	236	277	200	85	85	57	75	75

Gulf (Ver, Tab)								Southeast (Yuc, Camp, Q.Roo)		
POZA RICA	PEROTE	XALAPA	ESCAMELA	VERACRUZ	TIERRA BLANCA	MINATITLÁN	VILLAHERMOSA	PROGRESO	CAMPECHE	MÉRIDA
161	205	223	244	273	263	302	317	360	360	360
257	302	319	341	369	359	399	414	456	456	456
161	205	223	244	273	263	302	317	360	360	360
28	34	34	86	58	67	107	122	164	164	164
17	24	26	43	28	33	42	50	158	158	158
28	34	34	86	58	67	107	122	164	164	164
14	21	21	72	44	53	93	108	150	150	150
14	21	21	72	44	53	93	108	150	150	150
14	21	21	72	44	53	93	108	150	150	150
137	181	199	220	249	239	278	293	336	336	336
161	205	223	244	273	263	302	317	360	360	360
137	181	199	220	249	239	278	293	336	336	336
96	141	158	179	208	198	237	252	295	295	295
137	181	199	220	249	239	278	293	336	336	336
81	125	143	164	193	182	222	237	279	279	279
174	229	242	257	286	275	315	330	372	372	372
105	150	167	188	217	207	246	261	304	304	304
137	181	199	220	249	239	278	293	336	336	336
114	159	176	197	226	216	255	270	313	313	313
44	36	49	64	93	82	122	137	179	179	179
29	29	46	0	29	19	58	73	116	116	116
64	70	83	97	126	116	155	170	213	213	213
56	70	83	97	126	116	155	170	213	213	213
44	28	31	9	17	0	17	61	104	104	104
128	150	115	109	101	74	51	101	75	75	75
44	28	31	9	17	0	17	61	97	97	97
44	66	31	25	17	6	30	45	87	87	87
98	121	86	79	71	45	21	6	10	16	10
92	114	79	73	65	38	15	0	43	43	43
44	28	31	9	17	0	17	61	104	104	104
35	57	22	39	8	4	8	23	66	66	66
47	24	35	5	20	10	49	64	107	107	107
14	21	21	72	44	53	93	108	150	150	150
0	0	0	45	18	26	66	81	123	123	123
47	28	35	9	20	10	49	64	107	107	107
0	0	0	45	18	26	66	81	123	123	123
44	28	31	9	17	0	17	61	104	104	104
0	13	0	43	16	25	64	79	122	122	122
44	28	31	9	17	0	17	61	104	104	104
14	21	21	72	44	53	93	108	150	150	150
35	57	22	39	8	4	8	23	66	66	66
54	76	41	35	26	0	23	38	81	81	81
44	28	31	9	17	0	17	61	104	104	104
47	28	35	9	20	10	49	64	107	107	107
47	28	35	9	20	10	49	64	107	107	107
35	57	22	39	8	4	8	23	66	66	66
92	114	79	73	65	38	15	0	43	43	43
38	60	25	19	10	0	23	38	81	81	81
125	147	112	106	97	71	48	33	10	10	10
135	157	122	116	95	81	58	43	0	0	0

Figure A2. Toll cost matrix for hydrogen transportation.

References

1. Morales, A.; Pérez, M.; Pérez, J.; De León, S. Energías renovables y el hidrógeno: Un par prometedor en la transición energética de México. *Investig. Cienc.* **2017**, *25*, 92–101. [CrossRef]
2. Ehsan, S.; Abdul, M. Hydrogen production from renewable and sustainable energy resources: Promising green energy carrier for clean development. *Renew. Sustain. Energy Rev.* **2016**, *57*, 850–866. [CrossRef]
3. Orecchini, F.; Bocci, E. Biomass to hydrogen for the realization of closed cycles of energy resources. *Energy* **2007**, *32*, 1006–1011. [CrossRef]
4. De León Almaráz, S. Multi-Objective Optimization of a Hydrogen Supply Chain. Ph.D. Thesis, Toulouse Institute of Technology, Toulouse, France, 2014.
5. Parker, N. Optimizing the Design of Biomass Hydrogen Supply Chains Using Real-Word Spatial Distributions: A Case of Study Using California Rice Straw. Master's Thesis, University of California, Berkeley, CA, USA, 2007.
6. Rico, J. Desarrollo de una Red de Valor Con Base a la Gestión de Bioenergía, Para Determinar Estrategias de Negocios. Ph.D. Thesis, Instituto Tecnológico de Orizaba, Orizaba, Mexico, 2015.
7. Azzaro-Pantel, C. *Hydrogen Supply Chain Design, Deployment and Operation*; Elsevier: Amsterdam, The Netherlands, 2018; ISBN 9780128111987.
8. Kim, J.; Moon, I. Strategic design of hydrogen infrastructure considering cost and safety using multiobjective optimization. *Int. J. Hydrogen Energy* **2008**, *33*, 5887–5896. [CrossRef]
9. Almansoori, A.; Shah, N. Design and operation of a stochastic hydrogen supply chain network under demand uncertainty. *Int. J. Hydrogen Energy* **2012**, *37*, 3965–3977. [CrossRef]
10. Güler, M.G.; Geçici, E.; Erdoğan, A. Design of a future hydrogen supply chain: A multi period model for Turkey. *Int. J. Hydrogen Energy* **2021**, *46*, 16279–16298. [CrossRef]
11. Gabrielli, P.; Charbonnier, F.; Guidolin, A.; Mazzotti, M. Enabling low-carbon hydrogen supply chains through use of biomass and carbon capture and storage: A Swiss case study. *Appl. Energy* **2020**, *275*, 115245. [CrossRef]
12. Quarton, C.J.; Samsatli, S. The value of hydrogen and carbon capture, storage and utilization in decarbonizing energy: Insights from integrated value chain optimization. *Appl. Energy* **2020**, *257*, 113936. [CrossRef]
13. Li, L.; Manier, H.; Manier, M.-A. Hydrogen supply chain network design: An optimization-oriented review. *Renew. Sustain. Energy Rev.* **2019**, *203*, 342–360. [CrossRef]
14. Ochoa, J.; Azzaro, C.; Martinez, G.; Aguilar, A. Social cost-benefit assessment as a post-optimal analysis for hydrogen supply chain design and deployment: Application to Occitania (France). *Sustain. Prod. Consum.* **2020**, *24*, 105–120. [CrossRef]
15. Zakaria, I.H.; Ibrahim, J.A.; Othman, A.A. Waste biomass toward hydrogen fuel supply chain management for electricity: Malaysia perspective. In Proceedings of the AIP Conference Proceedings, Kedah, Malaysia, 11–13 April 2016; Volume 1761, p. 020111. [CrossRef]
16. Lam, H.L.; Ng, W.P.; Ng, R.T.; Ng, E.H.; Aziz, M.K.A.; Ng, D.K.S. Green strategy for sustainable waste-to-energy supply chain. *Energy* **2013**, *57*, 4–16. [CrossRef]
17. Gumte, K.; Pantula, P.; Miriyala, S.; Mitra, K. Achieving wealth from bio-waste in a nationwide supply chain setup under uncertain environment through data driven robust optimization approach. *J. Clean. Prod.* **2021**, *291*, 125702. [CrossRef]
18. Goodzarzian, F.; Wamba, S.; Mathiyazhagan, K.; Taghipour, A. A new bi-objective green medicine supply chain network design under fuzzy environment: Hybrid metaheuristic algorithms. *Comput. Ind. Eng.* **2021**, *160*, 107535. [CrossRef]
19. Abdolazimi, O.; Esfandarani, M.S.; Shishebori, D. Design of a supply chain network for determining the optimal number of items at the inventory groups based on ABC analysis: A comparison of exact and meta-heuristic methods. *Neural Comput. Appl.* **2021**, *33*, 6641–6656. [CrossRef]
20. Paul, A.; Shukla, N.; Paul, S.K.; Trianni, A. Sustainable supply chain management and multi-criteria decision-making methods: A systematic review. *Sustainability* **2021**, *13*, 7104. [CrossRef]
21. Tordecilla, R.; Juan, A.; Montoya, J.; Quintero, C.; Panadero, J. Simulation-optimization methods for designing and assessing resilient supply chain networks under uncertainty scenarios: A review. *Simul. Model. Pract. Theory* **2021**, *106*, 102166. [CrossRef]
22. Hosseini, S.; Ghatreh, M.; Abbasi, F. A novel hybrid approach for synchronized development of sustainability and resiliency in the wheat network. *Comput. Electron. Agric.* **2020**, *168*, 105095. [CrossRef]
23. Gital, Y.; Bilgen, B. Multi-objective optimization of sustainable biomass supply chain network design. *Appl. Energy* **2020**, *272*, 115259. [CrossRef]
24. Rasi, R.; Sohanian, M. A multi-objective optimization model for sustainable supply chain network with using genetic algorithm. *J. Model. Manag.* **2021**, *16*, 714–727. [CrossRef]
25. Zailan, R.; Lim, J.; Manan, Z.; Wan, S.; Mohammadi, B.; Jamaluddin, K. Malaysia scenario of biomass supply chain-cogeneration system and optimization modeling develpment: A review. *Renew. Sustain. Energy Rev.* **2021**, *148*, 111289. [CrossRef]
26. Nunes, L.; Causer, T.; Ciolkosz, D. Biomass for energy: A review on supply chain management models. *Renew. Sustain. Energy Rev.* **2020**, *120*, 109658. [CrossRef]
27. Seo, S.-K.; Yun, D.-Y.; Lee, C.-J. Design and optimization of a hydrogen supply chain using a centralized storage model. *Appl. Energy* **2020**, *262*, 114452. [CrossRef]
28. Yuen, S.; Shen, B.; Dong, W.; Yong, S.; Akbar, M.; Sunarso, J. Techno-economic analysis for biomass supply chain: A state-of-the-art review. *Renew. Sustain. Energy Rev.* **2021**, *135*, 110164. [CrossRef]

29. Rafique, R.; Jat, M.; Rehman, H.; Zahid, M. Bioenergy supply chain optimization for addressing energy deficiency: A dynamic model for large-scale network designs. *J. Clean. Prod.* **2021**, *318*, 128495. [CrossRef]
30. Li, L.; Manier, H.; Manier, M.-A. Integrated optimization model for hydrogen supply chain network design and hydrogen fueling station planning. *Comput. Chem. Eng.* **2020**, *134*, 106683. [CrossRef]
31. Ochoa, J.; Azzaro, C.; Aguilar, A. Optimization of a hydrogen supply chain network design under demand uncertainty by multi-objective genetic algorithms. *Comput. Chem. Eng.* **2020**, *140*, 106853. [CrossRef]
32. Debernardi, H.; Ortiz, H.; Rosas, D. Energía Disponible en el Campo Cañero Mexicano. Códoba, Veracruz. 2014. Available online: https://www.atamexico.com.mx/wp-content/uploads/2017/11/3-DIVERSIFICACI%C3%93N-2015.pdf (accessed on 3 April 2020).
33. SAGARPA. Planeación Agrícola Nacional 2017–2030. 2016. Available online: https://www.gob.mx/agricultura/acciones-y-programas/planeacion-agricola-nacional-2017-2030-126813 (accessed on 22 March 2020).
34. CONADESUCA. 6to. Informe Estadístico del Sector Agroindustrial de la Caña de Azúcar en México, Zafras 2009–2010/2018–2019, Comité Nacional para el Desarrollo Sustentable de la Caña de Azúcar. 2019. Available online: https://siiba.conadesuca.gob.mx/Archivos_Externos/6to_informe_estad%C3%ADstico.pdf (accessed on 17 February 2020).
35. Comisión Nacional de Hidrocarburos. Reservas de Hidrocarburos en México Conceptos Fundamentales y Análisis. 2018. Available online: https://www.gob.mx/cnh/documentos/analisis-de-informacion-de-las-reservas-de-hidrocarburos-de-mexico-al-1-de-enero-del-2018?idiom=es (accessed on 25 February 2020).
36. IRENA. *Hydrogen from Renewable Power: Technology Outlook for the Energy Transition*; International Renewable Energy Agency: Abu Dhabi, United Arab Emirates, 2018. Available online: www.irena.org (accessed on 30 March 2020).
37. Ferrero, D.; Gamba, M.; Lanzini, A.; Santarelli, M. Power-to-gas hydrogen: Techno-economic assessment of processes towards a multi-purpose energy carrier. *Energy Procedia* **2016**, *101*, 50–57. [CrossRef]
38. Mendoza, A.; Cadena, A.; de Buen, O. Estudio de Pesos y Trasnportes, Secretaría de Comunicaciones y Transporte. 2010. Available online: http://www.dof.gob.mx/nota_detalle.php?codigo=5508944&fecha=26/12/2017 (accessed on 18 April 2020).

Article

Applying the Crow Search Algorithm for the Optimal Integration of PV Generation Units in DC Networks

Luis Fernando Grisales-Noreña [1],*, Brandon Cortés-Caicedo [2], Gerardo Alcalá [3] and Oscar Danilo Montoya [4,5],*

[1] Department of Electrical Engineering, Faculty of Engineering, Universidad de Talca, Curicó 3340000, Chile
[2] Departamento de Mecatrónica y Electromecánica, Facultad de Ingeniería, Instituto Tecnológico Metropolitano, Medellín 050036, Colombia
[3] Centro de Investigación en Recursos Energéticos y Sustentables, Universidad Veracruzana, Coatzacoalcos, Veracruz 96535, Mexico
[4] Grupo de Compatibilidad e Interferencia Electromagnética (GCEM), Facultad de Ingeniería, Universidad Distrital Francisco José de Caldas, Bogotá 110231, Colombia
[5] Laboratorio Inteligente de Energía, Facultad de Ingeniería, Universidad Tecnológica de Bolívar, Cartagena 131001, Colombia
* Correspondence: luis.grisales@utalca.cl (L.F.G.-N.); odmontoyag@udistrital.edu.co (O.D.M.)

Abstract: This paper presents an efficient master–slave methodology to solve the problem of integrating photovoltaic (PV) generators into DC grids for a planning period of 20 years. The problem is mathematically formulated as Mixed-Integer Nonlinear Programming (MINLP) with the objective of minimizing the total annual operating cost. The main stage, consisting of a discrete-continuous version of the Crow search algorithm (DCCSA), is in charge of determining the installation positions of the PV generators and their corresponding power ratings. On the other hand, at the slave level, the successive approximation power flow method is used to determine the objective function value. Numerical results on 33- and 69-bus test systems demonstrate the applicability, efficiency and robustness of the developed approach with respect to different methodologies previously discussed in the scientific literature, such as the vortex search algorithm, the generalized normal distribution optimizer and the particle swarm optimization algorithm. Numerical tests are performed in the MATLAB programming environment using proprietary scripts.

Keywords: DC networks; PV generators; crow search algorithm; discrete-continuous codification; master–slave optimization; successive approximation power flow method; electrical systems planning

MSC: 65K05; 65K10; 68N99; 90C26; 90C59

Citation: Grisales-Noreña, L.F.; Cortés-Caicedo, B.; Alcalá, G.; Montoya, O.D. Applying the Crow Search Algorithm for the Optimal Integration of PV Generation Units in DC Networks. *Mathematics* **2023**, *11*, 387. https://doi.org/10.3390/math11020387

Academic Editors: Antonin Ponsich, Mariona Vila Bonilla, Bruno Domenech and Nicu Bizon

Received: 6 December 2022
Revised: 29 December 2022
Accepted: 5 January 2023
Published: 11 January 2023

1. Introduction

1.1. General Background

Due to technological advances in the field of power electronics, the implementation and use of DC networks have been growing in recent years, with which it is expected to bring electricity to end-users at medium and low voltage levels through DC transmission and sub-transmission systems [1,2]. In comparison with traditional AC systems, DC systems have the following advantages [3–5]: (i) higher efficiency, since the absence of reactive elements (i.e., inductive reactances and reactive power flows) reduces power losses and improves voltage profiles; (ii) reduced operating and investment costs associated with network maintenance; and (iii) simpler integration of distributed energy resources, such as distributed generation based on renewable resources and energy storage systems, since most of these devices operate in DC. This last advantage has enabled researchers around the world to develop strategies that allow the transition from classical, fossil fuel-based centralized generation systems to decentralized power generation systems based on renewable resources such as photovoltaic (PV) or wind energy [6,7]. This will not only help to meet the energy demands of end-users; it will also reduce dependence on fossil fuels and, at the same time, the environmental impact of their use.

The integration of distributed energy resources such as PV generation poses challenges for engineers in charge of the planning, design, and operation of electrical systems since an inadequate integration of these resources leads to the following problems [8]: (i) the degradation of voltage profiles, (ii) the overloading of distribution lines, (iii) increased energy losses, and (iv) the deterioration of power quality, among others. However, it is evident that appropriate integration of photovoltaic generators into power grids considerably improves the technical-operational conditions of the system and makes it possible to reduce the greenhouse gas emissions caused by fossil sources [9].

1.2. Motivation

Being a country located between the tropics of Cancer and Capricorn, Colombia has an abundant solar resource, which is the reason why, through legislation, as is the case of Law 1715 of 2014, the number of projects for the integration of PV generators into conventional electricity grids has increased in recent years [10,11]. However, the currently installed capacity of PV generation is far from the maximum usable capacity, representing 0.76% of the total energy generated in the country (these data come from the observation of the energy matrix reported by XM during 2019, before the pandemic). For this reason, the main motivation of this research study is to propose alternatives that allow taking advantage of the country's abundant solar resource, thus allowing an energy transition that reduces the emissions of polluting gases while providing a high-quality service to the end-users to be achieved that is as economical as possible. Note that the optimal integration of PV generators in DC networks is a complex problem from the point of view of mathematical modeling, as it is represented by a mixed-integer nonlinear programming (MINLP) model that combines discrete and continuous variables. As a result of the above, it is necessary to develop efficient solution strategies that address the problem with high-quality results and reduced computation times. Therefore, this research also aims to propose an optimization methodology to solve the problem under study by finding the best possible solution with low computational costs.

1.3. State of the Art

To address the problem of integrating generators based on renewable energy resources into DC networks, different methodologies have been recently reported in the specialized literature. Some of these research works are presented below.

In [12], a methodology is presented to assess the technical-economic feasibility of integrating and operating large-scale photovoltaic generators in AC/DC distribution networks. The objective functions considered are the minimization of operating costs and energy losses of the grid. To solve this problem, the non-dominant sorting genetic algorithm-II is used. Numerical performance achieved on the 33-bus IEEE test feeder demonstrates the feasibility of the suggested method. In [13], hybridization between the particle swarm optimization algorithm and the gravitational search algorithm is suggested to address the integration problem of renewable energy sources based mainly on photovoltaic and wind generation. The main goal for this work is to reduce energy losses in the grid and increase profits for renewable energy owners. Numerical performance on the 69-bus IEEE test system shows the effectiveness and applicability of the suggested methodology in terms of the solution compared to other population-based metaheuristic algorithms. In reference [14], the problem of integrating distributed generators (mainly based on solar and wind power) into DC grids is represented by a mixed-integer nonlinear linear programming (MINLP) model. This work aims to minimize the installation costs of distributed generators and save in power purchasing. The authors used the GAMS software to solve the mathematical model. Numerical performance on the 21-bus test feeder shows the application and effectiveness of the suggested approach. In [15], a method on the basis of the equilibrium optimization algorithm is proposed for the efficient location and size of PV generators and batteries in distribution grids. The purposes of this work include minimizing the cost of energy not supplied, the investment costs associated with the installation of the PV generators and batteries, their operating costs, the power losses through the distribution lines, and the CO_2 emissions produced in relation to the network and the systems. Numerical results on 30- and 69-bus feeders demonstrate the efficiency and robustness of the suggested methodology with respect to other algorithms reported in the specialized bibliography, such as genetic algorithms, particle swarm optimization, differential evolution and gray wolf optimization.

The authors of [16] address the problem of integrating of distributed generators in DC networks through a mixed-integer semi-definite programming model. This model is solved using the MATLAB CVX tool, with which the authors manage to minimize the power losses of the system. Numerical performance achieved on 21-bus and 69-bus test feeders demonstrates the efficiency of the suggested method in terms of solution quality compared to classical metaheuristic methods. In [17], a mixed-integer convex model is proposed to solve the problem of integrating generators based on renewable resources and energy storage systems in DC distribution networks. The aim of this work is to reduce the costs associated with energy losses. This model is solved using the MATLAB CVX tool, with excellent numerical results of 21- and 69-bus feeders showing the effectiveness and implementability of the suggested method. In reference [18], the problem regarding the integration of distributed generation sources in DC networks is addressed through a second-order conic programming model. This model seeks to minimize the power losses in the system lines. Numerical results on 21 and 69 bus feeders demonstrate the effectiveness and robustness of the suggested method compared to solutions representing the MINLP model of the problem.

Recently, the problem of integrating PV generators into DC networks has been solved by considering economic approaches based mainly on master–slave methodologies working with discrete-continuous coding. This type of codification allows siting and sizing problems to be solved in a unified manner, improving the exploration and exploitation of metaheuristic algorithms while reducing their computation times. An example of this is the work published by [19], which proposes a leader–follower optimization method consisting of the discrete-continuous version of the vortex search algorithm (DCVSA) and the successive approximations power flow method. The main idea of this work is to reduce the total annual operating costs, taking into account the investment, operation and maintenance costs of the power generation systems. The numerical performance on the 33-bus and 69-bus test feeders demonstrates the feasibility and effectiveness of the developed methodology. Finally, as in the previous case, the study by [20] uses the discrete-continuous version of the generalized normal distribution optimizer (DCGNDO) to solve the PV-generator integration problem. This work's main objective is to minimize the total annual operating costs. The results obtained in the 33- and 69-bus tests demonstrate the developed methodology's applicability and efficiency compared with the DCVSA.

1.4. Contributions and Scope

Considering the review literature review presented in the previous subsection, the main contributions of this document are presented as follows:

- A new optimization approach to solve the mathematical model representing the optimal integration of PV generators into DC grids. This methodology combines the discrete-continuous version of the crow search algorithm with the successive approximation power flow method within the framework of a master–slave optimization strategy.
- A solution strategy that finds the optimal global solution to the problem of integrating PV generators into DC networks, improving the results reported by the specialized literature regarding solution quality and repeatability.
- A new optimization approach based on the leader–follower operation scheme that allows solving a high-complexity optimization problem with reduced processing times

(less than 1.5 min) and consistent numerical performance in the DC versions of the 33- and 69-bus test feeders.

It is worth mentioning that this research is in the area of distribution system planning, which means that all the optimization algorithms (proposed and comparative methodologies) are evaluated based on simulations (offline validations) with the information provided by the distribution company regarding the generation and demand profiles being these data-averaged values obtained from historical information. In addition, once the expected size and location of the PV plants are determined, the distribution company will implement the physical stage, i.e., the construction of the PV plants; once these are ready to operate, then efficient day-ahead economic/technical/environmental dispatch methodologies for real-time operation must be implemented. This means that more research is required to plan and operate renewables in monopolar DC networks, which is an opportunity to continue contributing to the development of sustainable electrical networks in future works. However, numerical results in the studied test feeders demonstrated that in the case of the optimal location and sizing of PV plants in DC networks, all the combinatorial methods reach efficient numerical results, and the difference among them is minimal, which implies that with this research, the solution of the studied problem can be considered closed.

1.5. Document Organization

The rest of the paper is structured as follows: Section 2 provides a full description of the MINLP model representing the problem under study, i.e., the optimal integration of PV generators into DC distribution networks with the objective of reducing total annual operating costs; Section 3 describes the general implementation of the discrete-continuous version of the crow search algorithm and the successive approximation power flow method in order to evaluate the objective function; Section 4 presents the main characteristics of the DC versions of the 33- and 69-bus test feeders, the typical PV generation and demand curves for a Colombian region and the parametric information used to calculate the objective function value; Section 5 shows the numerical results, validations, analysis and discussion obtained for the optimal integration of PV generators for both test systems; and Section 6 lists the main conclusions of this study and future works.

2. Mathematical Formulation

The problem of the integration of PV generators into DC networks is presented here. The problem is mathematically formulated as Mixed-Integer Nonlinear Linear Programming (MINLP) where the decision variables are put in relation to the choice of the bus where the PV generator is placed, and the nonlinearity of the model arises in the power flow formulation due to the nonlinear nature of its general equation [21]. The objective function and the set of constraints of the optimization model representing the problem of integrating PV generators in DC distribution systems are described below.

2.1. Formulation of the Objective Function

The objective function corresponds to the minimization of the total annual operating cost of the DC network, which consists of three parts: annual power purchase costs for the substation bus, annual investment costs and maintenance costs for the PV generators. The components of the objective function are shown in (1) to (4).

$$\min \; A_{cost} = A_1 + A_2 + A_3, \tag{1}$$

$$A_1 = C_{kWh} T C_a C_c \left(\sum_{h \in \mathcal{H}} \sum_{k \in \mathcal{N}} p_{k,h}^{cg} \Delta h \right), \tag{2}$$

$$A_2 = C_{pv} C_a \left(\sum_{k \in \mathcal{N}} p_k^{pv} \right), \tag{3}$$

$$A_3 = C_{O\&M} T \left(\sum_{h \in \mathcal{H}} \sum_{k \in \mathcal{N}} p_{k,h}^{pv} \Delta h \right),$$ (4)

with

$$C_a = \left(\frac{t_a}{1 - (1 + t_a)^{-N_t}} \right),$$

$$C_c = \left(\sum_{t \in \mathcal{T}} \left(\frac{1 + t_e}{1 + t_a} \right)^t \right).$$

Here, the value of the objective function is given by A_{cost} and represents the total annual operating costs of the system. A_1 is the value of the annualized energy purchasing costs at the substation bus. A_2 is the value of the annualized investment costs, while A_3 is the value of the operation and maintenance costs of the PV generators. C_{kWh} is the average energy purchase price of the substation bus. T represents the average number of days per year. C_a is the annuity factor that allows finding the value of the periodic payments to be made by the network operator, depending on the expected internal interest rate t_a and the planning period N_t. C_c is a factor related to the increase in electricity costs during the planning period, which depends on the expected annual energy cost increase rate specified by the network operator t_e. $p_{k,h}^{cg}$ is the active power produced by each conventional generator connected to bus k during time period h. Δh is the length of the time period in which the electrical variables are assumed to be constant. C_{pv} is the average installation cost of 1 kW of solar power. p_k^{pv} is the nominal power of the PV generator connected to bus k. $C_{O\&M}$ is the maintenance and operation cost of the PV generator set, and $p_{k,h}^{pv}$ is the active power generated by each connected PV unit at bus k in the time interval h. Finally, $\mathcal{N}$, $\mathcal{H}$, and $\mathcal{T}$ are the sets containing all the buses in the network, the time periods in a daily operation scenario and the number of years in the planning horizon, respectively.

2.2. Set of Constraints

Equations (5) to (11) show the set of constraints representing the problem of integrating a PV generator into a DC grid.

$$p_{k,h}^{cg} + p_{k,h}^{pv} - P_{k,h}^{d} = \sum_{j \in \mathcal{N}} G_{kj} v_{k,h} v_{j,h}, \{\forall k \in \mathcal{N}, \quad \forall h \in \mathcal{H}\},$$ (5)

$$p_{k,h}^{pv} = p_k^{pv} C_h^{pv}, \{\forall k \in \mathcal{N}, \quad \forall h \in \mathcal{H}\},$$ (6)

$$P_k^{cg,\min} \leq p_{k,h}^{cg} \leq P_k^{cg,\max}, \{\forall k \in \mathcal{N}, \quad \forall h \in \mathcal{H}\},$$ (7)

$$y_k P_k^{pv,\min} \leq p_k^{pv} \leq y_k P_k^{pv,\max}, \{\forall k \in \mathcal{N}\},$$ (8)

$$v_k^{\min} \leq v_{k,h} \leq v_k^{\max}, \{\forall k \in \mathcal{N}, \quad \forall h \in \mathcal{H}\},$$ (9)

$$\sum_{k \in \mathcal{N}} y_k \leq N_{pv}^{ava},$$ (10)

$$y_k \in \{0,1\}, \{\forall k \in \mathcal{N}\}.$$ (11)

Here, $P_{k,h}^{d}$ is the active power required by the bus k in time period h. $v_{k,h}$ and $v_{j,h}$ denote the voltages of buses k and j in time period h, respectively, and G_{kj} is the conductance value associated with buses k and j. C_h^{pv} is the expected PV electricity production curve in the area of influence of the power system. $P_k^{cg,min}$ and $P_k^{cg,max}$ are the active power limits associated with each conventional generator connected at bus k. $P_k^{pv,min}$ and $P_k^{pv,max}$ are the active power limits associated with the PV generator connected at bus k. On the other hand, y_k is a binary variable responsible for finding the location of a PV generation unit at bus k. v_k^{min} and v_k^{max} are the minimum and maximum voltage regulation limits allowed by all the buses that make up the electrical system. Finally, N_{pv}^{disp} is a constant parameter related to the maximum number of PV generators that can be installed in the grid.

2.3. Model Analysis and Interpretation

The mathematical model displayed in (1) to (11) can be interpreted as follows. Equation (1) defines the objective function of the problem, which is the sum of the annual energy purchasing costs at the substation bus, as shown in Equation (2), with the annual investment costs of the PV generators, as proposed in (3), and maintenance and operating costs, as indicated in (4). Constraint (5) presents the active power balance for each system bus in each time period. This equation is the most challenging constraint of the problem studied, and since it is non-linear and non-convex, it must be solved adequately by numerical methods [4]. Equation (6) states that the active power generation of PV generators varies as a function of their rated power and the expected generation curve in the grid's influence zone. Inequality constraint (7) defines the lower and upper active power injection limits of conventional generators. Inequality (8) is also a constraint that determines the minimum and maximum active generation boundaries of the PV generators to be installed throughout the system. Similarly, (9) is a box constraint that defines the lower and upper bounds of voltage regulation for all busbars and time periods, while (10) defines the maximum number of PV generators available for installation in the grid. Finally, (11) shows the binary nature of the decision variable y_k.

The main complications of the presented model are (i) the existence of non-linearities and non-convexities in the active power balance equation and (ii) the mixture of integer and continuous variables. Therefore, to solve the problem under study, a master–slave optimization methodology based on the discrete-continuous version of the crow search algorithm (DCCSA) and the successive approximations method version is proposed, which has not been previously presented in the specialized literature and constitutes one of the main achievements of this work.

3. Proposed Methodology

This section presents a master–slave methodology applied to solve the problem of integrating PV generators into DC networks. In the master stage, the buses where the PV generators are placed are defined, along with their rated power. In the slave stage, the power flow constraints defined in the MINLP model are evaluated to determine the value of the objective function. Each component of the proposed methodology is presented in the following sections.

3.1. Master Stage: Discrete-Continuous Crow Search Algorithm

The DCCSA is a bio-inspired metaheuristic algorithm that is based on the rational behavior of crow flocks [22]. Crows are characterized by being ambitious birds, as they chase each other to stock up on the best food. In addition, crows watch where other birds hide their food in order to steal it when they are away [23]. Consequently, after having stolen the food, crows take the necessary measures to avoid becoming another victim, moving their hiding place or changing their route [24]. This behavior can be mathematically modeled by following simple rules that allow for a correct exploration and exploitation of the solution space [22]:

- ✓ Crows live in swarms
- ✓ Crows can remember where food sources are
- ✓ Crows chase each other to commit theft

✓   Crows guard their hideouts against robbery using stochastic behavioral factors

3.1.1. Initial Population

DCCSA is a population-based algorithm consisting of crows randomly placed in the environment, which allows the algorithm to begin the process of exploring and exploiting the solution space. The starting population of crows adopts the structure displayed in (12):

$$X^t = \begin{bmatrix} x_{11}^t & x_{12}^t & \cdots & x_{1N_v}^t \\ x_{21}^t & x_{22}^t & \cdots & x_{2N_v}^t \\ \vdots & \vdots & \ddots & \vdots \\ x_{N_i1}^t & x_{N_i2}^t & \cdots & x_{N_i,N_v}^t \end{bmatrix}, \tag{12}$$

where X^t is the population of crows in iteration t, N_i is the number of individuals in the population and N_v is the number of variables or the size of the solution space. To create the initial population of crows, (13) is used, which generates an array of random numbers in the lower and upper bounds that contain possible solutions for the PV generator integration problem.

$$X^0 = x_{\min}ones(N_i, N_v) + (x_{\max} - x_{\min})rand(N_i, N_v), \tag{13}$$

where $ones(N_i, N_v) \in \mathbb{R}^{N_i \times N_v}$ is an array containing ones; $rand(N_i, N_v) \in \mathbb{R}^{N_i \times N_v}$ is an array filled with random numbers between 0 and 1 that are generated by a uniform distribution; and $x_{\min} \in \mathbb{R}^{N_v \times 1}$ and $x_{\max} \in \mathbb{R}^{N_v \times 1}$ are vectors representing the lower and upper boundaries of the solution space, as shown below:

$$x_{\min} = \begin{bmatrix} x_{1,\min} \\ x_{2,\min} \end{bmatrix}, y_{\max} = \begin{bmatrix} x_{1,\max} \\ x_{2,\max} \end{bmatrix}.$$

Here, $x_{1,\min} \in \mathbb{R}^{N_{pv}^{ava} \times 1}$ and $x_{1,\max} \in \mathbb{R}^{N_{pv}^{ava} \times 1}$ represent the lower and upper limits of the decision variables associated with the location of the PV generators at the demand buses. On the other hand, $x_{2,\min} \in \mathbb{R}^{N_{pv}^{ava} \times 1}$ and $x_{2,\max} \in \mathbb{R}^{N_{pv}^{ava} \times 1}$ are the lower and upper bounds of the decision variables related to the size of the PV generation units.

Each individual generated by (13) must respect the coding shown in (14), which allows for determining the optimal location and size of PV generation units to be installed in the DC network.

$$X_i^t = [5, z, ..., 18 \mid 1.6593, p_z^{pv}, ..., 2.210]; i = 1, 2, ..., N_i. \tag{14}$$

Finally, in each iteration t, every crow in the population is able to memorize the position of the hiding place of its food, as presented in (15), which stores the location of the best food cache that each crow has found so far.

$$M^t = \begin{bmatrix} x_{11}^t & x_{12}^t & \cdots & x_{1N_v}^t \\ x_{21}^t & x_{22}^t & \cdots & x_{2N_v}^t \\ \vdots & \vdots & \ddots & \vdots \\ x_{N_i1}^t & x_{N_i2}^t & \cdots & x_{N_i,N_v}^t \end{bmatrix} \tag{15}$$

3.1.2. Crow Movement

To start the DCSSA, a crow j is supposed to want to visit its hideout, which is located at position M_j^t. On the other hand, crow i decides to follow j to approach its hiding place. Two situations may arise in this context: (i) search and (ii) evasion.

1. Case 1: Search

   From this situation, crow j does not know that crow i is chasing it. Therefore, crow i manages to approach crow j's hideout, thus switching its position in the solution space. The new location can be represented mathematically, as shown in (16).

$$X_i^{t+1} = X_i^t + rand\, fl\, (M_j^t - X_i^t),\qquad(16)$$

   where *rand* is a random number between 0 and 1 generated by uniform distribution, and fl is the flight length of the crow i.

2. Case 2: Evasion

   In this situation, crow j knows that crow i is chasing it. Consequently, the crow j tries to trick the crow i by moving to a random position in the solution space in order to protect its hiding place from being sacked.

   The two possible situations that may arise can be summarized as shown in (17).

$$X_i^{t+1} = \begin{cases} X_i^t + rand\, fl\, (M_j^t - X_i^t) & \text{Si}\, r_j \geq A_p \\ \text{random} & \text{otherwise} \end{cases}\qquad(17)$$

   Here, r_j is a random number between 0 and 1 that is generated by a uniform distribution, and A_p is the probability that crow j notices that crow i is following it.

3.1.3. Memory Updating

From the situations described above, the position of the crows is modified. Therefore, the new position of the food source must be updated. Therefore, if the new meal location adaptation function is better than the previously memorized position adaptation function, the crow updates its memory with the new position, as depicted in (18).

$$M_i^{t+1} = \begin{cases} X_i^{t+1} & \text{Si}\, F_F(X_i^{t+1}) < F(M_i^t) \\ M_i^t & \text{otherwise} \end{cases},\qquad(18)$$

where $F_f(\cdot)$ represents the adaptation function to be minimized.

3.2. Slave Stage: Successive Approximations Power Flow Method

The successive approximation method for solving the power flow in DC power systems was originally presented in [25]. This method allows iterative solving of the active power balance equation shown in (5). Therefore, it permits the slave phase to estimate the value of the adaptation function for each individual that composes the crow population, ensuring that the constraints presented in the MINLP model are respected, as previously mentioned in Section 2. The recursive formula that allows the solving of the power flow formulated in (5) is presented in (19).

$$\mathbb{V}_{d,h}^{m+1} = -\mathbf{G}_{dd}^{-1}\left[\mathbf{diag}^{-1}(\mathbb{V}_{d,h}^m)(\mathbb{P}_{d,h} - \mathbb{P}_{pv,h}) + \mathbf{G}_{ds}\mathbb{V}_{s,h}\right].\qquad(19)$$

Here, m is the iteration counter and $\mathbb{V}_{d,h}$ is the vector containing the voltage at the demand buses for each period h. $\mathbf{G}_{ds}$ is the component of the conductance matrix that associates the slack bus with the demand buses, while $\mathbf{G}_{dd}$ is the component of the conductance matrix that relates the demand buses to each other. $\mathbb{P}_{d,h}$ is the vector containing the active power consumed at the load buses for each period h. $\mathbb{P}_{pv,h}$ is the vector containing the active power generated by each PV generator for each period h. $\mathbb{V}_{s,h}$ is the vector containing the voltage at the substation bus terminals for each period h, which is a known parameter of the power flow solution. Finally, $\mathbf{diag}(z)$ is a diagonal matrix made up of the elements of vector z.

To determine the convergence of the iterative process, the criteria specified in (20) is employed, in which the maximum difference in the magnitudes of the demand voltages (i.e., $\mathbb{V}_{d,h}$) for each period h of two consecutive iterations is less than a given tolerance.

$$\max_h\left\{\left|\mathbb{V}_{d,h}^{m+1} - \mathbb{V}_{d,h}^m\right|\right\} \leq \zeta \tag{20}$$

In (20), ζ is defined as the convergence error, which takes the value 1×10^{-10} for this study.

Having solved the power flow in all time periods h using the successive approximation power flow method, the next step is to estimate the power generated at the terminals of the substation bus, as shown in (21).

$$\mathbb{P}_{s,h} = \mathbf{diag}(\mathbb{V}_{s,h})(\mathbf{G}_{ss}\mathbb{V}_{s,h} + \mathbf{G}_{sd}\mathbb{V}_{d,h}). \tag{21}$$

Here, $\mathbb{P}_{s,h}$ is the vector containing the active power generated at the slack bus for each period h. $\mathbf{G}_{ss}$ is the component of the conductance matrix associated with the slack bus, while $\mathbf{G}_{sd}$ is the component of the conductance matrix that relates the slack bus to the demand bus. Note that when solving (21), it is possible to obtain the value of A_1. The solution given by each individual in the master phase that follows the encoding given by (14) allows us to obtain the values A_2 and A_3. However, in order to discard potentially infeasible solutions that do not meet the constraints of the solution space, the objective function shown in (1) is substituted by the fitness function described in (22) [26,27].

$$\begin{aligned} F_f =&\, A_{cost} + \beta_1 \max_h\left\{0, \mathbb{V}_{d,h} - v^{\max}\right\} - \beta_2 \min_h\left\{0, \mathbb{V}_{d,h} - v^{\min}\right\} \\ &- \beta_3 \min_h\left\{0, \mathbb{P}_{s,h} - P_k^{gc,\min}\right\} \end{aligned} \tag{22}$$

In (22), F_f is the value of the adaptation function, and β_1, β_2 and β_3 are penalty factors applied to the objective function. These penalty factors are activated when the solution specified in the master stage does not meet the voltage regulation or generation capacity constraints of the slack bus. For this research article, the value of these penalty factors is taken as 1×10^6. One of the main advantages of using an adaptation function is that it allows the optimization approach to explore and exploit the solution space efficiently, given that if all the constraints specified in the MINLP model are satisfied, the final value of F_f is equal to the value of the objective function [28].

Figure 1 presents the general implementation of the proposed master–slave methodology for the integration of PV generation units into DC grids.

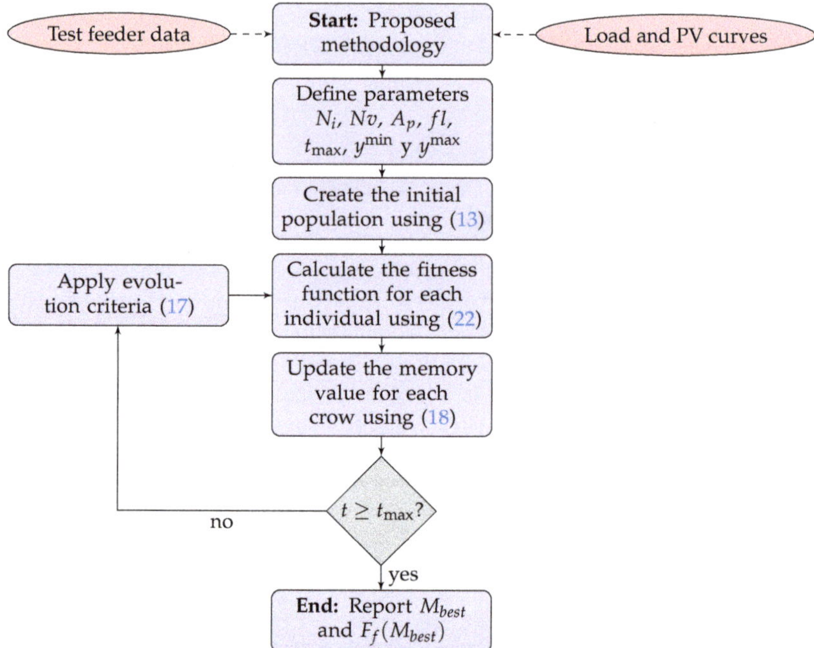

Figure 1. General application of the master–slave approach to address the problem of optimal integration of PV generating units in DC networks.

4. Test Feeders

To solve the problem under study, the DC versions of the 33- and 69-bus feeders (both with a radial topology) were used [29]. The main characteristics of each test feeder are presented below.

4.1. 33-Bus Test Feeder

This test feeder is an adaptation of the 33-bus ac test feeder commonly used to solve the problem of integrating PV generators into electrical systems. This feeder was originally proposed in [30]. Initially, it consists of 33 buses and 32 distribution lines, as shown in Figure 2a. To transform this feeder into a DC network, a voltage base of 12.66 kV and a power base of 100 kW are used. Additionally, the reactance component of all the distribution lines is neglected, as well as the reactive power consumption in all the buses that make up the feeder. In the maximum consumption scenario, the system loads consume 3715 kW. The parametric information for this system can be consulted in [31].

4.2. 69-Bus Test Feeder

This test feeder is an adaptation of the 69-bus AC test feeder commonly used to solve the problem of integrating PV generators into distribution systems, which was originally proposed in [32]. Initially, this feeder consists of 69 bus and 68 distribution lines, as shown in Figure 2b. To transform this feeder into a DC network, a voltage base of 12.66 kV and a power base of 100 kW are used. As in the previous feeder, the reactance component of all the distribution lines and the reactive power consumption in all the buses that make up the system are neglected. In the maximum consumption scenario, the system loads consume 3890.7 kW. The parametric information of this system can be found in [31].

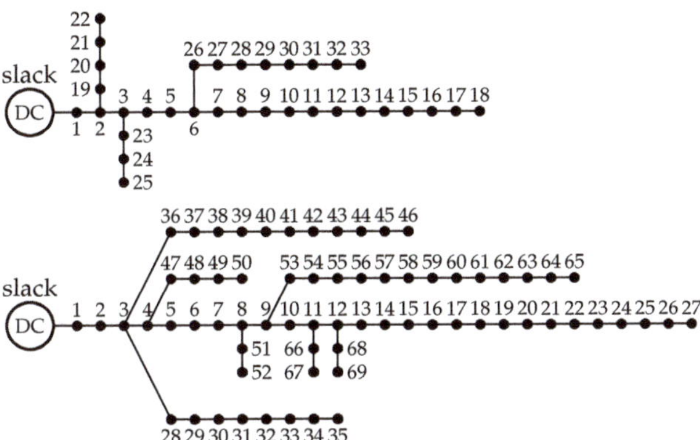

Figure 2. Electrical schematic of the test feeders: (**a**) 33-bus and (**b**) 69-bus.

Remark 1. *In this research work, the equivalent DC system electrical configuration is assumed to be monopolar, i.e., the voltage difference between the positive pole and the neutral conductor is the same as that assigned in the AC network [33].*

4.3. Calculating the Objective Function

To estimate the value of the adaptation function described in (22), the information shown in Table 1 was employed [34,35]. The cost of electricity considered for this study is a real value reported in [36], which corresponds to the average cost of energy reported by the utility company CODENSA of Bogota, Colombia, in May 2019. We have taken this value to have a fair comparison between the proposed methodology and the methodologies previously used to solve the problem under study in DC grids since this electricity price was used by the different comparison methodologies

Table 1. Information used to calculate the objective function value.

Parameter	Value	Unit	Parameter	Value	Unit
C_{kWh}	0.1390	USD/kWh	T	365	days
t_a	10	%	N_t	20	years
Δh	1	h	t_e	2	%
C_{pv}	1036.49	USD/kWp	$C_{O\&M}$	0.0019	USD/kWh
N_{pv}^{ava}	3	-	ΔV	± 10	%
$P_k^{pv,min}$	0	kW	$P_k^{pv,max}$	2400	kW
β_1	1×10^6	USD/V	β_2	1×10^6	USD/V
β_3	1×10^6	USD/W	-	-	-

To determine the effect of integrating PV generators into the test feeders presented in the previous subsection, the generation and demand curves for the city of Medellín, Colombia, were used (see Figure 3). These curves were first reported in [37].

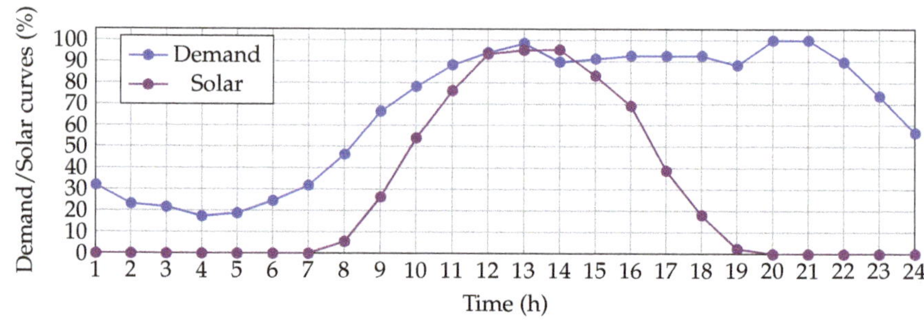

Figure 3. Daily demand and generation curves for Medellín, Colombia.

5. Numerical Results and Discussions

All simulations were implemented in the MATLAB programming environment, version 2022*a*, using proprietary scripts on a desktop computer with an Intel(R) Core(TM) i9-11900 processor CPU@2.50Ghz, 64.0 GB RAM and a 64-bit Windows 10 pro operating system. In order to demonstrate the performance of the suggested optimization approach, the DCCSA has been compared with the next methodologies, which were previously employed to address the problem under study in both AC and DC networks: (i) the BONMIN solver of the specialized GAMS software (exact solution of the MINLP model) [38], (ii) the discrete-continuous version of the Chu & Beasley Genetic Algorithm (DCCBGA) [38], (iii) the discrete-continuous version of the vortex search algorithm (DCVSA) [19], (iv) the discrete-continuous version of the generalized normal distribution optimizer (DCGNDO) [20], and (v) the discrete-continuous version of the parallelized particle swarm particle optimization algorithm (DCPPSO) [39]. Finally, for both test feeders, the installation of three PV generation units with a maximum size of 2400 kW was proposed.

5.1. DCCSA Parameters

To address the problem of the optimal integration of PV generators into DC networks, we used the information contained in Table 2.

Table 2. Parameters of the discrete-continuous crow search algorithm used in the master stage.

Parameter	DCCSA
Number of individuals (N_i)	62
Maximum iterations ($t_{\max}$)	622
Flight length (fl)	1.8468
Awareness probability (A_p)	0.0145

For the choice of the parameters listed in Table 2, the DCCSA was tuned using the CBGA [40] with an initial population of 50 individuals and a maximum number of iterations of 350. The tuning stage consists of using a metaheuristic algorithm in a previous stage, in the case of this study, the CBGA, to find the optimal parameters of the DCCSA to achieve a balance between thee exploration and exploitation of the algorithm. This is performed in order to guarantee the convergence of the algorithm and to ensure that the algorithm finds the global optimal solution (or very close optimal solutions) for the problem of integration of PV generators into DC networks. Similarly, another advantage of tuning metaheuristic algorithms is that it increases repeatability, i.e., each time the algorithm is run, it will always find the same or a very close solution. The parameters for tuning were: (i) the population size (N_i), with a range of [1–100] individuals; (ii) maximum number of iterations (t_{max}), with a range of [1–1000]; (iii) the flight length (fl), with a range of [0–3.5]; and (iv) the awareness probability (A_p), with a range of [0–1]. These parameters were selected due to the influence that each of them has on the performance of the algorithm since the modification of each of these parameters directly affects the exploration and exploitation of the DCCSA [22].

Figure 4 presents a flowchart where the operating principle of the tuning stage can be observed. In this stage, the CBGA is responsible for generating the optimal DCCSA parameters (i.e., N_i, t_{max}, f_l and A_p) represented by the red arrow. These parameters enter the master–slave stage to evaluate the value of the objective function of the problem under study represented by the green arrow. Once the iterative process is finished, the tuning stage finds the DCCSA parameters that allow finding the best value of the objective function. Regarding the parameters of the algorithms used for comparative purposes, they were taken from the original papers in which they were used for the first time to solve the problem of the location and sizing of PV generators in DC networks.

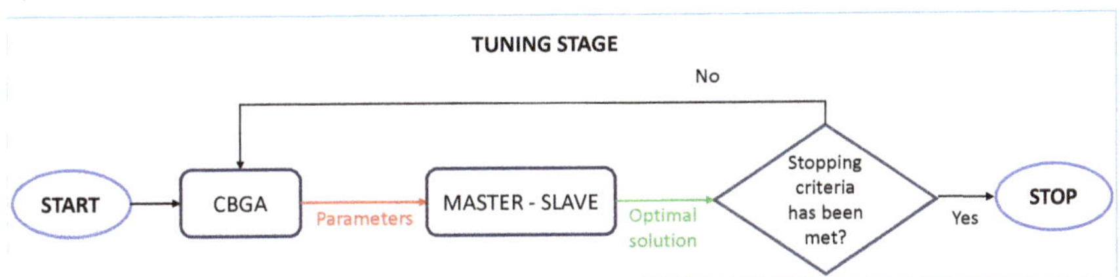

Figure 4. Operating principle of the tuning stage.

Likewise, it was also proposed to carry out 100 consecutive evaluations of the developed method to find the best, average and worst values of the objective function. In addition, the standard deviation of the two proposed test feeders and the average calculation time taken by the algorithm to find the optimal location and size of the PV generator were calculated.

5.2. Results from the First Test Feeder

5.2.1. Numerical Results

Table 3 shows the numerical values found by applying the proposed approach in the DC version of the 33-bus test feeder. The data in this table are listed from left to right as follows: the methodology used, the buses where the PV generators were installed and their rated power and the total annual operating costs.

Table 3. Numerical performance for the DC version of the 33-bus feeder.

Method	Site and Size (bus, MW)	A_{cost}(USD/Year)	Reduction(%)
Bench. Case	-	3,644,043.01	0
BONMIN	$\{18(1.4301), 32(2.0611), 33(0.1437)\}$	2,664,089.12	26.8919
DCCBGA	$\{11(1.1630), 14(0.9435), 31(1.4828)\}$	2,662,724.82	26.9294
DCVSA	$\{9(0.5803), 15(1.2914), 31(1.7156)\}$	2,662,425.32	26.9376
DCGNDO	$\{10(0.9743), 16(0.9202), 31(1.6925)\}$	2,662,371.59	26.9391
DCPPSO	$\{10(0.9680), 16(0.9189), 31(1.6999)\}$	2,662,371.59	26.9391
DCCSA	$\{10(0.9742), 16(0.9198), 31(1.6930)\}$	2,662,371.59	26.9391

The numerical results from the 33-bus test feeder show the following:

✓ All metaheuristic algorithms outperform the solution provided by BONMIN (i.e., the exact solution of the MINLP model), corroborating that the existence of binary variables leads the exact solution methods to be trapped in local optima.
✓ DCCSA, DCGNDO and DCPPSO are the methodologies that reach the best solution, achieving a reduction of 981,671.42 USD/year with respect to the benchmark case, confirming that the global optimal solution for the 33-bus test feeder is 2,662,371.59 USD/year

and is reached by locating the PV generation units at buses 10, 16 and 31, with a total rated capacity of 3586.97 kWp.

✓ The methodologies used to achieve the integration of PV generators allow savings of more than 26.85% compared to the benchmark case, with DCCSA, DCGNDO and DCPPSO being the methodologies that achieve the greatest reduction of 26.9391%. When these methodologies are compared to the other solution methodologies, there is a reduction in annual operating costs of approximately 0.0472% with respect to BONMIN, 0.0097% with respect to DCCBGA and 0.0015% with respect to DCVSA.

5.2.2. Statistical Analysis

Table 4 shows the results obtained by performing 100 consecutive evaluations of the master–slave approach in the 33-bus test feeder. The data in this table are listed from left to right as follows: the methodology used, the best solution found, the percent standard deviation and the average computation time.

Table 4. Numerical performance comparison in the 33-bus feeder after 100 consecutive evaluations.

Method	Best(USD/Year)	SDT(%)	Avg. Time(s)
BONMIN	2,664,089.12	0	1.29
DCCBGA	2,662,724.82	0.0557	2.43
DCVSA	2,662,425.32	0.0620	76.86
DCGNDO	2,662,371.59	0.0601	159.99
DCPPSO	2,662,371.59	0.0398	16.81
DCCSA	2,662,371.59	0.0058	20.86

The results in Table 4 show the following:

✓ DCCSA, DCGNDO and DCPPSO, in comparison with the methodologies reported in the literature, provide better results in terms of annual operating costs. They outperform BONMIN by 0.0645%, DCCBGA by 0.0133% and DCVSA by 0.0020%.
✓ Regarding computation times, BONMIN, DCCBGA and DCPPSO are faster than the proposed methodology, reducing processing times by 93.8016%, 88.3510% and 19.4160%, respectively, in comparison with DCCSA. However, as it is a planning problem, 20 s is a low processing time compared to the planning horizon, and DCCSA can solve the PV generator integration problem for DC networks while providing quality solutions with low computation times.
✓ Regarding the standard deviation, the superiority of the proposed DCCSA can be understood, achieving a reduction of 867.5515% with respect to DCCBGA, 977.5298% with respect to DCVSA, 943.9829% with respect to DCNGDO and 591.9624% with respect to DCPPSO. A comparison with respect to BONMIN was not made because when the MINLP model is exactly solved, its solution will always be the same.

The aforementioned demonstrates the effectiveness and robustness of DCCSA in solving the challenges of integrating PV generators into DC grids in order to reduce annual operating costs. The suggested method provides the best performance in terms of optimal solution and reproducibility with a short processing time. This makes it the best option to solve the problem of 33-bus test feeders, achieving an adequate solution not only from an economic point of view but also from a technical-operational point of view.

5.3. Results from the Second Test Feeder

5.3.1. Numerical Results

The numerical values found by the application of the proposed methodology on the DC version of the 69-bus test feeder are shown in Table 5. This table shows the same information as Table 3.

Table 5. Numerical results for the DC version of the 69-bus feeder.

Method	Site and Size (Bus, MVAr)	A_{cost}(USD/Year)	Reduction(%)
Bench. Case	-	3,817,420.38	0
DCCBGA	$\{19(0.7908),61(1.7891),64(1.1474)\}$	2,785,598.86	27.0293
DCVSA	$\{23(0.7720),62(2.3403)63(0.6185)\}$	2,785,538.58	27.0309
BONMIN	$\{27(0.4971),61(2.4000),65(0.8531)\}$	2,785,208.63	27.0395
DCGNDO	$\{19(0.4970),61(2.4000),64(0.8470)\}$	2,785,011.53	27.0447
DCPPSO	$\{22(0.5310),61(2.4000),64(0.8105)\}$	2,784,987.68	27.0453
DCCSA	$\{21(0.4855),61(2.4000),64(0.8598)\}$	2,784,979.35	27.0455

The following can be concluded from the information presented in Table 5:

✓ The developed DCCSA is the methodology that achieves the best solution for the 69-bus test feeder, i.e., a reduction of approximately 1,032,441.03 USD/year with respect to the benchmark case, which indicates that the optimal solution to the problem addressed in this study is 2,784,979.35 USD/year and is reached by locating the PV generators at buses 21, 61 and 64, with a total rated capacity of 3745.29 kWp.

✓ The comparison methodologies used to achieve the problem addressed allow savings of more than 27% with respect to the benchmark case, with DCCSA being the methodology that achieves the greatest reduction (i.e., 27.0455%). Similarly, when comparing the proposed methodology to the other solution methodologies reported in Table 5, reductions in the objective function value of approximately 0.0060% with respect to BONMIN, 0.0162% with respect to DCCBGA, 0.0146% with respect to DCVSA, 0.0008% with respect to DCGNDO and 0.0002% with respect to DCPPSO were obtained.

5.3.2. Statistical Analysis

As in the previous case, the effectiveness and robustness of the DCCSA for solving the problem under study were determined by performing 100 consecutive evaluations of the suggested approach in the 69-bus test feeder, whose results can be seen in Table 6. This table presents the same information as Table 4.

Table 6. Numerical performance comparison of the 69-bus feeder after 100 consecutive evaluations.

Method	Best(USD/Year)	SDT(%)	Avg. Time(s)
DCCBGA	2,785,598.86	0.1289	7.74
DCVSA	2,785,538.58	0.0975	269.22
BONMIN	2,785,208.63	0	2.03
DCGNDO	2,785,011.53	0.2384	376.88
DCPPSO	2,784,987.68	0.0226	28.24
DCCSA	2,784,979.35	0.0178	69.96

These results show the following:

✓ DCCSA, in comparison with all of the methodologies reported in the literature, provides better results in terms of the evaluation of the objective function. It outperforms BONMIN by 0.0082%, DCCBGA by 0.0222%, DCVSA by 0.0201%, DCNGDO by 0.0012% and DCPPSO by 0.0003%.

✓ As for the processing times, it can be seen that BONMIN, DCCBGA and DCPPSO are faster than the proposed methodology, reducing processing times by 97.0996%, 88.9359% and 59.6316%. It is also important to note that the DCCSA solves this highly complex optimization problem with the best results in less than 1.5 min.

✓ With regard to the standard deviation, the superiority of the proposed DCCSA is appreciated, achieving a reduction of 624.1573% with respect to DCCBGA, 447.7528%

with respect to DCVSA, 1239.3258% with respect to DCNGDO and 26.9663% with respect to DCPPSO. As mentioned in the previous subsection, no comparison was made with BONMIN.

The results presented above highlight that the developed DCCSA exhibits the best performance in terms of its solution and repeatability, with low processing times, making it the best choice to solve the problem of integrating PV generators into the 69-bus test feeder.

6. Conclusions and Future Work

This research document presents a master–slave approach to address the problem regarding the integration of PV generators into DC grids. In the master phase, the DCCSA is in charge of determining the bus where the photovoltaic generators will be installed, along with their rated powers, while the slave phase calculates the value of the adaptation function using the successive approximation power flow method. The DCCSA parameters were tuned using the CBGA. The numerical performance demonstrated the implementability and effectiveness of the optimization approach developed for the 33- and 69-bus DC test feeders when compared to different methodologies reported in the specialized literature, such as the BONMIN solver of GAMS, the Chu & Beasley genetic algorithm, the vortex search algorithm, the normal distribution optimizer and the particle swarm optimization algorithm.

In this regard, the following remarks can be made:

✓ The DCCSA achieves a reduction in the total annual operating costs of approximately 981,671.42 and 1,032,441.03 USD/year for each test feeder. These values represent reductions of 26.9391% and 27.0455%, respectively.

✓ The developed DCCSA obtains lower standard deviation values for the DC version of the 33- and 69-bus test feeders, showing improvements of 591.9624% and 26.9663% concerning the DCPPSO (i.e., second-best results), respectively. These standard deviation results confirm the repeatability and robustness of the proposed DCCSA in solving the PV generation unit integration problem, ensuring that, in each evaluation, the response is within a radius of 153 USD/year for the 33-bus feeder and 637 USD/year for the 69-bus feeder.

✓ The computation times taken by the proposed methodology to solve the MINLP model are 20.86 and 69.96 s for the 33 and 69-bus test feeders, respectively. This demonstrates that the developed methodology is a robust tool that allows the solving of highly complex mathematical models, ensuring quality answers when compared to other methods reported in the literature, as well as with low processing times. This allows the conclusion that the developed DCCSA is the best solution methodology to solve the problem herein addressed.

As future work, the following can be proposed: (i) reformulating the mathematical model of the problem under study while considering freeing the PV generators, i.e., disabling the maximum power point tracking of the PV generators; (ii) reformulating the mathematical model of the problem under study while considering the integration of batteries; (iii) including the problem of optimal conductor selection in the planning of DC networks, taking the investment costs of each conductor into account; and (iv) the reformulation of the studied problem via mixed-integer convex approximations by ensuring the global optimum finding with gradient-based algorithms combined with branch and cut optimization techniques.

Author Contributions: Conceptualization, methodology, software and writing (review and editing): B.C.-C., L.F.G.-N., G.A. and O.D.M. All authors have read and agreed to the published version of the manuscript.

Funding: This research received no external funding.

Acknowledgments: This work was developed with the colaboration of the Universidad de Talca, Instituto tecnológico Metropolitano, Universidad Veracruzana and Universidad Distrital Francisco José de Caldas.

References

1. Valencia, A.; Hincapie, R.A.; Gallego, R.A. Optimal location, selection, and operation of battery energy storage systems and renewable distributed generation in medium–low voltage distribution networks. *J. Energy Storage* **2021**, *34*, 102158. [CrossRef]
2. Abdelgawad, H.; Sood, V.K. A comprehensive review on microgrid architectures for distributed generation. In Proceedings of the 2019 IEEE Electrical Power and Energy Conference (EPEC), Montréal, QC, Canada, 16–18 October 2019; IEEE: New York, NY, USA, 2019; pp. 1–8.
3. Li, J.; Liu, F.; Wang, Z.; Low, S.H.; Mei, S. Optimal power flow in stand-alone DC microgrids. *IEEE Trans. Power Syst.* **2018**, *33*, 5496–5506. [CrossRef]
4. Garcés, A. On the convergence of Newton's method in power flow studies for DC microgrids. *IEEE Trans. Power Syst.* **2018**, *33*, 5770–5777. [CrossRef]
5. Alassi, A.; Bañales, S.; Ellabban, O.; Adam, G.; MacIver, C. HVDC transmission: Technology review, market trends and future outlook. *Renew. Sustain. Energy Rev.* **2019**, *112*, 530–554. [CrossRef]
6. Saeed, M.H.; Fangzong, W.; Kalwar, B.A.; Iqbal, S. A Review on Microgrids' Challenges & Perspectives. *IEEE Access* **2021**, *9*, 166502–166517.
7. González, D.M.L.; Rendon, J.G. Opportunities and challenges of mainstreaming distributed energy resources towards the transition to more efficient and resilient energy markets. *Renew. Sustain. Energy Rev.* **2022**, *157*, 112018. [CrossRef]
8. Abdul Kadir, A.F.; Khatib, T.; Elmenreich, W. Integrating photovoltaic systems in power system: Power quality impacts and optimal planning challenges. *Int. J. Photoenergy* **2014**, *2014*, 1–7.. [CrossRef]
9. Lamb, W.F.; Grubb, M.; Diluiso, F.; Minx, J.C. Countries with sustained greenhouse gas emissions reductions: An analysis of trends and progress by sector. *Clim. Policy* **2022**, *22*, 1–17. [CrossRef]
10. Moreno, C.; Milanes, C.B.; Arguello, W.; Fontalvo, A.; Alvarez, R.N. Challenges and perspectives of the use of photovoltaic solar energy in Colombia. *Int. J. Electr. Comput. Eng.* **2022**, *12*, 4521–4528. [CrossRef]
11. López, A.R.; Krumm, A.; Schattenhofer, L.; Burandt, T.; Montoya, F.C.; Oberländer, N.; Oei, P.Y. Solar PV generation in Colombia—A qualitative and quantitative approach to analyze the potential of solar energy market. *Renew. Energy* **2020**, *148*, 1266–1279. [CrossRef]
12. Gao, S.; Jia, H.; Marnay, C. Techno-economic evaluation of mixed AC and DC power distribution network for integrating large-scale photovoltaic power generation. *IEEE Access* **2019**, *7*, 105019–105029. [CrossRef]
13. Radosavljević, J.; Arsić, N.; Milovanović, M.; Ktena, A. Optimal placement and sizing of renewable distributed generation using hybrid metaheuristic algorithm. *J. Mod. Power Syst. Clean Energy* **2020**, *8*, 499–510. [CrossRef]
14. Cardona Isaza, J.A. *Ubicación y Dimensionamiento óptimo de Generadores Distribuidos en Redes DC*; Universidad Tecnológica de Pereira: Pereira, Colombia, 2022.
15. Abou El-Ela, A.A.; El-Seheimy, R.A.; Shaheen, A.M.; Wahbi, W.A.; Mouwafi, M.T. PV and battery energy storage integration in distribution networks using equilibrium algorithm. *J. Energy Storage* **2021**, *42*, 103041. [CrossRef]
16. Gil-González, W.; Molina-Cabrera, A.; Montoya, O.D.; Grisales-Noreña, L.F. An mi-sdp model for optimal location and sizing of distributed generators in dc grids that guarantees the global optimum. *Appl. Sci.* **2020**, *10*, 7681. [CrossRef]
17. Basto-Gil, J.D.; Maldonado-Cardenas, A.D.; Montoya, O.D. Optimal Selection and Integration of Batteries and Renewable Generators in DC Distribution Systems through a Mixed-Integer Convex Formulation. *Electronics* **2022**, *11*, 3139. [CrossRef]
18. Molina-Martin, F.; Montoya, O.D.; Grisales-Noreña, L.F.; Hernández, J.C. A Mixed-Integer conic formulation for optimal placement and dimensioning of DGs in DC distribution Networks. *Electronics* **2021**, *10*, 176. [CrossRef]
19. Cortés-Caicedo, B.; Molina-Martin, F.; Grisales-Noreña, L.F.; Montoya, O.D.; Hernández, J.C. Optimal Design of PV Systems in Electrical Distribution Networks by Minimizing the Annual Equivalent Operative Costs through the Discrete-Continuous Vortex Search Algorithm. *Sensors* **2022**, *22*, 851. [CrossRef]
20. Montoya, O.D.; Gil-González, W.; Grisales-Noreña, L.F. Solar Photovoltaic Integration in Monopolar DC Networks via the GNDO Algorithm. *Algorithms* **2022**, *15*, 277. [CrossRef]
21. Yang, Z.; Xie, K.; Yu, J.; Zhong, H.; Zhang, N.; Xia, Q. A general formulation of linear power flow models: Basic theory and error analysis. *IEEE Trans. Power Syst.* **2018**, *34*, 1315–1324. [CrossRef]
22. Askarzadeh, A. A novel metaheuristic method for solving constrained engineering optimization problems: Crow search algorithm. *Comput. Struct.* **2016**, *169*, 1–12. [CrossRef]
23. Jain, M.; Rani, A.; Singh, V. An improved crow search algorithm for high-dimensional problems. *J. Intell. Fuzzy Syst.* **2017**, *33*, 3597–3614. [CrossRef]
24. Hussien, A.G.; Amin, M.; Wang, M.; Liang, G.; Alsanad, A.; Gumaei, A.; Chen, H. Crow search algorithm: Theory, recent advances, and applications. *IEEE Access* **2020**, *8*, 173548–173565. [CrossRef]
25. Montoya, O.D.; Garrido, V.M.; Gil-González, W.; Grisales-Noreña, L.F. Power flow analysis in DC grids: Two alternative numerical methods. *IEEE Trans. Circuits Syst. II Express Briefs* **2019**, *66*, 1865–1869. [CrossRef]

26. Sahoo, R.R.; Ray, M. PSO based test case generation for critical path using improved combined fitness function. *J. King Saud Univ.-Comput. Inf. Sci.* **2020**, *32*, 479–490. [CrossRef]

27. Zhang, X.; Beram, S.M.; Haq, M.A.; Wawale, S.G.; Buttar, A.M. Research on algorithms for control design of human–machine interface system using ML. *Int. J. Syst. Assur. Eng. Manag.* **2021**, *13*, 462–469. [CrossRef]

28. Harman, M.; Jia, Y.; Zhang, Y. Achievements, open problems and challenges for search based software testing. In Proceedings of the 2015 IEEE 8th International Conference on Software Testing, Verification and Validation (ICST),Graz, Austria, 13–17 April 2015; pp. 1–12.

29. Kaur, S.; Kumbhar, G.; Sharma, J. A MINLP technique for optimal placement of multiple DG units in distribution systems. *Int. J. Electr. Power Energy Syst.* **2014**, *63*, 609–617. [CrossRef]

30. Baran, M.E.; Wu, F.F. Network reconfiguration in distribution systems for loss reduction and load balancing. *IEEE Power Eng. Rev.* **1989**, *9*, 101–102. [CrossRef]

31. Sahoo, N.; Prasad, K. A fuzzy genetic approach for network reconfiguration to enhance voltage stability in radial distribution systems. *Energy Convers. Manag.* **2006**, *47*, 3288–3306. [CrossRef]

32. Baran, M.E.; Wu, F.F. Optimal capacitor placement on radial distribution systems. *IEEE Trans. Power Deliv.* **1989**, *4*, 725–734. [CrossRef]

33. Monteiro, V.; Monteiro, L.F.; Franco, F.L.; Mandrioli, R.; Ricco, M.; Grandi, G.; Afonso, J.L. The Role of Front-End AC/DC Converters in Hybrid AC/DC Smart Homes: Analysis and Experimental Validation. *Electronics* **2021**, *10*, 2601. [CrossRef]

34. Castiblanco-Pérez, C.M.; Toro-Rodríguez, D.E.; Montoya, O.D.; Giral-Ramírez, D.A. Optimal Placement and Sizing of D-STATCOM in Radial and Meshed Distribution Networks Using a Discrete-Continuous Version of the Genetic Algorithm. *Electronics* **2021**, *10*, 1452. [CrossRef]

35. Wang, P.; Wang, W.; Xu, D. Optimal sizing of distributed generations in DC microgrids with comprehensive consideration of system operation modes and operation targets. *IEEE Access* **2018**, *6*, 31129–31140. [CrossRef]

36. Montoya, O.D.; Gil-González, W.; Grisales-Noreña, L.; Orozco-Henao, C.; Serra, F. Economic dispatch of BESS and renewable generators in DC microgrids using voltage-dependent load models. *Energies* **2019**, *12*, 4494. [CrossRef]

37. Grisales-Noreña, L.F.; Montoya, O.D.; Ramos-Paja, C.A. An energy management system for optimal operation of BSS in DC distributed generation environments based on a parallel PSO algorithm. *J. Energy Storage* **2020**, *29*, 101488. [CrossRef]

38. Montoya, O.D.; Grisales-Noreña, L.F.; Perea-Moreno, A.J. Optimal Investments in PV Sources for Grid-Connected Distribution Networks: An Application of the Discrete–Continuous Genetic Algorithm. *Sustainability* **2021**, *13*, 13633. [CrossRef]

39. Grisales-Noreña, L.F.; Montoya, O.D.; Marín-García, E.J.; Ramos-Paja, C.A.; Perea-Moreno, A.J. Integration of PV Distributed Generators into Electrical Networks for Investment and Energy Purchase Costs Reduction by Using a Discrete–Continuous Parallel PSO. *Energies* **2022**, *15*, 7465. [CrossRef]

40. Beasley, J.E.; Chu, P.C. A genetic algorithm for the set covering problem. *Eur. J. Oper. Res.* **1996**, *94*, 392–404. [CrossRef]

 *mathematics*

Article

Optimization-Based Energy Disaggregation: A Constrained Multi-Objective Approach

Jeewon Park, Oladayo S. Ajani and Rammohan Mallipeddi *

Department of Artificial Intelligence, Kyungpook National University, Daegu 37224, Republic of Korea
* Correspondence: mallipeddi.ram@gmail.com

Abstract: Recently, optimization-based energy disaggregation (ED) algorithms have been gaining significance due to their capability to perform disaggregation with minimal information compared to the pattern-based ED algorithms, which demand large amounts of data for training. However, the performances of optimization-based ED algorithms depend on the problem formulation that includes an objective function(s) and/or constraints. In the literature, ED has been formulated as a constrained single-objective problem or an unconstrained multi-objective problem considering disaggregation error, sparsity of state switching, on/off switching, etc. In this work, the ED problem is formulated as a constrained multi-objective problem (CMOP), where the constraints related to the operational characteristics of the devices are included. In addition, the formulated CMOP is solved using a constrained multi-objective evolutionary algorithm (CMOEA). The performance of the proposed formulation is compared with those of three high-performing ED formulations in the literature based on the appliance-level and overall indicators. The results show that the proposed formulation improves both appliance-level and overall ED results.

Keywords: energy disaggregation; non-intrusive load monitoring; optimization-based energy disaggregation; constrained multi-objective optimization; evolutionary algorithms

MSC: 68T20

Citation: Park, J.; Ajani, O.S.; Mallipeddi, R. Optimization-Based Energy Disaggregation: A Constrained Multi-Objective Approach. *Mathematics* **2023**, *11*, 563. https://doi.org/10.3390/math11030563

Academic Editors: Antonin Ponsich, Mariona Vila Bonilla and Bruno Domenech

Received: 21 December 2022
Revised: 11 January 2023
Accepted: 17 January 2023
Published: 20 January 2023

1. Introduction

In the modern world, the residential sector accounts for nearly one-third of global energy consumption [1]. Unlike traditional indirect feedback, such as monthly bills, the provision of appliance-based consumption feedback is projected to result in 12% energy savings per year [2] combined with additional features, such as the identification of faulty and/or energy-inefficient devices [2]. In order to provide appliance-level consumption feedback, it is essential to monitor the power consumption of each appliance directly (intrusive) or indirectly (non-intrusive) referred to as appliance load monitoring (ALM). Therefore, ALM can be classified as intrusive ALM (IALM) or non-intrusive ALM (NIALM) [1]. In IALM, one or more sensors are used to measure the consumption of each appliance, resulting in accurate measurements, but it is costly due to the amount of hardware required. On the other hand, NIALM, or energy disaggregation (ED), employs a single sensor to measure the consumption of the whole house, and appliance-level consumption is estimated using artificial-intelligence-based techniques. In the last few decades, the combined growth of artificial intelligence and smart meters led to an exponential growth of ALM [2–4] because of its capability to promote energy awareness with minimal infrastructure.

Given the aggregated measurements, $y(t)$, from the smart meter [1,2] over time, $t = 1, 2, \ldots, T$, the goal of ED is to estimate the energy consumption, $y_i(t)$, of each device, $i \in 1, 2, \ldots, n$, such that

$$y(t) = \sum_{i=1}^{n} y_i(t) + \sigma(t), \tag{1}$$

where $y(t)$ denotes the aggregate active power (P) [5] and $\sigma(t)$ represents the measurement noise.

From Equation (1), it is evident that ED is an over-parameterized and highly ill-posed problem. Furthermore, ED gets complicated as the number, types, and similarity between the devices increases [1], coupled with measurement errors [2]. Frameworks proposed for ED can be classified as (a) unsupervised or (b) supervised [1,2,6].

Unsupervised ED approaches [7–9] leverage unsupervised and generic learning features; however, they often fail when appliances with similar operating characteristics are featured in the network or when the power rating of one appliance is a linear combination of two or more appliances [10]. Supervised ED frameworks require representative labeled datasets to facilitate training of the components of the model. Furthermore, the type and amount of the training dataset depend on the components present. The challenges associated with machine-learning-based approaches are summarized in [10–14]. Among them, the main challenges are the ones associated with the data required for feature extraction and model training, such as

1. Exponential increase in data requirement as the number of appliances increases.
2. Depending on feature extraction, the sampling rate of data collection needs to be changed.
3. Data are household-specific due to unique device combinations and their usage patterns.
4. Class imbalance is inherent due to infrequent operation of some devices.
5. To incorporate new devices, the processes of data collection and training need to be repeated.

Optimization-based ED approaches alleviate the need for a training process that demands large amounts of data. Contrary to machine learning approaches, optimization-based ED approaches employ simple and readily available information corresponding to electrical devices such as different modes of operation and their associated power ratings. Additionally, new appliances can be integrated easily into the network by appending the appliance-specific information (states and ratings). Given the above information, ED can be formulated as a single-objective or multi-objective optimization problem with/without constraints [4,15–17]. The performances of optimization-based ED algorithms depend on various factors [14]. However, the main ones among them are the objective function(s) and constraints. In other words, the performance strongly depends on how the problem is formulated. In the literature, the objective and constraint functions are based on energy disaggregation error, sparsity of switching events, and some constraints regarding device operation depending on how the problem is formulated. Recently, in [13,18], ED is formulated as a multi-objective optimization problem. However, these formulations are unconstrained and do not consider the device's operation characteristics.

Motivated by the need for more efficient ED problem formulations that take into account the associated constraints in order to realize good ED results, this work formulates ED as a constrained multi-objective problem (CMOP), where sparsity and disaggregation error are considered as the two objectives. In addition, device-specific operational characteristics are considered as constraints. The formulated CMOP is solved using the constrained multi-objective evolutionary algorithm (CMOEA), and its performance is compared with those of state-of-the-art optimization-based ED formulations. The main contributions of this paper are highlighted as follows:

1. A novel constrained multi-objective formulation of energy disaggregation is proposed.
2. In the formulation, sparsity and disaggregation error are considered as the objectives to be optimized.
3. The constraints are formulated based on the device-specific operation characteristics of each appliance.
4. The performance of the proposed CMOP is evaluated using a constraint multi-objective evolutionary algorithm (CMOEA); it compares favorably with other methods in the literature.

The remainder of the paper is organized as follows. In Section 2, a review of the different formulations of optimization-based ED existing in the literature is presented. Section 3 presents the formulation proposed in the current work, where ED is formulated as a constrained multi-objective problem (CMOP). Section 4 presents the simulation results and a comparison with state-of-the-art optimization-based ED algorithms.

2. Literature Review on Optimization-Based Energy Disaggregation

Electrical devices, generally, operate in one of the predefined modes that are associated with estimated power-consumption levels, as depicted in Table 1. Given the information on the number of devices (n) in the network, the operational modes, and the associated power consumption corresponding to each device, ED can be formulated as an optimization problem as a constrained/unconstrained single or multi-objective problem [16]. In the literature, most of the optimization-based ED algorithms [15,19,20] represent ED as a binary optimization problem where a device i with l_i non-off modes is decomposed into l_i virtual two-state (on/off (1/0)) devices. For appliance i, let $P_i = \left[p_i^1, \ldots, p_i^{l_i} \right]^T$ represent a power rating corresponding to l_i virtual devices. Then, for n devices, the power rating corresponding to the $m = \sum_{i=1}^n l_i$ virtual devices is given by an ($m \times 1$) vector $P = [P_1, P_2, \ldots, P_i, \ldots, P_n]^T$. At time t, the operational status of m virtual on/off devices is given by the binary vector

$$S(t) = \left[s_1^{(1)}(t), \ldots; s_1^{(l_1)}(t), \ldots, s_i^{(1)}(t), \ldots, s_i^{(l_i)}(t), \ldots, s_n^{(1)}(t), \ldots, s_n^{(l_n)}(t) \right]^T, \qquad (2)$$

where $s^{(j)}(t) = \{0,1\}$ for $j = \{l_1, l_2, \ldots, l_n\}$.

Table 1. Details of appliances, their modes of operation with associated power ratings, and their power deviations [4].

No. of Appliances	Appliance	Maximum No of Modes	Power Rating (p)			Power Deviation (Θ)		
n		l_i	p_i^1	p_i^2	p_i^3	Θ_i^1	Θ_i^2	Θ_i^3
D1	LCD-Dell	1	25	-	-	5	-	-
D2	LCD-LG	1	22	-	-	5	-	-
D3	Coffee Maker	3	700	900	1100	100	100	100
D4	iMac	2	35	50	-	5	10	0
D5	Desktop	2	40	50	-	15	20	-
D6	Server	1	130	-	-	20	-	-
D7	Water Cooler	3	65	380	450	5	10	10
D8	Laptop	3	15	30	70	5	10	10
D9	Microwave	3	1000	1200	1700	100	100	100
D10	Printer	3	400	700	900	50	80	100
D11	Refrigerator	2	115	350	-	15	10	-

The aim of any ED algorithm is to find the operational state of each device in the network at each time instance given by ($S(t)$), so that estimated power consumption $\hat{y}(t)$ resembles the aggregated measurements, $y(t)$, from the smart meter [1,2], over time $t = 1, 2, \ldots, T$. In addition, $\hat{y}(t)$ is a combination of $\hat{y}_i(t)$, where $i = 1, 2, \ldots, n$. Therefore, during the estimation of ($S(t)$), the estimation of $\hat{y}_i(t)$, where $i = 1, 2, \ldots, n$, should match the true power-consumption levels of the individual appliances.

In order to approximate $(S(t))$, the intuitive and the most commonly employed objective function in optimization-based ED is the least-square error between $y(t)$ and $\hat{y}(t)$, as shown below [15,19,20].

$$\text{minimize} f = \sum_{t=1}^{T} (y_i(t) - \hat{y}_i(t))^2, \tag{3}$$

where $\hat{y}(t) = S(t)^T P$.

To handle optimization-based energy disaggregation, as formulated in (3), integer programming [17], mixed integer programming [19], evolutionary algorithms [4,15,16,21], etc., have been employed. The search space associated with the binary optimization problem given by (3) increases drastically with the increase in the number of devices and their associated operational modes. Furthermore, the energy disaggregation given by (3) is over-parameterized. Hence, the solutions obtained may fail to represent the practical operation of an appliance. The different issues associated with optimization-based ED algorithms are summarized in [14]. In other words, it is essential to improve the problem formulation considering additional objectives and/or constraints.

Due to the binary representation of the ED problem, where appliance i with l_i non-off operating modes is represented as l_i virtual devices, during the estimation of S, the appliance i might operate in more than one of the possible modes, which is impractical. To address this problem, the authors of [19] considered an inequality constraint that forces the device to operate in only one of the l_i modes or switches off all the l_i two state devices.

As shown in Table 1, the power rating of one on/off device can be similar to those of others, or the power rating of one device can be represented as a linear combination of multiple devices. This results in a situation where there exist multiple possible solutions for a given aggregate value. To address this issue, in [19], it has been experimentally demonstrated that choosing a combination of appliances with the lowest number of devices being on at a given time would result in better performance.

Currently, the smart meters provide high-frequency data. In other words, consecutive measurements of $y(t)$ are obtained at significantly shorter intervals (say 10 s). Therefore, minimizing the least-square error (3) alone may result in frequent appliance switching (on/off). To enforce temporal sparsity, in [3], ED is expressed as a constrained single-objective problem. In this framework, Sparse Switching Event Recovering (SSER), the goal is to minimize the total number of on/off switchings (4) subject to power-limit constraints given by (5).

$$\text{minimize} TSE(\triangle S) = \sum_{j=1}^{m} \sum_{t=1}^{T} \left| \triangle S^{(j)}(t) \right|, \tag{4}$$

subject to

$$S'(t)(P - \Theta) \leq y(t) \leq S'(t)(P + \Theta). \tag{5}$$

where $S = [S(1), ..., S(i), ..., S(T)]$ is the $(m \times T)$ matrix. $(\Theta = [\Theta_1, \Theta_2, \ldots, \Theta_m]^T)$ is the approximate power deviation variation corresponding to each power state $(P = [P_1, P_2, \ldots, P_m]^T)$. $TSE(.)$ denotes the total switching events in $\triangle S$ given by

$$\triangle S = S.D,$$

where differential matrix (D) of size $T \times (T-1)$ is given by:

$$D = \begin{bmatrix} -1 & & & & \\ 1 & -1 & & & \\ & 1 & \ddots & & \\ & & \ddots & -1 & \\ & & & 1 & -1 \\ & & & & 1 \end{bmatrix}$$

In other words, corresponding to each operational mode, the deviation from the rated power (Θ) is assumed to be provided. It is challenging to estimate (Θ) corresponding to every operational mode resulting in serious degradation in the performance [3].

The over-parameterized formulation in Equation (3) is regularized in [22], which is referred to as sparse optimization (Sopt), as shown below.

$$\text{minimize} f = \sum_{t=1}^{T} (y(t) - \hat{y}(t))^2 +$$

$$\Rightarrow \lambda_1 \sum_{i=1}^{n} \sum_{t=1}^{T} \left\| \begin{bmatrix} w_i^{(1)}(t) \\ \cdot \\ \cdot \\ \cdot \\ w_i^{(l_i)}(t) \end{bmatrix} \odot \begin{bmatrix} s_i^{(1)}(t) \\ \cdot \\ \cdot \\ \cdot \\ s_i^{(l_i)}(t) \end{bmatrix} \right\|_1 +$$

$$\Rightarrow \lambda_1 \sum_{i=1}^{n} \sum_{t=1}^{T} \left\| k_i \begin{bmatrix} s_i^{(1)}(t) - s_i^{(1)}(t-1) \\ \cdot \\ \cdot \\ \cdot \\ s_i^{(l_i)}(t) - s_i^{(l_i)}(t-1) \end{bmatrix} \right\|_\infty , \tag{6}$$

subject to

$$\sum_{j=1}^{l_i} s_i^{(j)}(t) = 1, i = 1, \ldots, n, and \ t = 1, \ldots, T \tag{7}$$

The equality constraint (7) is to enforce that continuous operating devices operate in at least one of the l_i non-off states. In (6), the penalty terms are expected to provide the temporal sparsity. However, the performance significantly varies based on the non-negative weight vector $[w_i^{(1)}(t), \ldots, w_i^{(l_i)}(t)]^T$ and hyperparameters $(\lambda_1, \lambda_2,$ and $k_i(i = 1, \ldots, n))$.

Recently, ED is modeled as a multi-objective optimization problem in [18], where objectives are

$$\text{minimize} \begin{cases} f_1 = |y(t) - \hat{y}(t)| \\ f_2 = \phi_o d_o(s(t), s(t-1)) + \phi_s d_s(s(t), s(t-1)), \end{cases} \tag{8}$$

where function $d_s(s(t), s(t-1))$ represents the number of mode changes, and function $d_o(s(t), s(t-1))$ represents the number of on/off changes. Generally, solving a multi-objective optimization problem leads to a number of trade-off solutions where each solution is a prospective energy disaggregation. Therefore, it is essential to select a solution from the set to estimate the power consumption profile of devices. In [18], a decision-maker (DM) function defined by the following equation is employed to select the optimal ED solution from the set of trade-off solutions.

$$DM = f_1(s(t)) + \left[(1 + f_2(s(t))) \sqrt{|f_1(s(t)) - f_1(s(t-1))|} \right] \tag{9}$$

In [13], it was observed that minimization of least-square error (f_1 in (8)) maximizes the sum of the variations in switching events (f_2 in (8)) and vice versa. This is because of the featured inherent noise and similarity between the appliances in terms of power ratings. In other words, minimization of least-square error and total variation of switching events are conflicting. In [13], the problem is solved as a multi-objective problem. However, instead of employing the decision function, once the trade-off set is obtained, a solution where the disaggregated individual device operations match the practical device operation is selected (using some reference signals). The reference signals are considered to available or given by the manufacturer. In addition, in [13], the ED is solved as an discrete optimization problem instead of binary optimization problem where the state matrix (S) is represented as

$$SP = \begin{bmatrix} sp_1(1) & \cdots & sp_1(T) \\ \vdots & \ddots & \vdots \\ sp_n(1) & \cdots & sp_n(T) \end{bmatrix} \tag{10}$$

where SP is a state matrix of size $n \times T$ and $sp_i(t)$ is the consumption of device $i = 1, 2, \ldots, n$ at time instance $t = 1, 2, \ldots, T$. The objective functions considered are

$$Minimize : E = \sum_{t=1}^{T} (y(t) - \sum_{i=1}^{n} sp_i(t))^2 \tag{11}$$

$$Minimize \quad \begin{matrix} \sum_{i=1}^{n} \sum_{t=2}^{T} [(sp_i(t) \neq sp_i(t-1))(sp_i(t)sp_i(t-1) \neq 0)] \\ + \\ \sum_{i=1}^{n} \sum_{t=2}^{T} [(sp_i(t) \neq sp_i(t-1))(sp_i(t)sp_i(t-1) = 0)] \end{matrix} \tag{12}$$

Equation (11) is similar to f_1 in (8), and Equation (12) is similar to f_2 in (8). In addition, to effectively solve the multi-objective ED using the multi-objective evolutionary algorithm, problem-specific mutation and crossover operators were proposed.

Based on the review, it can be concluded that to improve the performance of optimization-based ED algorithms, novel problem formulations in terms of objectives and constraints are very crucial. Hence, more efficient formulations and algorithms are needed to address the ED problem.

3. Energy Disaggregation as a Constrained Multi-Objective Optimization Problem

In [13], the ED problem is formulated as an unconstrained multi-objective optimization problem given by Equations (11) and (12). In the second objective related to temporal sparsity (12), the sum of appliance on/off switching is combined with appliance state switching. It is to be remembered that the appliance on/off switching and appliance state changing strongly depend on the type of device. For instance, a refrigerator is a continuous operational device that rarely switches on/off and also switches operational modes with less frequency. However, a printer is a device that is regularly switched on/off, and during a certain period of operation, the number of state switches is high compared to the number in devices such as refrigerators. In other words, it is essential to take the device-specific operational constraints into account. In this work, appliance-specific operational constraints are incorporated, and ED is formulated as a constrained multi-objective optimization problem (CMOP). It is solved using a constrained multi-objective evolutionary algorithm (CMOEA). The appliance-specific operational constraints include a number of state switches per unit time of operation. This is specific to devices and the way in which they are designed to be operated. In addition, this information can be easily obtained from the manufacturer or through some data collection regarding how the particular device is operated in a network.

In the current framework, the objectives considered are same as (11) and (12). However, the minimization of (11) and (12) is subjected to n constraints, one corresponding to each device, represented as follows.

$$\frac{\sum_{t=2}^{T}[(sp_i(t) \neq sp_i(t-1))(sp_i(t)sp_i(t-1) \neq 0)]}{\sum_{t=1}^{T}[(sp_i(t) \neq 0)]} \leq b_i \qquad i = 1, ..., n \qquad (13)$$

In the constraints given by (13), the left-hand side represents the number of state switching events corresponding to a device per unit time of operation in a prospective energy disaggregation vector. The right-hand side b_i represents the numerical value specific to the device. In other words, continuously operating devices such as refrigerators have low values of b_i, as the number of state switches is significantly low for a large period of operation. On the other hand, for devices such as a coffee maker, the number of state switching events is significantly higher over a shorter period of time. It has to be remembered that obtaining the values of b_i corresponding to device operation is not difficult to do.

To solve the CMOP defined by (11)–(13), any existing state-of-the-art CMOEA can be employed. However, in the current work, I_{SDE^+} [23], which is an evolutionary multi-objective algorithm, is used. I_{SDE^+} is effective at handling multi-objective problems with a variety of landscapes and is computationally efficient. I_{SDE^+} is combined with superiority of feasible (SF) to handle the constraints. In addition, to effectively solve the ED problem formulated as a CMOP, application-specific variation operators (crossover and mutation) proposed in [13] are employed. The overall framework used to solve the ED, formulated as a CMOP—CMOEA (I_{SDE^+} with superiority of feasible)—is shown in Algorithm 1.

Algorithm 1: General framework of the CMOEA employed to solve the ED formulated as a CMOP.

---

1 **Input**: N (population size)
2 $\boldsymbol{P} \leftarrow$ Initialization
3 $\boldsymbol{ISDE+} \leftarrow$ Evaluation (P)
4 **while** *predefined termination criteria not satisfied* **do**
5     $\boldsymbol{M} \leftarrow$ Mating selection (P, N, ISDE+)
6     $\boldsymbol{O} \leftarrow$ Variation (M, N)
7     $\boldsymbol{Q} \leftarrow P \cup O$
8     $\boldsymbol{ISDE+} \leftarrow$ Evaluate (Q)
9     $[\boldsymbol{P}, \boldsymbol{ISDE+}] \leftarrow$ Environmental selection $(Q, N, ISDE+)$
10 **end**
11 **Output**: P

In the proposed framework, the CMOEA starts with random initialization of a set of solutions (N) for the given ED problem, where each prospective solution is represented as shown in (10). The ISDE+ indicator value that depends on the two objectives given by ((11) and (12)) and constraint violation given by (13) is evaluated for individual solution candidates in the population (as outlined in line 2 of Algorithm 1). Later, mating selection is carried out, in which the population members with superior ISDE+ values are prioritized in a probabilistic manner (line 4 in Algorithm 1). The solutions selected during mating selection (M) are then used to produce new solutions, namely, the offspring population (O) (line 5 in Algorithm 1). The process of producing new solutions using the solutions and their objective values in the population is referred to as variation. In the current work, we employ the problem-specific variation operators proposed in [13]. The population (P) and offspring population (O) are combined (Q) (line 6 in Algorithm 1) and evaluated (line 7 in Algorithm 1). Finally, environmental selection is performed, where the best NP candidates of Q are chosen to be the population (P) for the next generation (line 8 in Algorithm 1). The steps mating selection, variation to produce new solutions, evaluation, and environmental selection (Algorithm 1, lines 4–8) are repeated until a predefined stopping criterion is met. After the termination, the final population (P) which contains trade-off solutions that satisfy the objectives and constraints are considered as the output. In other words, each solution in the trade-off set represents a possible ED. From the set of trade-off solutions, the solution with the lowest value of disaggregation error is selected as the best possible energy disaggregation result.

4. Simulation Results and Analysis

To evaluate the performance of the proposed framework, we considered some instances of ED problems from the benchmark suite proposed in [14]. Specifically, we selected instances I_1, I_{12} and I_{18}, which are problem instances that feature cases where almost all devices are in operation, the power rating of one appliance is a linear combination of multiple appliances, and simultaneous switching of appliances with similar states or multiple devices whose linear combinations are similar to each other. These instances were chosen because they represent the different challenges posed by optimization problems formulated as ED.

Furthermore, as shown in [14], the performances of ED algorithms must be evaluated by a number of metrics, including both appliance-level and overall performance metrics. Therefore, we employ standard metrics such as per-appliance accuracy (AC_i), estimated energy fraction index (EEFI) ($\hat{h}_i$), and relative squared error (RSE_i) at the appliance level; and overall accuracy (ACC), overall state prediction accuracy (SPA), and fraction of total energy assigned correctly (FTEAC) at the overall level to compare the performance of the proposed framework with the baseline results from the literature. A better-performing ED algorithm is expected to have higher values for overall performance indicators—ACC, SPA, and FTEAC. Among the appliance-level indicators, AC_i is expected to be higher, and RSE_i is expected to be lower. However, ($\hat{h}_i$) is expected to be as close as possible to (h_i).

All the simulations were performed in MATLAB 2020a installed on a PC with 64-bit Windows 10, a 3.30 GHz CPU, and 24 GB of RAM. Based on the aforementioned problem instances and metrics, we first evaluated the ED performance with and without the constraints defined by Equation (13). In Tables 2–4, the effects of the appliance-specific constraints on the energy disaggregation performance are evaluated considering problem instances I_1, I_{12}, and I_{18}. Tables 5–7, present a comparative analysis of the proposed framework with state-of-the-art energy disaggregation frameworks, such as ALIP [19], MONILM [18], and SOPT [22].

In Tables 2–4, it can be observed that with respect to most of the devices, the energy disaggregation performance with constraints is better than that without constraints in most of the per-appliance metrics. In addition, a similar observation can be made with respect to overall performance metrics, such as SPA and FTEAC. However, in instance I_{12}, the ACC of the proposed framework with constraints is less, but the performance is drastically improved in terms of SPA. This is because the use of constraints helped the framework perform better on D_{11}, which was in operation for significant amount of time and consumed significant amount of power (h). Therefore, it justified the use of appliance-specific constraints defined by (13).

Table 2. Effect of appliance-specific constraints on the performance of energy disaggregation considering the I_1 problem instance.

No of Appliances	AC_i		h	$\hat{h}_i$		RSE_i	
n	Without	With	Ground Truth	Without	With	Without	With
D1	1	1	0.05	0.0584	0.0542	0	0
D2	0.94	0.94	0.0328	0.0514	0.0477	0.4669	0.4469
D3	1	1	0	0	0	0	0
D4	0.8570	0.9072	0.0728	0.1074	0.0680	0.1589	0.0994
D5	0.5394	0.8796	0.0734	0.0149	0.0867	1	0.0713
D6	0.9358	0.9358	0.2267	0.3039	0.2816	0.0766	0.0766
D7	0.5809	0.7407	0.3025	0.1359	0.2147	0.6718	0.4313
D8	0.6825	0.7576	0.0489	0.0933	0.0398	0.1001	0.3891
D9	1	1	0	0	0		
D10	0.5423	0.5	0.0759	0.0987	0	1	1
D11	0.9823	0.9711	0.1158	0.1359	0.2074	0.3824	2.0084

Overall Metrics		
	Without	With
Overall Energy Disaggregation Accuracy (ACC (%))	87.6556	90.7188
State Prediction Accuracy (SPA (%))	56.4899	64.899
Fraction of Total Energy assigned correctly (FTEAC)	0.7749	0.8222

Table 3. Effect of appliance-specific constraints on the performance of energy disaggregation considering the I_{12} problem instance.

No of Appliances	AC_i		h	$\hat{h}_i$		RSE_i	
n	without	With	Ground Truth	Without	With	Without	With
D1	1	1	0.0501	0.0550	0.0650	0.0876	0.0060
D2	1	1	0	0.0523	0		
D3	1	1	0	0	0		
D4	0.7218	0.8644	0.0706	0.1101	0.0867	0.1256	0.3309
D5	0.8285	0.5618	0.0797	0.1174	0.0259	1	0.1811
D6	0.9167	0.9150	0.2567	0.2896	0.3134	0.0826	0.0777
D7	0.7303	0.5945	0.2779	0.2738	0.1548	0.7184	0.5988
D8	0.5214	0.6224	0.0204	0.1018	0.0552	2.2408	8.0803
D9	1	1	0	0	0		
D10	0.5	0.5	0.0061	0	0.0000	1	1
D11	0.5	0.9601	0.2387	0	0.2989	0.1101	1.0000

Overall Metrics		
	without	With
Overall Energy Disaggregation Accuracy (ACC (%))	89.4814	86.1355
State Prediction Accuracy (SPA (%))	33.6869	60.2778
Fraction of Total Energy assigned correctly (FTEAC)	0.7512	0.8172

Table 4. Effect of appliance-specific constraints on the performance of energy disaggregation considering I_{18} problem instance.

No of Appliances	AC_i		h	$\hat{h}_i$		RSE_i	
n	Without	With	Ground Truth	Without	Proposed	Without	Proposed
D1	0.7583	1	0.0652	0.0363	0.0696	0.4833	0.0000
D2	0.6561	0.94	0.0056	0.0349	0.0612	5.4466	8.2330
D3	1	1	0	0	0		
D4	0.5	0.8730	0.0848	0.0000	0.1135	1	0.1172
D5	0.6138	0.5354	0.1076	0.0504	0.0142	0.7165	1.0000
D6	0.9408	0.9408	0.2939	0.3658	0.3617	0.0586	0.0586
D7	0.6373	0.6373	0.1684	0.1468	0.1658	0.6727	0.7272
D8	0.6687	0.7534	0.0926	0.0422	0.0497	0.4666	0.3115
D9	1	1	0	0	0		
D10	0.5	0.5	0.0277	0.00000	0.0000	1	1
D11	0.9809	0.9744	0	0.32356	0.1644	0.9485	0.1166

Overall Metrics		
	without	With
Overall Energy Disaggregation Accuracy (ACC (%))	88.4347	88.8737
State Prediction Accuracy (SPA (%))	53.4091	54.0657
Fraction of Total Energy assigned correctly (FTEAC)	0.7293	0.8333

In Tables 5–7, it can be observed that the performance of the proposed framework, in terms of SPA and FTEAC, is better than the state-of-the-art methods for instances I_1 and I_{18}, but slightly worse for I_{12}. However, in ACC, the performance of the proposed framework is worse. As mentioned in the literature [13], this is not a concern, because a high value of ACC does not signify superior performance, as each mode of the device is represented with a discrete value, and thus achieving an ACC close to 100% is not possible. In other words, even accurate energy disaggregation does not result in an ACC close to 100%. Therefore, the performance of the proposed framework seems to be superior for instances I_1 and I_{18}. However, for instance I_{12}, the performance of ALIP seems better than that of the proposed framework. For instance I_{12}, nearly 80% of the total energy is consumed by continuously operating devices, such as D_6, D_7, and D_{11}. In ALIP, an equality constraint is specifically employed to handle continuously operating devices, resulting in superior performance.

Table 5. Comparison of the proposed framework with the state-of-the-art methods in terms of energy disaggregation on the I_1 problem instance.

No of Appliances	AC_i				h	$\hat{h}_i$				RSE_i			
n	ALIP	MONILM	SOPT	Proposed	Ground Truth	ALIP	MONILM	SOPT	Proposed	ALIP	MONILM	SOPT	Proposed
D1	0.6648	0.8722	0.8403	1	0.05	0.0169	0.0381	0.0348	0.0542	0.6704	0.2556	0.3194	0
D2	0.6162	0.841	0.94	0.94	0.0328	0.0144	0.0359	0.045	0.0477	0.9208	0.5983	0.4469	0.4469
D3	1	1	1	1	0	0.0221	0.0131	0.0233	0	-	-	-	-
D4	0.6612	0.8093	0.8073	0.9072	0.0728	0.0387	0.0714	0.1023	0.0680	0.614	0.2651	0.1778	0.0994
D5	0.6663	0.8039	0.8226	0.8796	0.0734	0.0403	0.0842	0.1023	0.0867	0.6019	0.2493	0.2055	0.0713
D6	0.9358	0.8605	0.5012	0.9358	0.2267	0.2663	0.2113	0.0007	0.2816	0.0764	0.2175	0.9972	0.0766
D7	0.7923	0.7633	0.9194	0.7407	0.3025	0.2307	0.1895	0.286	0.2147	0.3687	0.4554	0.1109	0.4313
D8	0.5763	0.4203	0.4363	0.7576	0.0489	0.0398	0.095	0.0898	0.0398	1.0479	2.1133	1.9037	0.3891
D9	1	1	1	1	0	0.0057	0	0	0	-	-	-	-
D10	0.6593	0.7096	0.664	0.5	0.0759	0.055	0.0307	0.0358	0.0000	0.8208	0.4959	0.6601	1
D11	0.9656	0.8919	0.9767	0.9711	0.1158	0.2729	0.232	0.242	0.2074	2.2818	2.5971	1.2418	2.0084

Overall Metrics	ALIP	MONILM	SOPT	Proposed
Overall Energy Disaggregation Accuracy (ACC (%))	99.8051	99.6126	96.5757	90.7188
State Prediction Accuracy (SPA (%))	60.0758	49.899	42.2475	64.899
Fraction of Total Energy assigned correctly (FTEAC)	0.7785	0.7769	0.7011	0.8222

Table 6. Comparison of the proposed framework with state-of-the-art methods in terms of energy disaggregation in the I_{12} problem instance.

No of Appliances	AC_i				h	$\hat{h}_i$				RSE_i			
n	ALIP	MONILM	SOPT	Proposed	Ground Truth	ALIP	MONILM	Proposed	SOPT	ALIP	MONILM	SOPT	Proposed
D1	0.6949	0.9018	0.5	1	0.0501	0.0197	0.0434	0.0650	0	0.6133	0.2598	1	0.0060
D2	1	1	1	1	0	0.016	0.0356	0	0.029	-	-	-	-
D3	1	1	1	1	0	0.0042	0	0	0	-	-	-	-
D4	0.6332	0.8008	0.833	0.8644	0.0706	0.0364	0.0861	0.0867	0.0942	0.6694	0.2413	0.1654	0.3309
D5	0.6292	0.8163	0.8243	0.5618	0.0797	0.0347	0.0933	0.0259	0.1087	0.6795	0.2202	0.1911	0.1811
D6	0.9465	0.8697	0.9127	0.9150	0.2567	0.2833	0.2337	0.3134	0.2613	0.0126	0.1836	0.0878	0.0777
D7	0.8121	0.7788	0.5	0.5945	0.2779	0.2571	0.2289	0.1548	0	0.3326	0.3994	1	0.5988
D8	0.5574	0.6015	0.527	0.6224	0.0204	0.0416	0.0849	0.0552	0.0887	2.511	6.8094	6.1865	8.0803
D9	1	1	1	1	0	0	0.0061	0	0.0061	-	-	-	-
D10	0.5	0.5	0.5	0.5	0.0061	0.0218	0.046	0.0000	0	2.44	4.04	1	1
D11	0.8915	0.6184	0.7128	0.9601	0.2387	0.2961	0.1436	0.2989	0.39	0.5494	0.9286	1.2979	1

Overall Metrics	ALIP	MONILM	SOPT	Proposed
Overall Energy Disaggregation Accuracy (ACC (%))	99.6766	99.5509	97.2710	86.1355
State Prediction Accuracy (SPA (%))	65.8081	47.0202	54.2929	60.2778
Fraction of Total Energy assigned correctly (FTEAC)	0.8697	0.8561	0.666	0.8172

Table 7. Comparison of the proposed framework with state-of-the-art methods in terms of energy disaggregation in the I_{18} problem instance.

No of Appliances	AC_i				h	$\hat{h}_i$				RSE_i			
n	ALIP	MONILM	SOPT	Proposed	Ground Truth	ALIP	MONILM	SOPT	Proposed	ALIP	MONILM	SOPT	Proposed
D1	0.6389	0.8139	0.9903	1	0.0652	0.0181	0.0409	0.064	0.0696	0.7222	0.3722	0.0194	0.0000
D2	0.5852	0.7129	0.94	0.94	0.0056	0.0179	0.047	0.0571	0.0612	3.4572	7.5177	8.1831	8.2330
D3	1	1	1	1	0	0.0065	0	0.0065	0	-	-	-	-
D4	0.6079	0.7868	0.7338	0.8730	0.0848	0.033	0.1168	0.1277	0.1135	0.7384	0.2318	0.292	0.1172
D5	0.6226	0.8231	0.8591	0.5354	0.1076	0.0412	0.1046	0.1305	0.0142	0.6951	0.207	0.1219	1.0000
D6	0.9408	0.8385	0.9041	0.9408	0.2939	0.3392	0.2591	0.31	0.3617	0.0586	0.2731	0.1332	0.0586
D7	0.7751	0.8421	0.6236	0.6373	0.1684	0.2041	0.1655	0.024	0.1658	0.521	0.3469	0.7221	0.7272
D8	0.5641	0.7003	0.682	0.7534	0.0926	0.0319	0.0857	0.0501	0.0497	0.8537	0.5692	0.4666	0.3115
D9	1	1	1	1	0	0	0	0	0	-	-	-	
D10	0.7638	0.7491	0.6947	0.5	0.0277	0.0261	0.0232	0.0464	0.0000	0.5213	0.4129	1.2459	1
D11	0.9483	0.8001	0.9369	0.9744	0	0.3222	0.157	0.1995	0.1644	1.401	0.8337	0.535	0.1166

Overall Metrics

	ALIP	MONILM	SOPT	Proposed
Overall Energy Disaggregation Accuracy (ACC (%))	98.8239	99.3654	95.3903	88.8737
State Prediction Accuracy (SPA (%))	55.2778	50.4545	46.9949	54.0657
Fraction of Total Energy assigned correctly (FTEAC)	0.7723	0.8655	0.8117	0.8333

5. Conclusions and Future Work

In this work, ED was formulated as a constrained multi-objective optimization problem, where the objectives are minimizing energy disaggregation error and temporal sparsity, and constraints related to the practical operation of the devices were proposed. Specifically, in the proposed formulation, the constraints make sure that each device operation during the ED process adheres to the associated practical operational characteristics. Results from the experiments conducted in this work show that the incorporation of the constraints enhanced the ED performance in various metrics (appliance-level and overall) compared to the case where the constraints were not considered. Furthermore, when compared with state-of-the-art ED algorithms, the proposed constrained multi-objective framework was able to demonstrate superior performance.

Author Contributions: Conceptualization, J.P., O.S.A. and R.M.; methodology, O.S.A. and R.M.; formal analysis, O.S.A. and R.M.; data curation, J.P., O.S.A. and R.M.; writing—original draft preparation, J.P., O.S.A. and R.M.; writing—review and editing, J.P., O.S.A. and R.M.; supervision, R.M.; All authors have read and agreed to the published version of the manuscript.

Funding: This research received no external funding.

Institutional Review Board Statement: Not applicable.

Informed Consent Statement: Not applicable.

Data Availability Statement: Not applicable.

Acknowledgments: This work partly was supported by the Basic Science Research Program through the National Research Foundation of Korea (NRF) funded by the Ministry of Education (2021R1I1A3049810), and the National Research Foundation (NRF), Korea, under project BK21 FOUR.

Conflicts of Interest: The authors declare no conflict of interest.

References

1. Zoha, A.; Gluhak, A.; Imran, M.A.; Rajasegarar, S. Non-Intrusive Load Monitoring Approaches for Disaggregated Energy Sensing: A Survey. *Sensors* **2012**, *12*, 16838–16866. [CrossRef]
2. Zeifman, M.; Roth, K. Nonintrusive appliance load monitoring: Review and outlook. *IEEE Trans. Consum. Electron.* **2011**, *57*, 76–84. [CrossRef]
3. Tang, G.; Wu, K.; Lei, J.; Tang, J. Plug and play! a simple, universal model for energy disaggregation. *arXiv* **2014**. arXiv:1404.1884.
4. Tang, G.; Wu, K.; Lei, J.; Tang, J. A simple model-driven approach to energy disaggregation. In Proceedings of the 2014 IEEE International Conference on Smart Grid Communications (SmartGridComm), Venice, Italy, 3–6 November 2014; IEEE: Piscataway, NJ, USA, 2014; pp. 566–571.
5. Laughman, C.; Lee, K.; Cox, R.; Shaw, S.; Leeb, S.; Norford, L.; Armstrong, P. Power signature analysis. *IEEE Power Energy Mag.* **2003**, *1*, 56–63. [CrossRef]
6. Pereira, L.; Nunes, N. Performance evaluation in non-intrusive load monitoring: Datasets, metrics, and tools—A review. *Wiley Interdiscip. Rev. Data Min. Knowl. Discov.* **2018**, *8*, e1265. [CrossRef]
7. Gonçalves, H.; Ocneanu, A.; Bergés, M. Unsupervised disaggregation of appliances using aggregated consumption data. In Proceedings of the 1st KDD Workshop on Data Mining Applications in Sustainability (SustKDD), San Diego, CA, USA, 21 August 2011.
8. Johnson, M.J.; Willsky, A.S. Bayesian Nonparametric Hidden Semi-Markov Models. *J. Mach. Learn. Res.* **2013**, *14*, 673–701.
9. Winkler, P.; Le Ray, G.; Pinson, P. Unsupervised Energy Disaggregation: From Sparse Signal Approximation to Community Detection. *IEEE Trans. Smart Grid* **2019**, 1–8. Available online: http://pierrepinson.com/docs/Lerayetal2019-unsupnilm.pdf (accessed on 19 June 2022).
10. Hart, G. Nonintrusive appliance load monitoring. *Proc. IEEE* **1992**, *80*, 1870–1891. [CrossRef]
11. Srinivasan, D.; Ng, W.; Liew, A. Neural-network-based signature recognition for harmonic source identification. *IEEE Trans. Power Deliv.* **2006**, *21*, 398–405. [CrossRef]
12. Tsai, M.S.; Lin, Y.H. Modern development of an Adaptive Non-Intrusive Appliance Load Monitoring system in electricity energy conservation. *Appl. Energy* **2012**, *96*, 55–73. [CrossRef]
13. Ghorbanpour, S.; Pamulapati, T.; Mallipeddi, R.; Lee, M. Energy disaggregation considering least square error and temporal sparsity: A multi-objective evolutionary approach. *Swarm Evol. Comput.* **2021**, *64*, 100909, . [CrossRef]
14. Ajani, O.S.; Kumar, A.; Mallipeddi, R.; Das, S.; Suganthan, P.N. Benchmarking Optimization-Based Energy Disaggregation Algorithms. *Energies* **2022**, *15*, 1600. [CrossRef]
15. Egarter, D.; Elmenreich, W. EvoNILM: Evolutionary appliance detection for miscellaneous household appliances. In Proceedings of the GECCO '13 Companion, Amsterdam, The Netherlands, 6–10 July 2013.
16. Ghorbanpour, S.; Pamulapati, T.; Mallipeddi, R. Swarm and evolutionary algorithms for energy disaggregation: challenges and prospects. *Int. J. Bio Inspired Comput.* **2021**, *17*, 215–226. [CrossRef]
17. Suzuki, K.; Inagaki, S.; Suzuki, T.; Nakamura, H.; Ito, K. Nonintrusive appliance load monitoring based on integer programming. In Proceedings of the 2008 SICE Annual Conference, Tokyo, Japan, 20–22 August 2008; pp. 2742–2747. [CrossRef]
18. Machlev, R.; Belikov, J.; Beck, Y.; Levron, Y. MO-NILM: A multi-objective evolutionary algorithm for NILM classification. *Energy Build.* **2019**, *199*, 134–144. [CrossRef]
19. Bhotto, M.Z.A.; Makonin, S.; Bajić, I.V. Load Disaggregation Based on Aided Linear Integer Programming. *IEEE Trans. Circuits Syst. II Express Briefs* **2017**, *64*, 792–796. [CrossRef]
20. Shen, Q.; Wang, X. An analysis of the optimization disaggregation algorithm in the estimation related to energy consumption of appliances in buildings. *Appl. Math. Comput.* **2014**, *234*, 506–519. [CrossRef]
21. Egarter, D.; Sobe, A.; Elmenreich, W. Evolving Non-Intrusive Load Monitoring. In Proceedings of the 16th European Conference, EvoApplications 2013, Vienna, Austria, 3–5 April 2013.
22. Piga, D.; Cominola, A.; Giuliani, M.; Castelletti, A.; Rizzoli, A.E. Sparse Optimization for Automated Energy End Use Disaggregation. *IEEE Trans. Control Syst. Technol.* **2016**, *24*, 1044–1051. [CrossRef]
23. Pamulapati, T.; Mallipeddi, R.; Suganthan, P.N. I_{SDE}+—An Indicator for Multi and Many-Objective Optimization. *IEEE Trans. Evol. Comput.* **2018**, *23*, 346–352. [CrossRef]

Article

Searching for a Unique Exciton Model of Photosynthetic Pigment–Protein Complexes: Photosystem II Reaction Center Study by Differential Evolution

Denis D. Chesalin and Roman Y. Pishchalnikov *

Prokhorov General Physics Institute of the Russian Academy of Sciences, 119991 Moscow, Russia; genoa-and-pittsburgh@mail.ru
* Correspondence: rpishchal@kapella.gpi.ru; Tel.:+7-499-503-8777

Abstract: Studying the optical properties of photosynthetic pigment–protein complexes (PPCs) in the visible light range, both experimentally and theoretically, is one of the ways of gaining knowledge about the function of the photosynthetic machinery of living species. To simulate the PPC optical response, it is necessary to use semiclassical theories describing the effect of external fields–matter interaction, energy migration in molecular crystals, and electron–phonon coupling. In this paper, we report the results of photosystem II reaction center (PSIIRC) linear optical response simulations. Applying the multimode Brownian oscillator model and the theory of molecular excitons, we have demonstrated that the absorption, circular and linear dichroism, and steady-state fluorescence of PSIIRC can be accurately fitted with the help of differential evolution (DE), the multiparametric evolutionary optimization algorithm. To explore the effectiveness of DE, we used the simulated experimental data as the target functions instead of those actually measured. Only 2 of 10 DE strategies have shown the best performance of the optimization algorithm. With the best tuning parameters of DE/rand-to-best/1/exp strategy determined from the strategy tests, we found the exact solution for the PSIIRC exciton model and fitted the spectra with a reasonable convergence rate.

Keywords: differential evolution; evolutionary computations; chlorophyll; absorption; cumulant expansion; multimode Brownian oscillator model; inhomogeneous broadening; photosystem II reaction center

Citation: Chesalin, D.D.; Pishchalnikov, R.Y. Searching for a Unique Exciton Model of Photosynthetic Pigment–Protein Complexes: Photosystem II Reaction Center Study by Differential Evolution. *Mathematics* **2022**, *10*, 959. https://doi.org/10.3390/math10060959

Academic Editors: Antonin Ponsich, Mariona Vila Bonilla and Bruno Domenech

Received: 22 February 2022
Accepted: 15 March 2022
Published: 17 March 2022

Publisher's Note: MDPI stays neutral with regard to jurisdictional claims in published maps and institutional affiliations.

1. Introduction

Among the numerous proteins of living organisms, pigment–protein complexes (PPCs) are perhaps the most interesting object for numerical simulations of the optical response of proteins [1,2] simply because they control the light-driven reactions of the photosynthetic process in organisms that transform light energy into chemical energy. With chlorophylls, bacteriochlorophylls, and carotenoids as the main pigment molecules, PPCs actively absorb from 300 to 900 nm, providing effective light harvesting in the visible range and subsequent energy transport within a complex and between different complexes [3]. The optical properties of individual pigments usually determine those of the whole PPC; however, in some cases, the interaction energies between pigments in PPC and the local binding proteins have much greater effect on the PPC's optical response [1]. The number of pigments in PPCs is crucial as well; it varies from a single carotenoid, such as in the orange carotenoid protein (OCP) complex [4]; dozens of chlorophylls, such as in the trimeric Fenna–Matthews–Olson complex [5–7]; hundreds of chlorophylls, such as in photosystem I [8,9]; and thousands of bacteriochlorophylls, such as in chlorosomes [10].

The main feature of the electronic absorption bands of photosynthetic pigments is a phonon wing, the shape and intensity of which depends on the electron–phonon interaction [11]. Chlorophylls and bacteriochlorophylls are characterized by a set of several dozen vibronic states and a relatively weak electron–phonon interaction, while carotenoids have four pronounced vibronic states and strongly interact with the electronic states of a pigment [12]. All of these features can be taken into account within the framework of a theory called the multimode Brownian oscillator model (MBOM) [13]. This theory allows for simulating a realistic absorption lineshape of an electronic transition of any pigment by introducing the spectral density function [14]. The shape of a pigment absorption spectrum is modeled by considering that an electronic transition is coupled to a set of effective vibronic modes. Each mode is described by three parameters: the frequency, the damping factor, and the Huang–Rhys factor (otherwise the electron–phonon coupling) [15]. The first two parameters can be estimated experimentally, while the only way to obtain the Huang–Rhys factor is by modeling the pigment's optical response. Thus, by combining the results of experimental analysis and theoretical modeling, we can determine a characteristic set of microparameters for a pigment molecule and use them in further modeling.

In order to model the linear optical properties in the case of an assembly of interacting pigment molecules, it is necessary to use the theory of molecular excitons [16] in addition to MBOM. According to this theory, any system of interacting molecules of an arbitrary geometry and dimensionally is described by the Frenkel exciton Hamiltonian, which is the basis for the theory of molecular crystals [17]. Thus, combining the MBOM and the exciton theory, the system's optical response can be simulated with a high degree of accuracy. There are many studies in which the combination of these two theories was used to obtain realistic simulated spectra and kinetics of various PPCs. Nevertheless, the main disadvantage of these works is the lack of optimization of the experimental data fitting procedure; evidence of the uniqueness of the theoretical models are usually not given.

The use of evolution optimization algorithms [18,19], in particular differential evolution (DE) [20,21], has shown that the search for optimal quantum models of primary photosynthesis processes is possible [22,23]. As opposed to genetic evolutionary algorithms, DE creates a new generation of model parameters, perturbing the current generation with the scaled difference of randomly selected population members. The detailed introduction to DE can be found in different surveys of the topic [24,25]; including descriptions of some modifications of the classical version of the algorithm designed to improve the convergence of DE [26–29].

The aim of our study is to explore, for the first time, the potential of DE to be an effective optimization routine for the fitting of the PPC optical response. We have chosen the reaction center of photosystem II (PSIIRC) as an example of PPCs (Figure 1), the linear spectra of which should be simulated and an appropriate exciton model created [30–33]. PSIIRC is a rather small protein. It contains only eight cofactors embedded in the protein matrix: six chlorophylls (Chl) and two pheophytins [34]. The eight pigments in PSIIRC give us an optically active, eight-level excitonic manifold; however, PSIIRC also has three so-called charge separation states [31], which are optically inactive and will not be considered in our simulations. For the sake of simplicity and clarity, we will use the pre-calculated linear spectra of PSIIRC as target functions instead of the actually measured ones. The absence of noise in the spectra will allow us to estimate the DE convergence with great accuracy.

The statement of the optimization problem is considered in the second section. The DE algorithm and the references that describe its applications for the modeling of the optical response of photosynthetic pigments are discussed in the third section. In the fourth section, we briefly survey some important aspects of quantum theory on the basis of which the simulation procedures were written. The quantum model of energy transfer in PSIIRC that was used to generate the target functions is explained in the fifth section. The results of the strategy test for different settings of DE and full datasets of the fitting procedure for two different strategies are given in section six. Finally, some features of the strategy test, the algorithm convergence at different DE settings, and further perspectives of DE application for the modeling of primary photosynthetic processes are discussed in section seven.

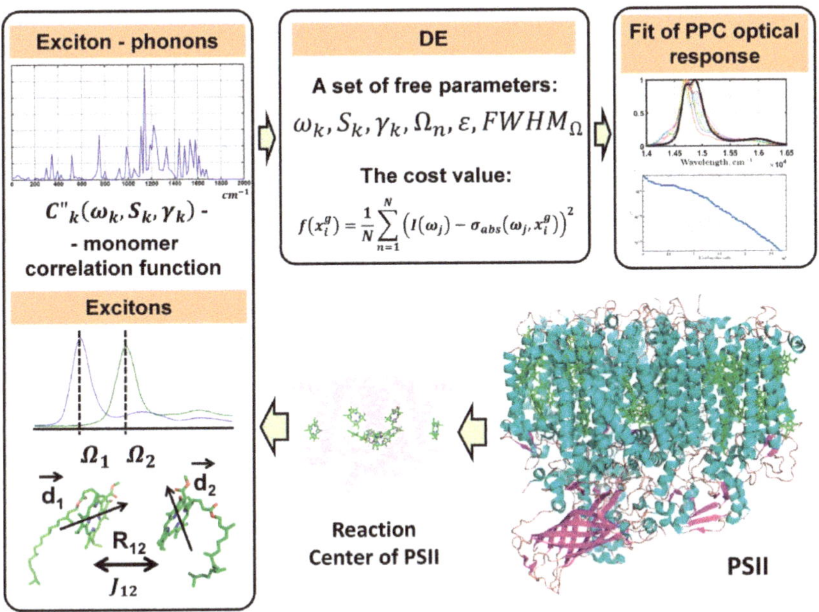

Figure 1. Scheme of the PPC linear optical response fitting procedure. The crystal structure of photosystem II and the isolated reaction center are shown at the bottom of the figure. The left block represents the employed theories in the simulation: the exciton theory (transition energies Ω_n, transition moments d_n, distances between the centers of transition moments R_{nm}, coupling energies J_{nm}, the dielectric constant ε, the full width at half maximum of inhomogeneous broadening $FWHM_\Omega$) and the multimode Brownian oscillator model ($\left\{ \omega_j, S_j, \gamma_j \right\}$ are frequencies, the Huang–Rhys factors, and damping factors of a vibronic mode). The upper blocks symbolize the differential evolution fitting procedures ($I(\omega_n)$ is a measured spectrum, $\sigma_{abs}\left(\omega_n, x_i^g \right)$ is a simulated spectrum). See more detailed explanations in the text.

2. Statement of the Optimization Problem

Using of the molecular exciton theory in modeling the optical properties of a PPC from a mathematical point of view is the sequential implementation of computational procedures such as matrix diagonalization, fast Fourier transform, and numerical integration of time- and frequency-dependent functions. Depending on the characteristics of the vibronic modes of the correlation functions, the number of pigments in the complex, and the interaction energies between them, the speed and quality of the calculated spectra can vary significantly. In general (Figure 1), a set of parameters $x_j = \{\omega_k, S_k, \gamma_k, \Omega_n, \varepsilon, FWHM_\Omega\}$ is fed to the input of the simulation program, which entirely determines the simulated spectra of the complexes. This set, with which the spectra are calculated, we will hereafter refer to as a solution of the PSIIRC optical response modeling. Since the exciton theory is semiclassical and does not assume ab initio calculations, in order to find x_j, it is necessary to compare the calculated spectra with those measured experimentally. Thus, by varying the values of x_j, one can try to find a solution for which the calculated spectra most accurately describe the measured ones. Ideally, the best solution is the set that corresponds to the exact coincidence of the calculated and measured spectra.

The dependence of the calculated spectra on x_j is very complex and cannot be factorized. Many publications are devoted to the search "manually" for a set of x_j for the PSIIRC exciton model and almost always use the simultaneous simulation of several spectra obtained by different experimental techniques. For example, in the paper by Novoderezhkin et al. [32], four exciton models of energy transfer in PSIIRC are considered, which correspond to four different sets of x_j. It is clear that the process of finding a set of model parameters that would allow the most accurate fit of the calculated and measured spectra can be optimized. Evolutionary algorithms are applicable for this purpose if we consider the squared difference between the calculated and experimental spectra as an objective function to be minimized. The use of DE in this case is preferable to genetic algorithms, since it allows for varying the parameters continuously instead of discretely. Moreover, DE allows us to classify the found solutions x_j with respect to the value of the objective function. The smallest value of the objective function corresponds to the smallest difference between the calculated and experimental spectra. Of course, the algorithm may become stuck in the local minimum, when any changes in x_j within certain values x_j^{local} make the objective function only worse and convergence stagnates.

Thus, the combined software implementation of the optical response modeling procedures and the differential evolution algorithm will make it possible to find the exciton model parameters that will provide the best match between the experimental and computational data.

3. Differential Evolution

The algorithm of DE has four data processing steps: initialization, mutation, crossover, and selection. The initialization of DE runs once at the beginning of the fitting, while three other steps sequentially repeat themselves as many times as required to obtain the appropriate simulated spectra (Figure 1).

At the initialization of DE, a matrix of model parameters, **X**, is filled with random values within specified limits, and then the objective function values are estimated. The size of the matrix is $D \times N_p$, where D is equal to the number of model parameters, and N_p is the size of the population. Assuming that x_i are the parameters to find, $j = 1, 2, \ldots, D$ and $x = x_j$, then the matrix of model parameters is written as $\mathbf{X} = x_i$, where $i = 1, 2, \ldots, N_p$. The elements of the matrix are chosen taking into account the boundary conditions, which consider the physical limits of the parameters to find.

After the initialization step, the main cycle of DE starts with the generation of a new matrix, $\mathbf{X}^g$, where $g = 0, 1, \ldots, g_{max}$ is a generation index. In the classical version of DE, a mutant vector, $\mathbf{v}_i^g$, is calculated according to one of the following five expressions:

$$\mathbf{v}_i^g = \mathbf{x}_{r0}^g + F\left(\mathbf{x}_{r1}^g - \mathbf{x}_{r2}^g\right), \tag{1}$$

$$\mathbf{v}_i^g = \mathbf{x}_{best}^g + F\left(\mathbf{x}_{r1}^g - \mathbf{x}_{r2}^g\right), \tag{2}$$

$$\mathbf{v}_i^g = \mathbf{x}_i^g + F\left(\mathbf{x}_{best}^g - \mathbf{x}_{r0}^g\right) + F\left(\mathbf{x}_{r1}^g - \mathbf{x}_{r2}^g\right), \tag{3}$$

$$\mathbf{v}_i^g = \mathbf{x}_{best}^g + F\left(\mathbf{x}_{r1}^g - \mathbf{x}_{r2}^g\right) + F\left(\mathbf{x}_{r3}^g - \mathbf{x}_{r4}^g\right), \tag{4}$$

$$\mathbf{v}_i^g = \mathbf{x}_{r0}^g + F\left(\mathbf{x}_{r1}^g - \mathbf{x}_{r2}^g\right) + F\left(\mathbf{x}_{r3}^g - \mathbf{x}_{r4}^g\right), \tag{5}$$

where F is the weighting factor and $F \in [0, 1]$; $\mathbf{x}_i^g$, $\mathbf{x}_{r0}^g$, $\mathbf{x}_{r1}^g$, and $\mathbf{x}_{r2}^g$ are randomly chosen vectors from the current population; and $(i \neq r0 \neq r1 \neq r2) \in [0, Np]$. $\mathbf{x}_{best}^g$ is a vector corresponding to the best solution (minimum of the objective function).

The diversity of the trial vector population can be increased by applying a crossover procedure. In this case, a new trial vector, $\mathbf{u}_j^g$, is created by exchanging the elements of each target vector of the current population with those of a mutant one. The crossover rate, $Cr \in [0, 1]$, determines the number of exchanged values in the trial vector. There are two types of crossovers: binomial and exponential. The combination of Equations (1)–(5) and 2 crossovers provides us with 10 different strategies to create a new generation of model parameters. The names of the strategies are formed as follows: DE/x/y/z, where x is a base vector (*rand, best, rand-to-best*), y is the number of differences (1 or 2), and z is the crossover type (*exp* or *bin*). Thereby, the convergence of the algorithm can be controlled by choosing the optimal strategy and varying the weighting factor and crossover rate.

Thus, the target vector of a new generation, g + 1, is determined by comparing $f\left(\mathbf{u}_i^g\right)$ and $f\left(\mathbf{x}_i^g\right)$. The expression for the objective function $f\left(\mathbf{x}_i^g\right)$ is:

$$f\left(\mathbf{x}_i^g\right) = \frac{1}{N}\sum_{n=1}^{N}\left(I(\omega_n) - \sigma_{abs}\left(\omega_n, \mathbf{x}_i^g\right)\right)^2,\tag{6}$$

where $I(\omega_n)$, for example, is a measured absorption spectrum of PSIIRC at frequency ω_n, $\sigma_{abs}\left(\omega_n, \mathbf{x}_i^g\right)$ is a simulated absorption spectrum of PSIIRC, and N is the number of points in the spectra. After the objective functions are evaluated, $\mathbf{x}_i^{g+1}$ vector allocation is made according to the following conditions:

$$\mathbf{x}_i^{g+1} = \begin{cases} \mathbf{u}_i^g, & f\left(\mathbf{u}_i^g\right) \le f\left(\mathbf{x}_i^g\right) \\ \mathbf{x}_i^g, & f\left(\mathbf{u}_i^g\right) > f\left(\mathbf{x}_i^g\right) \end{cases},\tag{7}$$

When a new population is completed, the next cycle of DE starts, and the optimization runs until the predetermined minimum of the objective function is reached or the number of generations reaches a specified maximum.

4. Theory

4.1. Multimode Brownian Oscillator Model

According to quantum theory of the radiation interaction with matter, the optical response of any pigment molecule can be estimated by expanding the expression for the polarization of a system, $P(t)$, in powers of the radiation field assuming this field as a perturbation. Consider a system of two electronic states: a ground state $|g\rangle$ and an excited one $|e\rangle$, and let $\mu_{eg}(\mathbf{q})$ be a transition dipole moment between states. Then, the polarization can be written as the expectation value of $\mu(\mathbf{q})$:

$$P(\mathbf{r}, t) = Tr\left[(\mu_{eg}(\mathbf{q})|e\rangle\langle g| + \mu_{ge}(\mathbf{q})|g\rangle\langle e|)\rho(t)\right],\tag{8}$$

where $\rho(t)$ is a density matrix whose time evolution is determined by the Hamiltonian of the system. The expansion of $\rho(t)$ in powers of the field results in the decomposition of polarization $P(\mathbf{r}, t) = P(\mathbf{r}, t)^{(1)} + P(\mathbf{r}, t)^{(2)} + P(\mathbf{r}, t)^{(3)} + \cdots$. The first term $P(\mathbf{r}, t)^{(1)}$ of this decomposition is responsible for the linear absorption:

$$P(\mathbf{r}, t)^{(1)} = -\frac{i}{\hbar}\int_0^\infty dt_1 E(\mathbf{r}, t - t_1)S^{(1)}(t_1),\tag{9}$$

where $S^{(1)}(t_1) = \frac{i}{\hbar}\theta(t_1)\langle\mu_{eg}(t_1)\mu_{eg}(0)\rho(-\infty)\rangle + c.c.$ is the linear response function in the Liouville representation, $E(\mathbf{r}, t)$ is the radiation field, $\theta(t_1)$ is the Heaviside step function, and $\langle\ldots\rangle$ denotes the averaging over nuclear degrees of freedom, $t_1 = \tau_2 - \tau_1$, where τ_i are the ordered points on $[t_0, t]$ used in the decomposition of polarization. A general expression for an absorption spectrum can be written in an integral form:

$$\sigma_{abs}(\omega) = \int_{-\infty}^{\infty} dt\, S^{(1)}(t_1)e^{i\omega t},\tag{10}$$

Introducing the effective operator of the electronic energy gap, $U(\tau) = \exp\left(\frac{i}{\hbar}H_g\tau\right)U\exp\left(-\frac{i}{\hbar}H_g\tau\right)$, where $U = H_e(\mathbf{q}) - H_g(\mathbf{q}) - \hbar\omega_{eg}$, $H_e(\mathbf{q})$, and $H_g(\mathbf{q})$ are Hamiltonians of the electronic excited and ground states, and ω_{eg} is an arbitrary parameter. Considering the time evolution of $U(\tau)$, the first order response function [13] can be expressed as:

$$S^{(1)}(t_1) = \frac{i}{\hbar}\theta(t_1)e^{-i\omega_{eg}t_1 - g(t_1)} + c.c.,\tag{11}$$

$$g(t) = \int_0^t d\tau_2 \int_0^{\tau_2} d\tau_1 C(\tau_1),\tag{12}$$

$$C(\tau_1) = \frac{1}{\hbar^2}\langle U(\tau_1)\,U(0)\rho_g\rangle,\tag{13}$$

where $g(t)$ is the lineshape function; $C(\tau_1)$ is the two-time correlation function of $U(\tau)$. Since the correlation function is complex, it can be expressed in the time domain as $C(t) = C'(t) + C''(t)$, and as $C(\omega) = \int_{-\infty}^{\infty} dt\, e^{i\omega t}C(t) = C'(\omega) + C''(\omega)$ in the frequency domain. Considering the fluctuation dissipation theorem, we obtain the expression for $C(t)$ in the following form:

$$C(t) = \int_{-\infty}^{\infty} d\omega\,\cos(\omega t)\coth(\beta\hbar\omega/2)C''(\omega) + i\int_{-\infty}^{\infty} d\omega\,\sin(\omega t)C''(\omega),\tag{14}$$

where $C''(\omega)$ is the imaginary part of $C(\omega)$ and can be treated classically. This feature of $C''(\omega)$ makes it quite suitable for the modeling of the optical response. Thus, the equation for $g(t)$ in terms of $C''(\omega)$ is written as:

$$g(t) = \frac{1}{2\pi}\int_{-\infty}^{\infty} d\omega\frac{1 - \cos\omega t}{\omega^2}\coth(\beta\hbar\omega/2)C''(\omega) - \frac{i}{2\pi}\int_{-\infty}^{\infty} d\omega\frac{\sin(\omega t) - \omega t}{\omega^2}C''(\omega).\tag{15}$$

Taking into account Equations (10) and (11), the final expression for numerical simulation of the absorption lineshape is given by:

$$\sigma_{abs}(\omega) = \frac{1}{\pi}Re\int_0^{\infty} dt\, e^{i(\omega - \omega_{eg})t}e^{-g(t)}.\tag{16}$$

The combination of Equations (15) and (16) allow for the modeling of the linear optical response of a single electronic transition interacting with an arbitrary set of vibronic modes.

In order to evaluate $C''(\omega)$, the theory of MBOM must be applied [13]. In terms of MBOM, a system consisting of an electronic state interacting with a set of vibronic states is described by the following Hamiltonians:

$$H_{sys} = H_g + H_e + H_{VB},\tag{17}$$

$$H_g = \sum_j^N \left(\frac{p_j^2}{2m_j} + \frac{1}{2}m_j\omega_j^2 q_j^2\right),\tag{18}$$

$$H_e = \hbar\omega_{eg}^0 + \sum_j^N \left(\frac{p_j^2}{2m_j} + \frac{1}{2}m_j\omega_j^2 (q_j + d_j)^2\right),\tag{19}$$

$$H_{VB} = \sum_n^M \left[\frac{p_n^2}{2m_n} + \frac{1}{2}m_n\omega_n^2 x_n^2 - x_n\sum_j c_{nj}q_j + \frac{\sum_j c_{nj}^2 q_j^2}{2m_n\omega_n^2}\right],\tag{20}$$

where H_g and H_e are Hamiltonians of the ground $|g\rangle$ and the excited $|e\rangle$ states. Vibronic states of the system are modeled by introducing a certain number of effective vibronic modes. Each mode is characterized by frequency ω_j, mass m_j, momentum p_j, coordinate q_j, and displacement d_j of the excited state potential curve. j is the index of a mode, and N is the number of modes. The influence of the local environment is represented by the H_{VB} part in the system Hamiltonian, which depends on another set of oscillators, the bath modes, and their parameters $\{p_n, x_n, \omega_n, m_n\}$. The coupling between electronic and vibronic states is set by microparameters c_{nj}. Finally, the MBOM correlation function is calculated using the path integral method. The imaginary part of $C(t)$ is written in the form:

$$C''(\omega) = \sum_j \frac{2S_j \omega_j^3 \omega \gamma_j}{\left(\omega_j^2 - \omega^2\right)^2 + \omega^2 \gamma_j^2},$$
(21)

where $S_j = d_j^2/2$ are the Huang–Rhys factors, and γ_j are the damping factors for each ω_j that are determined empirically.

So, to calculate the absorption or the fluorescence spectrum of a monomeric pigment molecule, such as chlorophyll, bacteriochlorophyll, or carotenoid, both in solvent and in protein, one has to evaluate the spectral density (Equation (21)) then the lineshape function (Equation (15)), and the absorption lineshape is simulated according to Equation (16).

4.2. Excitons

In the previous section, it was shown how to model the linear optical response of a monomeric pigment. The theory of molecular excitons considering electronic transition and interaction energies between pigment molecules of PSIIRC allows calculating the contributions of each Chl molecule to the resulting spectra and population kinetics [16,30,34,35]. We consider that PSIIRC consist of eight two-level Chl molecules; each molecule can be either in a ground $|0\rangle$ or in an excited $|n\rangle$ state. n runs from 1 to N, where N is the number of pigments in PSIIRC. Denoting $B_n^+ = |n\rangle\langle 0|$ as the exciton creation operator and $B_n = |0\rangle\langle n|$ as that of annihilation, the PSIIRC exciton Hamiltonian is then written in the form:

$$H_{ext} = \sum_n \Omega_n B_n^+ B_n + \frac{1}{2} \sum_{n \neq m} J_{mn} \left(B_m^+ B_n + B_n^+ B_m\right),$$
(22)

where Ω_n is the transition energy between the ground and the excited states of a pigment. B_n^+ and B_n obey the commutation rules $[B_n, B_n^+] = 1$. J_{mn} is a matrix of coupling energies calculated employing the extended dipole approximation [36]. This method of calculating the interaction energies using the values of partial charges is much more accurate than the classical dipole–dipole approximation.

Diagonalizing the Hamiltonian (Equation (22)), we obtain the eigenstates c_n^α and eigenvalues ϵ_α that allow for the transformation of the system parameters from the site representation to the exciton representation. Thereby, the lineshape function (Equation (15)) in the exciton representation is $g_{\mu\nu\alpha\beta}(t) = \sum_{mnkl} c_m^\mu c_n^\nu c_k^\alpha c_l^\beta g_{mnkl}(t)$, where $\alpha, \beta, \ldots = 1 \ldots N$ are indices of the exciton states. Finally, the expressions for exciton absorption, circular and linear dichroism, and fluorescence spectra will be presented as a sum over exciton states [9,36]:

$$\sigma_{abs}^{ext}(\omega) \approx \frac{\omega}{\pi} \sum_\alpha^N \mathbf{d}_\alpha^2 Re \int_0^\infty dt e^{i(\omega - \epsilon_\alpha)t} e^{-g_{\alpha\alpha\alpha\alpha}(t)} e^{-0.5K_{\alpha\alpha}t}$$
(23)

$$\sigma_{CD}^{ext}(\omega) \approx \frac{\omega}{\pi} \sum_\alpha^N \mathbf{R}_\alpha Re \int_0^\infty dt e^{i(\omega - \epsilon_\alpha)t} e^{-g_{\alpha\alpha\alpha\alpha}(t)} e^{-0.5K_{\alpha\alpha}t}$$
(24)

$$\sigma_{LD}^{ext}(\omega) \approx \frac{\omega}{\pi} \sum_\alpha^N \left[d_\alpha^{z\,2} - \frac{1}{2}\left(d_\alpha^{x\,2} + d_\alpha^{y\,2}\right)\right] Re \int_0^\infty dt e^{i(\omega - \epsilon_\alpha)t} e^{-g_{\alpha\alpha\alpha\alpha}(t)} e^{-0.5K_{\alpha\alpha}t}$$
(25)

$$\sigma_{fl}^{ext}(\omega) \approx \frac{\omega^3}{\pi} \sum_\alpha^N \frac{(\mathbf{nd}_\alpha)^2 e^{\epsilon_\alpha\beta}}{\sum_n e^{\epsilon_\alpha\beta}} Re \int_0^\infty dt e^{i(\omega - \epsilon_\alpha + 2\lambda_{\alpha\alpha\alpha\alpha})t} e^{-g_{\alpha\alpha\alpha\alpha}^*(t)} e^{-0.5K_{\alpha\alpha}t},$$
(26)

where $\lambda_{\alpha\alpha\alpha\alpha} = -\lim_{\tau\to\infty} Im\left[\frac{dg_{\alpha\alpha\alpha\alpha}(\tau)}{d\tau}\right]$ is the reorganization energy of an exciton state α; $K_{\alpha\alpha} = \sum_\beta K_{\alpha\beta}$ are the exciton relaxation rates; $\mathbf{d}_\alpha = \sum_n c_n^\alpha \mathbf{d}_n$ is the Q$_y$ transition moments of Chl transformed to the exciton representation; and $\mathbf{R}_\alpha = \sum_{nm} c_n^\alpha c_m^\alpha r_{nm}(\mathbf{d}_n \times \mathbf{d}_m)$ is a matrix of the rotational strength necessary for CD spectra simulation.

5. Exciton Model of the Photosystem II Reaction Center

PSIIRC is an important PPC of photosystem II of higher plants and cyanobacteria. All of the light quanta energy absorbed by photosynthetic PPC with minimum losses eventually transfers to the reaction centers where chemical reactions of charge separation occur. In addition to being the location of chemical reactions, reaction centers actively serve as a light-harvesting complex, too. PSIIRC has six Chls, two pheophytins, and two carotenoids, which absorb at the visible range. In this study, we are going to model the PSIIRC optical response only in the so-called Q$_y$ region of Chl absorption which corresponds to the 650–750 nm range. It means that the excited states of carotenoids will not be taken into account in the exciton Hamiltonian (Equation (22)). Since our simulations are focused only on the linear spectroscopy (absorption, steady-state fluorescence, circular and linear dichroism), the radical pair states are not considered. Thus, the PSIIRC exciton Hamiltonian in our modeling will include contributions of eight pigments: two Chls of the special pair (PD1 and PD2), two accessory Chls, two pheophytins, and two peripheral Chls (Figure 2B).

To explore the potential of DE as an effective optimization procedure for fitting the PSIIRC spectra, we will use the simulated experimental data instead of the measured ones. The real data are always noisy and may contain some inconspicuous contributions that can only worsen the convergence of the optimization. The simulated experimental data as target functions will allow the algorithm, in the case of successful configuration, to converge almost to machine zero and determine the local minima for parameters of the PSIIRC quantum model.

To simulate the optical response of Chl, we used the results of our previous studies. The spectral density, absorption, and fluorescence spectra of monomeric Chl are shown in Figure 2A. However, to take into account that the surrounding of Chls in PSIIRC is different from that of it in solution, we used special values of $\{\omega_{low}, S_{low}, \gamma_{low}\}$ determined previously for the lowest vibronic mode. The total number of vibronic modes in Equation (21) was 39. The number of points in the time and frequency arrays was defined as $n = 2^{11} = 2048$. The time step of integration was 0.0042 ps. The full set of $\{\omega_j, S_j, \gamma_j\}$ for Chl can be found in our previous publications [9,23].

Parameters of the PSIIRC exciton model are shown in Table 1. The energies of the Q$_y$ transition of Chls and pheophytins were chosen in such a way that the simulated linear spectra, according to Equations (16)–(20), were as close as possible to the measured ones at room temperature. The interaction energies between PSIIRC cofactors were calculated in the extended dipole approximation, except for the coupling between Chls in the special pair; according to the previous studies, it was set as 150 cm^{-1}. It must be stressed that the inhomogeneous broadening $FWHM_\Omega$ [9] was not taken into account in the simulations, since it requires averaging over random perturbations of diagonal elements of the exciton Hamiltonian (Equation (22)). We deliberately made such a simplification in order to allow the algorithm of DE to converge to machine zero. The Q$_y$ transition moments and the spectral densities of Chl and pheophytin are slightly different and when modeling the real experimental data, these distinctions must be accounted for; however, for the purposes of this study, it is enough to consider them to be equal. As a result, the calculated spectra of absorption, steady-state fluorescence, and linear and circular dichroism we used for the strategy tests and for optimization with the maximum number of free parameters are presented in Figure 2B.

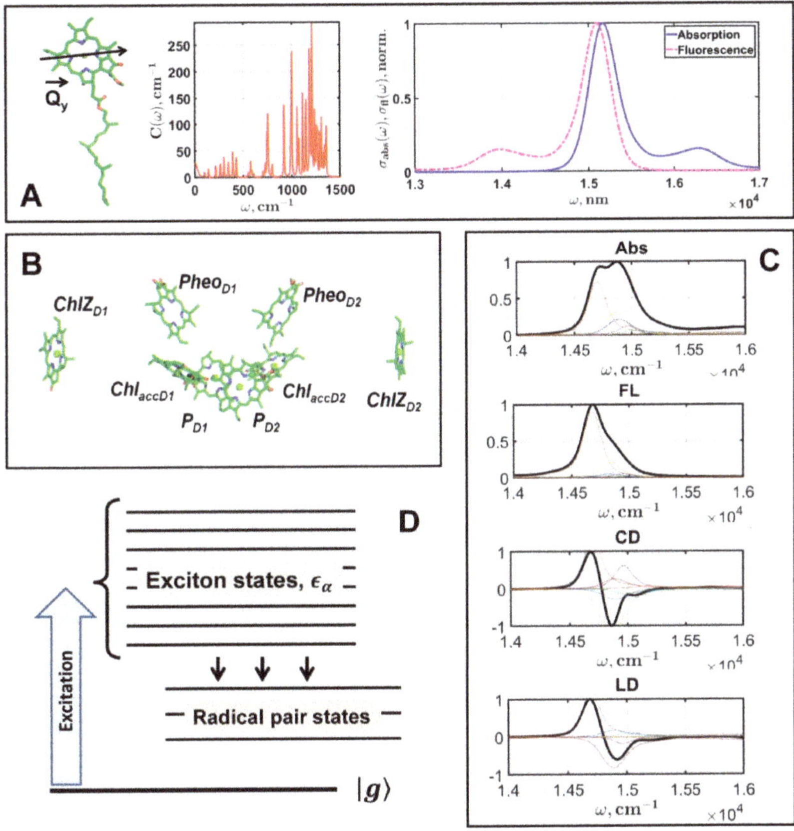

Figure 2. Optical properties of monomeric Chl, the main pigment of Photosystem II reaction center: the spectral density (red) of Chl and the simulated absorption (blue) and fluorescence spectra (magenta) of Chl (**A**); the mutual orientation of Photosystem II reaction center cofactors (**B**) is a key factor in the exciton theory (a scheme (**D**) of the energy levels of PSIIRC) that was applied to simulate absorption, circular and linear dichroism, and steady-state fluorescence spectra. These spectra were used as the target functions (**C**).

Table 1. Material Hamiltonian of the PSIIRC exciton model used for simulation of target functions.

	PD1	PD2	$Chl_{acc}D1$	$Chl_{acc}D2$	PheoD1	PheoD2	Chl_ZD1	Chl_ZD2	Ω_n
PD1	0	150.00	−30.94	−100.96	−3.91	19.01	0.74	0.96	14,960.0
PD2		0	−96.75	−23.53	24.53	−4.22	1.11	1.06	15,070.0
$Chl_{acc}D1$			0	12.43	60.92	−4.96	2.98	0.03	15,045.0
$Chl_{acc}D2$				0	−5.80	54.97	−0.02	2.71	15,080.0
PheoD1					0	3.10	−4.06	−0.32	15,100.0
PheoD2						0	−0.29	−4.44	15,120.0
Chl_ZD1							0	0.24	15,180.0
Chl_ZD2								0	15,170.0

6. Results

6.1. Strategy Test

According to our previous modeling of the linear absorption of monomeric chlorophylls, bacteriochlorophylls, and carotenoids, *DE/rand-to-best/1/exp* and *DE/best/1/bin* strategies have demonstrated the best convergence rates. Preliminary trial runs of the optimization algorithm for the PSIIRC complex showed that, in general, the results of convergence are similar to those we obtained for monomeric pigments. Thus, in the case of PSIIRC, it was decided to run a strategy test only for the two best DE classical strategies. The strategy control parameters varied from 0.55 to 0.85 for F and from 0.8 to 1.0 for Cr with a discrete step of 0.05.

The number of free parameters required to simulate the linear optical response of PSIIRC without inhomogeneous broadening is 12:8 Q_y transition energies Ω_n for Chls and pheophytins, the effective dielectric constant, which is used to calculate the coupling energies between pigments, and 3 parameters $\{\omega_{low}, S_{low}, \gamma_{low}\}$ for the lowest vibronic mode in the spectral density. To reduce the calculation time and to more clearly demonstrate the effect of convergence, only 5 of the 12 PSIIRC model parameters were set as free during the fitting procedure: 4 transition energies and the dielectric constant. Moreover, fewer free parameters allow the optimization algorithm to converge in fewer generations. So, we set $g_{max} = 50$ and performed 30 runs of the program for each $\{F, Cr\}$ pair. The results of the strategy test are shown in Figure 3A,B.

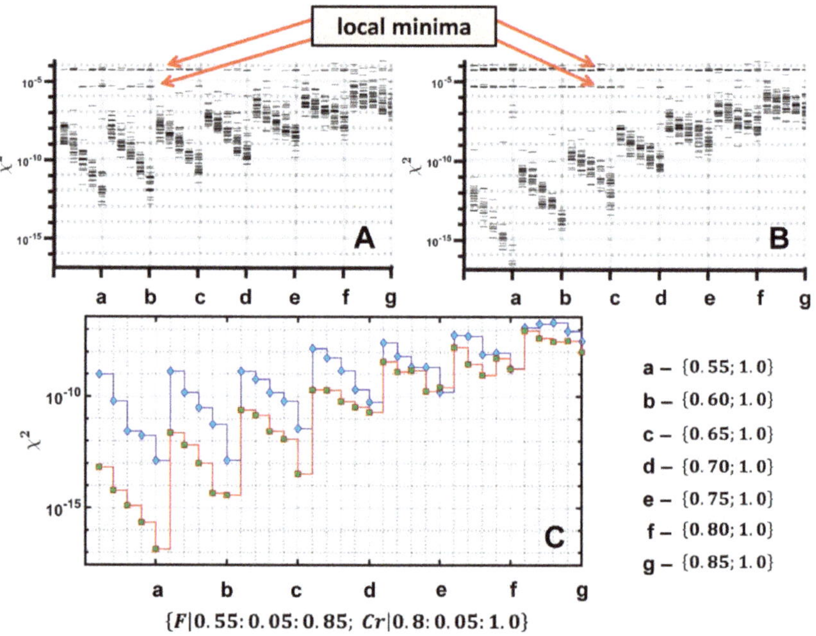

Figure 3. Strategy test. Distributions of the objective functions values obtained for *DE/best/1/bin* (**A**) and *DE/rand-to-best/1/exp* (**B**) strategies after 30 runs of DE for each $\{F, Cr\}$ pair. Comparison of the best results for *DE/best/1/bin* (blue) and *DE/rand-to-best/1/exp* (red) are shown in plot (**C**). Latin letters correspond to $\{F, Cr\}$ pairs. Red lines indicate the local minima of optimization.

Figure 3C demonstrates the best values of the objective function obtained for two strategies. The plots show that the optimization becomes stuck in at least in two minima. Thus, it can be argued that if the value of objective function is less than the lowest local minimum, which is equal to 4.37472×10^{-7}, the algorithm finds the best solution and does not stick at any local minima. Taking into account this criterion, the number of successful optimizations for all $\{F, Cr\}$ pairs and two strategies was calculated. These data are shown in Tables 2 and 3.

Table 2. The results of the strategy test. The number of successful optimizations of PSIIRC fitting after 30 runs of DE for each $\{F;\ Cr\}$ pair and the *DE/rand-to-best/1/exp* strategy.

F	Cr				
	0.8	**0.85**	**0.9**	**0.95**	**1**
0.55	29	25	27	24	21
0.60	27	27	25	24	23
0.65	29	27	28	26	25
0.70	28	30	28	28	28
0.75	30	28	30	28	25
0.80	27	29	28	26	25
0.85	19	23	25	24	27

Table 3. The results of the strategy test. The number of successful optimizations of PSIIRC fitting after 30 runs of DE for each $\{F;\ Cr\}$ pair and the *DE/best/1/bin* strategy.

F	Cr				
	0.8	**0.85**	**0.9**	**0.95**	**1**
0.55	20	14	9	18	9
0.60	24	24	23	22	17
0.65	21	23	22	17	24
0.70	22	26	23	24	22
0.75	27	29	26	26	23
0.80	25	24	24	24	25
0.85	24	28	26	25	26

To compare the effectiveness of the two strategies, we calculated the percentage of convergence, taking into account the criterion of the lowest local minimum. The convergence probability for *DE/rand-to-best/1/exp* strategy is 87.9% (923 of 1050) and 74.9% (786 of 1050) for *DE/best/1/bin*.

6.2. PSIIRC Linear Optical Response Modeling

To perform the optimization of PSIIRC linear optical response modeling, we tuned the DE algorithm based on the values of convergence rates and convergence probabilities obtained in the strategy test. Considering the results of the strategy test, the corresponding values $\{F = 0.55, Cr = 0.9\}$ were chosen. Unlike the strategy test, the number of free parameters was nine. All of the excitation energies Ω_n were set free, as well as the dielectric constant. The initial boundaries for Ω_n were from 14,500 cm^{-1} to 15.300 cm^{-1} and from 0.5 to 2 for the dielectric constant, ε.

For greater accuracy and more detailed information about the convergence dynamics, the maximum number of generations g_{max} chosen was 300. The results of the PSIIRC linear optical response fitting for the two strategies are shown in Figures 4 and 5. As we can see, the *DE/rand-to-best/1/exp* strategy found the best solution as opposed to *DE/best/1/bin* which stuck at local minimum.

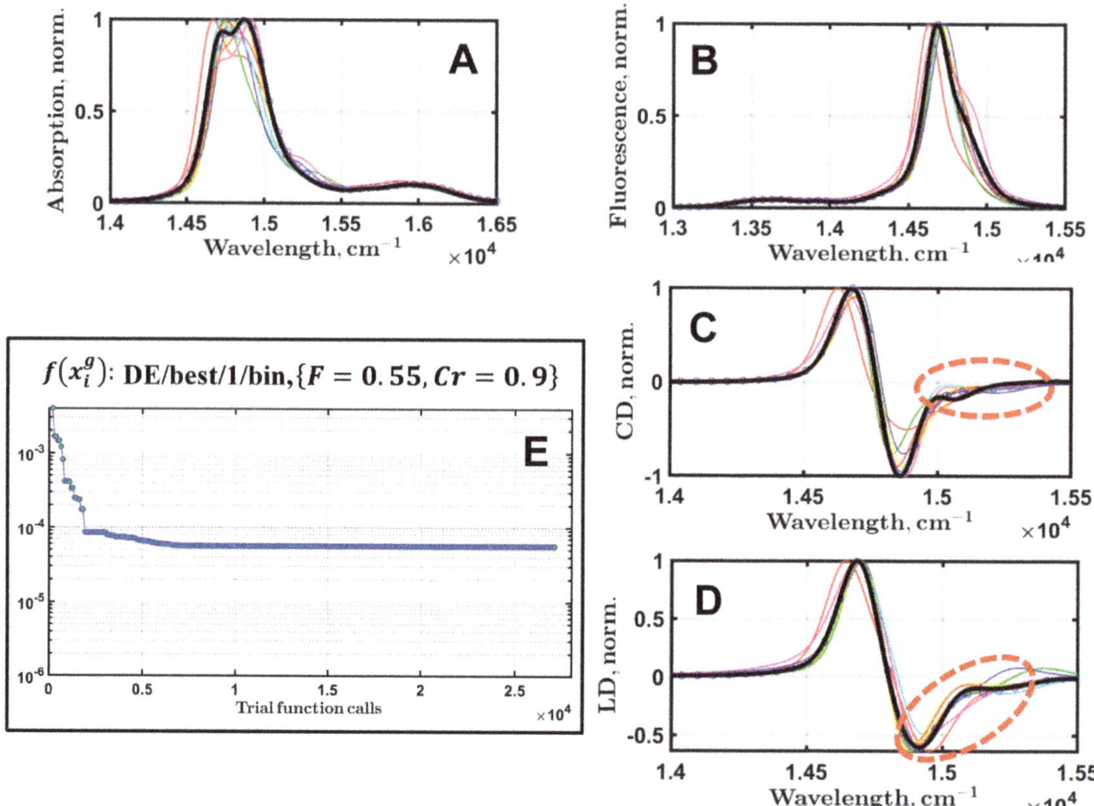

Figure 4. Results of the PSIIRC linear optical response fitting. The case when DE is stuck in a local minimum. *DE/best/1/bin* strategy. The best simulated spectra (black lines with circular markers) of absorption (**A**), steady-state fluorescence (**B**), circular (**C**), and linear (**D**) dichroism are shown. Color lines are the spectra found by DE after first 7 generations. Black thick lines are the target spectra. Red ovals indicate those parts of the spectra where there are significant discrepancies between the target spectra and those found by DE. The objective function $f\left(x_i^g\right)$ dependence on the number of the trial function (linear response simulation) calls (**E**).

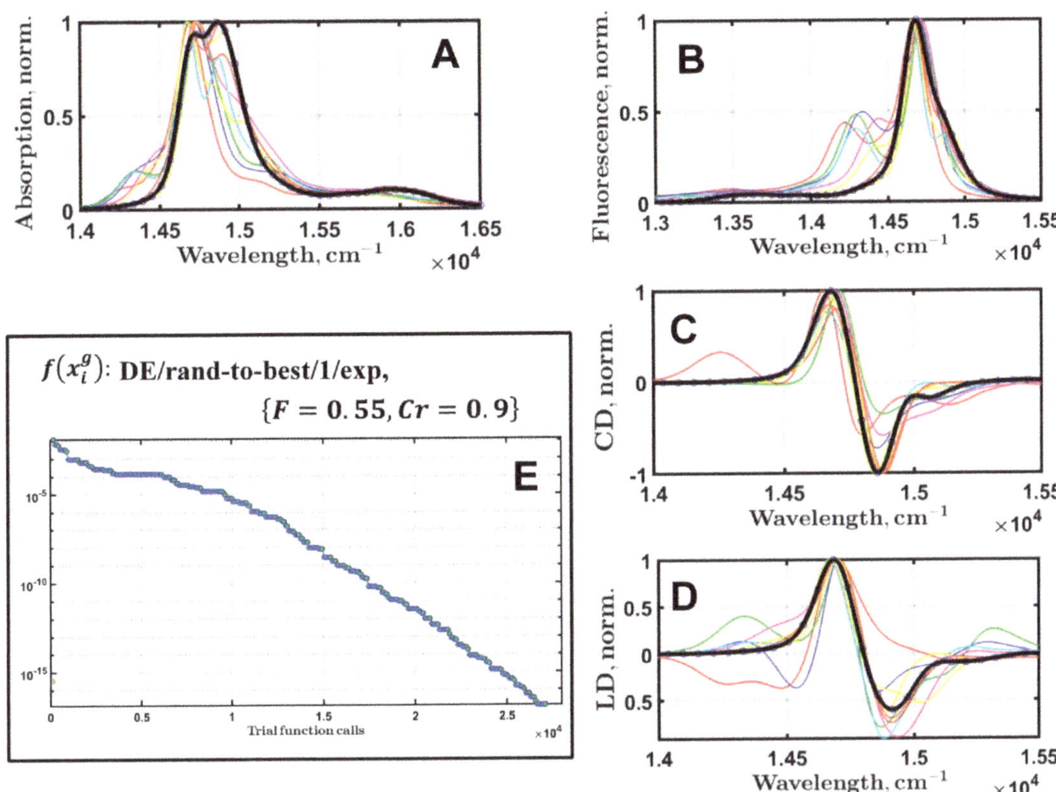

Figure 5. Results of the PSIIRC linear optical response fitting. Successful DE run. *DE/rand-to-best/1/exp* strategy. The best simulated spectra (black lines with circular markers) of absorption (**A**), steady-state fluorescence (**B**), circular (**C**), and linear (**D**) dichroism are shown. Color lines are the spectra found by DE after first 7 generations. Black thick lines are the target spectra. The objective function $f\left(x_i^g\right)$ dependence on the number of the trial function (linear response simulation) calls (**E**).

7. Discussion

Some attempts to use the evolutionary multiparametric optimization for modeling the optical properties of PPC have already been made. The genetic algorithm was used to estimate the charge and energy transfer rates in photosystem I core complexes [37]. However, this first effort was not really successful: linear spectra and kinetics were not simulated; only transfer rates were calculated to compare with those measured. Moreover, the protein crystal structure was not available at that moment, which significantly limited the proper use of exciton theory. The later attempt employing the same approach was used for the light-harvesting complex II from higher plants [38]. In this case, some parameters of the exciton model were estimated, taking into account the existing crystal structure, but the overall optimization approach based on two-dimensional lattice model appeared to be not very effective.

PPCs with a large number of pigments are the perfect objects for testing the effectiveness of optimization algorithms. Simulating PPC optical properties requires considering the electronic transition energies of all pigments in PPCs [2]. The exciton spectra of photosystem I from *Synechococcus elongatus* [39] were simulated with the help of the genetic algorithm. The monomeric complex of photosystem I is characterized by a large number of Chls, about 100 per complex. The main drawback of this study is the simulation of linear optical response as "stick" spectra, which is a rough approximation of the spectrum widths for such PPCs. However, transition energies were modeled using a variation of genetic algorithm. For each pigment, the Q_y transition energy could vary discretely within the interval of 660–715 nm with a step of 0.25 nm. Such a discretization of the parameters to search is the main disadvantage of the genetic algorithm in comparison with DE.

Another interesting example of the modeling of the exciton dynamic, linear spectra and time-resolved fluorescence of the monomeric photosystem I, applying a certain mutation strategy of genetic algorithm, is described in [40]. The spectra of individual excitonic states were simulated in terms of the Redfield relaxation theory, which already makes it possible to estimate the width of each exciton state depending on its energy. A different protein environment around Chls creates variations in the Q_y transition energies, which cannot be estimated theoretically. The use of evolutionary optimization in this case is one of the ways to solve the problem of finding Q_y transition energies without resorting to tedious quantum mechanical calculations. The authors of [40] proposed several mutation strategies of the genetic algorithm for searching for transition energies; however, due to the huge search space, the algorithm did not achieve a correct assignment of most Chl energies.

In our simulations of the linear spectra of PSIIRC, we applied the modified Redfield theory [35] (actually, a combination of MBOM and the exciton theory) to calculate the absorption profile for each exciton state (Equations (23)–(26)). This approach allows us to reproduce very realistic spectra for each exciton state of PSIIRC. Chls and Pheos in PSIIRC are characterized by a set of *39* vibronic modes, which gives the advantage in modeling the effect of electron–phonon interaction compared with the standard Redfield approach. However, in applying this complex approach, we increase the number of free model parameters, which in turn increases the computational costs of the experimental data fitting. Our previous studies of the linear optical response of Chl, BChl, and carotenoids in solutions were modeled considering MBOM theory [4,22,23]. To overcome the complexity of the fitting procedure with several dozen free parameters, the DE algorithm was used. All of the classical strategies and their tuning parameters were tested, and the best DE strategy for fitting of monomeric pigments had been found.

PSIIRC is a system of interacting pigments fixed in the protein skeleton and, in comparison with monomers, needs additional computational procedures such as matrix diagonalization (eigenstate problem) and integrations over a time scale to assess the relaxation rates. As in the case of monomers, we perform the strategy tests for PSIIRC modeling. The test results showed that the *DE/rand-to-best/1/exp* and *DE/best/1/bin* strategies are the best choices for the system; however, after thoroughly testing the strategy parameters $\{F, Cr\}$, it was found that each of those two has its own advantages and disadvantages.

For such computational algorithms, sticking at local minima for high-dimensional multimodal function is a rather serious problem. The global solution to this problem will sufficiently simplify the calculations because when the algorithm passes through the last local minimum, the rate of convergence becomes clearly exponential (Figure 5). In fact, with an increase in the number of free parameters, the convergence probability decreases greatly. Even with optimal values of the parameters, the percentage of convergence has a certain limit, which is determined by the initial conditions and the statement of a specific problem.

For example, with the best tuning parameters $\{F = 0.55, Cr = 0.9\}$, the convergence rate for the *DE/rand-to-best/1/exp* strategy with nine unknown parameters is equal about 27%. *DE/best/1/bin* cannot find the best solution and is always stuck. It is worth noting that with an increase in the number of unknowns, the value of the best tuning parameters is retained. Therefore, the next step in solving the problem will be the creation (development) of the algorithm, which can determine local minimum and after it can get out of there. A more flexible selection of tuning parameters [27,28,41] (for example, an adaptation of the SADE algorithm [29,42]) or a way to get out from the local minima could make a wide step towards solving this problem.

8. Conclusions

We have shown that the use of a heuristic evolutionary algorithm such as DE in modeling the optical properties of PPCs allows us to obtain high-quality calculated spectra and, at the same time, to assess the uniqueness of the obtained parameters of the exciton model of the energy transfer in PPCs.

In this study, the DE algorithm was used for simulation of the linear optical response of a system of interacting chlorophyll and pheophytin pigments. Applying the MBOM and the theory of molecular excitons, we have demonstrated that the linear optical response of the PSIIRC (absorption, circular and linear dichroism, and steady-state fluorescence) can be fitted by DE with high accuracy. To explore the effectiveness of DE, we used the simulated experimental data as the target functions instead of those of actually measured. After the strategy test was performed, it appeared that only 2 of the 10 DE strategies have shown the best performance of the optimization algorithm. The best tuning parameters were determined to run the full optimization of PSIIRC linear optical response modeling. Finally, using the *DE/rand-to-best/1/exp* strategy, we found the exact solution for the PSIIRC exciton model and fitted the spectra with a reasonable convergence rate.

However, the chosen "optimal" strategies and their settings still do not allow us to find the desired exciton models of pigment–protein complexes with 100% probability. The main problem is that the optimization algorithm becomes stuck in the local minima (Figure 4 is a typical example). Thus, the development of modified DE strategies that can detect local minima and allow the algorithm to find ways to bypass them in the parameter space that minimizes the objective function is the immediate goal of our further research in this field.

Author Contributions: Validation, formal analysis, investigation, software, writing—original draft preparation, D.D.C.; conceptualization, methodology, resources, software, writing—original draft preparation, writing—review and editing, visualization, supervision, project administration, funding acquisition, R.Y.P. All authors have read and agreed to the published version of the manuscript.

Funding: This research was funded by a grant of the Ministry of Science and Higher Education of the Russian Federation for large scientific projects in priority areas of scientific and technological development (subsidy identifier 075-15-2020-774).

Institutional Review Board Statement: Not applicable.

Informed Consent Statement: Not applicable.

Data Availability Statement: No additional data available.

Acknowledgments: This study was carried out using equipment of the shared research facilities of the HPC computing resources at Moscow State University.

Conflicts of Interest: The authors declare no conflict of interest. The funders had no role in the design of the study, in the collection, analyses, or interpretation of the data, in the writing of the manuscript, or in the decision to publish the results.

References

1. Jang, S.J.; Mennucci, B. Delocalized excitons in natural light-harvesting complexes. *Rev. Mod. Phys.* **2018**, *90*, 035003. [CrossRef]
2. Mirkovic, T.; Ostroumov, E.E.; Anna, J.M.; van Grondelle, R.; Scholes, G.D. Light Absorption and Energy Transfer in the Antenna Complexes of Photosynthetic Organisms. *Chem. Rev.* **2017**, *117*, 249–293. [CrossRef] [PubMed]
3. Croce, R.; van Amerongen, H. Natural strategies for photosynthetic light harvesting. *Nat. Chem. Biol.* **2014**, *10*, 492–501. [CrossRef] [PubMed]
4. Pishchalnikov, R.Y.; Yaroshevich, I.A.; Slastnikova, T.A.; Ashikhmin, A.A.; Stepanov, A.V.; Slutskaya, E.A.; Friedrich, T.; Sluchanko, N.N.; Maksimov, E.G. Structural peculiarities of keto-carotenoids in water-soluble proteins revealed by simulation of linear absorption. *Phys. Chem. Chem. Phys.* **2019**, *21*, 25707–25719. [CrossRef]
5. Chaillet, M.L.; Lengauer, F.; Adolphs, J.; Muh, F.; Fokas, A.S.; Cole, D.J.; Chin, A.W.; Renger, T. Static Disorder in Excitation Energies of the Fenna-Matthews-Olson Protein: Structure-Based Theory Meets Experiment. *J. Phys. Chem. Lett.* **2020**, *11*, 10306–10314. [CrossRef]
6. Adolphs, J.; Renger, T. How proteins trigger excitation energy transfer in the FMO complex of green sulfur bacteria. *Biophys. J.* **2006**, *91*, 2778–2797. [CrossRef] [PubMed]

7. Higgins, J.S.; Lloyd, L.T.; Sohail, S.H.; Allodi, M.A.; Otto, J.P.; Saer, R.G.; Wood, R.E.; Massey, S.C.; Ting, P.C.; Blankenship, R.E.; et al. Photosynthesis tunes quantum-mechanical mixing of electronic and vibrational states to steer exciton energy transfer. *Proc. Natl. Acad. Sci. USA* **2021**, *118*, e2018240118. [CrossRef] [PubMed]

8. Croce, R.; van Amerongen, H. Light-harvesting in photosystem I. *Photosynth. Res.* **2013**, *116*, 153–166. [CrossRef] [PubMed]

9. Pishchalnikov, R.Y.; Shubin, V.V.; Razjivin, A.P. The role of vibronic modes in formation of red antenna states of cyanobacterial PSI. *Photosynth. Res.* **2020**, *146*, 75–86. [CrossRef] [PubMed]

10. Gunther, L.M.; Jendrny, M.; Bloemsma, E.A.; Tank, M.; Oostergetel, G.T.; Bryant, D.A.; Knoester, J.; Kohler, J. Structure of Light-Harvesting Aggregates in Individual Chlorosomes. *J. Phys. Chem. B* **2016**, *120*, 5367–5376. [CrossRef]

11. Ueno, S.; Tanimura, Y. Modeling and Simulating the Excited-State Dynamics of a System with Condensed Phases: A Machine Learning Approach. *J. Chem. Theory Comput.* **2021**, *17*, 3618–3628. [CrossRef] [PubMed]

12. Jansen, T.L.C. Computational spectroscopy of complex systems. *J. Chem. Phys.* **2021**, *155*, 170901. [CrossRef] [PubMed]

13. Mukamel, S. *Principles of Nonlinear Optical Spectroscopy*; Oxford University Press: New York, NY, USA; Oxford, UK, 1995; Volume 6, p. 543.

14. Kubo, R. Generalized Cumulant Expansion Method. *J. Phys. Soc. Jpn.* **1962**, *17*, 1100–1120. [CrossRef]

15. Lax, M. The Franck-Condon Principle and Its Application to Crystals. *J. Chem. Phys.* **1952**, *20*, 1752–1760. [CrossRef]

16. Abramavicius, D.; Palmieri, B.; Voronine, D.V.; Sanda, F.; Mukamel, S. Coherent Multidimensional Optical Spectroscopy of Excitons in Molecular Aggregates; Quasiparticle versus Supermolecule Perspectives. *Chem. Rev.* **2009**, *109*, 2350–2408. [CrossRef] [PubMed]

17. Brixner, T.; Hildner, R.; Kohler, J.; Lambert, C.; Wurthner, F. Exciton Transport in Molecular Aggregates—From Natural Antennas to Synthetic Chromophore Systems. *Adv. Energy Mater.* **2017**, *7*, 1700236. [CrossRef]

18. Holland, J.H. *Adaptation in Natural and Artificial Systems: An Introductory Analysis with Applications to Biology, Control, and Artificial Intelligence*; University of Michigan Press: Ann Arbor, MI, USA, 1975; p. 183.

19. Goldberg, D.E. *Genetic Algorithms in Search, Optimization and Machine Learning*, 13th ed.; Addison-Wesley Publishing Company, Inc.: Reading, MA, USA, 1989; p. 432.

20. Storn, R.; Price, K. Differential evolution—A simple and efficient heuristic for global optimization over continuous spaces. *J. Glob. Optim.* **1997**, *11*, 341–359. [CrossRef]

21. Storn, R. System design by constraint adaptation and differential evolution. *IEEE Trans. Evol. Comput.* **1999**, *3*, 22–34. [CrossRef]

22. Pishchalnikov, R.Y.; Bondarenko, A.A.; Ashikhmin, A.A. Optimizing the Multimode Brownian Oscillator Model for the Optical Response of Carotenoids in Solution by Fine Tuning of Differential Evolution. *Lobachevskii J. Math.* **2020**, *41*, 1545–1553. [CrossRef]

23. Pishchalnikov, R. Application of the differential evolution for simulation of the linear optical response of photosynthetic pigments. *J. Comput. Phys.* **2018**, *372*, 603–615. [CrossRef]

24. Das, S.; Mullick, S.S.; Suganthan, P.N. Recent advances in differential evolution—An updated survey. *Swarm Evol. Comput.* **2016**, *27*, 1–30. [CrossRef]

25. Rocca, P.; Oliveri, G.; Massa, A. Differential Evolution as Applied to Electromagnetics. *IEEE Antennas Propag. Mag.* **2011**, *53*, 38–49. [CrossRef]

26. Wu, G.H.; Mallipeddi, R.; Suganthan, P.N.; Wang, R.; Chen, H.K. Differential evolution with multi-population based ensemble of mutation strategies. *Inf. Sci.* **2016**, *329*, 329–345. [CrossRef]

27. Qin, A.K.; Huang, V.L.; Suganthan, P.N. Differential Evolution Algorithm with Strategy Adaptation for Global Numerical Optimization. *IEEE Trans. Evol. Comput.* **2009**, *13*, 398–417. [CrossRef]

28. Paterlini, S.; Krink, T. Differential evolution and particle swarm optimisation in partitional clustering. *Comput. Stat. Data Anal.* **2006**, *50*, 1220–1247. [CrossRef]

29. Liu, J.; Lampinen, J. A fuzzy adaptive differential evolution algorithm. *Soft Comput.* **2005**, *9*, 448–462. [CrossRef]

30. Gelzinis, A.; Abramavicius, D.; Ogilvie, J.P.; Valkunas, L. Spectroscopic properties of photosystem II reaction center revisited. *J. Chem. Phys.* **2017**, *147*, 115102. [CrossRef] [PubMed]

31. Raszewski, G.; Saenger, W.; Renger, T. Theory of optical spectra of photosystem II reaction centers: Location of the triplet state and the identity of the primary electron donor. *Biophys. J.* **2005**, *88*, 986–998. [CrossRef] [PubMed]

32. Novoderezhkin, V.I.; Andrizhiyevskaya, E.G.; Dekker, J.P.; van Grondelle, R. Pathways and timescales of primary charge separation in the photosystem II reaction center as revealed by a simultaneous fit of time-resolved fluorescence and transient absorption. *Biophys. J.* **2005**, *89*, 1464–1481. [CrossRef]

33. Muh, F.; Plockinger, M.; Renger, T. Electrostatic Asymmetry in the Reaction Center of Photosystem II. *J. Phys. Chem. Lett.* **2017**, *8*, 850–858. [CrossRef] [PubMed]

34. Renger, G.; Renger, T. Photosystem II: The machinery of photosynthetic water splitting. *Photosynth. Res.* **2008**, *98*, 53–80. [CrossRef]

35. Renger, T. Semiclassical Modified Redfield and Generalized Forster Theories of Exciton Relaxation/Transfer in Light-Harvesting Complexes: The Quest for the Principle of Detailed Balance. *J. Phys. Chem. B* **2021**, *125*, 6406–6416. [CrossRef] [PubMed]

36. Pishchalnikov, R.Y.; Shubin, V.V.; Razjivin, A.P. Spectral differences between monomers and trimers of photosystem I depend on the interaction between peripheral chlorophylls of neighboring monomers in trimer. *Phys. Wave Phenom.* **2017**, *25*, 185–195. [CrossRef]

37. Trinkunas, G.; Holzwarth, A.R. Kinetic modeling of exciton migration in photosynthetic systems. 3. Application of genetic algorithms to simulations of excitation dynamics in three-dimensional photosystem core antenna reaction center complexes. *Biophys. J.* **1996**, *71*, 351–364. [CrossRef]
38. Trinkunas, G.; Connelly, J.P.; Muller, M.G.; Valkunas, L.; Holzwarth, A.R. Model for the excitation dynamics in the light-harvesting complex II from higher plants. *J. Phys. Chem. B* **1997**, *101*, 7313–7320. [CrossRef]
39. Vaitekonis, S.; Trinkunas, G.; Valkunas, L. Red chlorophylls in the exciton model of photosystem I. *Photosynth. Res.* **2005**, *86*, 185–201. [CrossRef] [PubMed]
40. Bruggemann, B.; Sznee, K.; Novoderezhkin, V.; van Grondelle, R.; May, V. From structure to dynamics: Modeling exciton dynamics in the photosynthetic antenna PS1. *J. Phys. Chem. B* **2004**, *108*, 13536–13546. [CrossRef]
41. Gong, W.Y.; Cai, Z.H. Differential Evolution With Ranking-Based Mutation Operators. *IEEE Trans. Cybern.* **2013**, *43*, 2066–2081. [CrossRef] [PubMed]
42. Noman, N.; Iba, H. Accelerating differential evolution using an adaptive local search. *IEEE Trans. Evol. Comput.* **2008**, *12*, 107–125. [CrossRef]

Article

Multi-Objective Multi-Scale Optimization of Composite Structures, Application to an Aircraft Overhead Locker Made with Bio-Composites

Xavier Martínez [1,2], Jordi Pons-Prats [2,3], Francesc Turon [1,2], Martí Coma [2], Lucía Gratiela Barbu [2,4] and Gabriel Bugeda [2,4,*]

1. Department of Nautical Science and Engineering, Universitat Politècnica de Catalunya (UPC), Pla de Palau 18, 08003 Barcelona, Spain
2. Centre Internacional de Mètodes Numèrics a l'Enginyeria (CIMNE), Campus Nord, Gran Capità s/n, 08034 Barcelona, Spain
3. Department of Physics, Aeronautical Division, Universitat Politècnica de Catalunya (UPC), Campus Baix Llobregat, c/Esteve Terrades 5, 08860 Castelldefels, Spain
4. Department of Civil and Environmental Engineering, Universitat Politècnica de Catalunya (UPC), Campus Nord, Gran Capità s/n, 08034 Barcelona, Spain
* Correspondence: bugeda@cimne.upc.edu

Citation: Martínez, X.; Pons-Prats, J.; Turon, F.; Coma, M.; Barbu, L.G.; Bugeda, G. Multi-Objective Multi-Scale Optimization of Composite Structures, Application to an Aircraft Overhead Locker Made with Bio-Composites. *Mathematics* 2023, 11, 165. https://doi.org/10.3390/math11010165

Academic Editors: Antonin Ponsich, Mariona Vila Bonilla and Bruno Domenech

Received: 14 November 2022
Revised: 15 December 2022
Accepted: 24 December 2022
Published: 28 December 2022

Abstract: The use of composite materials has grown exponentially in transport structures due to their weight reduction advantages, added to their capability to adapt the material properties and internal micro-structure to the requirements of the application. This flexibility allows the design of highly efficient composite structures that can reduce the environmental impact of transport, especially if the used composites are bio-based. In order to design highly efficient structures, the numerical models and tools used to predict the structural and material performance are of great importance. In the present paper, the authors propose a multi-objective, multi-scale optimization procedure aimed to obtain the best possible structure and material design for a given application. The procedure developed is applied to an aircraft secondary structure, an overhead locker, made with a sandwich laminate in which both, the skins and the core, are bio-materials. The structural multi-scale numerical model has been coupled with a Genetic Algorithm to perform the optimization of the structure design. Two optimization cases are presented. The first one consists of a single-objective optimization problem of the fibre alignment to improve the structural stiffness of the structure. The second optimization shows the advantages of using a multi-objective and multi-scale optimization approach. In this last case, the first objective function corresponds to the shelf stiffness, and the second objective function consists of minimizing the number of fibres placed in one of the woven directions, looking for a reduction in the material cost and weight. The obtained results with both optimization cases have proved the capability of the software developed to obtain an optimal design of composite structures, and the need to consider both, the macro-structural and the micro-structural configuration of the composite, in order to obtain the best possible solution. The presented approach allows to perform the optimisation of both the macro-structural and the micro-structural configurations.

Keywords: multi-objective optimization; multi-scale homogenization; bio-composite materials

MSC: 74P05

1. Introduction

One of the main motivations for the exponential increase in the use of composite materials, especially in transport structures such as airplanes, ships, and automobiles, is their excellent ratio between mechanical performance and weight. Any weight reduction in these structures directly relates to the reduction of energy consumption of the transportation mode. The other main advantage of composites is that they offer to adapt the material properties, and its internal micro-structure, to the strength requirements of the structure in which they are used. An example of this specific composite customization is found in the development of 3-Dimensional Laminate (3DL) sails for the yacht industry [1] in which the different composite fibres are placed individually in the sail to obtain optimal performance.

Several factors allow for having highly efficient composites such as the aforementioned 3DL sails. One of them is the development of manufacturing techniques, such as it is Automatic Fibre Placement (AFP) [2], that allows allocating, with extreme precision, a single fibre filament in the structural component. Another factor that has been key for the development of these new composites is the numerical formulations and tools required to predict the structural performance that will be obtained with such composites. As composites have become more complex, these tools have to be improved to account for the new material configurations. Therefore, in most cases, it is not sufficient to evaluate the composite as an orthotropic material and to apply a failure threshold criterion such as the ones reported in [3]. Instead, it is necessary to use more complex formulations that obtain the composite performance by means of its constituent materials. One of these approaches is the serial-parallel mixing theory (S/P RoM) [4], which acts as a constitutive equation manager and is based on the definition of a set of compatibility equations between the composite components, usually fibre and matrix, relating their stress-strain performance. The validity of the S/P RoM to accurately characterize the composite stiffness, as well as different composite failure modes, has already been demonstrated in several references [5–8].

However, when the composite architecture becomes more complex, for instance when having woven laminates, the serial/parallel mixing theory is not enough to account for all the micro-structural interactions between the composite constituents, being necessary the use of more complex formulations. Good candidates for such purposes are numerical multiscale procedures [9,10], which are based on solving the structural problem at hand by splitting it into two different scales. Namely, a macro-scale that discretizes the global problem, and a micro-scale that defines the material micro-structure. With this approach, the macro-scale deformation gradient tensor is used for the solution of the micro-scale problem, and then, using the microscopic results, for the obtainment of the macro-scale stress tensor. The microscopic problem is solved on a Representative Volume Element (RVE) model, which must be periodic and representative of the material [11]. This approach is of special relevance for the analysis of woven composites, which micro-structural performance cannot be properly captured by the serial-parallel mixing theory. Examples of multiscale analysis on woven composites can be found in [12,13]

The vast number of possibilities that offer composite materials, regarding their internal configuration and their disposition in the structure, represents also a challenge to obtain the optimal configuration for a given structural application. One of the most common solutions used to customize the composite for a given application is by giving a preferred orientation to the composite laminate, which is done by placing more fibres in a given direction compared to the other ones. The definition of the optimal direction can be done based on a deep understanding of the structural performance of the element considered or, as it is done with more recent technologies, by coupling the structural analysis with an optimization procedure. In this regard, Nikbakt et al. [14] have made an exhaustive review of the work that has been conducted by the scientific community regarding composite optimization. In their review they divide the work into the different types of structures analysed: beams, plates, shells, and other types; and, for each one of them, they provide the optimizations made based on the different objective functions considered, i.e., weight minimization or buckling load maximization. In their work, they realized that most of the analyses made are based on finding the best stacking sequence to obtain the desired structure performance. Examples of these types of analyses can be found in the work conducted by Ehsani and Rezaeepazhand [15], Wei et al. [16], and Zhou et al. [17]. In the last work mentioned, the authors not only provide the optimal composite orientation, but they also find the best structural topology to handle the applied loads.

Optimization procedures are based on the modification of different structural parameters to obtain the desired structural performance, which is represented by an objective function. There are also optimization strategies that seek to obtain the improved performance of different objective functions. This is, for instance, maximizing the structure stiffness while minimizing its weight. Those are the so-called multi-objective optimization procedures [18]. Several authors have already applied multi-objective optimization strategies to obtain the best composite configuration and, therefore, take maximum advantage of the multiple configuration capabilities of composite materials. Recent examples of this approach can be found in [19,20]. In these analyses, since they are based on a multi-objective optimization, the outcome is a Pareto Front, which provides the solutions showing an equilibrium of the optimized functions. The points on the Pareto Front are those which dominates the other solutions, which means that no solution improves the dominant ones without getting one of the objective function worse. A further step has been done in the works of Li et al. [21] and Coelho et al. [22,23] in which the optimization method is applied to a multiscale procedure for the characterization of the composite structures. In the last one of these works, the parameters that can be modified to obtain the optimized structure performance are at the macro-level by modifying the fibre orientation, and at the micro-level by changing the volumetric participation of the different components in the composite. Other examples of multi-scale optimization can be found on references [24–29], which focus on topological optimization. In [26–28], the authors propose a two-scale optimization process using pre-computed microstructure format, while in [27–29] the authors propose a multi-scale approach for structures and the materials of the components.

Current work enhances the path started by the abovementioned authors by coupling a multi-objective optimization code with a multi-scale structural code. The proposed approach focusses on shape optimization, not on topological one. The requirements from the selected applications better fit with shape optimization. With this approach, it will be possible to find the optimal micro-structure of the material and the optimal configuration of the composite in the structure, complying with several objective functions and both the macro- and the micro-level. This brings the opportunity, at the design stage, of tailoring the material to obtain the desired structural performance. For instance, the optimization process can improve the design characteristics at the micro-level (stacking sequence or composite micro-structural configuration) and at the macro-level (composite orientation or shape optimization) simultaneously. It will be shown that this new approach brings different solutions than a simple macro-scale optimization, which can result in a more efficient and structure with an improved performance.

The first section of this manuscript includes a brief description of the numerical tools developed to conduct the analysis: the multi-scale procedure used to characterize composite materials, and the multi-objective optimization code that is coupled to the structural one. Afterwards, the optimization procedure proposed is defined and validated by solving a case study. The case considered consists in the optimization made of an aircraft secondary structure, an overhead locker (or hatrack), designed to be manufactured using eco-composite materials. The eco-composites considered in this work have been identified in the context of the ECO-COMPASS project [30,31] (European Union's Horizon 2020 research and innovation programme under grant agreement No 690638) as being a renewable and ecologically improved solution compared to traditional ones.

The numerical tools developed in this work, together with the use of eco-composite materials, are expected to facilitate the incorporation of these new materials in transportation structures. The optimization made will optimize the structure and the material micro-structure, obtaining the best possible configuration for the application considered. Finally, the incorporation of eco-composites in an optimal configuration is one of the important actions that can be taken to minimize the carbon footprint associated with transportation structures.

2. Multi-Scale and Multi-Objective Procedures for the Analysis and Optimisation of Composite Materials

The multi-scale, multi-objective optimization procedure proposed in this work is obtained by coupling a composite structural solver, PLCd [32], in which the composite performance is obtained by means of a numerical homogenization, with a multi-objective optimization software, RMOP [33], capable of interacting with any other software solving a physical problem. This section describes briefly the formulation used by both codes, as well as the procedure used to couple them.

2.1. Numerical Multi-Scale Model

Numerical multi-scale models are of special relevance for the analysis of structures with a complex micro-structural behaviour, such as composites. In these cases, it is very difficult to define a constitutive law that captures accurately the mechanical performance of the material. Instead, a multi-scale approach uses a micro-structural model, defined by means of a Representative Volume Element (RVE), to obtain the material response. The material performance provided by the micro-model is used afterward by the macro-model of the structure to be analysed.

Figure 1 presents a schematic representation of a multiscale approach. This figure shows that the solution of the beam model (macro-structure) produces a set of strain fields that are converted to displacements in the micro-model in order to analyse its response. The results provided by the micro-model are a set of forces that are then transformed to stresses in order to feed the macro-model. In other words, the micro-model works like a constitutive equation of the material, as it provides the stress field associated with a given strain. With this approach, it is possible to obtain the elastic performance of composites with complex micro-structures (such as honeycombs, woven composites, etc.) as well as to predict the complex failure modes associated to these materials.

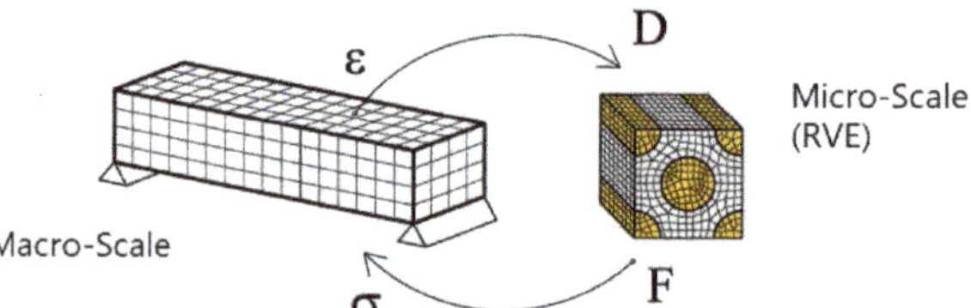

Figure 1. Schematic representation of a multiscale analysis.

The mathematical foundations on which the multiscale analysis is based, as well as its detailed numerical implementation, can be obtained from the reference Otero et al. [9]. In this work the authors show that the displacement to be applied to the micro-scale (RVE) can be obtained as the displacement of the macro-scale and the addition of a micro-fluctuation in the micro-model. This can be written as:

$$u_\mu = u + w(X_\mu) \tag{1}$$

being u_μ the displacement of the micro-scale, u the displacements of the macro-scale, and $w(X_\mu)$ the micro-fluctuations on the micro-model, which vary for each point in the microscale X_μ.

Besides describing the displacement field of the micro-model, a multiscale method also need to relate the different scales considered. This is done with the average theorems [34]. These theorems state that a given parameter in the macro-model can be obtained as the integral over the volume of this same parameter in the micro-model. If the average theorem is applied to the deformation gradient, the expression obtained is:

$$F(X_0) = \frac{1}{V_\mu} \int_{V_\mu} F_\mu(X_0, X_\mu)\, dV \tag{2}$$

where $F(X_0)$ is the deformation gradient, in X_0, of the macro-structure, V_μ is the RVE volume, and F_μ are the deformation gradients in all RVE points. Equation (2) can be used to obtain the different displacement fields that are kinematically admissible in the micro-scale. In order to solve the equilibrium of the RVE at the micro-level, it is also necessary to define a set of boundary conditions. Among the different possible sets of boundary conditions that can be applied, this work will use periodic boundary conditions, which can be written as:

$$w\left(X_\mu^+\right) = w\left(X_\mu^-\right) \ \forall \ X_\mu \in \left\{\partial V_\mu^+, \partial V_\mu^-\right\}. \tag{3}$$

being $V_\mu^+ \ V_\mu^-$ the parallel periodic boundaries in the RVE. With these boundary conditions, the kinematical constraint defines a periodic displacement fluctuation on parallel faces of the RVE.

Once solved the Boundary Value Problem (BVP) at the Representative Volume Element, the stresses in the macro-model can be obtained from the stresses computed at the micro-scale using the average theorem:

$$\sigma = \frac{1}{V} \int_V \sigma_\mu \, dV \tag{4}$$

where, σ and σ_μ are the stresses in the macro-model and in the micro-model respectively, and V is the volume of the RVE.

The multiscale procedure previously summarized is described in detail in the work of Otero et al. [9] and implemented in the CIMNE in-house finite element software PLCd [32]. The abovementioned work also describes an enhanced approach that incorporates the second-order displacements of the macro-model in the displacement field of the RVE. A full second-order homogenization is proposed by Geers et al. in [35]. Although current work will be limited to the linear-elastic performance of the structure, multiscale methods can account for material non-linearities. In this regard, Otero et al. [36], and Zaghi et al. [37] proposed different strategies to incorporate material-damage at the micro-structural level, with an affordable computational cost.

2.2. Multi-Objective Optimization Procedures

The multi-objective optimisation tool used in this work is RMOP [33], the Robust Multi-Objective Optimization platform. It is a CIMNE in-house tool used along with the optimization analysis. It is an optimization platform that implements Genetic Algorithms, Particle Swarm Optimization methods and, Gradient Based methods.

RMOP is implemented as a set of libraries to open the possibility to implement new developments resulting from research. This coupling is done with a script, which can be defined using any programming or scripting language. The basic concept is that the solver works as a black box, so the optimizer is sending the request to the solver and this is answering the requested evaluation for both the objective functions and the constraints. This coupling can be done through the command line and ASCII files, but also using directly the RAM memory of the computer.

The Genetic Algorithm implementation in RMOP for the solution of multi-objective optimisation problems is based on the NSGA-II [38]. It uses a $\lambda + \mu$ strategy and Crowded-Comparison Operator for the selection operator [38], a Simulated Binary Crossover [39] for the crossover operator, and a Polynomial Mutation [18] for the mutation operator. These operators define how the evaluated individuals are selected to become parents and how the new offspring are created. The results of a multi-objective problem are normally represented using the Pareto optimality or non-dominated individuals concept [18]. Figure 2 shows this concept for a problem with two conflicting objective functions. For a given multi-objective problem, the solution is the Pareto optimal set resulting from the used optimization method. This gives a representation of all compromised designs between the two conflicting objectives. Real-world problems normally involve different conflicting objectives without any unique optimum design for all of them. In this case, a set of compromised solutions known as Pareto optimal (or non-dominated) solutions, can be obtained. A solution of a multi-objective problem is considered Pareto optimal if there are no other solutions improving all the objectives simultaneously. This means that an improvement in any of the objective functions implies a deterioration of any of the rest. The goal of solving the corresponding optimisation problem is then to provide a set of Pareto optimal solutions representing a trade-off of information amongst the objectives.

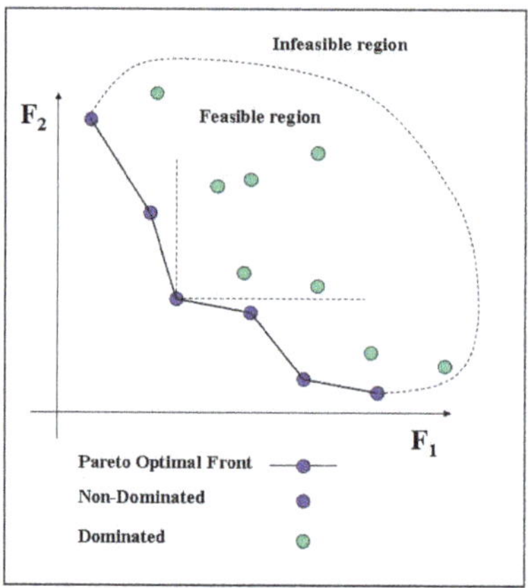

Figure 2. Pareto Optimality.

For a minimisation problem, a vector x_1 is said partially less than vector x_2 if, and only if:

$$\forall i: \quad f_i(x_1) \leq f_i(x_2) \quad and \quad \exists i: \quad f_i(x_1) < f_i(x_2) \tag{5}$$

In this case the solution x_1 dominates the solution x_2.

As Evolutionary Algorithms (EAs) consider multiple points simultaneously, they are capable of finding a number of different solutions in a Pareto set. A comprehensive theory, literature review, and implementation of Multi-objective EAs (MOEAs) including the NSGAII and VEGA algorithms is given by Deb in reference [18].

The RMOP platform is also enabling a pure hybrid approach. Two or more populations can be defined, assigning exploration and exploitation roles to each of them. The roles can be defined through the values of the probability of cross-over and mutation, or through the assignment of different objective functions and constraints to each population [39]. Hybridization through the combination of optimization methods, as well as the mixing between multiple populations and a combination of methods is an ongoing development of the platform [40–42].

2.3. Multi-Objective Multi-Scale Optimization Procedure

Once having defined the multi-scale procedure to be used in this work, which is implemented in the finite element code PLCd [32], and the multi-objective optimization software RMOP [33], this section describes the procedure used to couple both numerical tools, in order to perform a multi-objective, multi-scale optimization of eco-composite structures.

This coupling is conducted using an interface program that reads the results obtained from the numerical analysis made by PLCd, and sends them to the optimization code. Then RMOP takes the results obtained and uses them to generate new populations that can be analysed in order to obtain the optimal structure configuration. It also draws the Pareto front (Figure 2) that can be used afterward by the user to select the most convenient solution for the structural problem solved.

The analysis of the results obtained from the different populations is made based on the objective functions to be optimized (either minimized or maximized). These objective functions are defined in the interface program, and can take data from the input data file of the structural solver (for instance, minimization of the stiffness required from the material) or from the output file (i.e., minimization of the deformation or of the maximum stresses).

Once the new populations have been defined, the interface program modifies the input data files required by the structural solver to run the new analyses. When the mechanical simulation is conducted using a multiscale analysis, the definition of the new simulation populations can be made by modifying either the parameters of the macro-model, the parameters of the micro-model or both.

The whole procedure is described schematically in Figure 3.

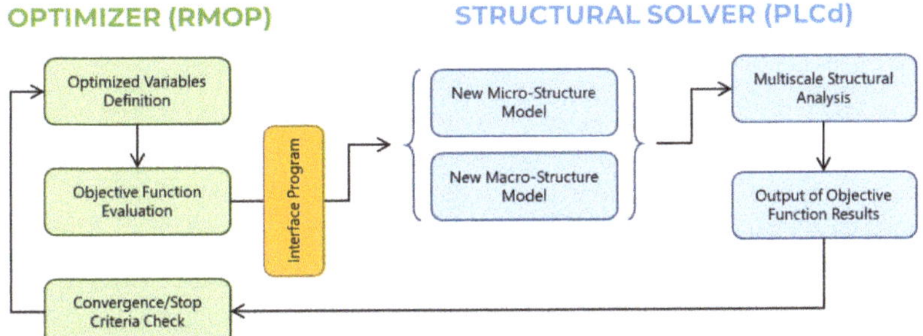

Figure 3. Flow diagram describing the process developed and the interaction between the optimizer and the structural solver.

3. Optimization Example, Application to an Aircraft Overhead Locker (Hatrack)

The structure chosen to validate the optimization procedure is the cabin stowage bin located above the seats in an airplane, this is also called hatrack. The hatrack is usually made with a sandwich material in which the core is a honeycomb and the skins are made of a woven glass-fibre embedded in a phenolic resin. In this work, the honeycomb core is made of aramid and cellulose, and the skins are made with a ramie woven embedded in an eco-epoxy matrix.

This section describes, first, the considered hatrack and the numerical model developed. Afterwards, two different optimization analyses are presented. The first one looks for the best possible orientation of the woven skins in the hatrack shelf. This optimization affects only the macro-structural model of the hatrack. The second optimization analysis conducted affects both, the macro- and the micro-scales of the model, as it looks for the best woven orientation as well as for the optimal micro-structural configuration of the laminate. The comparison of both analyses shows that, in order to obtain the best possible solution, the composite microstructure is as important as its orientation in the structure.

3.1. Aircraft Overhead Locker to Be Optimized

3.1.1. Geometry

A hatrack is the cabin stowage bin located above the seats in an airplane. The hatrack considered has been obtained from the patent of Welch and Roth [43]. It is subjected to three fuselage frames, which are at a distance of 20 inches. Therefore, the hatrack has a length of 40 inches (1016 mm), a width of 583 mm, and a height of 251 mm. Its exact geometry is shown in Figure 4.

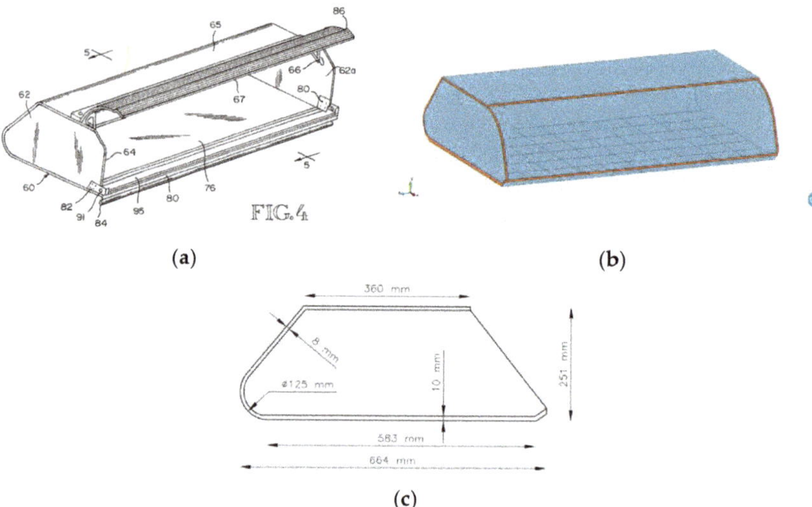

(a)   (b)

(c)

Figure 4. Hatrack considered for the design optimization. (**a**) Original geometry obtained from a patent template [43]. (**b**) Numerical model developed reproducing the original geometry. (**c**) Cross section dimensions.

The main interest of the current analysis is focussed on the performance of the hatrack shelf (element 76 in Figure 4a), as this is where the largest stresses and strains are found. To minimize the computational cost of the model, all non-structural parts have not been included (e.g., the hatrack door and its lock). Another simplification made on the geometry is in the areas where the hatrack attachments to the fuselage are located. These regions have large stress concentrations and require of specific reinforcements to sustain the loads. Usually, in these spots, the core material is filled with resin to create a monolithic region. For the sake of simplicity, these reinforcements are not defined in the model. Therefore, stress concentrations in these regions will be ignored.

3.1.2. Boundary Conditions

As it has been already mentioned, the hatrack is attached to the three fuselage frames that are located along its length in the upper and lower panels. This defines a total of six supports, as shown in Figure 5. In the model, this boundary condition is applied by restricting the displacements in the three spatial directions, along a line with a length of 30 mm.

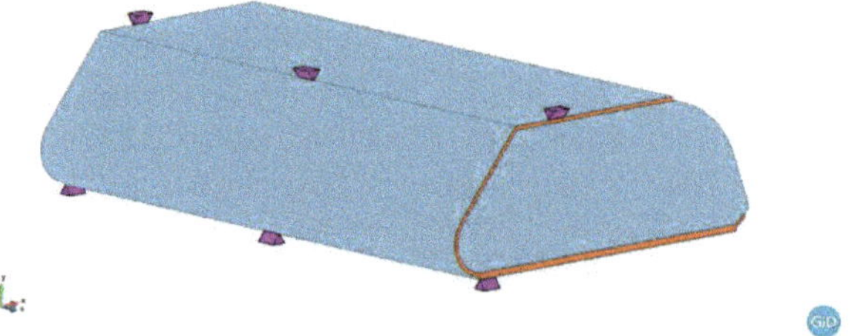

Figure 5. Boundary conditions on displacements defined for the numerical model.

As for the loads applied, a hatrack with these dimensions is allowed to support a total of 50 kg of weight due to the luggage stored in it (live-load). A dead-load of 20 kg has to be added to the live-load, which accounts for the actual weight of the hatrack, plus the weight of different systems attached to it such as the intercommunication elements, the lights, the oxygen masks, etc. The total load of 70 kg is applied to the shelf of the bin. This is correct for the luggage and the systems attached to the hatrack, but not for the self-weight, which is distributed along the whole geometry. Applying the load in such a manner has been done for the sake of simplicity and provides an extra safety factor.

The structural analysis will consider two different limit states. The Service Limit State (SLS) will apply the weight of 70 kg multiplied by gravity and divided by the area in which it is distributed. It defines a pressure of 1.14×10^{-3} MPa to be applied at the shelf. This load will be used to verify the maximum displacement suffered by the hatrack in service conditions. The other case corresponds to an Ultimate Limit State (ULS). In this case, the load is affected by a gravity acceleration of 8 g, as it is defined in Certification Specification CS-25 provided by EASA [44]. In this case, the applied pressure is 9.15×10^{-3} Mpa and this load will be used to evaluate the stresses found in the structural elements.

3.1.3. Materials

The hatrack is made with a sandwich laminate in which both, the skins and the core, contain eco-materials. The skins of the sandwich are made of woven ramie fibres embedded in a partly rosin-based epoxy matrix (AGMP 3600 [45]). The core is made with a honeycomb that contains a high percentage of cellulose.

Two different sandwich laminates are used in the hatrack. The shelf, in which the luggage is placed, has a total thickness of 14 mm, with a core of 10 mm and two skins of 2 mm each. The rest of the hatrack has a thickness of 8 mm, with a core of 6 mm and skins of 1 mm each. The mechanical performance of the skin and core materials is obtained using a multiscale strategy, with the definition of a RVE to represent the woven skin and another RVE to characterize the honeycomb core.

Sandwich skins

This section describes the Representative Volume Element Model developed to characterize the woven ramie laminate used for the sandwich skins. The geometry of the representative volume element has been generated using a micro-photograph of the composite. Both, the micro-photograph and the geometry of the model developed, are shown in Figure 6. The mesh of the RVE contains 9547 linear tetrahedral elements and 2287 nodes.

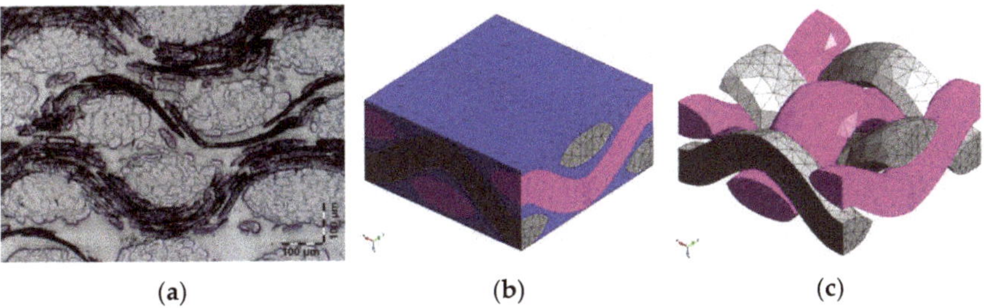

(a)   (b)   (c)

Figure 6. (**a**) Micro-photograph of the woven ramie laminate. (**b**) Representative volume element developed to simulate the material. (**c**) Detail of the fibres waviness defined in the RVE.

As it is shown in Figure 6b, the representative volume element has been defined with three different bulk materials, fibres in X direction (grey), fibres in Y direction (pink) and matrix (blue). Fibres in X and Y directions have exactly the same mechanical properties and differentiation has been made in order to have the possibility of changing them in the optimization process. The mechanical properties of ramie fibres have been obtained from average values shown in publications [46,47], and the properties of the epoxy resin come from the manufacturer's specifications. These properties are described in Table 1. The fibre volume content in the prepeg is 51.4%.

Table 1. Mechanical properties of ramie fibres and epoxy resin considered in the simulation.

Material	Young Modulus [GPa]	Poisson Modulus	Shear Modulus [Gpa]
Ramie	24.0	0.2	11.8
Epoxy	3.5	0.25	1.4

Sandwich core

The green-honeycomb core used for the analysis is an aramid paper core with a high percentage of cellulose. The material has been characterized with a Representative Volume Element which geometry has been defined from available photographs of the actual material. The material and the RVE model developed are shown in Figure 7. The mesh of the RVE has 9600 linear hexahedral elements and 15,666 nodes. A detail of the mesh is included in Figure 7c, where it is shown that there are two elements along the thickness of the honeycomb in order to capture the possible (although not probable) bending of the honeycomb paper.

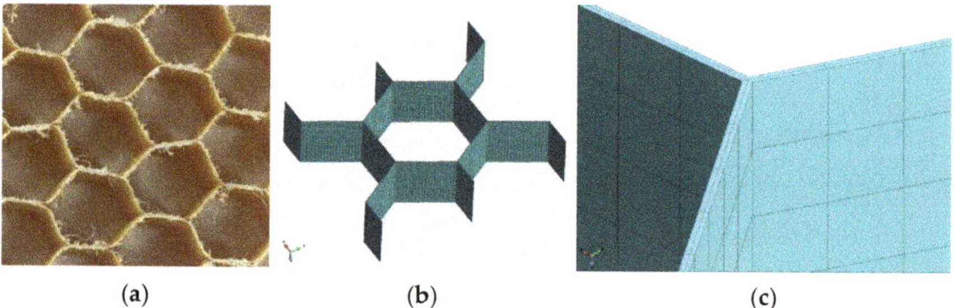

(a)   (b)   (c)

Figure 7. (**a**) Green honeycomb. (**b**) RVE developed to simulate the material. (**c**) Detail of the mesh defined.

The material properties defined for the hybrid cellulose-aramid films that constitute the honeycomb are described in Table 2.

Table 2. Mechanical properties of ramie fibres and epoxy resin considered in the simulation.

Material	Young Modulus [GPa]	Poisson Modulus [Gpa]	Shear Modulus [Gpa]
Green Honeycomp	18.0	0.3	6.9

3.1.4. Hatrack Finite Element Mesh

The hatrack is discretized with linear hexahedral solid elements. It has 79,173 elements and 106,276 nodes. Figure 8 shows a detail of the mesh. The laminate has three finite elements along its thickness, one corresponding to the core (pink elements in Figure 8) and two, one at each side of the core, corresponding to the woven skins. The mesh defined corresponds to a compromise between results accuracy and computational cost, as the structural model has to be analyzed many times during the optimization procedure.

Figure 8. Detail of the finite element mesh developed to analyse the hatrack.

3.1.5. Laminar Orientation

When working with composite structures, one of the most important aspects is the orientation of the composite in the structural element, as composites are highly orthotropic in terms of strength and stiffness. Figure 9 shows the original axis defined for the different hatrack elements. In this drawing the X, Y, and Z axes are represented in blue, red and green respectively. This orientation has been applied to the woven ramie prepreg and to the honeycomb. Figure 9 also shows that the hatrack shelf is divided into 6 × 10 rectangular elements. This division has been made to facilitate the definition of different woven orientations in the optimization problem.

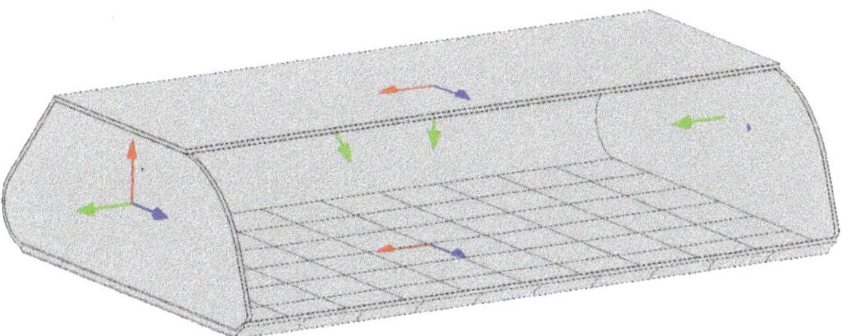

Figure 9. Laminate original orientation.

3.1.6. Hatrack Structural Performance

The hatrack has been analysed with the materials and boundary conditions previously described. This analysis provides the performance of the hatrack in its original configuration and will be compared with the improved response after conducting the optimization analysis.

Deformation under Service Limit State (SLS) loads

The deformation of the hatrack under SLS conditions is depicted in Figure 10. In this case, the load applied is a pressure load at the shelf surface, with a value of 9.15×10^{-3} Mpa. Under this load, the larger deformations are found at the centre of the free border of the shelf, and the value of the vertical displacement is 4.22 mm. The deformations of the hatrack under Ultimate Limit State (ULS) loads is eight times bigger, as the loads are also eight times bigger and the model is linear. In this case, the maximum displacement is 33.7 mm.

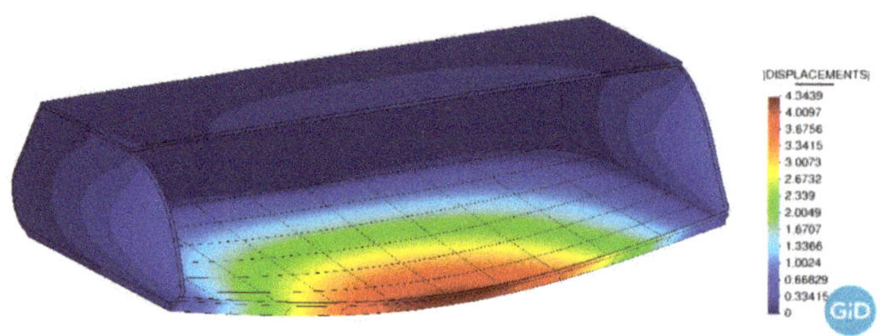

Figure 10. Hatrack deformation under SLS loads.

3.1.7. Stresses under Ultimate Limit State (ULS) Loads

The structure validation also requires that the maximum stresses under ULS loads do not reach the threshold value. The maximum tensile stress for the sandwich skin is limited to 59 Mpa, and the maximum shear stress allowed in the honeycomb is limited to 0.73 Mpa. Figure 11a shows the maximum tensile stress (first principal stress) in the skins of the hatrack. These stresses are larger at the back, close to the support area, and on the sides, close to the opened area. It is possible to see also stress concentrations around the different support regions. Figure 11b shows these same stresses in the hatrack shelf. These two figures show that at no point the maximum stress of 50 Mpa is exceeded, except in the support regions (where the hatrack is fixed to the airplane frames). The stress concentration in this regions is not plotted to avoid hiding the rest of the results.

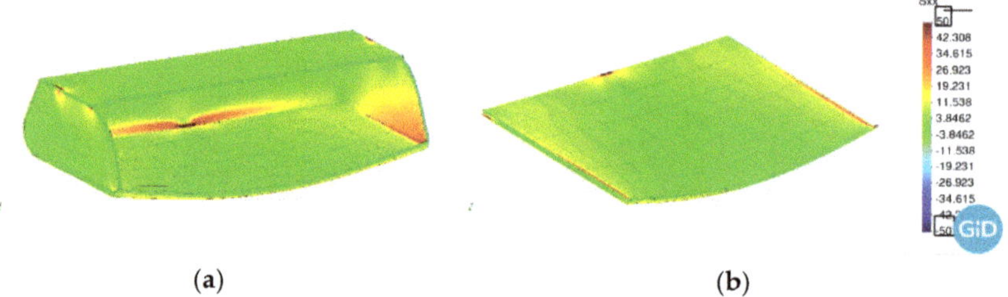

(a) (b)

Figure 11. Hatrack maximum normal stresses (principal stress) in the skin, under ULS loads. (**a**) Stresses in all hatrack skins. (**b**) Detail image of the stresses at the shelf.

Figure 12 shows the shear stresses in the honeycomb core at the shelf. In this image it is shown that they do not exceed the maximum stress value of 0.73 MPa at any point except, again, at the regions where the hatrack is fixed to the airplane frames.

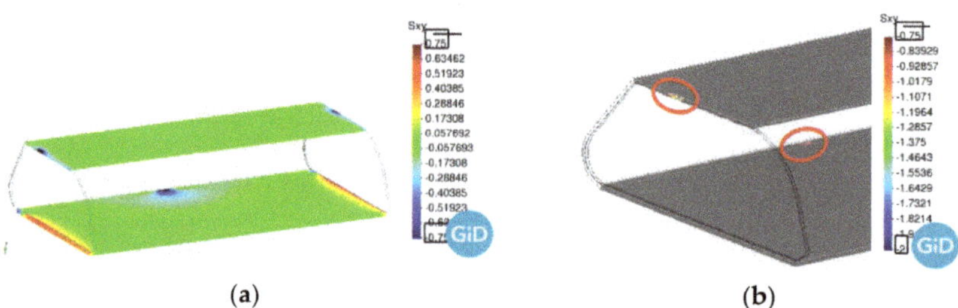

(a) (b)

Figure 12. Hatrack maximum shear stress in the honeycomb core, under ULS loads. (**a**) Elements with stresses in the range of $(-0.75, 0.75)$ MPa. (**b**) Elements with stresses larger than 0.75 Mpa in absolute value.

The stress results shown in previous figures prove the validity of the designed model to sustain the applied loads under the Ultimate Limit State. The only regions where the maximum loads are exceeded are those in which the hatrack is fixed. To handle these stress concentrations, in these regions the honeycomb core is filled with resin, obtaining a stiffer material that can easily handle the applied loads. The design of this structural detail is not within the objectives of this work and, for this reason, these larger stresses are disregarded.

3.2. First Optimization Analysis: Optimal Fibre Alignment for an Improvement of Structure Stiffness

The first optimization analysis conducted aims to improve the stiffness of the hatrack shelf by finding the optimal orientation of the woven skins. With this aim, the numerical model has divided the shelf into a total of 60 regions, 6 rows by 10 columns (see Figure 9), in which it is possible to define independent orientations. The stiffness of the shelf is evaluated with the displacement obtained at the centre of the free edge, where the maximum displacements are found, as shown in Figure 10. Therefore, the objective function here is the maximum vertical displacement in a given point (centre of the free edge of the shelf), which will be minimized. In order to achieve this, the algorithm can modify 60 different angles, each one corresponding to the skin orientation at each of the rectangular elements shown in Figure 9.

The initial displacement, under ULS loads and with the orientation following the X and Y axis shown in Figure 8, is 33.7 mm. With this initial deformation, the coupled PLCd-RMOP software is run, and the maximum displacement is evaluated after each structural solver analysis. Table 3 describes the set-up parameters for the RMOP optimizer, applied to the described application. The evolution of this displacement with the different evaluations made of the model is shown in Figure 13. This figure shows that the optimization software has been able to reduce the displacement of the structure with successive iterations, reaching an asymptotic value of 32.8 mm

Table 3. RMOP parameters configuration for the single objective analysis.

Set-Up Parameter for the RMOP Optimizer	Value
Objective functions	1, displacement
Design variables	70, fiber orientations
Range for the design variables	[−90, +90]
Population size	200
Probability of crossover	0.9
Mutation probability	0.1
Number of CPU	8

The woven skin orientation that sh"uld 'e defined to obtain the minimum displacement is shown in Figure 14. This picture shows that centre skins should maintain their original orientation, which is perpendicular to the shelf lip, while the outer skins must be oriented looking approximately towards the point of maximum deformation. This analysis validates the procedure to optimize the design of the hatrack.

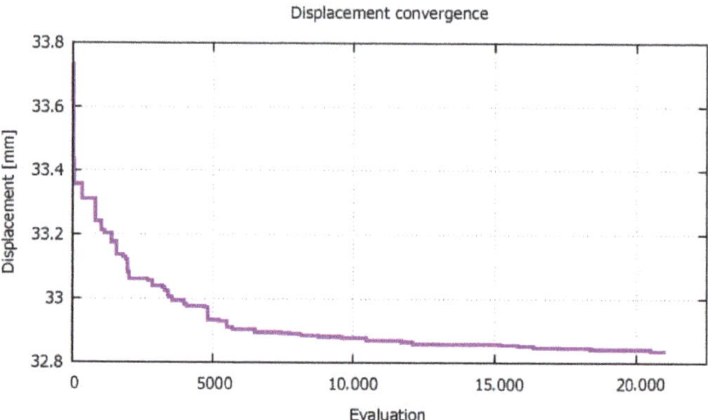

Figure 13. Maximum shelf displacement for the different evaluations of the model made by the optimization software.

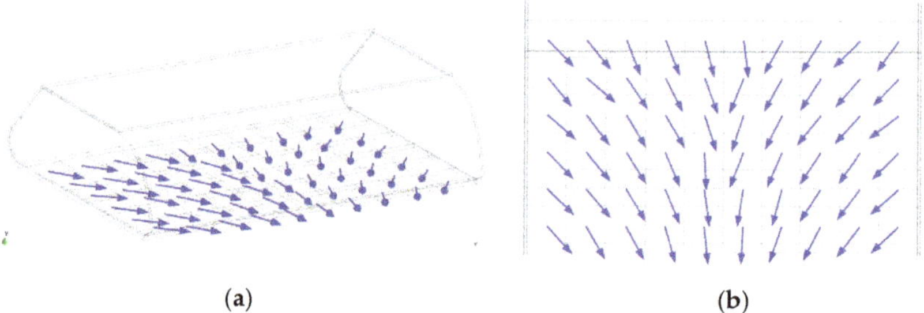

(a) (b)

Figure 14. Optimal woven skin orientation (X direction) required to obtain the minimum shelf displacement. (**a**) isometric view. (**b**) plane view.

The results also show that the improvement obtained in the structure stiffness with the optimal woven disposition is only 0.9 mm for the ULS loads, which is only 2.7% of the original deformation. Therefore, it can be concluded that in this case, the orientation of the woven composite in the shelf does not play a significant role in the hatrack shelf stiffness.

As for the structural performance of the cabin bin, it is very similar to the performance previously seen. The displacement obtained with the numerical model is shown in Figure 15, where it is seen that the displacement pattern is identical to the one obtained with the woven without an optimized alignment. Figure 16a,b shows the maximum tensile stresses in the sandwich skins and the shear stresses in the core, respectively. In both cases, the stress threshold is not exceeded except in the support regions. The results shown in these figures prove that the optimized design is valid and that it is possible to increase the stiffness of the hatrack shelf with a specific orientation of the woven skins in the sandwich laminate.

Figure 15. Hatrack displacements with the optimal woven skin design.

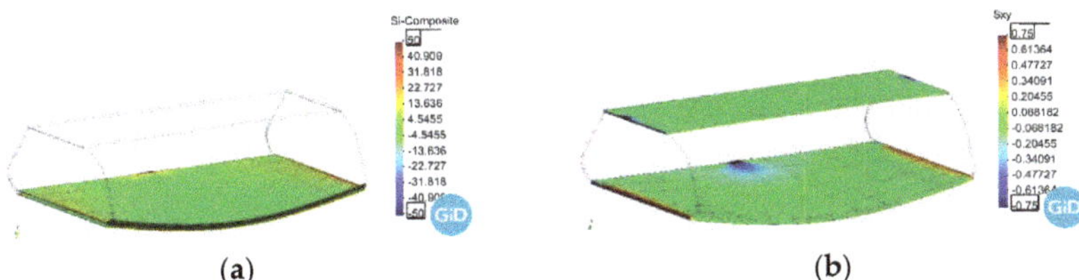

Figure 16. Hatrack shelf stresses with the optimal woven skin design. (**a**) Tensile stresses at the woven skins. (**b**) Shear stresses at the honeycomb.

3.3. Second Optimization Analysis: Multi-Objective, Multi-Scale Optimization

The second analysis made is a multi-objective, multi-scale optimization. Two different objective functions are defined to obtain the optimal design of the structure. The first one corresponds to the shelf stiffness, which is measured as the in previous case, by minimizing the vertical displacement at the centre of the free edge. The second objective function consists on minimizing number of fibres placed in one of the woven directions, which seeks a reduction on the material cost and weight (flax fibres density are about 1.5 g/cm^3 [48], while epoxy resins have densities around 1.2 g/cm^3 [3]). To obtain the optimal design, the software will modify the woven skin orientation, as has been done in the previous case, and will also modify the material configuration by reducing the amount of fibres in one of the directions of the woven material. Table 4 describes the set-up parameters for this second and multi-objective test case.

Table 4. RMOP parameters configuration for the multi-objective analysis.

Set-Up Parameter for the RMOP Optimizer	Value
Objective functions	2, displacement and amount of fibre layers
Design variables	71, fiber orientations and number of layers
Range for the design variables	[0, 180]
Population size	200
Probability of crossover	0.9
Mutation probability	0.1
Number of CPU	4

The fibre reduction is achieved by varying the stiffness of fibres in the Y direction of the Representative Volume Element defined to characterize the ramie woven prepreg (7b). This stiffness reduction is defined using the parallel mixing theory [49] to obtain the fibre yarn stiffness, as it is shown in Equation (6), assuming that the maximum stiffness is obtained if all the yarn contains ramie fibre, and that stiffness is reduced by adding some percentage of rosin epoxy resin in the fibre yarn. If this percentage reaches 100% there will be no fibres in the yarn, which provides a material similar to a unidirectional composite. In Equation (6), E_B, E_f and E_m correspond to the stiffness of the bundle, the ramie fibres and the rosin epoxy matrix, respectively, and k_f is the volumetric participation of fibre.

$$E_B = k_f \cdot E_f + \left(1 - k_f\right) \cdot E_m \tag{6}$$

In a multi-objective optimization analysis, there is not an optimal case, but a set of cases that provide an optimal result for one of the objective functions, for a given value of the other objective function. This set of results are represented in the Pareto front, which shows the best candidates obtained by combining both variables. Figure 17 shows the Pareto front after several evaluations. This figure shows that the evolution of this front from the first analysis (purple line) to the evaluation 30th (green line) is quite visible, and that after evaluation 30th the difference is minimal. The final result is obtained after 150 evaluations and it is shown with a red line.

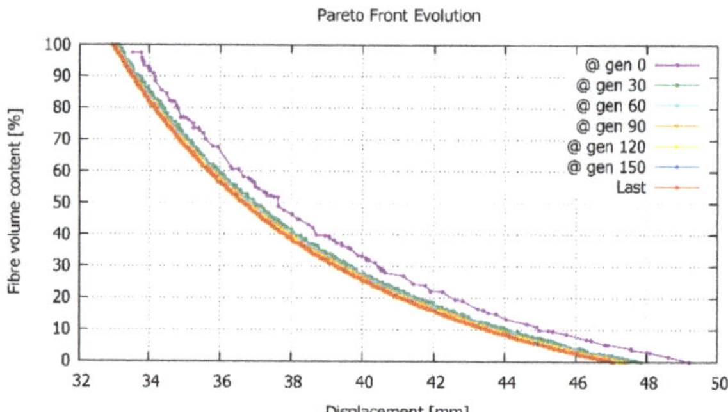

Figure 17. Pareto front evolution showing the performance of the different objective functions defined for the analysis.

The results shown in Figure 17 for a fibre volume content of 100% are the same that were obtained in the first optimization analyses conducted, in which the original displacement is close to 34 mm. This displacement can be reduced to 33 mm with a correct fibre disposition. On the other hand, if we consider the case in which we remove all fibres in Y direction, the deformation of the hatrack shelf becomes substantially larger, being of 47 mm if fibres are oriented in the optimal situation, and 49 mm otherwise. Between these two limits, there is a wide range of solutions, providing different displacements for different fibre volume contents.

Figure 17 also shows that the slope of the Pareto front is substantially steeper for the fibre volume contents between 100% and 60% than for lower fibre volume contents. In other words, a reduction of 10% in fibre volume for higher fibre volume fractions has a lower implication in the shelf displacement, than this same reduction is applied in lower fibre volume fractions. A limit between these two tendencies is found around a fibre volume content of 60%, in which we have an increase from 33 mm to 35.5 mm (7.5%) in the maximum vertical deformation of the shelf, with a reduction of a 40% of fibre volume content. Therefore, from an engineering point of view, this can be an optimal design for the hatrack.

Considering this case as the optimal design, in the following the results obtained for this case are described. The first result shown is the orientation of the woven skin in the hatrack shelf. Figure 18 shows the directions of the stiffer skin yarns that are required to obtain the best structural performance. Now it is important to bear in mind that the woven skins are orthotropic, as the yarns in the X direction (the one represented in Figure 18) contain 40% more fibres than in the perpendicular direction. If the orientation obtained for this case is compared with the orientation in the case of a regular woven laminate, with the same amount of fibres in both directions (Figure 14), it can be seen that in the central region of the hatrack, the yarns with a larger amount of fibres have to be placed perpendicularly to the direction obtained in the case of having the same yarns in both directions. On the other hand, in the region close to the edges, the stiffer yarns are perpendicular to the central ones, stitching them.

The comparison of fibre distribution displayed in Figures 14 and 18 show that the optimal pattern has a large dependence on the structural configuration of the woven laminate. Which justifies the convenience of multi-objective, multi-scale optimization methods to obtain the optimal structural design of eco-composite structures.

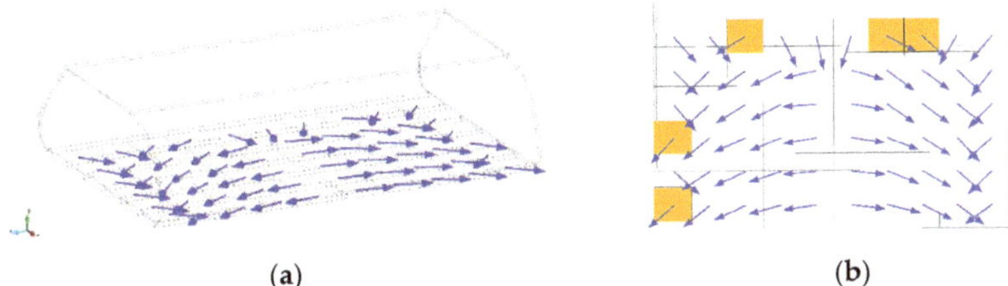

(a)   (b)

Figure 18. (**a**) Optimal woven skin orientation (X direction) required to obtain the minimum shelf displacement. In (**b**) the elements for which the optimizer has not found the optimal orientation are marked in orange.

With this new fibre distribution, the structural performance of the hatrack is very similar to the one observed in previous analyses. Deformations follow the same pattern shown by previous models (Figure 19), although in this case the maximum displacement obtained is a bit larger, as it has a value of 32 mm under Ultimate Limit State loads. As for the stresses, they are below the failure threshold (Figure 20a,b). Therefore, the hatrack is feasible under this configuration, although its stiffness has been reduced due to the reduction of fibre content.

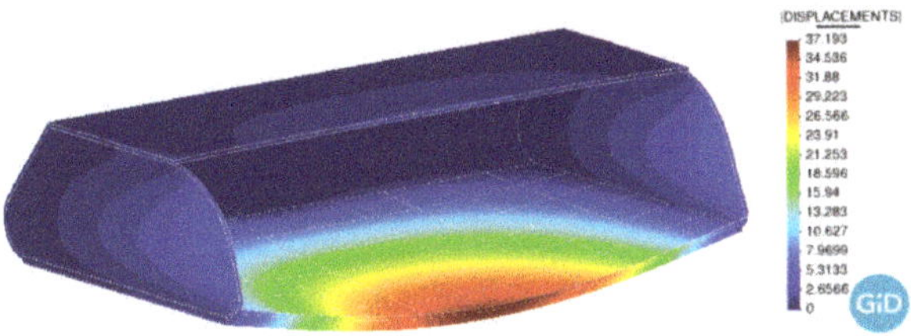

Figure 19. Hatrack displacements with the optimal woven skin design and 60% of fibres in Y direction.

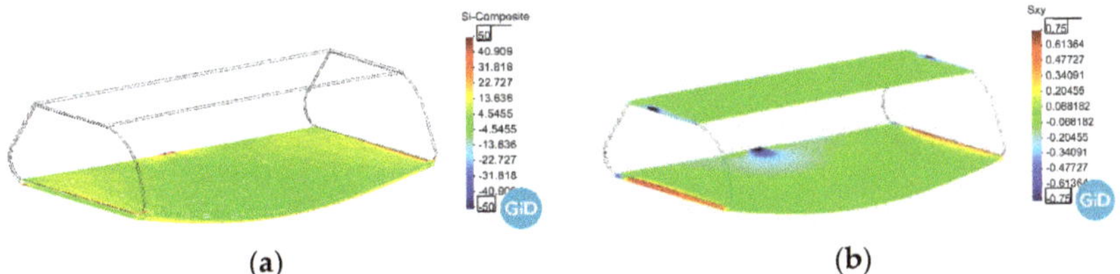

(a)   (b)

Figure 20. Hatrack shelf stresses with the optimal woven skin design and 60% of fibres in Y direction. (**a**) Tensile stresses at the woven skins. (**b**) Shear stresses at the honeycomb.

4. Discussion

The three numerical models analysed in Sections 3.1–3.3 have shown the performance of an aircraft overhead locker with three different material configurations. The first analysis shows the performance of the locker when the woven composite of the skins is oriented following the length and width directions of the structure (as it is shown in Figure 9), while in the other two simulations these orientations are defined using an optimization procedure. The first optimization analysis (Section 3.2) been made with a single objective function and modifying the structural configuration only at a macroscale level. This analysis has provided the optimal fibre orientation of the laminate skins in order to improve the stiffness of the hatrack shelf. The second optimization analysis (Section 3.3) has been conducted with two different objective functions, one consisting in obtaining the larger stiffness of the hatrack shelf, and the other one that seeks using the minimum amount of fibres in one of the directions of the woven skins (reducing the cost and the weight of the woven material).

The modifications made on the structure design to obtain the optimal performance have been the orientation of the shelf skins and the number of fibres of the woven composite, in one of its directions. This last modification is made at a micro-structural level, as it modifies the configuration of the Representative Volume Element.

The results obtained with the solution of both optimization problems have proved the capability of the software developed to obtain an optimal design of eco-composite structures. In both analyses, the optimal fibre orientations differ from the orientations considered in the original hatrack (analysis made in Section 3.1). However, the authors consider that the most relevant result is obtained from the comparison of the two optimization analyses conducted, as these have shown that a multi-objective optimization that modifies both, the macrostructure and the micro-structure, can provide solutions not considered by a regular optimization procedure. This opens a new framework in the design of composite structures, as now the material can be optimized at different scales, in order to provide the best possible material configuration, material disposition and structural configuration, for a given application. This framework is not found in other optimization analyses, in which either apply an optimization procedure to a multi-scale problem, or they use a multi-objective optimization to a structural problem, but not both.

Although the results demonstrated the capabilities of the proposed methodology, it has some limitations and drawback that the authors must further research. The use of a population-based optimizer requires a large number of simulations, which means a large computational cost. Parallelization is already in place, but further research is required to ensure the applicability to larger structural problems. The coupling between the optimizer and the multi-scale simulator is using a weak form, but hierarchical optimization or hybridization strategies could be developed to further benefit from all the scales. In this same line, any strategy aimed to reduce the computational cost of multiscale structural models will benefit the whole procedure, as multiscale methods are rather expensive and this limits the complexity of the models that can be considered in the optimization process.

The overhead locker analysed in this work is used mainly as an example of the capabilities of the proposed formulation. However, looking at the specific application, additional improvements of the design could be obtained by acting on the detailed geometry of the micro-structure, e.g., by considering a different weave pattern, or by modifying the hatrack shape at the macro-level.

5. Conclusions

The introduction of eco-composites in engineering structures and, more specifically in aeronautic structures, requires improving the existing knowledge about the material performance, and also to develop analysis tools capable of predicting the response of the material when it is in service. An optimal design of eco-composite structures needs the developed numerical tools to evaluate the structural performance and to correct its design, if necessary, in order to improve it.

Optimization tools provide a systemized solution for design improvement. Instead of relying on the designer experience to obtain a new structural configuration that provides better performance, it is possible to rely on an optimization code to do the work. In this case, the software analyses the variables that define the structure performance and, based on different procedures (e.g., genetic algorithms), defines new structure configurations that are expected to improve the structural performance.

Current work has coupled a multi-scale structural finite element software, PLCd, with a multi-objective optimization software, RMOP, to obtain a multi-objective multi-scale optimization package that can be used to obtain the optimal design of eco-composite structures. With this approach, it is possible to optimize the structural performance by modifying the configuration of the structure at a macro-level, but also having the possibility of modifying the configuration of the internal micro-structure of the composite material. The analysis conducted with the developed tools has shown that with the multi-scale multi-objective approach proposed, it is possible to obtain improved structural configurations, not reachable by the optimizer otherwise. These tools are expected to facilitate the use of these new materials in different structural applications, and especially in aeronautical structures, in order to reduce the environmental impact of transportation.

Author Contributions: Conceptualization, X.M., J.P.-P., F.T., M.C., L.G.B. and G.B.; methodology, X.M., J.P.-P., F.T., M.C. and L.G.B.; software, X.M., J.P.-P., F.T., M.C. and L.G.B.; validation, X.M., J.P.-P., F.T., M.C., L.G.B. and G.B.; formal analysis, X.M., J.P.-P., F.T., M.C., L.G.B. and G.B.; investigation, X.M., J.P.-P., F.T., M.C. and L.G.B.; resources, X.M., J.P.-P., F.T., M.C., L.G.B. and G.B.; data curation, X.M., J.P.-P., F.T., M.C., L.G.B. and G.B.; writing—original draft preparation, X.M., J.P.-P., F.T., M.C., L.G.B. and G.B.; writing—review and editing, X.M., J.P.-P., F.T., M.C., L.G.B. and G.B.; visualization, X.M., J.P.-P., F.T., M.C., L.G.B. and G.B.; supervision X.M., J.P.-P. and G.B. All authors have read and agreed to the published version of the manuscript.

Funding: This work has received funding from the European Union's Horizon 2020 research and innovation programme under grant agreement No 690638, from the Special Research Plan on Civil Aircraft of Ministry for Industry and Information of the People's Republic of China (MIIT) under Grant No MJ-2015-H-G-103, and from the Spanish Ministerio de Economia y Competividad through the project MAT2014-60647-R, Multi-scale and multi-objective optimization of composite laminate structures (OMMC). The authors also acknowledge the Severo Ochoa Centre of Excellence (2019–2023), which financially supported this work under the grant CEX2018-000797-S funded by MCIN/AEI/10.13039/501100011033.

Acknowledgments: Jordi Pons-Prats and Lucía Gratiela Barbu acknowledge the support of the Serra Hunter programme by the Catalan Government.

Conflicts of Interest: The authors declare no conflict of interest.

References

1. Pearson, W.E. Textiles to composites: 3D moulding and automated fibre placement for flexible membranes. In *Marine Applications of Advanced Fibre-Reinforced Composites*; Graham-Jones, J., Summerscales, J., Eds.; Woodhead Publishing Series in Composites Science and Engineering; Woodhead Publishing: Sawston, UK, 2016; pp. 305–334.
2. Crosky, A.; Grant, C.; Kelly, D.; Legrand, X.; Pearce, G. Fibre placement processes for composites manufacture. In *Advances in Composites Manufacturing and Process Design*; Elsevier Inc.: Amsterdam, The Netherlands, 2015; pp. 79–92.
3. Barbero, E.J. *Introduction to Composite Materials Design*, 3rd ed.; CRC Press: Boca Raton, FL, USA, 2017.
4. Rastellini, F.; Oller, S.; Salomón, O.; Oñate, E. Composite materials non-linear modelling for long fibre-reinforced laminates: Continuum basis, computational aspects and validations. *Comput. Struct.* **2008**, *86*, 879–896. [CrossRef]
5. Granados, J.J.; Martinez, X.; Nash, N.; Bachour, C.; Manolakis, I.; Comer, A.; Di Capua, D. Numerical and experimental procedure for material calibration using the serial/parallel mixing theory, to analyze different composite failure modes. *Mech. Adv. Mater. Struct.* **2021**, *28*, 1415–1433. [CrossRef]
6. Martinez, X.; Rastellini, F.; Oller, S.; Flores, F.; Oñate, E. Computationally optimized formulation for the simulation of composite materials and delamination failures. *Compos. Part B Eng.* **2011**, *42*, 134–144. [CrossRef]
7. Solis, A.; Sanchez-Saez, S.; Martinez, X.; Barbero-Pozuelo, E. Numerical analysis of interlaminar stresses in open-hole laminates under compression. *Compos. Struct.* **2019**, *217*, 89–99. [CrossRef]
8. Barbu, L.; Oller, S.; Martinez, X.; Barbat, A. High-cycle fatigue constitutive model and a load-advance strategy for the analysis of fiber reinforced composites. *Compos. Struct.* **2019**, *220*, 622–641. [CrossRef]
9. Otero, F.; Oller, S.; Martinez, X. Multiscale Computational Homogenization: Review and Proposal of a New Enhanced-First-Order Method. *Arch. Comput. Methods Eng.* **2018**, *25*, 479–505. [CrossRef]
10. Geers, M.G.D.; Kouznetsova, V.G.; Brekelmans, W.A.M. Multi-scale computational homogenization: Trends and challenges. *J. Comput. Appl. Math.* **2010**, *234*, 2175–2182. [CrossRef]
11. Hernández, J.A.; Oliver, J.; Huespe, A.E.; Caicedo, M.A.; Cante, J.C. High-performance model reduction techniques in computational multiscale homogenization. *Comput. Methods Appl. Mech. Eng.* **2014**, *276*, 149–189. [CrossRef]
12. Ricks, T.M.; Pineda, E.J.; Bednarcyk, B.A.; McCorkle, L.S.; Miller, S.G.; Murthy, P.L.N.; Segal, K.N. Multiscale Progressive Failure Analysis of 3D Woven Composites. *Polymers* **2022**, *14*, 4340. [CrossRef]
13. Lansiaux, H.; Soulat, D.; Boussu, F.; Labanieh, A.R. Development and Multiscale Characterization of 3D Warp Interlock Flax Fabrics with Different Woven Architectures for Composite Applications. *Fibers* **2020**, *8*, 15. [CrossRef]
14. Nikbakt, S.; Kamarian, S.; Shakeri, M. A review on optimization of composite structures Part I: Laminated composites. *Compos. Struct.* **2018**, *195*, 158–185. [CrossRef]
15. Ehsani, A.; Rezaeepazhand, J. Stacking sequence optimization of laminated composite grid plates for maximum buckling load using genetic algorithm. *Int. J. Mech. Sci.* **2016**, *119*, 97–106. [CrossRef]

16. Wei, R.; Pan, G.; Jiang, J.; Shen, K.; Lyu, D. An efficient approach for stacking sequence optimization of symmetrical laminated composite cylindrical shells based on a genetic algorithm. *Thin-Walled Struct.* **2019**, *142*, 160–170. [CrossRef]

17. Zhou, Y.; Nomura, T.; Saitou, K. Multi-component topology and material orientation design of composite structures (MTO-C). *Comput. Methods Appl. Mech. Eng.* **2018**, *342*, 438–457. [CrossRef]

18. Deb, K. *Multi-Objective Optimization Using Evolutionary Algorithms*; John Wiley & Sons: Hoboken, NJ, USA, 2001.

19. Serhat, G.; Basdogan, I. Multi-objective optimization of composite plates using lamination parameters. *Mater. Des.* **2019**, *180*, 107904. [CrossRef]

20. Fagan, E.M.; De La Torre, O.; Leen, S.B.; Goggins, J. Validation of the multi-objective structural optimisation of a composite wind turbine blade. *Compos. Struct.* **2018**, *204*, 567–577. [CrossRef]

21. Li, H.; Luo, Z.; Xiao, M.; Gao, L.; Gao, J. A new multiscale topology optimization method for multiphase composite structures of frequency response with level sets. *Comput. Methods Appl. Mech. Eng.* **2019**, *356*, 116–144. [CrossRef]

22. Coelho, P.G.; Fernandes, P.R.; Guedes, J.M.; Rodrigues, H.C. A hierarchical model for concurrent material and topology optimisation of three-dimensional structures. *Struct. Multidiscip. Optim.* **2008**, *35*, 107–115. [CrossRef]

23. Coelho, P.G.; Guedes, J.M.; Rodrigues, H.C. Multiscale topology optimization of bi-material laminated composite structures. *Compos. Struct.* **2015**, *132*, 495–505. [CrossRef]

24. Ferrer, A.; Cante, J.C.; Oliver, J. On multi-scale structural topology optimization and material design. In Proceedings of the Congress on Numerical Methods in Engineering (CMN_2015), Lisboa, Portugal, 29 June–2 July 2015.

25. Ferrer, A.; Oliver, J.; Cante, J.C.; Lloberas-Valls, O. Vademecum-based approach to multi-scale topological material design. *Adv. Model. Simul. Eng. Sci.* **2016**, *3*, 1–22. [CrossRef]

26. Ferrer, A.; Cante, J.C.; Hernández, J.A.; Oliver, J. Two-scale topology optimization in computational material design: An integrated approach. *Int. J. Numer. Methods Eng.* **2018**, *114*, 232–254. [CrossRef] [PubMed]

27. Kim, Y.; Yoon, G. Multi-resolution multi-scale topology optimization—A new paradigm. *Int. J. Solids Struct.* **2000**, *37*, 5529–5559. [CrossRef]

28. Sivapuram, R.; Dunning, P.; Kim, H.A. Simultaneous material and structural optimization by multiscale topology optimization. *Struct. Multidiscip. Optim.* **2016**, *54*, 1267–1281. [CrossRef]

29. Zhang, Y.; Xiao, M.; Zhang, X.; Gao, L. Topological design of sandwich structures with graded cellular cores by multiscale optimization. *Comput. Methods Appl. Mech. Eng.* **2020**, *361*, 112749. [CrossRef]

30. Bachmann, J.; Yi, X.; Gong, H.; Martinez, X.; Bugeda, G.; Oller, S.; Tserpes, K.; Ramon, E.; Paris, C.; Moreira, P.; et al. Outlook on ecologically improved composites for aviation interior and secondary structures. *CEAS Aeronaut. J.* **2018**, *9*, 533–543. [CrossRef]

31. Bachmann, J.; Yi, X.; Tserpes, K.; Sguazzo, C.; Barbu, L.G.; Tse, B.; Soutis, C.; Ramón, E.; Linuesa, H.; Bechtel, S. Towards a Circular Economy in the Aviation Sector Using Eco-Composites for Interior and Secondary Structures. Results and Recommendations from the EU/China Project ECO-COMPASS. *Aerospace* **2021**, *8*, 131. [CrossRef]

32. CIMNE-Composites. PLCd. PLastic Crack Dynamic Code. Available online: www.cimne.com/PLCd (accessed on 23 July 2019).

33. CIMNE-Optimization Group. RMOP. Robust Multi-Objective and Multidisciplinary Optimization Platform. Available online: http://tts.cimne.com/RMOP/ (accessed on 24 July 2019).

34. Nemat-Nasser, S. Averaging theorems in finite deformation plasticity. *Mech. Mater.* **1999**, *31*, 493–523. [CrossRef]

35. Geers, M.; Kouznetsova, V.G.; Brekelmans, W.A.M. MultiScale First-Order and Second-Order Computational Homogenization of Microstructures towards Continua. *Int. J. Multiscale Comput. Eng.* **2003**, *1*, 371–386. [CrossRef]

36. Otero, F.; Martinez, X.; Oller, S.; Salomón, O. An efficient multi-scale method for non-linear analysis of composite structures. *Compos. Struct.* **2015**, *131*, 707–719. [CrossRef]

37. Zaghi, S.; Martinez, X.; Rossi, R.; Petracca, M. Adaptive and off-line techniques for non-linear multiscale analysis. *Compos. Struct.* **2018**, *206*, 215–233. [CrossRef]

38. Deb, K.; Pratap, A.; Agarwal, S.; Meyarivan, T. A fast and elitist multiobjective genetic algorithm: NSGA-II. *IEEE Trans. Evol. Comput.* **2002**, *6*, 182–197. [CrossRef]

39. Deb, K.; Agrawal, R.B. Simulated Binary Crossover for Continuous Search Space. *Complex Syst.* **1995**, *9*, 115–148.

40. Coma, M.; Pons-Prats, J.; Bugeda, G. Hybrid optimization methods. In Proceedings of the CMN 2019—Congreso de Métodos Numéricos en Ingeniería, Guimarães, Portugal, 1–3 July 2019.

41. Pons-Prats, J.; Coma, M.; Bugeda, G. Optimization hybridization with multiple populations and optimization methods. In Proceedings of the EUROGEN 2019—International Conference on Evolutionary and Deterministic Methods for Design, Optimization and Control with Applications to Industrial and Societal Problems 2019, Guimarães, Portugal, 12–14 September 2019.

42. Coma, M.; Tousi, N.M.; Pons-Prats, J.; Bugeda, G.; Bergada, J.M. A New Hybrid Optimization Method, Application to a Single Objective Active Flow Control Test Case. *Appl. Sci.* **2022**, *12*, 3894. [CrossRef]

43. Welch, J.; Roth, R.H. Enhanced Luggage Bin System. U.S. Patent 6398163 B1, 30 May 2000.

44. EASA. Certification Specifications for Large Aeroplanes CS-25. 2003. Available online: https://www.easa.europa.eu/sites/default/files/dfu/decision_ED_2003_02_RM.pdf (accessed on 12 July 2019).

45. Yi, X.S.; Zhang, X.; Ding, F.; Tong, J. Development of Bio-Sourced Epoxies for Bio-Composites. *Aerospace* **2018**, *5*, 65. [CrossRef]

46. Ishikawa, A.; Okano, T.; Sugiyama, J. Fine structure and tensile properties of ramie fibres in the crystalline form of cellulose I, II, IIII and IVI. *Polymer* **1997**, *38*, 463–468. [CrossRef]

47. Nam, S.; Netravali, A.N. Green composites. I. physical properties of ramie fibers for environment-friendly green composites. *Fibers Polym.* **2006**, *7*, 372–379. [CrossRef]
48. Kozlowski, R. *Handbook of Natural Fibres: Types, Properties and Factors Affecting Breeding and Cultivation*; Textile Institute, Woodhead Publishing: Manchester, UK, 2012.
49. Truesdell, C.; Toupin, R. *The Classical Field Theories*; Springer: Berlin/Heidelberg, Germany, 1960; pp. 226–858.

MDPI
St. Alban-Anlage 66
4052 Basel
Switzerland
Tel. +41 61 683 77 34
Fax +41 61 302 89 18
www.mdpi.com

Mathematics Editorial Office
E-mail: mathematics@mdpi.com
www.mdpi.com/journal/mathematics

9 783036 579788